Dimorphic Fungi in Biology and Medicine

Dimorphic Fungi in Biology and Medicine

Edited by

Hugo Vanden Bossche
Frank C. Odds
Janssen Research Foundation
Beerse, Belgium

and

David Kerridge
University of Cambridge
Cambridge, United Kingdom

Springer Science+Business Media, LLC

Library of Congress Cataloging in Publication Data

Dimorphic fungi in biology and medicine / edited by Hugo Vanden Bossche, Frank
C. Odds, and David Kerridge.
    p.   cm.
    "Proceedings of the Fourth Symposium on Topics in Mycology on Fungi Dimor-
phism, held September 1–4, 1992, in Cambridge, United Kingdom"—T.p. verso..
    Includes bibliographical references and index.
    ISBN 978-1-4613-6226-5   ISBN 978-1-4615-2834-0 (eBook)
    DOI 10.1007/978-1-4615-2834-0
    1. Mycoses—Congresses. 2. Fungi—Congresses. I. Vanden Bossche, H. II. Odds, F.
C. III. Kerridge, David. IV. Symposium on Topics in Mycology on Fungal Dimorphism,
(1992: Cambridge, England) [DNLM: 1. Fungi—pathogenicity—congresses. 2. Fungi—
cytology—congresses. 3. Morphogenesis—physiology—congresses. 4. Mycoses—
therapy—congresses. 5. Mycoses—genetics—congresses. 6. Genetics, Biochemical—
congresses. QW 180 D582 1992]
QR245.D55   1993
616'.015—dc20
DNLM/DLC   93-24076
for Library of Congress   CIP

Proceedings of the Fourth Symposium on Topics in Mycology on Fungal Dimorphism,
held September 1–4, 1992, in Cambridge, United Kingdom

ISBN 978-1-4613-6226-5

© 1993 Springer Science+Business Media New York
Originally published by Plenum Press, New York in 1993

PREFACE

Fungal dimorphism is a topic that sounds inherently too rarified to attract more than a specialist audience. Yet some 230 individuals representing an eclectic mixture of interests, from basic science to medical practice, gathered in Churchill College, Cambridge in Semptember 1992 for a meeting devoted only to this subject. The symposium was the fourth in a series "Topics in Mycology" to be jointly organized by the Janssen Research Foundation and the International Society for Human and Animal Mycology. The participants enjoyed a rich and varied diet of oral presentations and poster displays in the field of fungal morphogenesis. This book sets down in print the material presented at the dimorphism symposium. We think that the high quality of these papers conveys very well the flavor of what was an excellent meeting.

The selection of contributions in this volume covers very wide ground indeed. Chapters devoted to some non-pathogenic fungi are included, because the scientific basis of morphological development belongs to the fields of cellular and molecular biology: it does not recognize the boundary imposed by considerations of virulence of a fungus for a human host. Yet morphogenetic change in those fungi that do cause human disease frequently appears to be a component of the pathological process: many important pathogens change from a hyphal form in the external environment to a round form in infected tissues. This relationship between dimorphism and pathogenicity is the point of contact between pure biology and medicine.

The intrests of clinicians are indeed well catered for, with chapters on diagnosis and treatment of mycoses caused by dimorphic pathogens. The reader will even find articles devoted to fungi not usually thought of as dimorphic – for example, *Aspergillus* species. The reason for this is simple: when such moulds spout spore-bearing structures from their vegetative hyphae they produce cells with a non-hyphal shape. The true unifying theme for this book is therefore cell shape diversity and its regulation. "Dimorphism" is, like that distinction between pathogenic and non-pathogenic fungi, a notion devised by people for their own convenience. By bringing together elements of morphogenesis that cross the customary man-made boundaries we hope we have produced a book that will give every reader a better than normal chance of finding something new within its pages. Certainly, those who attended the Cambridge symposium felt they had benefitted from the exposure to areas of biology and medicine beyond their normal experience.

Hugo Vanden Bossche
Frank C. Odds
David Kerridge

ACKNOWLEDGEMENTS

The Organizers are grateful to the Janssen Research Foundation for providing the opportunity to organize this symposiumn.

Thanks are due to Drs. G. Cauwenbergh, J.R. Graybill, D.W.R. Mackenzie, M. Shepherd, D. Soll, and J. Van Cutsem for their help in finding topics and speakers.

We are grateful to chairmen, speakers and all participants and especially to those who presented posters and contributed so much to the discussions.

We have been greatly assisted in our organizational duties by Mrs. H. Dergent, Mrs. A. Siegers, and Mrs. P. Preston.

As editors of this volume we would like to thank the publisher, Plenum Press, and extend special thanks to Mrs. Leen Geentjens and Ms. Cindy Michielsen for retyping the manuscripts.

CONTENTS

INTRODUCTORY PAPER

MOLECULAR GENETICS OF MORPHOGENETIC PROCESSES

CELL BIOLOGY AND BIOCHEMISTRY

DIMORPHISM AND PATHOGENESIS

DIAGNOSIS AND TREATMENT OF MYCOSES

Introductory Paper

FUNGAL DIMORPHISM : A SIDEWAYS LOOK

David Kerridge

Department of Biochemistry
Tennis Court Road
Cambridge, CB2 1QW
United Kingdom

ABSTRACT

The majority of important systemic fungal pathogens are pleomorphic growing as mycelia in the environment and as yeasts in humans.

Candida albicans is an exception to this and occurs as both yeasts and mycelia in the saprobic and pathogenic states. The ability of this organism to form hyphae is often considered an important virulence attribute, but the evidence is not conclusive and over emphasis on this one factor may detract attention from other properties of this important opportunistic pathogen. The molecular basis of the morphological switch is not understood. The transition from the yeast to mycelial form is induced by changes in the environmental conditions and much of the confusion in the literature stems from a failure to separate those changes in the cell directly associated with the morphological switch from other phenotypic effects.

The most significant difference between growth of hyphae and budding yeasts relates to sites of deposition of newly synthesized cell wall constituents. It is suggested that this should provide a focus for future research into this phenomenon.

INTRODUCTION

"Few fields of biological science based on so simple an observation, can have generated such a confused and contradictory literature as that of dimorphism in *Candida albicans*".[1]

In the four years that have elapsed since the publication of "Candida and Candidosis" from which that quotation is taken, interest in fungal dimorphism, in particular in fungi pathogenic to humans, has increased considerably. We are some way from understanding the molecular basis of these morphological changes, but application of the techniques of molecular genetics to the problem is

Dimorphic Fungi in Biology and Medicine, Edited by
H Vanden Bossche *et al*, Plenum Press, New York, 1993

already providing an insight into this phenomenon. This symposium is timely and bringing together scientists all with a deep interest in this phenomenon will provide a stimulus for future research. Writing the "scene setting" paper is a daunting task, but I hope to highlight the problems that we must address and provide a stimulus for future discussions.

THE CLINICAL IMPORTANCE OF DIMORPHIC FUNGI

There are over 300,000 species of fungi throughout the world of which some 200 have been shown to cause infections in humans. These human pathogens range from disfiguring dermatophytic fungi to organisms responsible for life threatening systemic mycoses. During the last half century changes in medical practice and in diseases to which humans are exposed have resulted in a significant increase in the importance of mycotic infections and in particular those caused by opportunistic pathogens. Such organisms are normally harmless but capable of causing life-threatening diseases in the compromised host i.e. one where the normal defence mechanisms are impaired or the microenvironment in which the fungus can grow is altered. The most important of these are *Aspergillus fumigatus*, *Candida albicans* and *Pneumocystis carinii* and it is these three fungi that have been responsible for the upsurge in interest.

In addition there is another group of fungi present in the environment capable of infecting normal healthy humans causing severe systemic mycoses. These fungi are classified in Hazard Group 3.[2] Such organisms "may cause severe human disease and present a severe hazard to laboratory workers. They may present a risk of spread to the community but there is usually effective prophylaxis or treatment available". All pathogenic fungi in this group have one other feature in common; they exhibit structural dimorphism, existing as a typical branching mycelium in the environment, but in an infected host these fungi grow as unicellular microorganisms.

Structural dimorphism does not occur in the systemic pathogens *A. fumigatus*, *Cryptococcus neoformans* and *P. carinii*. *C. albicans*, the most important opportunistic fungal pathogen, exhibits structural dimorphism but is anomalous in that it occurs as yeasts and mycelia in both commensal and pathogenic states. It is the most studied of all dimorphic fungal pathogens and there are a number of reasons for this. It is clinically important - for example, approximately 75% of women will suffer from vaginal candidosis at some stage in their lifetime - and prevalent in compromised patients. There is great interest in the possible role of dimorphism in fungal pathogenicity. It is also a potential model for studying morphogenesis in eukaryotic microorganisms.

In contrast to *Candida albicans*, other dimorphic fungi show a clear cut distinction between saprobic and pathogenic forms but these fungi have a restricted geographical distribution area and have not been so extensively studied.[3]

TERMINOLOGY

The terminology that has developed for the dimorphic fungi, in particular *Candida albicans*, is at its best confusing and at its worse potentially misleading. The title of this symposium is Fungal Dimorphism yet how many of the fungi under discussion exist only in two morphological forms? (Dimorphic: existing or occurring in two distinct forms. The Shorter Oxford English Dictionary, 3rd ed.). *C. albicans*, the most common "dimorphic' fungus, exists in four different morphological forms; yeasts, pseudomycelia, mycelia and chlamydospores. Polymorphic or pleomorphic would be correct, but 'dimorphic' is too well established for it to be changed now.

It is more confusing when we consider the names used for the unicellular form of *C. albicans*. There is a choice of yeast, blastospore and blastoconidium. By analogy to bacterial spores and fungal conidia the terms blastospore and blastoconidia imply resting cells, resistant to environmental extremes, rather than growing unicellular organisms. Perhaps now is the time to agree to eliminate these terms and restrict ourselves to using the word "yeast" to describe the unicellular form of *C. albicans*. The outgrowth of mycelia from yeasts is often referred to as 'germination' but again by analogy to other biological systems this suggests a transition from a nongrowing or resting cell to a growing cell. Whereas with *C. albicans* we are dealing with an environmentally induced morphological change in growing cells. It would be less misleading to refer to hyphal development or formation, and eliminate germination from common usage. The morphological change is reversible but how should the production of yeasts from hyphae be described?

THE CLINICAL IMPORTANCE OF FUNGAL DIMORPHISM

"The capacity to form a hypha is without doubt a major pathogenic attribute of infectious yeasts like *C. albicans*."[4]

"Little doubt that the hypha has evolved as a mechanism for tissue penetration".[4]

Virulence and pathogenicity are synonymous and defined as the capacity to cause disease. *C. albicans* is a commensal and normally harmless, only causing disease in a compromised host. It is therefore somewhat incongruous to discuss virulence factors in this organism. After all it is the changes in the host that are largely responsible for infections. There are properties of *C. albicans* which enhance its ability to cause disease in the compromised host, but they should not be over emphasised at the expense of those changes in the compromised host that allow exploitation of this specific environment by opportunistic pathogens.

The majority of fungi capable of causing disease in normal healthy humans are also dimorphic, but for these fungi it is the unicellular yeast and not mycelial morphology that is found in the host (Table 1). Two exceptions are the ubiquitous *A. fumigatus* which shows hyphal growth both in its normal habitats in the environment and in the host, and *C. albicans* where both yeasts and mycelia can be found in lesions. It is clear that the ability to grow in the hyphal form is not a prerequisite for infection by all pathogenic fungi. In fact if we consider Group 3 fungal pathogens, it would appear that an ability to grow as a unicellular yeast is a "major virulence factor".

Other *Candida* species are also opportunistic pathogens and many like *C. glabrata* and *C. krusei* are distinguished from *C. albicans* by their inability to produce hyphae either in culture or in lesions, but cause disease in the compromised host. The frequency of these infections is less than those caused by *C. albicans* and may reflect not only a reduced pathogenicity but also their presence as commensals in humans. However, there is an implicit assumption that inability to produce hyphae is responsible for the reduced pathogenicity of these *Candida* species.

Table 1. Systemic fungal infections

Fungus	Morphology	Infectious form	Hazard group
*Aspergillus fumigatus**	Mycelial	Mycelia	2
Blastomyces dermatitidis	Pleomorphic	Yeast	3
*Candida albicans**	Pleomorphic	Yeast and mycelium	2
*Candida spp.**	Unicellular	Yeast	2
Cryptococcus neoformans	Unicellular	Encapsulated yeast	2
Coccidioides immitis	Pleomorphic	Spherules	3
Histoplasma capsulatum	Pleomorphic	Yeast	3
Paracoccidioides brasiliensis	Pleomorphic	Yeast	3
Sporothrix schenkii	Pleomorphic	Yeast	3

* opportunistic pathogen

Pathogenicity or virulence is a multifactorial phenomenon and over emphasis on the specific observation, that both yeasts and mycelia of *Candida albicans* occur in lesions and that in the commensal state there would appear to be a great proportion of yeasts, may divert attention away from other potentially important properties of this fungus. In considering the role of cell shape in enhancing the ability of a fungus to grow in a human host it is important to examine the stages in disease development. These have been discussed in detail by Smith[5] and are:

i. colonization of the mucous surfaces;
ii. penetration through these surfaces;
iii. growth within the host;
iv. resistance to the host's defence mechanisms;
v. the pathogen must damage the host's tissues.

Hyphae can be distinguished from unicellular yeasts by three distinct features, the site of cell wall growth, cell shape, and the structure of the cell wall. If hyphae are to be assessed as a potential major virulence factor then these specific features must be considered in relationship to the individual stages in establishment of an infection.

Both yeasts and hyphae of *C. albicans* adhere to epithelial cells and there is evidence for enhanced adhesion by hyphae. This could result from an increased surface to volume ratio in elongated hyphal cells compared to the ovoid yeasts, changes in surface hydrophobicity and/or specific "adhesins" on the surface.[6] Of these various factors, it is only the difference in the surface to volume ratio that is unique. This initial stage in colonization may be aided by hyphal growth, but it is not essential.

The evidence for penetration of hyphae into epithelial cells is largely microscopic. It may be a purely physical phenomenon with the growing hyphal tip forcing its way through the plasma membrane of an epithelial cell or it might result from damage caused by exocellular lytic enzymes secreted at the hyphal tip. This could be important in the establishment of topical candidosis. All systemic fungal pathogens, with the exception of *C. albicans*, infect via the lung. *C. albicans* is found both on mucous surfaces and in the gut. Systemic candidosis results from the passage of yeasts rather than hyphae from the gut into the circulatory system.[7]

What advantages will a mycelial fungus have over a unicellular one growing in the host? An obvious one is the difference in the surface to volume ratio with the mycelial form having a larger surface area over which it can transport nutrients from the environment. Also by growing linearly it can spread throughout the host.

C. albicans is a commensal and normally nonpathogenic in the noncompromised host, i.e. humans are remarkably effective in defending themselves against infections by this organism.

Tissue damage is exemplified by the damage caused by mycelia in systemic aspergillosis where growing hyphae apparently stop at nothing in their spread through the host's tissues. A similar effect could occur with *C. albicans*.

Each stage in the establishment of an infection has been considered in relationship to hyphal development in *C. albicans*. Although there are stages where hyphal growth apparently aids the establishment of candidosis it cannot be considered a "major" virulence factor. The azole antimycotic drugs are effective in treating patients with systemic mycoses. The concentrations required are less than the MIC values for the same drugs determined *in vitro*. This apparently results from inhibition of hypha initiation and growth at drug concentrations lower than those required to prevent yeast growth and is taken as evidence for the importance of hyphal development in *C. albicans* as a virulence factor.[8] Hyphal growth is not essential for systemic fungal infections and over emphasis on this one feature of dimorphism may detract attention from other possibly more important virulence factors. Studies by Ryley and Ryley[9] have shown that although hyphal formation is not an essential virulence factor in the establishment of systemic candidosis in mice it may be important in the localization of the fungus in specific organs.

THE MOLECULAR BASIS OF THE DIMORPHIC TRANSITION

"When such a complex self regulating system is in the process of steady exponential growth, its basal elements must be in a state of dynamic equilibrium with the internal concentration of the numerous small molecules - amino acids, purines, pyrimidines, etc., from which they are synthesised. Changing the external environment, as by placing a cell in a different medium, will inevitably cause a complex shift in the concentration of all components of the internal environment, which in turn will react upon the rates of synthesis of the various large molecules, decreasing some rates and increasing others; in other words changing the chemical composition of the cell".[10]

The phenomenon under discussion is an environmentally induced reversible transition from growth as a unicellular yeast to that of an extending mycelium. Studies on the majority of the important fungal pathogens (Table 1) have concentrated on the mycelial to yeast transformation[3] since it is the unicellular form that is found in lesions whereas for *C. albicans* attention has been directed to the transformation from yeast to mycelium, even though both morphological forms occur in lesions. Little or no work has been done on the reverse outgrowth of yeasts from mycelia in this organism. The general approach used by the majority of research workers has been to choose a method for induction of hyphal growth and then examine one or more specific properties. If changes occur in the specific parameters being measured, it is usually assumed that these are linked to hypha initiation and development. The experimental methods used to induce the transition are as many and various as the research groups employing them (Table 2). The parameters examined range from cell envelope composition and function to the intracellular concentrations of intermediary metabolites, from localization of cytoskeletal proteins to macromolecular biosynthesis and all have been implicated in the morphological switch. The failure to appreciate the fact that environmental

Table 2. Factors affecting the dimorphic transition in *Candida albicans*

Environmental			
Physical	pH value, temperature		
Chemical	i.	Inorganic	Zinc, etc.
	ii.	Organic	Vitamins, amino acids, carbohydrates, etc.

Cellular			
Metabolic state of the cells			
Cell density			
Genetic	i.	Species	
	ii.	Ploidy	
	iii.	Mutations	

changes have significant effects on cell composition, and that in batch culture the environment is constantly changing has resulted in a confused and conflicting literature. The media used to initiate hyphal development range from serum to buffered *N*-acetyl glucosamine and it is unlikely that the composition of cells incubated under such diverse conditions will be identical. If we wish to compare the composition of the yeast and mycelial forms, it is important to grow them under identical environmental conditions and the only way to do that is in a chemostat. Given the absence of well characterized mutants of *C. albicans* this is difficult. Inevitably induction of hyphal growth involves a sudden shift from one environment to another, and will also influence the changes that are observed.[12]

The basic problem is to disassociate changes responsible for the dimorphic transition from those merely resulting from an altered environment. If this is to be done the problem must be clearly defined. Essentially we are dealing with two cell types, one of which grows as a ovoid budding yeast and the other as a filamentous hypha. Cell shape is controlled by the structure of the cell wall; the removal of which results in formation of spherical protoplasts. Knowledge of the molecular fine structure of the cell wall is still rudimentary and we lack a detailed understanding of the structural changes responsible for the differences in shape. Cell walls can be isolated by mechanical disintegration of fungi and exhaustive washing. But are these structures identical to the cell wall *in situ*? Models for the structure of fungal cell walls are relatively simple, and give little insight into the determinants of cell shape.[13,14]

Cell wall growth requires the coordinated synthesis of all its constituents, for *C. albicans* these will be β-glucans, chitin and mannoproteins. The first two are synthesized by plasma membrane enzymes, mannoproteins are synthesized in the endoplasmic reticulum and Golgi apparatus and subsequently transported to the cell envelope in association with small vesicles. These macromolecules are inserted into the growing cell wall and finally crosslinked to preexisting wall constituents. In both hyphae and budding yeasts the driving force for cell growth is the turgor pressure.[15]

Examination of the sites of wall growth in both hyphae and yeasts has demonstrated significant differences. In growing hyphae approximately 90% of wall extension occurs in the vicinity of the hyphal tip and only 10% over the remaining cell surface, whereas in budding yeasts only 70% occurs at the apex of the bud with the remaining 30% over the remaining bud surface. This distribution of wall growth continues until the bud reaches a certain size and then over 90% of wall growth takes place over the entire surface.[15] So here is one fundamental difference between the two morphological forms which provides a focal point for research. Two questions to be answered are:- what determines the location of the site(s) of cell wall growth? and how is this spatial location of cellular function affected by environmental changes? One feature of dimorphism in C. albicans is that pseudomycelia are an intermediate structural form between yeasts and hyphae. This structure comprises linearly arranged elongate cells. It is not clear if this is a discrete stage between budding and hyphal growth or if there is a continuous structural change between the two extremes.

Location of the sites of cell wall synthesis in fungi is controlled by the spatial organization of the cytoskeletal components, in particular actin granules and microfilaments.[16] There is no evidence for a role for cytoplasmic microtubules. Actin granules and associated microfilaments and calmodulin[17] accumulate at the site of cell wall evagination. These structures remain at the apex if hyphal growth occurs. In budding yeasts the granules do not remain at the apex of the bud but disperse throughout the cell. The cytoskeleton is clearly important in determining the sites of cell wall synthesis. The next fundamental question is how is the cytoskeletal organization affected by environmental shifts? Unfortunately we are some way from answering this question.

The emphasis throughout all the studies has been on the factors resulting in hyphal growth in C. albicans rather than growth of the yeast whereas for other dimorphic pathogens the reverse is true. The factors that induce hyphal development in C. albicans are many and various (Table 2). In addition to a wide range of environmental factors, the physiological state of the cells and their genomic composition are important. The mechanism(s) by which environmental factors are perceived by cells and affect cell wall biosynthesis are not understood.

Studies of the genetic control of the morphological switch have not been helped by the fact that C. albicans is diploid without a sexual stage in its life cycle. However progress is being made and studies of the yeast pseudomycelial switch in Saccharomyces cerevisiae[18] may well provide an insight into similar phenomena in C. albicans and other dimorphic fungi.

C. albicans rather than other dimorphic fungi has been used to illustrate the underlying problems. Synthesis of cell walls is fundamentally similar in all fungi and even though the system is complex the ultimate aim must be to develop a unifying hypthesis to account for the environmentally induced dimorphism.

As a final thought: can we avoid teleology* in our studies of fungal dimorphism?

---

* Teleology; the doctrine or study of ends or final causes, esp. as related to the evidences of design or purpose in Nature, Shorter Oxford English Dictionary 3rd Edition.

REFERENCES

1. F. C. Odds, "Candida and Candidosis", Balliere Tindall, London (1988).
2. Categorisation of pathogens according to hazard and categories of containment, HMSO, London (1988).
3. G. San Blas, Fungi pathogenic to humans: molecular aspects of dimorphism, in: "Handbook of Applied Mycology Vol.2, Humans, Animals and Insects", D.K. Orora, L.Ajello and K.J.Mukurji, eds., Marcel Dekker Inc. New York, Basel, Hong Kong (1991).
4. D. R. Soll, Current status of the molecular basis of Candida pathogenicity, in: "The fungal spore and disease in plants and animals", G. T. Cole and H. P. Hoch, eds., Plenum Press, New York and London (1991).
5. H. Smith, The chemotherapeutic potential of inhibition or circumvention of the determinants of microbial pathogenicity, *Symp. Soc. Gen. Microbiol.* 38: 367 (1985).
6. R. A. Calderone and P. C. Braun, Adherence and receptor relationships of *Candida albicans*, *Microbiol. Rev.* 55: 1 (1991).
7. M. J. Kennedy, Candida blastospore adhesion, association and invasion of the gastrointestinal tract of vertebrates, in: "The fungal spore and disease initiation in plants and animals", G. T. Coles and H. C. Hoch, eds., Plenum Press, New York and London (1991).
8. M. Borgers, M. De Brabander, H. Vanden Bossche and J. Van Cutsem, Promotion of pseudomycelial formation of *Candida albicans* in culture: a morphological study of the effects of miconazole and ketoconazole, *Postgrad. Med. J.* 55: 687 (1979).
9. J. F. Ryley and N. G. Ryley, *Candida albicans* - do mycelia matter? *J. Med. Vet. Mycol.* 23: 225 (1990).
10. D. Herbert, The chemical composition of microorganisms as a function of their environment, *Symp. Soc. Gen. Microbiol.* 11: 391 (1961).
11. B. Maresca and G. S. Kobayashi, Dimorphism in *Histoplasma capsulatum*; a model for the study of cell differentiation in pathogenic fungi, *Microbiol. Rev.* 53: 186 (1989).
12. O. Maale and N. O. Kjeldgaard, "Control of Macromolecular synthesis", W. A. Benjamin, New York (1966).
13. M. Shepherd, Cell envelope of *Candida albicans*, *CRC Crit. Rev. Microbiol.* 15: 7 (1987).
14. R. Sentandreu, E. Herrera, J.P. Martinez and M.V. Elorza, Yeast cell wall glycoproteins, in "Fungal cell wall and immune response", J-P. Latge and D. Boucias, eds., NATO ASI series H, Cell Biology Vol. 53 (1991).
15. J.H. Sietsma and J.G.H. Wessels, Cell wall assembly in fungal morphogenesis, in: "Fungal cell wall and immune response", J-P. Latge and D. Boucias, eds., NATO ASI series H. Cell Biology, Vol. 53 (1991).
16. K. Yokayama, H. Kaji, K. Nishimura and M. Miyaji, The role of microfilaments and microtubules in apical growth and dimorphism of *Candida albicans*, *J. Gen. Microbiol.* 136: 1067 (1990).
17. S.E. Brockerhoff and J.N. Davis, Calmodulin concentration at regions of cell growth in *Saccharomyces cerevisiae*, *J. Cell Biol.* 118: 619 (1992).
18. C.J. Gimeno, P.O. Ljungolahi, C.A. Styles and G.R. Fink, Unipolar cell divisions in the yeast *Saccharomyces cerevisiae* lead to filamentous growth; regulation by starvation and RAS, *Cell* 68: 1077 (1992).

Molecular Genetics of Morphogenetic Processes

MOLECULAR CONTROLS OF CONIDIOGENESIS IN *ASPERGILLUS NIDULANS*

William E. Timberlake

Department of Genetics
Department of Plant Pathology
University of Georgia
Athens, GA 30602-7223, USA

ABSTRACT

Three fungal species, *Aspergillus nidulans*, *Saccharomyces cerevisiae*, and *Neurospora crassa*, are providing insights into the mechanisms controlling fungal morphogenesis because they offer highly tractable molecular genetic systems and undergo morphogenetic processes representative of most fungal taxa. A common feature of these organisms is the ability to produce morphologically distinct cell types through alterations in the polarities and patterns of cell divisions. It is possible that homologous or analogous regulatory pathways exist that control cell division polarity to promote formation of hyphae or pseudohyphae or, alternatively, production of nearly spherical spores or yeast cells.

A. nidulans grows vegetatively by producing classical hyphae that are divided into multinucleate compartments by perforate septa. Hyphae show polarized growth, elongating by the addition of new cell wall material to the hyphal apex. By contrast, the asexual spores, conidia, are spherical cells produced on conidiophores by a specialized budding process that is controlled by at least three regulatory genes, *brlA*, *abaA*, and *wetA*. In *brlA* mutants, conidiophore initials are produced, but grow indeterminately as aerial hyphae; spherical expansion of the hyphal tip to form the conidiophore vesicle does not occur. Temperature shift experiments with thermosensitive *brlA* mutants have further shown that the *brlA* product is needed for formation of two tiers of conidiophore buds, metulae and phialides, and for conidium differentiation by the phialides. Thus, a primary activity of *brlA* appears to be regulation of the conversion from the hyphal growth form to a specialized budding growth form. Forced expression of *brlA* in hyphae under conditions that normally prevent conidiation leads to cessation of hyphal growth, and the production of conidia at the hyphal apices and cells reminiscent of blastospores subapically. In *abaA* mutants, metulae, which unlike phialides proliferate by axialary budding, undergo supernumerary divisions with no phialidic differentiation. In *wetA* mutants, conidia fail to undergo normal spore maturation and autolyse. The phenotypes of mutants suggest that these regulatory genes can be viewed to act by modifying the division patterns of the various specialized conidiophore cells. I discuss how the interactions of the products of the regulatory genes control expression of target structural genes and the regulatory genes themselves.

INTRODUCTION

Conidiation, the asexual reproductive pathway of the filamentous fungus *Aspergillus nidulans*, serves as a model for the study of the molecular and genetic controls of fungal morphogenesis. Conidia are uninucleate, mitotically-derived, environmentally resistant spores produced by multicellular conidiophores that project above the plane of the fungal colony. The controls of conidiophore development and spore differentiation have been subjected to extensive analysis, initially by classical mutagenesis and subsequently by use of recombinant DNA technology. The information gained has begun to provide a detailed understanding of how specific genes control the morphogenetic transitions that ultimately lead to differentiation of conidia. In this brief review of the molecular controls of conidiogenesis I will attempt to make three points. First, conidiophore development involves several types of cellular differentiations all of which are essential for normal spore production. Second, transitions from one developmental stage to the next are controlled by three major regulatory genes, designated *bristle (brlA)*, *abacus (abaA)*, and *wet-white (wetA)* that appear to encode pathway-specific transcriptional regulators. The products of these genes singly and in combinations activate responder genes whose products directly contribute to the form or specialized physiological activities of the conidiophore and conidia and also control their own activities. Third, these three regulatory genes can be manipulated to produce dramatic and informative changes in the growth form of the organism. Some of the modified growth forms suggest a linkage between the controls of the conidiation pathway and the controls of dimorphic switching. The reader is referred to Timberlake[1], Clutterbuck and Timberlake[2] and Timberlake and Clutterbuck[3] for more comprehensive reviews of the molecular genetic controls of conidiation in *A. nidulans*.

PATTERNS OF CELL DIFFERENTIATION DURING CONIDIOPHORE DEVELOPMENT

A. nidulans hyphae grow by apical extension involving deposition of new cell wall material at or near the hyphal tip. Cells near the periphery of a colony are incapable of producing conidiophores but, as they age and are internalized in the growing colony, they become developmentally competent, that is, they acquire the ability to respond to the environmental signals that induce conidiation.[4] Although what these signals are and how they are sensed and transduced are poorly understood, they almost certainly include exposure of the cells to an air interface[5] and to light[6] and probably localized depletion of nutrients. The first step of conidiophore development entails differentiation of a vegetative cell into an anchor or foot cell. Foot cells produce aerial branches that grow away from the plane of the colony to a height of approximately 100 μm, during which time multiple mitotic divisions occur. These conidiophore precursors, or stalk cells, are very similar to aerial hyphae that do not go on to form conidiophores and may indeed not be committed to this developmental pathway. Subsequent developmental steps, however, lead to production of differentiated cells that have a limited or no capacity to return to vegetative growth except through the production and germination of conidia.

Figure 1 shows the later stages of conidiophore development as observed by scanning and transmission electron microscopy. Apical extension of the stalk cell ceases (Figure 1A,F) and is followed by swelling of the tip to form a globose, multinucleate vesicle (Figure 1B,G). This is a major growth transition in which polarized (apical) growth is replaced by depolarized expansion. Formation of all subsequent cell types entails budding processes reminiscent of those displayed by budding yeasts. Buds, called metulae, form in close spacing on the surface of the

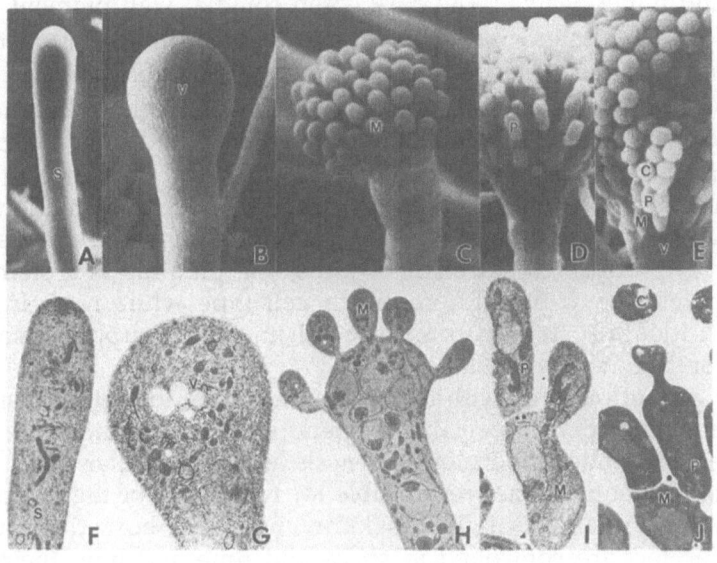

Fig. 1 Stages of conidiophore development. Panels A-E show scanning electron micrographs of the major steps leading to formation of a mature *Aspergillus* conidiophore. Panels F-J show transmission electron micrographs of cells at stages equivalent to those shown in A-E. An aerial hypha (A,F) is produced by a foot cell embedded in the growth medium and grows to a height of ~100 μm by addition of new cell wall material to the tip. The conidiophore stalk thus formed contains numerous nuclei due to multiple mitotic divisions that occur during stalk elongation. Once apical growth has been completed, the tip swells (B,G) to form the conidiophore vesicle. Numerous buds (metulae) form on the surface of the vesicle (C,H). These are subtended by nuclei that undergo a synchronous division, with one daughter nucleus entering each differentiating metula. Metulae undergo one or more divisions (D,I) to produce sporogenous phialides. Phialides form conidia (C) by undergoing repeated mitotic divisions, with one daughter nucleus being incorporated into the nascent spore and the other being retained in the phialide (J; Figure 2). The mature conidiophore (E) contains numerous clonal spore chains. Conidia continue to be produced by phialides for several days. Cell types are indicated as follows: S, conidiophore stalk; V, conidiophore vesicle; M, metula; P, phialide; C, conidium.

vesicle (Figure 1C,H) and the vesicular nuclei undergo a synchronous, oriented mitosis such that a single daughter nucleus enters each metula. Metulae then become separated from the vesicle by deposition of a septum, which is perforate, providing cytoplasmic connection with the remainder of the conidiophore. Metulae undergo from one to several divisions to produce phialides (Figure

1D,I).[7,8] Phialides, the sporogenous cells of the conidiophore, are uninucleate and retain cytoplasmic connection with metulae through perforate septa as shown in Figure 1I, where Woronin bodies, which are associated with septal pores in ascomycetes, can be seen within the metula just below the phialide septum. Thus, there is cytoplasmic continuity between the most distal conidiophore cells and the substrate mycelium. This continuity probably provides an essential source of elaborated nutrients to the conidiophore.

Phialides produce conidia by undergoing repeated mitotic divisions. Mitosis in the phialide is oriented along its axis. Following division, the prespore nucleus enters the swollen neck of the phialide, which is formed from the inner phialide wall layer,[7,9] and becomes arrested in the G1 phase of the cell cycle. The nascent spore then becomes separated from the phialide by formation of a complete septum, and the spore wall undergoes several maturation steps,[9] becoming dense, pigmented and impermeable. The dark green conidial wall pigment is derived from acetyl CoA by the stepwise action of a polyketide synthase encoded by wA [10] and a p-diphenol oxidase encoded by yA [11-13] and probably serves a structural function as well as protecting the spore from ultraviolet light.[14] In addition, the conidiophore sterigmata and conidia are covered by a hydrophobic rodlet layer composed in part of the product of the $rodA$ gene.[15] The phialide nucleus undergoes another division to repeat the process of spore formation. As this continues, newly formed spores displace previously formed spores, without detachment, to produce clonal chains. Thus, the phialide is a stem cell that repeatedly gives rise to one differentiated cell type while retaining its own differentiated identity. It is also a nurse cell in that it supplies many of the components of the outer spore wall.

A major goal of research with this system has been to identify genes whose products contribute to the specialized phenotypes of conidiophore cells and conidia to begin to understand morphogenesis at the molecular level. A second goal has been to identify genes responsible for regulation of the morphogenetic loci. Ultimately, one would like to know precisely how the activities of morphogenetic loci are controlled in space and time and how their products execute the assembly of this complex, multicellular, spore producing apparatus. Mutational studies[16] have identified genes of both types and have thus provided the basis for more recent molecular studies.

GENES CONTROLLING CONIDIOPHORE DEVELOPMENT

Clutterbuck and his colleagues subjected conidiation in *A. nidulans* to systematic mutational analysis, looking for mutants that were specifically altered in some aspect of the process. The mutants they identified can be considered to fall into three classes. The first class consists of mutants that are unable to begin conidiogenesis or produce decreased numbers of normal conidiophores and are often altered in patterns of vegetative growth. Some of these mutations probably identify genes that mediate environmental sensing and thus, when inactivated, prevent developmental initiation.[17] These genes are only now beginning to be analyzed in detail. Others, however, are more likely to identify genes whose activities are required for vegetative growth at low levels but are also required for conidiation but at higher levels. One such example is *trpC*, encoding a trifunctional enzyme needed for tryptophan biosynthesis. Loss of function *trpC* mutants grow in the presence of low supplementing levels of tryptophan, but fail to conidiate. High levels of tryptophan support both growth and conidiation.[18]

Interestingly, the *trpC* gene is developmentally regulated and tryptophan is the precursor of a dark brown pigment present in conidiophore cells.[19] Thus, some genes may play important roles both in vegetative growth and sporulation. On the other hand, many auxotrophs, for example arginine auxotrophs, conidiate poorly or not at all in the presence of low supplement levels that nevertheless support normal growth rates, but there is no evidence for specific involvement of the product of the pathway or a pathway intermediate in development. Thus, the effects of such auxotrophic mutations on development may be quite indirect.

The second class of mutants have deficiencies that are clearly restricted to the conidiogenesis pathway but produce essentially complete conidiophores and wild type numbers of viable spores. Some of these were identified soon after *A. nidulans* had been selected as a genetic model system: yellow- and white-spored strains (*yA* and *wA* mutants) that are defective in synthesis of the mature green conidial wall pigment (see above) were among the first *Aspergillus* mutants. Others were identified later in particular mutant backgrounds that were defective in synthesis of the tryptophan-derived conidiophore pigment, called ivory (*ivoA* and *ivoB*) mutants. Later cloning of these genes has shown that they are developmentally regulated, being inactive during vegetative growth and induced in conidiophores during development. In an alternative approach, recombinant DNA clones were selected containing genes that were specifically induced during conidiation.[20] Subsequent sequence analysis of these clones permitted identification of some (M.A. Stringer and W.E. Timberlake, unpublished results). Directed mutation of one such anonymous gene showed that it was essential for formation of the conidiophore and spore rodlet layer, and the gene was named *rodletless* (*rodA*) to reflect the mutant phenotype.[15] The identification of these genes is useful in at least two regards. First, the functions of their products provide insights into the physiological and structural alterations that are required for normal conidiophore development. Second, the genes represent molecular targets for loci that regulate gene activity during development and can be used to identify these loci and to characterize the activities of their products.

The final class of mutants have defects that result in the formation of highly abnormal or incomplete conidiophores and were found in extensive searches for strains that were able to initiate development but were asporogenous or oligosporogenous.[16] Two genes identified in these searches, bristle and abacus (*brlA* and *abaA*), are absolutely required for spore formation. In *brlA* mutants conidiophore stalks are produced, but no later cell types. In *abaA* mutants, metulae are produced but fail to form phialides. Instead, in the mutants, the metulae proliferate to produce abnormal chains of cells with properties similar to metulae.[8] A third gene, *wet-white*, is required for normal spore differentiation: mutants produce autolytic spores, which reduces spore viability by several orders of magnitude.[9] As will be discussed below, these genes are major regulators of conidiation-specific gene expression. In addition, oligosporogenous mutants have been identified, such as stunted and medusa (*stuA* and *medA* mutants) that produce deformed conidiophores but are nevertheless able to produce fairly normal, viable conidia, although inefficiently. These loci are likely to encode ancillary regulators of gene expression,[21] but are less well understood than *brlA*, *abaA*, and *wetA* and will not be discussed further here. The fact that even in extensive mutant searches a very limited number of loci was identified that when inactivated produced mutant phenotypes consistent with a regulatory role indicates that the actual number of such loci is low, almost certainly less than ten.

The *brlA*, *abaA* and *wetA* genes have been cloned and characterized in considerable detail.[22-25] Several lines of evidence support the conclusion that they are sequentially activated genes encoding transcription factors that alone and in combinations control expression of morphogenetic loci and the regulatory pathway itself. First, RNA blot analysis showed that each gene is developmentally regulated, their transcripts being undetectable in hyphae and accumulating during synchronously induced conidiation.[22] The developmental specificity of expression of the putative regulatory genes was confirmed in experiments in which their promoters were fused to the *lacZ* reporter gene[26,27] (M.A. Frizzell and W.E. Timberlake, unpublished results) and further showed that *brlA* and *abaA* are transcribed specifically in conidiophores, whereas *wetA* is transcribed mainly in differentiating spores. Second, double mutant studies[28] showed that *brlA* mutations are epistatic to *abaA* and *wetA* mutations and *abaA* mutations are epistatic to *wetA* mutations, and these relationships are manifested at the molecular level.[22] Furthermore, transcript accumulation patterns were consistent with *brlA* acting first, followed by *abaA* and then *wetA*. These results suggest that the putative regulatory genes define a dependent pathway: *brlA* --> *abaA* --> *wetA*. Third, the results of forced expression studies support this idea. The *brlA* structural gene was fused to the promoter from the *alcA* gene, encoding catabolic alcohol dehydrogenase, and the fusion gene was introduced into *Aspergillus* by transformation to produce strain TTA292.[23] Transcription from this promoter is repressed by glucose and induced by alcohols. This provided a mechanism for rapidly inducing *brlA* expression under conditions that normally completely suppress conidiation. Induction of *brlA* in TTA292 led to activation of morphogenetic loci such as *wA*, *yA* and *rodA* and of *abaA* and *wetA*. Similar experiments with an *alcA(p)::abaA* fusion gene also led to activation of some morphogenetic loci and *wetA* .[24] Somewhat unexpectedly, *abaA* induction in addition led to *brlA* activation, indicating that *abaA* is a positive feedback regulator of *brlA*. Moreover, these experiments showed that *wetA* mRNA accumulation is dependent upon integrity of the *wetA* gene, suggesting that the product of *wetA* is required for activation of the gene, that is *wetA* is positively autoregulatory. Forced expression of *wetA* induced transcription of mainly spore-specific genes but not of *brlA* or *abaA*. These results were used to formulate the regulatory model shown in Figure 2.

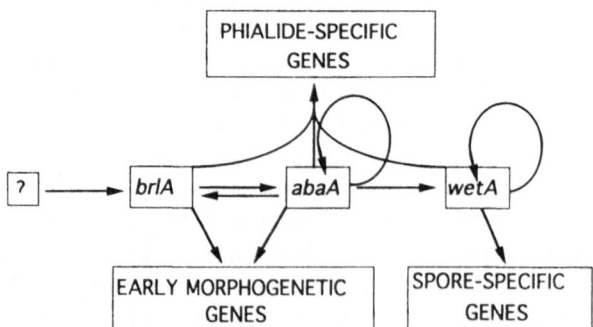

Fig. 2 A model for genetic control of *A. nidulans* conidiation.

The molecular genetic analyses of *brlA*, *abaA* and *wetA* provide no information about the biochemical functions of their products. However, the results from several other types of analysis are consistent with the hypothesis that each gene encodes a positively acting transcription factor. First, in the case of *brlA* and *abaA* , the peptides predicted from conceptual translation of the genes' mRNAs have highly significant similarities with the DNA binding domains of known classes of transcription factors.[23,29] The sequence of *wetA* also revealed features suggestive of a transcription factor, although no explicit sequence similarities were found in data base searches.[25] Second, *brlA* and *abaA* were capable of activating transcription in yeast of a minimal yeast promoter to which potential *cis*-acting regulatory regions from *Aspergillus* had been fused[30] (R. Aramayo, A. Andrianopoulos and W.E. Timberlake, unpublished results) and specific *bristle* and *abacus* response elements could be mapped by directed mutations. These elements were also capable of mediating gene-specific *trans* activation in *Aspergillus*. Third, in the case of *abaA*, polypeptide produced in *E. coli* binds *in vitro* specifically and with high affinity to the *abacus* response elements identified in yeast (A. Andrianopoulos and W.E. Timberlake, unpublished results). These elements occur upstream of transcription start in all known *abaA* inducible genes, including *brlA*. Interestingly, they also occur in the *abaA* promoter itself, indicating that *abaA* is subject to positive autoregulation, as shown in Figure 2. The positive feedback loops shown in Figure 2 could be responsible for making the core regulatory pathway independent of the environmental signals that are initially responsible for *brlA* activation, that is, represent the molecular basis for developmental commitment.

In summary, three regulatory genes have been identified that define a core pathway that appears to be responsible for controlling gene expression during conidiophore development. These genes probably encode positively acting transcription factors that induce expression of morphogenetic loci whose products contribute directly to the form and function of conidiophores and spores. They also regulate the activity of the pathway through a series of feedback loops that may act to control the levels of the regulatory proteins precisely as development proceeds, independent of the external queues that initially triggered the process. The model presented in Figure 2 predicts that molecular intervention in the core regulatory pathway might enable one to induce conidiation under conditions that are normally fully suppressive of the process. This prediction is met as described in the following section.

EFFECTS OF MISSCHEDULED EXPRESSION OF REGULATORY GENES

Forced expression of *brlA, abaA* or *wetA* in hyphae strongly inhibits growth and probably diminishes or eliminates the ability of the cells to acquire and metabolize nutrients from the medium.[3,23-25] As shown in Figure 3, forced expression of *brlA* leads to the most dramatic changes in the growth form of the fungus. In strain TTA292 (*alcA(p)::brlA, abaA$^+$;wetA$^+$*), induction of *brlA* leads to conversion of hyphal tips to reduced conidiophores that begin to produce conidia (Figure 3A). The conidia formed are fairly normal in shape, size, and ability to germinate, but show some interesting abnormalities in wall composition. They appear to lack outer wall layers so that the inner, chitinous wall can be stained with calcofluor, unlike the situation with conidia produced by normal, aerial conidiophores. It is probable that this defect is due to diffusion away from the spore of the outer wall components, for example pigment precursors and *p*-diphenol oxidase and rodlet proteins, because the genes encoding these products are activated normally. Prolonged incubation of TTA292 under inductive conditions leads to continued formation of conidia, including formation of conidia

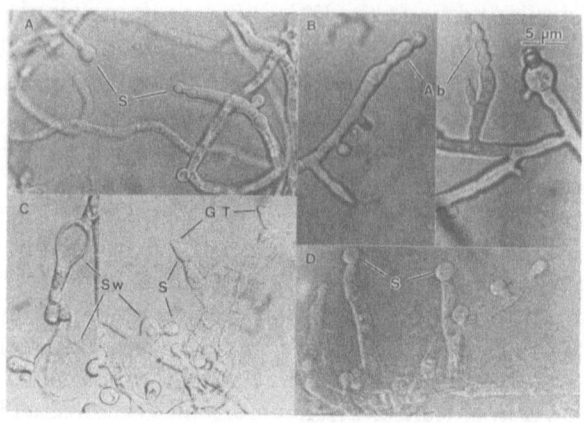

Fig. 3 Effects of developmental mutations on conidiation in submerged culture. The *alcA*(p)::*brlA* strain (A) was crossed with strains containing the *abaA14* (B), *wetA6* (C), or *stuA1* (D) morphogenetic mutations, and doubly mutant progeny were isolated. Development of double mutants in submerged culture was induced by growing them in the presence of L-threonine. Characteristic mutant structures are indicated by arrows: hyphae (H); spores (S); abacus-like structures (Ab); swollen regions (Sw); and germ tubes (GT).

by subapical budding (see Figure 1A) so that the medium becomes densely filled with spores. Simultaneously the mycelium becomes depleted of cytoplasm and highly vacuolated, suggesting that elaborated cellular components are being utilized for spore production. These results show that activation of one regulatory gene, *brlA*, is sufficient to drive the core pathway leading to spore differentiation. However, the reduced conidiophores lack most of the complexity of normal conidiophores. Thus, other gene activities are probably needed for the development of morphological complexity. In addition, formation of conidiophores and conidia in submerged culture may prevent normal assembly of structures due to loss of outer wall components. Activation of *abaA* or *wetA* is insufficient to cause sporulation.[24,25] With *abaA* , this was perhaps unexpected as forced expression leads to activation of the core regulatory pathway (see Figure 2), as does forced *brlA* expression. Thus, the order of gene activation may be crucial for morphogenesis. This would not be surprising as substrates for later developmental steps need to be elaborated by earlier steps in order for development to proceed in an orderly manner.

Forced expression of *brlA* in strains lacking *abaA* or *wetA* function also results in interesting alterations in growth form. As shown in Figure 1B, activation of *brlA* in an *abaA⁻* strain leads to formation of abacus like structures, not conidia. Prolonged stimulation of *brlA* expression causes continued proliferation of these structures so that the culture becomes densely populated with constricted cells that resemble yeast pseudohyphae. As shown in Figure 1C, activation of *brlA* in a *wetA⁻* strain permits spore formation, but the spores undergo premature germination as evidenced by the germ tubes indicated in the figure. Some of these germ tubes elongate slightly and then produce another spore. Prolonged stimulation of *brlA* expression continues this process so that the culture becomes densely populated by germinated and ungerminated conidia. This process is reminiscent of budding growth in some fungi. Thus, experimental intervention in the conidiation control pathway can lead to dramatic, and unexpected, changes in growth form, with the novel growth forms resembling normal growth forms displayed by some other fungi.

CONCLUSIONS

Several genes have been identified that control conidiophore development in *A. nidulans*. These appear to encode positively acting transcription factors that control expression of one another through a series of feedback loops and expression of morphogenetic loci. As we gain an understanding of how these genes are regulated and how their products act and interact to control gene expression it has become possible to utilize them to alter the growth form of the organism. As more information is acquired concerning additional regulators, such as *stuA* and *medA*, it may become possible to convert the normally filamentous growth form of *Aspergillus* to forms characteristic of truly pleomorphic fungi such as *Candida albicans*. It is possible that homologous or analogous regulatory pathways operate to control sporulation and pleomorphic switching. We have found that other *Aspergillus* and *Penicillium* species utilize regulatory pathways similar to that identified in *A. nidulans* to control conidiophore development. It will be of considerable interest to see if more distantly related organisms, such as *Candida* and *Saccharomyces*, have homologous genes and, if so, what their functions are in controlling growth and development.

REFERENCES

1. W. E. Timberlake, Molecular genetics of *Aspergillus* development, *Annu. Rev. Genet.* 24: 5 (1990).
2. A. J. Clutterbuck and W. E. Timberlake, Genetic regulation of sporulation in the fungus *Aspergillus nidulans*,, in: "Development, The Molecular Genetic Approach", V. E. A. Russo, S. Brody, D. Cove, and S. Ottolenghi, eds., Springer-Verlag, Berlin (1992).
3. W. E. Timberlake and A. J. Clutterbuck, Genetic regulation of conidiation, in: "Physiology and Genetics of *Aspergillus*", D. Martinelli and J. Kinghorn, eds., Chapman and Hall, London. In press. (1993).
4. D. E. Axelrod, M. Gealt and M. Pastushok, Gene control of developmental competence in *Aspergillus nidulans, Dev. Biol.* 34: 9 (1973).
5. D. E. Axelrod, Kinetics of differentiation of conidiophores and conidia by colonies of *Aspergillus nidulans. J. Gen. Microbiol.* 73 : 181 (1972).
6. J. L. Mooney and L. N. Yager, Light is required for conidiation in: *Aspergillus nidulans,* Genes Dev. 4: 1473 (1990).
7. C. W. Mims, E .A. Richardson and W. E. Timberlake, Ultrastructural analysis of conidiophore development in the fungus *Aspergillus nidulans* using freeze-substitution, *Protoplasm* 44: 132 (1988).
8. T. C. Sewall, C. W. Mims and W. E. Timberlake, *abaA* controls phialide differentiation in *Aspergillus nidulans, Plant Cell* 2: 731 (1990).
9. T. C. Sewall, C. W. Mims and W. E. Timberlake, Conidial differentiation in wild type and *wetA⁻* strains of *Aspergillus nidulans, Dev. Biol.* 138 : 499 (1990).
10. M. E. Mayorga and W. E. Timberlake, The developmentally regulated *Aspergillus nidulans wA* gene encodes a polypeptide homologous to polyketide and fatty acid synthases, *Mol. Gen. Genet.* In press (1992).
11. A. J. Clutterbuck, Absence of laccase from yellow-spored mutants of *Aspergillus nidulans, J. Gen. Microbiol.* 70: 423 (1972).
12. D. J. Law and W. E. Timberlake, Developmental regulation of laccase levels in *Aspergillus nidulans, J. Bacteriol.* 144: 509 (1980).
13. R. Aramayo and W. E. Timberlake, Sequence and molecular structure of the *Aspergillus nidulans yA* (laccase I) gene, *Nucl. Acids Res.* 18: 3415 (1990).
14. R. Aramayo, T.H. Adams and W.E. Timberlake, A large cluster of highly expressed genes is dispensable for growth and development in *Aspergillus nidulans, Genet.* 122: 65 (1989).

15. M. A. Stringer, R.A. Dean, T.C. Sewall and W.E. Timberlake, *Rodletless,* a new *Aspergillus nidulans* developmental mutant induced by directed gene inactivation, *Genes Dev.* 5: 1161 (1991).

16. A. J. Clutterbuck, A mutational analysis of conidial development in: *Aspergillus nidulans, Genet.* 63 : 317 (1969).

17. T. H. Adams, W. A. Hide, L. N. Yager and B. N. Lee, Isolation of a gene required for programmed initiation of development by *Aspergillus nidulans, Mol. Cell. Biol.* In press (1992).

18. M. M. Yelton, J. E. Hamer, E. R. de Souza, E. J. Mullaney and W. E. Timberlake, Developmental regulation of the *Aspergillus nidulans trpC* gen, *Proc. Natl. Acad. Sci. USA* 80: 7576 (1983).

19. C. E. Birse and A. J. Clutterbuck, N-acetyl-6-hydroxytryptophan oxidase, a developmentally controlled phenol oxidase from *Aspergillus nidulans, J. Gen. Microbiol.* 136: 1725 (1990).

20. C. R. Zimmermann, W. C. Orr, R. F. Leclerc, E. C. Barnard and W. E. Timberlake, Molecular cloning and selection of genes regulated in *Aspergillus* development, *Cell* 21: 709 (1980).

21. K. Y. Miller, J. Wu and B. L. Miller, StuA is required for cell pattern formation in *Aspergillus, Genes Dev.* In press (1992).

22. M. T. Boylan, P. M. Mirabito, C. E. Willett, C. R. Zimmermann and W. E. Timberlake, Isolation and physical characterization of three essential conidiation genes from *Aspergillus nidulans, Mol. Cell. Biol.* 7: 3113 (1987).

23. T. H. Adams, M. T. Boylan and W. E. Timberlake, *brlA* is necessary and sufficient to direct conidiophore development in *Aspergillus nidulans, Cell* 54: 353 (1988).

24. P. M. Mirabito, T. H. Adams and W. E. Timberlake, Interactions of three sequentially expressed genes control temporal and spatial specificity in *Aspergillus* development, *Cell* 57: 859 (1989).

25. M. A. Marshall and W. E. Timberlake, *Aspergillus nidulans wetA* regulates spore-specific gene expression, *Mol. Cell. Biol.* 11: 55 (1991).

26. J. Aguirre, T. H. Adams and W. E. Timberlake, Spatial control of developmental regulatory genes in *Aspergillus nidulans, Exp. Mycol.* 14: 290 (1990).

27. T. H. Adams and W. E. Timberlake, Upstream elements repress premature expression of an *Aspergillus* developmental regulatory gene, *Mol. Cell. Biol.* 10: 4912 (1990).

28. S. D. Martinelli, Phenotypes of double conidiation mutants of *Aspergillus nidulans, J. Gen. Microbiol.* 114: 277 (1979).

29. A. Andrianopoulos and W. E. Timberlake, ATTS, a new and conserved DNA binding domain, *Plant Cell* 3: 747 (1991).

30. Y. C. Chang and W. E. Timberlake, Identification of *Aspergillus brlA* response elements (BREs) by genetic selection in yeast, *Genet.* In press (1992).

31. T. H. Adams and W. E. Timberlake, Developmental repression of growth and gene expression in *Aspergillus, Proc. Natl. Acad. Sci. USA* 87: 5405 (1990).

PHYSICAL AND GENETIC MAPPING OF *CANDIDA ALBICANS*

B.B. Magee[1], Wen-Shen Chu[2], Bernhard Hube[3], Rachel J. Wright[4],
E.H.A. Rikkerink[5], and P.A. Sullivan[4]

[1] Department of Genetics and Cell Biology, 250 Bio Sciences
Center, University of Minnesota, St. Paul, MN 55108
[2] Present address: Food Industry Research and Development
Institute, P.O. Box 246, Hsinchu, 30099 Taiwan, Republic of
China
[3] Department of Molecular and Cell Biology Marischal College,
University of Aberdeen, AB9 1AS, Scotland
[4] Biochemistry Department, University of Otago, Box 56, Dunedin,
New Zealand
[5] New Zealand Department of Scientific and Industrial Research,
Mt. Albert Research Centre, 120 Mt. Albert Road, Private Bag,
Auckland

ABSTRACT

Pulsed-field gel electrophoresis was used to analyze several aspects of the
genetics of *Candida albicans*. A restriction map of the genome of strain 1006 was
prepared using the eight-base-pair specific enzyme *Sfi*I. The previously
determined genetic map was superimposed on this physical map. Preparation of a
similar map for the white-opaque strain WO-1 demonstrated the presence of three
reciprocal chromosomal translocations in this organism. Mapping the cloned
genes for the extracellular aspartyl proteinase to the electrophoretic karyotype
under different stringencies demonstrated that there are at least four genes
homologous to both the cloned genes.

INTRODUCTION

The importance of *Candida albicans* in human disease has led to a great deal of
work on this organism in the last 10 years. As the single most important fungal
human pathogen, this opportunistic yeast has been studied from a large number
of aspects. Much work has been done to try to identify virulence factors, such as
its extracellular acid proteinase, its ability to grow at 37 °C and above, and its
special ability to adhere to mammalian cells. In addition, such properties as

complement and steroid receptors have been examined. Characteristics of the cell cycle, such as the yeast to hyphal dimorphism and phenotypic colony morphology and cell shape transitions have been studied as well.[1]

Although much information has been gathered about these properties, the difficulty in carrying out a genetic analysis in *C. albicans* has greatly hindered our ability to resolve the question of the importance of these characteristics in pathogenesis of the organism. The demonstration of transformation and the construction of a replicating plasmid has greatly advanced the genetics,[2,3] but it is still quite difficult to correlate the results of parasexual genetic analysis and molecular biology.

Parasexual genetics is effective for determining linkage and gene order for genes relatively far apart,[4-7] but it is not particularly effective in determining genetic fine structure. The appropriate multiple auxotrophs are cumbersome to construct, and getting sufficient data on the frequency of rare recombinants is also difficult. It therefore seems that physical mapping may be a better way to look at gene order when the genes are close together. This approach, however, will require much more than a simple assignment to chromosomes, especially since the chromosomes of *C. albicans* range in size from 1 to 4.3 megabases.[8-10]

Poulter and his colleagues[7] have led in the development of the linkage map of *C. albicans*. They have demonstrated the existence of 7 linkage groups, to each of which at least two genes are assigned. We have classified many of the auxotrophic mutations isolated in our laboratory and Poulter's into complementation groups. It is therefore possible to use linkage data from both laboratories to expand the genetic map. We will present some the these data below.

The advent of pulsed-field gel electrophoresis,[11] which enables one to separate megabase-sized DNA molecules, suggests a way to correlate the two kinds of analysis. Establishment of a definite electrophoretic karyotype allowed the unequivocal demonstration of 8 as the chromosome number;[9,10] together with the fact that seven linkage groups had been demonstrated, this suggested that correlation of the electrophoretic karyotype (the physical map) with the genetic map must be quite straightforward.

The general approach to this correlation is simple: several of the genes whose mutation helped to define the genetic map have been cloned; hybridizing these to a southern blot of electrophoretic karyotype shows which electrophoretic band corresponds to a particular linkage group.[8]

Cloning by complementation of auxotrophies in *C. albicans* is complicated by the fact that the plasmids presently available tend to replicate as oligomers in this organism;[3] hence, complementation of a particular auxtrophy by transformation with a library usually yields a large plasmid composed of several different small ones, and determining which of the plasmids from the original library carries the complementary sequence is cumbersome although not impossible.[12] Far easier is cloning by complementation in *Saccharomyces cerevisiae* or *Escherichia coli*. As a result, the most frequent approach is first to isolate *Candida* genes, by complementation of a specific *S. cerevisiae* auxotrophy (for example, *ura3*[13]), then to use the gene to transform various *C. albicans* Ura⁻auxotrophs to determine which corresponds to *ura3*. If the gene complementary to a mapped auxotrophy can be found in this way, it can be assigned to the electrophoretic map and the linkage group assigned to a particular chromosome.

In order to refine this physical assignment of the genetic loci to expand the linkage data, we decided to construct a restriction map of the *C. albicans* genome, using the enzyme *Sfi*I.

In this paper we describe the construction of this map. This map was then used to analyze the karyotypic abnormalities of the white-opaque strain, WO-1; these abnormalities turned out to result from three reciprocal chromosomal translocations. Finally, in assigning the genes encoding the extracellular acid proteinase to the electrophoretic karyotype, we have obtained evidence that there is a family of such proteinase genes.

MATERIALS AND METHODS

Strains: The strains used were 1006 (*ura3 lys1 arg57 ser57*),[12] ATCC10261 and ATCC10231 (both obtained from the American Type Culture Collection), SC5314,[13] and WO-1.[14] Pulsed field electrophoresis conditions were as follows: for karyotype separation and for the largest *Sfi*I bands, 120-300 s(linear ramp) 30 h, 420-900 s(linear ramp) 66$_h$, 900-1200 s(linear ramp) 30 h at 65 V in 0.4% chromosome grade agarose (Bio-Rad); for the medium-sized *Sfi*I bands, 60 s/15 h, 90 s/8 h, 120 s/10 h at 200 V in a 1% agarose gel; and for the smallest *Sfi*I bands, 10-45 s (linear ramp) 30-34 hr/1% agarose gel/200 V. All the PFGEs were run at 12-15° C in 0.5 X TBE buffer (0.045 M Tris-base, 0.045 M boric acid, 1 mM EDTA). Gels were transferred to MSI magnagraph filters as described previously[8] and probed with sequences labeled with ^{32}P by oligo priming. Stringency washes were performed according to MSI protocols. The enzyme *Sfi*I (New York Biolabs) was used in accordance with the instructions of the manufacturer.

RESULTS

Construction of a *Sfi*I Map of the Genome of *C. albicans* Strain 1006

Trial digestion of the *C. albicans* chromosome preparation with the enzyme *Sfi*I (of nucleotide specificity GGCCNNNNNGGCC) suggested that it generated a reasonable number of fragments with a broad range of sizes (from 45 Kb to 1500 Kb). We therefore chose this enzyme to prepare our map.

Separation of all the *Sfi*I bands required three sets of pulsed-field gel electrophoresis conditions, as shown in Fig. 1. It is evident from the brightness of the various bands on the stained gel that some of these bands contain more than one molecular fragment, for example bands S and O. Separations like these in Fig. 1 were blotted and hybridized with 54 sequences which had previously been assigned to chromosomes. Table 1 shows these results. Several of the electrophoretic bands clearly contain fragments of similar size from different chromosomes, since probes from more than one chromosome hybridize to them. A similar blot was probed with a plasmid containing a subtelomeric repeated DNA fragment. The fragments which hybridize to this plasmid are telomeric. The assignment of fragments to chromosomes and the identification of the ones which are telomeric does not by itself allow us to determine fragment order. Furthermore, fragments for which we happen to have no homologous sequences cannot be assigned to a particular chromosome using these techniques. Therefore, we decided to use Smith-Bernsteil mapping to order fragments. This technique involves digestion for various times with *Sfi*I. These samples are then subjected to pulsed-field gel electrophoresis and the blots probed with a sequence which is on a telomeric fragment. Fig. 2 shows an informative experiment of this sort. The band that hybridizes with the sequence GAL1 at early times is chromosome 1. At later

Table 1. The results of hybridization of the chromosomes and the *Sfi*I fragments of the strain 1006 with cloned sequences.

Probe	Chromosome	*Sfi*I Fragment
l609 (crDNA)	R_1, R_2	U, S
p1055 (*ADE1*)	R_1, R_2	M, S
pCDC10	R_1, R_2	K, S
pYSK210 (*MGL1*)	R_1, R_2	B_2
cos19BF$_1$	R_1, R_2	K, S
pYSK231 (*SOR9*)	R_1, R_2	U, S
pYSK208 (*GAL1*)	1	S
p1002 (*ACT1*)	1	L
pHS100 (*TUB2*)	1	S
pCDC3	1	S
pJG-T1 (*TRP1*)	1	S
p1077 (*SER57*)	1	L
pKC1 (*LYS2*)	1	S
pAL3-1 (*SEC18*)	1	L
pAR84-3 (*HIS3*)	2	U
p1926 (*TS*)	2	U
pHPT1	2	U
p3e-6	2	U
pCHR2	2	U
p6b-1	2	U
pMK3 (*ADE2*)	3	P
pYSK230 (*SOR2*)	3	P
pET18 (*URA3*)	3	O
p190	3	P
P3e-4	3	P
cos33BF$_1$	3	P
pCA1 (*CA*)	3	P
p79	3	P
p1123 (*LYS1*)	4	F_2
STC7 (*HIS4*)	4	O, BB
pCHR4	4	F_2
p3g-1	4	F_2
p99	4	F_2
p316	4	O
p356	4	H
pTK2-9-1	5	M
snc3	5	I
snc5	5	I

Table 1. (Continued)

Probe	Chromosome	*Sfi*I Fragment
p2002 (MTX^R)	6	O
p3-167 (BEN^R)	6	O
pMK22 (CARS)	6	C_2, C_3
pAR84-1	6	O
pTK2-25-1	6	C_2, C_3
pJB4 (*ADH*)	6	O
pCAP	6	O
p1879 (*DHFR*)	7	G
p1129 (ARG57)	7	G
pAL1 (*LEU2*)	7	D_2, C_1, C_2
pCHR7	7	F_1
p212	7	D_2, A_1
p282	7	D_2, A_1

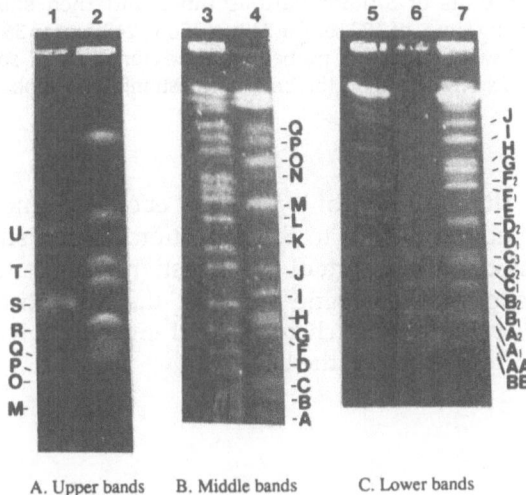

A. Upper bands   B. Middle bands   C. Lower bands

Fig. 1 Electrophoretic separation of *Sfi*I digestion fragments. Chromosome preparations in agarose beads were digested with the restriction enzyme *Sfi*I. The beads were then subjected to one of three electrophoresis conditions, according to the size of the fragments to be separated. A. Large bands: 120-300 sec switching time on a linear ramp for 30 h, 420-900 sec on a linear ramp for 66 h, and 900-1200 sec on a linear ramp for 30 h at 65 v in 0.4 % agarose. B. Middle-sized bands: 60 sec for 15 h, 90 sec for 8 h and 120 sec for 10 h at 200 v in 1 % agarose. C. Small bands: 10-45 sec on a linear ramp for 30-34 h at 200 v in 1 % agarose. The instrument was a Bio-Rad Chef II; all runs were at 12 C.

times, smaller bands appear, first at 1900Kb and then at 1400. We can then infer that the *Sfi*I map of chromosome 1 is in part S-J. Since S is a telomeric fragment, and since LYS1 and TRP1, which are on the other side of the centromere from *GAL1*, also hybridize to this fragment, these two pieces constitute one arm, the centromere, and part of the second arm of chromosome 1.

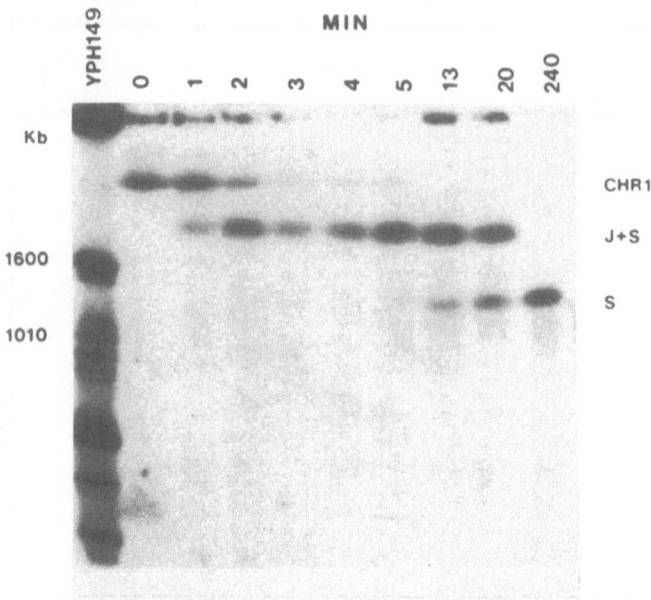

Fig. 2 Partial digestion of the genome to determine band order. The beads prepared as in Fig. 1 were digested with 2.5 units of *Sfi*I for various times and then subjected to pulsed-field electrophoresis at switching times of 120 sec for 24 h at 150 v, 240 sec for 36 h at 150 v, and 1200 sec for 24 h at 65 v. The gel was blotted and probed with the cloned *GAL1* sequence. YPH149 is the karyotype of an *S. cerevisiae* strain; MIN is the time of digestion; Kb is kilobases.

To determine the number of bands in each chromosome we isolated chromosomes and subjected them to *Sfi*I digestion. Fig. 3 shows an example of these results. The electropherogram of the digestion was blotted and probed with the telomeric sequences to determine which of the chromosomal fragments are telomeric. In addition, individual chromosomal digestions were blotted with the labeled chromosome to identify all the bands.

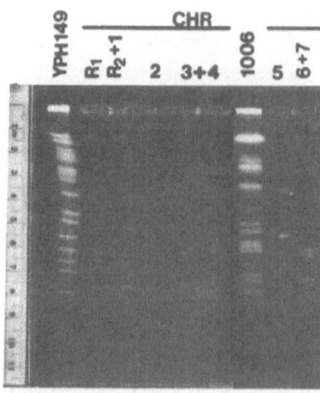

Fig. 3 Isolated chromosome analysis to determine telomeric fragments. Chromosomes were excised from a standard pulsed-field electrophoretic separation. They were then digested with *Sfi*I and subjected to pulsed-field electrophoresis as in Fig. 1B, except that the 120 sec pulse for 10 h was omitted. The resulting gel was probed with a telomere-associated sequence.

From these experiments we have determined that these are 33 *Sfi*I fragments of 40 Kb or more generated by complete digestion of the genome of *C. albicans* strain 1006. Fig. 4 shows the *Sfi*I map of this strain.

The chromosomes which carry the rDNA genes in *C. albicans* are highly variable in size and, depending on the strain, can migrate on a pulsed-field gel in any position from well above chromosome 1 to at or below chromosome 2. Since the chromosomes are numbered from the largest to the smallest, we have chosen to call the rDNA-bearing chromosome "R", making no assumptions about where it migrates. Fragments V and S carry the rDNA genes from chromosomes R_1 and R_2,

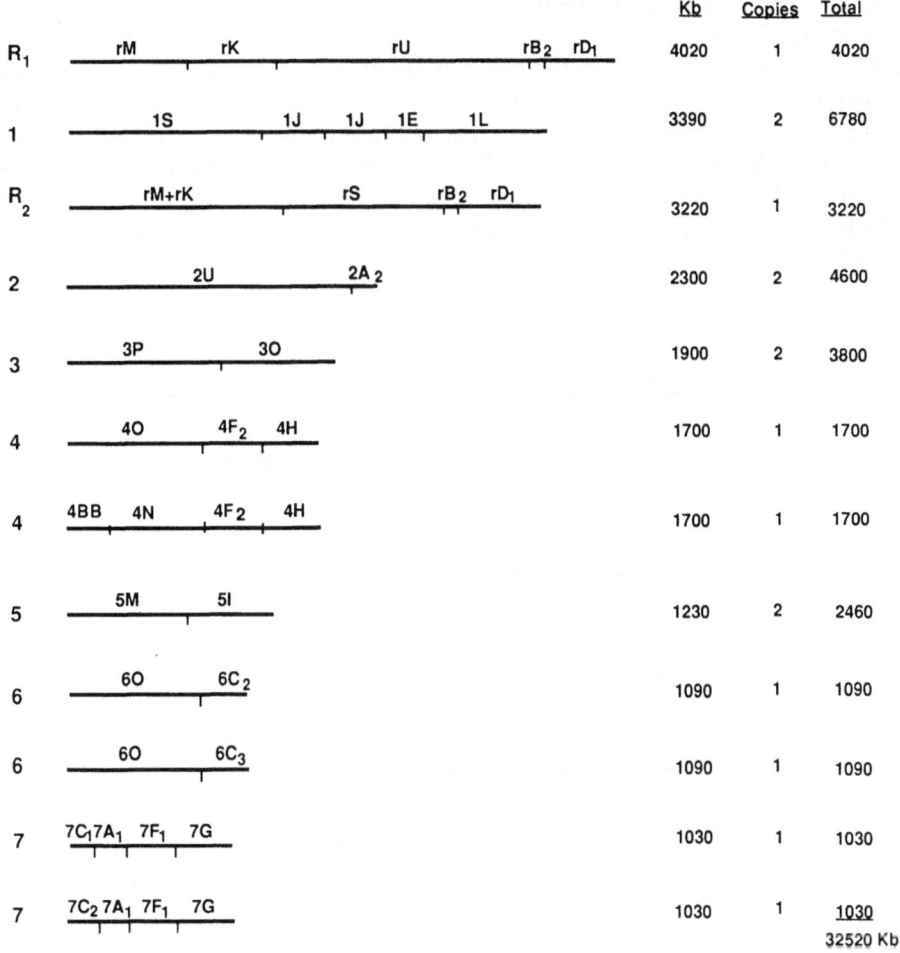

Fig. 4 *Sfi*I restriction map of the genome of *Candida albicans* strain 1006. The numbers on the right are the sizes of the chromosomes in kilobases.

respectively. The difference in size of these fragments accounts for the difference in size between the two homologues.

Some of the largest chromosomes (2 and 3) of strain 1006 apparently have only one *Sfi*I site, while smaller ones, like 4, have several. Thus, to a first approximation, the *Sfi*I sites are not randomly distributed throughout the chromosomes. The sum of the chromosome sizes adds up to 32 Mb, well within the range of estimated sizes for the *C. albicans* diploid genome. As described below, there are certainly small *Sfi*I fragments (<2 Kb) which we have not accounted for, and there may be some in the 1-40 Kb range which we have overlooked.

Correlation of the Physical Map with the Genetic Map

Figure 5 shows the genetic map superimposed on the physical map; the relevant *Sfi*I sites are indicated. Linkage group 1 corresponds to chromosome R; we have no clones of other genes to allow us to develop a fine structure map. Chromosome 1 contains the genes assigned to linkage group 7. We have made most progress in mapping this chromosome. *GAL1* is on one arm; the other arm contains, beginning with the centromere-proximal gene, *LYS2* ,*TRP1*, and *SER57*. Interestingly, *GAL1*, *LYS2* and *TRP1* all hybridize to the same *Sfi*I fragment, 1S, so this 800 Kb piece must contain the centromere as well. Chromosome 3 corresponds to linkage group 3, but since it has only 1 *Sfi*I site, the physical map is not very informative. However, *ade2* and *ura3*, which are on separate *Sfi*I fragments, map on opposite sides of the centromere. Chromosome 4 corresponds to linkage group 2; it contains *lys1* and *his4*. Chromosomes 2, 5, 6, and 7 have not been assigned to any linkage groups, although chromosome 7 has two auxotrophic markers *arg57* and *leu2*, assigned to it.

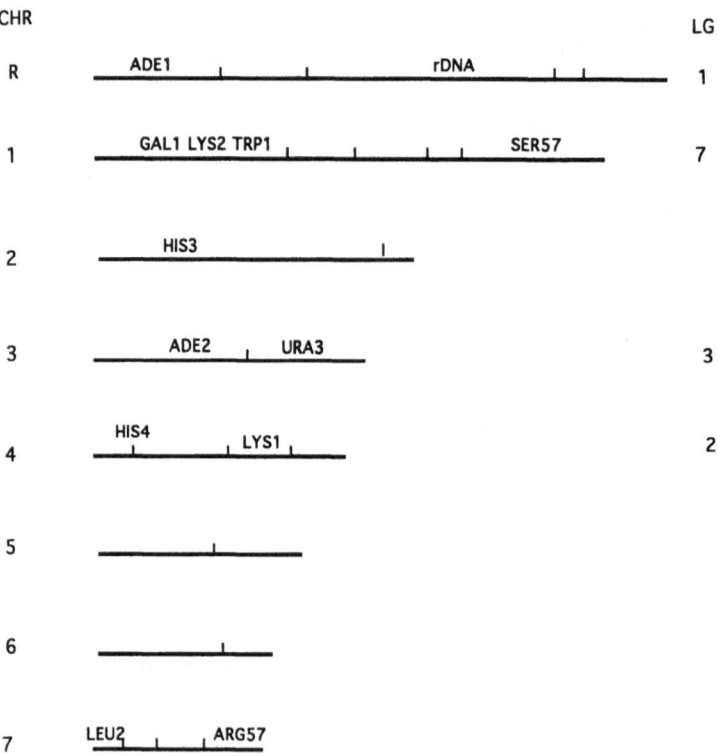

Fig. 5 Relationship of the genetic and physical maps of strain 1006.

The Switching Strain WO-1 Has 3 Chromosomal Translocations

Strain WO-1 carries out the white-opaque transition, which involves a change in cell shape, as well as in many other properties.[14-16] The electrophoretic karyotype of strain WO-1 differs from that of strain 1006 in several ways: the relative intensity of some bands differs between the two (e.g., chromosome 2) and there are extra bands in WO-1 compared to 1006 (chromosomes 4.5 and snc [for supernumerary chromosome]). Since WO-1 is the "type strain" for the white-opaque transition, it is important to determine the nature of these differences to determine whether they play a role in switching. We therefore set out to construct a *SfiI* map of this strain to see how it differed from that of 1006.

We first asked whether the gene-chromosome assignments were the same in WO-1 as in 1006. Many genes which hybridized to single band on the electrophoretic karyotype of 1006 hybridized to two bands in WO-1. This behavior has been observed previously for the G-protein isolated from *C. albicans* [17] and for the acid proteinase which seems to be differentially expressed in opaque cells.[18] Although this result suggests that the genomic organization of WO-1 might be very different from that of 1006, examination of the bands produced from the two strains by *SfiI* digestion shows very few differences. Furthermore, hybridization of the cloned sequences to blots of the separation of the *SfiI* fragments demonstrates that although many of the clones hybridize to two "chromosomes," none hybridize to more than one *SfiI* band. We conclude that although the two strains may differ at the level of organization of chromosomes, at the next lower level of organization, that probed by *Sfi* I digestion, they are very similar.

The *SfiI* map of strain WO-1 was prepared by using the methods described for strain 1006. Figure 6 shows the *SfiI* map of the genome of this organism. There are several important aspects to this map. The first is that WO-1 has undergone three chromosomal translocations: one homologue of chromosome 5 has undergone a reciprocal translocation with one homologue of chromosome 1, and both products now migrate at the same rate as chromosome 2. The other homologue of chromosome 5 has translocated with one homologue of chromosome 6. One of these translocation products migrates with chromosome 4 and the other is the 800 Kb supernumerary chromosome. The third reciprocal translocation occurred between one homologue of chromosome 4 and one homologue of chromosome 7. These products run as chromosome 4.5 and chromosome 5 on the WO-1 electrophoretic karyotype.

The second important point is that WO-1, at the level of the *SfiI* map, is diploid. That is, all *SfiI* fragments are present in two copies, and the total amount of genetic information in the strain is the same, within error, as that in 1006. The third important point is that all the translocations appear to occur at or near *SfiI* sites. Although only three translocations have occurred, it may be significant that none occurred except at *SfiI* sites. A possible explanation for this behavior is the existence of a repeated sequence which contains 4 *SfiI* sites within about 2 Kb.

This repeated sequence, RPS-1, studied by Iwaguchi *et al.* [19] hybridized to the 27A sequence identified by Scherer and Stevens.[20] Like 27A, RPS-1 occurs on every chromosome but chromosome 3. Figure 7 shows a schematic depiction of this sequence. We postulate that RPS-1 and other parts of 27A constitute regions of homology which lead to ectopic pairing, recombination between non-homologous chromosomes, and thus translocation. If this is true, and if this repeated sequence accounts for most of the *SfiI* sites in the genome, translocations would occur preferentially at or near *SfiI* sites. We are presently cloning a chromosomal break point from strain WO-1 to test this hypothesis.

Mapping the Acid Proteinase Demonstrates the Presence of a Gene Family

The extracellular acid proteinase, first observed by Staib[21] and studied intensively by Ruchel,[22] has been considered a likely candidate for a virulence factor in *C. albicans*.[23,24] The gene has recently been cloned by Hube *et al.*,[25] by

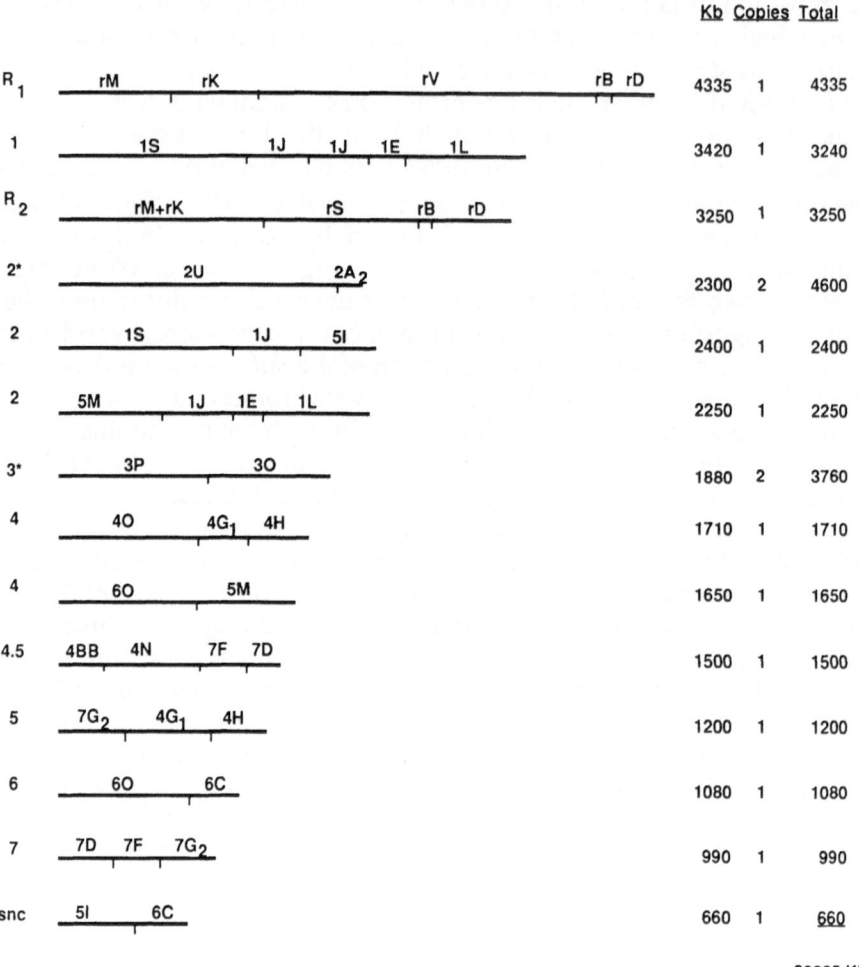

			Kb	Copies	Total
R₁	rM rK rV rB rD		4335	1	4335
1	1S 1J 1J 1E 1L		3420	1	3240
R₂	rM+rK rS rB rD		3250	1	3250
2*	2U 2A₂		2300	2	4600
2	1S 1J 5I		2400	1	2400
2	5M 1J 1E 1L		2250	1	2250
3*	3P 3O		1880	2	3760
4	4O 4G₁ 4H		1710	1	1710
4	6O 5M		1650	1	1650
4.5	4BB 4N 7F 7D		1500	1	1500
5	7G₂ 4G₁ 4H		1200	1	1200
6	6O 6C		1080	1	1080
7	7D 7F 7G₂		990	1	990
snc	5I 6C		660	1	660

32805 Kb

Fig. 6 *Sfi*I restriction map of the genome of *Candida albicans* strain WO-1.

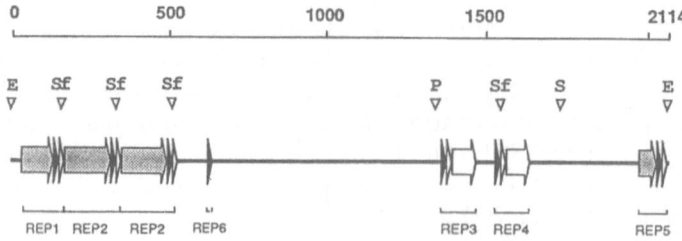

Fig. 7 A schematic representation of the repeated sequence RPS-1 characterized by Iwaguchi, *et al.*[19]

Wright *et al.*,[26] and by Ganesan *et al.*[27] The three groups used different strains to prepare their libraries, and the clones from the first two groups are only 77% homologous at the DNA level. (The sequence of the clone from Ganesan *et al.* is unavailable.) To determine whether this was due to strain differences, we used a portion of the coding sequence of the two clones to probe electrophoretic karyotypes to determine where each mapped. Fig. 8B shows that the clone from Hube *etal.*, *PRA10*, maps to chromosome 6, while Sullivan's clone, *PRA11*, maps to chromosome R. Thus, these clones represent two different genes, and both genes are present in each of our strains.

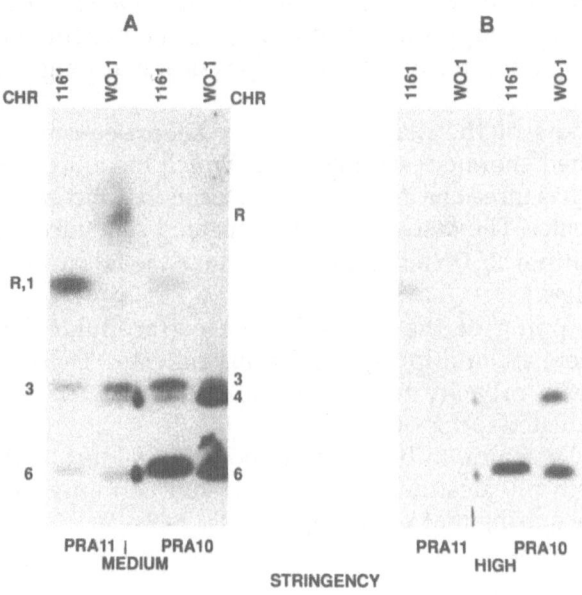

Fig. 8 Probe of CHEF chromosome separations with the clone *PRA10* and *PRA11* genes. A. The blot washed at medium stringency. B. The blot washed at high stringency.

The blots described in the previous paragraph were washed at high stringency. The presence of two unlinked genes suggested to us that there might be more genes in the *C. albicans* genome related to this putative virulence factor. We therefore probed the blots in Fig. 8B with the same probes, but this time we washed at medium stringency. Figure 8A shows that when probed in this fashion, the blot revealed that several chromosomes apparently contain homologous genes. In addition to chromosomes R and 6, chromosomes 3 and 4 hybridize to the probes with a strong signal.

This mapping is supported by southern hybridizations of restriction enzyme digests of genomic DNA. At high stringency each of the two clones hybridizes only with its cognate fragments; this is true both for the strains from which the genes were cloned and for three other unrelated strains. At low stringency each probe lights up the fragments corresponding to the other gene, and in addition several new bands appeared. We infer that these bands come from the genes located on chromosomes 3 and 4.

It is extremely interesting that there should be multiple genes for the extracellular acid proteinase in *C. albicans* since this protein has been implicated as an important virulence factor in these organism. Multiple genes might function as a sort of redundancy, to be sure that the capacity to make the enzyme is not lost.

Alternatively, they might be differentially regulated, depending upon the niche in which the organism finds itself (different hosts, different organs, differing immune responses). Morrow et al. [18] have recently shown that the PRA10 gene is expressed in opaque but not white cells of WO-1, so that regulation of proteinase appearance appears also to be connected with the phenotypic transition. Characterization of these multiple genes will no doubt clarify their roles if they do have different ones.

CONCLUSIONS

*Sfi*I is a very useful enzyme for mapping the *Candida albicans* genome. This map will be useful in supporting and clarifying the genetic map. It will also be useful as a starting point for fine-structure physical mapping of the genome of the organism.

The usefulness of the map has already been demonstrated by its use in characterizing the genomic rearrangements which have occurred in WO-1. This switching strain has three chromosomal translocations, and each has occurred at or very near a *Sfi*I site. The association of a repeated sequence cloned by Iwaguchi et al.[19] with multiple *Sfi*I sites may account for elevated recombination in the region of these sites.

Physical mapping of the genes for the extracellular acid proteinase has indicated that there are multiple genes for this enzyme. These genes may provide redundancy in the capacity to secrete the proteinase, or they may represent differentially regulated forms of the protein.

A great deal remains to be learned about *C. albicans*, and the expansion of genetic tools both physical and parasexual, will certainly help us move more rapidly toward acquiring that knowledge.

ACKNOWLEDGMENTS

This work was supported by USPHS grant AI16567 awarded to P.T. Magee. We are grateful to Dr. Magee for editorial assistance and to Dr. David Soll for helpful discussions and for the gift of strain WO-1.

REFERENCES

1. S. Scherer and P. T. Magee, Genetics of *Candida albicans, Microbiol. Rev.* 54 : 226 (1990).
2. M. B. Kurtz, M. W. Cortelyou and D. R. Kirsch, Integrative transformation using a cloned *Candida ADE2* gene, *Mol. Cell Biol.* 6: 142 (1986).
3. M. B. Kurtz, M. W. Cortelyou, S. M. Miller, M. Lai and D. R. Kirsch, Development of autonomously replicating plasmids for *Candida albicans,*, *Mol. Cell Biol.* 7: 209 (1987).
4. A. K. Goshorn and S. Scherer, Genetic analysis of prototrophic natural variants of *Candida albicans., Genetics* 123: 667 (1989).
5. P.T. Magee, E.H.A. Rikkerink and B.B. Magee, Methods for the genetics and molecular biology of *Candida albicans, Anal. Biochem.* 175 : 361 (1988).
6. S. N. Kakar, R. M. Partridge and P. T. Magee, A genetic study of *Candida albicans:* isolation of a wide variety of auxotrophs and demonstration of linkage and complementation, *Genetics* 104: 241 (1983).
7. R. T. Poulter, V. Hanrahan, K. Jeffrey, D. Markie, M.G. Shepherd and P.A. Sullivan, Recomination analysis of naturally occurring diploid *Candida albicans, J. Bacteriol.* 152: 969 (1982).
8. B.B. Magee, Y. Koltin, J. Gorman and P.T. Magee, Assignment of cloned *Candida albicans* genes to bands on the electrophoretic karyotype, *Mol. Cell Biol.* 8: 4721 (1988).

9. B. A. Lasker, Carle, G. F., Kobayashi, G. S. and G. Medoff, Comparison of the separation of *Candida albicans* chromosome-sized DNA by pulsed-field gel electrophoresis techniques, *Nucleic Acids Research* 17 : 3783 (1989).

10. B.L. Wickes, J. Staudinger, B.B. Magee, K.-J. Kwon-Chung, P.T. Magee and S. Scherer, Physical and genetic mapping of *Candida albicans*: several genes previously assigned to chromosome 1 map to chromosome R, the rDNA-containing linkage group, *Infect. Immun.* 59: 2480 (1991).

11. G. F. Carle and M. V. Olson, Separation of chromosomal DNA molecules from yeast by orthogonal-field-alteration gel electrophoresis, *Nuc. Acid Res.* 12: 5647 (1984).

12. A.K. Goshorn, S. Grindle and S. Scherer, Gene isolation by complementation in *Candida albicans* and application to physical and genetic mapping, *Infect. Immun.* 60: 876 (1992).

13. A. M. Gillum, E. T. H. Tsay and D. R. Kirsch, Isolation of the *Candida albicans* gene for orotidine-5'-phosphate decarboxylase by complementation of *S. cerevisiae* and *E. coli pyrF* mutations, *Mol. Gen. Genet.* 198: 79 (1984).

14. B. Slutsky, M. Staeball, J. Anderson, L. Risen, M. Pfaller and D.R. Soll, "White-opaque transition" : a second high-frequency switching system in *Candida albicans, J. Bacteriol.* 169: 189 (1987).

15. J. M. Anderson and D. R. Soll, Unique phenotype of opaque cells in the white-opaque transition of *Candida albicans, J. Bacteriol.* 169: 5579 (1987).

16. E.H.A. Rikkerink, B.B. Magee and P.T. Magee, Opaque-white transition: a programmed morphological transition in Candida albicans, *J. Bacteriol.* 170: 895 (1988).

17. C. Sadhu, D. Hoekstra, M.J. McEachern, S.I. Reed and J.B. Hicks, A G-protein a subunit from asexual *Candida albicans* functions in the mating signal transduction pathway of *Saccharomyces cerevisiae* and is regulated by the a1 - a 2 repressor, *Molec. Cell. Biol.* 12: 1977 (1992).

18. B. Morrow, T. Srikantha and D.R. Soll, Transcription of the gene for pepsinogen, PEP1, is regulated by white-opaque switching in *Candida albicans*, Molec. Cell. Biol. 12: 2997 (1992).

19. S. Iwaguchi, M. Homma and K. Tanaka, Isolation and characterization of a repeated sequence (RPS1) of *Candida albicans, J. Gen. Microbiol.* in press (1992).

20. S. Scherer and D. A. Stevens, A *Candida albicans* dispersed, repeated gene family and its epidemiologic applications, *Proc. Natl. Acad. Sci. USA.* 85: 1452 (1988).

21. F. Staib, Proteolysis and pathogenicity of *Candida albicans* strains, *Mycopathol. Mycol. Appl.* 37: 345 (1969).

22. R. Rüchel, Properties of a purified proteinase from the yeast *Candida albicans, Biochem. Biophys. Acta.* 659: 99 (1981).

23. F. Macdonald and F.C. Odds, Virulence for mice of a proteinase-secreting strain of *Candida albicans* and a proteinase deficient mutant, *J. Gen. Microbiol.* 129: 431 (1983).

24. K.-J Kwon-Chung, D. Lehman, C. Good and P.T. Magee, Genetic evidence for role of extracellular proteinase in virulence of *Candida albicans, Infect. Immun.* 49: 571 (1985).

25. B. Hube, C. J. Turver, F. C. Odds, H. Eiffert, G. J. Boulnois, H. Kochel and R. Rüchel, Sequence of the *Candida albicans* gene encoding the secretory aspartate proteinase, *J. Med. Vet. Mycol.* 29: 129 (1991).

26. R.J. Wright, H. Carne, A.D. Hieber, I.L. Lamont, G.W. Emerson and P.A. Sullivan, Two genes for secreted aspartate proteinases in *Candida albicans*, In press *J. Bacteriol.* (1992).

27. K. Ganesan, A. Banerjee and A. Datta, Molecular cloning of the secretory acid proteinase gene from *Candida albicans* and its use as a species specific probe, *Infect. Immun.* 59 : 2972 (1991).

GENETIC BASIS FOR DIMORPHISM AND PATHOGENICITY IN *CANDIDA ALBICANS**

William A. Fonzi, Susan Saporito-Irwin, Jiang-Ye Chen, and
Paul Sypherd

Department of Microbiology and Molecular Genetics
College of Medicine
University of California
Irvine, CA 92717, USA

ABSTRACT

Our research is aimed at understanding how genetic information is differentially used in the yeast-hyphal morphogenesis of *C. albicans*, and the biochemical role such gene products play in the process. Toward this end, we have focused on genetic elements that are activated by pH and by temperature, two conditions that regulate the decision to produce yeast or hyphae. One gene, called *PHR1*, is actively transcribed only at pH's near neutrality. The inferred amino acid sequence of this gene is 56% identical to a protein of *S. cerevisiae* that is anchored to the membrane by GPI (glycosylphosphatidylinositol). When *PHR1* is deleted on both chromosomes, the double mutant is unable to form hyphal cells. A temperature regulated genetic element was found to have the characteristics of a retrotransposon and to be moderately repeated in the genome, with copies on several chromosomes. The distribution of this element was also found to be strain-specific. We have speculated on a role such an element could play in the pathogenesis of *C. albicans*.

INTRODUCTION

A molecular genetic approach to studying dimorphism in *Candida albicans* is based on the notion that the process can be comprehended through understanding the genes involved and their regulation. This notion is difficult to refute, but in the past has been easy to ignore, largely because of the intractability of the organism

---

* Part of paper is adapted from J.Y. Chen and W.A. Fonzi, J. Bacteriol. 174 (1992) 5624-5632 with kind permission from the American Society for Microbiology, Journals Division, the copyright holder.

for genetic studies. In more recent years, however, the development of molecular genetic techniques, especially transformation and recombinant DNA methods[1,2] has forged a new arena for research on dimorphism in this important organism.

A role for hyphal production in pathogenicity has been extensively studied. As reviews by Odds[3] have concluded, the weight of the evidence leads one to the view that the production of hyphae is intimately related to the process of invasion and disease.

One of the most compelling aspects of yeast-hyphal morphogenesis is the possibility that these changes are accompanied by changes in gene expression. A clear resolution to this question would come from genetic studies. However, the absence of a mating system and diploid nature of *C. albicans* genome[4] has made it difficult to use traditional mutant approaches to detect genes that may be involved in dimorphism. Our approach, therefore, has been to use differential cDNA hybridization to screen for genes that are developmentally regulated. The experimental design of the screen was to prepare a λ phage cDNA library from cells in the process of hyphae formation and to hybridize plaques of the plated library with labeled first strand cDNAs prepared from yeast or hyphal cells. In the course of these studies, we succeeded in recovering genes expressed only in the hyphal phase, and we have identified a repeat sequence that has the characteristics of a retrotransposon. The nature of this latter genetic element, and a possible role for it are presented in detail.

MATERIALS AND METHODS

Strains and Culture Conditions

The *C. albicans* strain employed in these studies were SC5314,[1] SGY243,[2] 3153A,[5] ATCC38696, CAF3-1, a ura⁻ derivative of SC5314 (Fonzi and Irwin, unpublished results), and CAS8 (phr::*his*G/Δphr1). The cells were routinely maintained on yeast extract-peptone-dextrose (YPD) medium at 30 °C. When examining the effects of pH or temperature on gene expression, the medium of Lee et al.[6] was used.

DNA Cloning and Sequencing

In vitro manipulations of plasmid DNA were conducted by standard procedures.[7] Plasmids were amplified and maintained in *E. coli* strains HB101 or XL1-Blue (Stratagene). Hybridization screening of phage λ libraries was performed as described by Carlock.[8] Nucleotide sequences were determined with the dideoxy chain-termination method[9] using Sequenase (United States Biochemical). Nucleotide sequence analyses were performed using the Wisconsin Genetics Computer Group Sequence Analysis Software Package, Version 7.0.[10] Homology searches of the Genebank and EMBL databases were conducted using the FASTA Program of Pearson and Lipman.[11]

Southern and Northern Blot Analysis

Southern and Northern blot hybridizations were conducted using standard protocols.[7] *C. albicans* genomic DNA was prepared by the method of Scherer and Stevens[12] and RNA was prepared by the method of Langford and Gallwitz.[13]

CHEF gel electrophoresis for chromosome analysis was conducted essentially as described by Magee et al.[14]

Construction of *PHR1* Null Mutant

To construct a null mutant of *PHR1*, a portion of the *PHR1* gene was deleted and replaced with a DNA fragment consiting of *hisG*-URA3-*hisG*. A DNA fragment containing the disrupted *PHR1* gene was used to transform the ura⁻ *C. albicans* strain CAF-3. Transformation of *C. albicans* was conducted as described by Kelly et al.[2] Transformed cells were selected as ura⁺ and the homologous integration event was verified by Southern blot analysis (data not shown). Spontaneous ura⁻ derivatives of the transformed cells were selected on medium containing 5-fluoroorotic acid.[15] These clones were screened by Southern blot analysis to identify those which had undergone intragenic recombination between the *hisG* sequences. The second copy of the *PHR1* gene was deleted by the eviction and transplacement method of Winston et al.[16] creating an 850 bp deletion within the coding region. This strain was designed CAS-8.

RESULTS AND DISCUSSION

We set out to isolate genes whose expression was correlated with the process of hypha formation. A cDNA library was prepared from polyA⁺ RNA isolated from strain SC5314 which had been induced to form hyphae for one hour. Yeast cells were grown in Lee's medium at pH 4.5 and 25 °C and hyphae were induced in the same medium at pH 6.8 and 37 °C. The extracted mRNA from these cells was used to synthesize radioactive first strand cDNA probes. These probes were used to hybridize to duplicate copies of the cDNA library. General housekeeping genes would hybridize with both probes, while those genes expressed specifically in hyphae would only hybridize with the cDNA probe made from RNA extracted from hyphal cells. These probes would not only be sensitive to differences to morphology changes but also be responsive to the environmental conditions used to induce hypha formation; i.e. pH and temperature.

By using this technique, 21 unique clones were identified. Northern blot analysis allowed us to classify the genes into the following groups:
1. ECE, or extent of cell elongation
2. HSG, or hyphal-specific genes
3. PHR, or pH-responsive genes
4. Temperature-responsive genes were also identified.

Isolation and Characterization of the pH-Responsive Gene, *PHR1*

The expression of one of the cDNA clones, #73, was shown to be regulated by the pH of the external medium. Total RNA was extracted from cells grown in Lee's synthetic medium pH 4.5 and 6.8, at both 25 °C and 37 °C. Northern blot analysis using the entire cDNA isert as a hybridization probe is shown in Fig. 1. A 1.95 kb transcript was detected in cells that were grown at pH 6.8 but not pH 4.5 irrespective of temperature or morphology of the cells.

The isolation of a pH responsive clone from *C. albicans* was of considerable interest to us because pH may be a significant factor in determining the pathogenic status in *C. albicans*. The ability to respond and adapt to changes in pH of the external environment would be of particular significance in organisms that inhabit the gut. In fact, differential gene expression in response to changes in pH has also been detected in *E. coli*,[17,18] *Shigella*[19] and *Salmonella*.[20] In addition, pH has been shown to affect directly the expression of virulence determinants in *Vibro cholerae*.[21]

The pH of the external environment has been shown to be an important parameter in the decision to make hyphae in *C. albicans*. All *in vitro* conditions used to induce hyphae require a pH around neutrality. By affecting the morphological form in *C. albicans*, pH may in turn affect the ability of the fungus to adhere to host cells and/or invade them. pH may also directly affect adhesion of *C. albicans* to host cells. A pH optimum of 6-8 has been reported to be necessary for cell adhesion *in vitro*.[22,23]

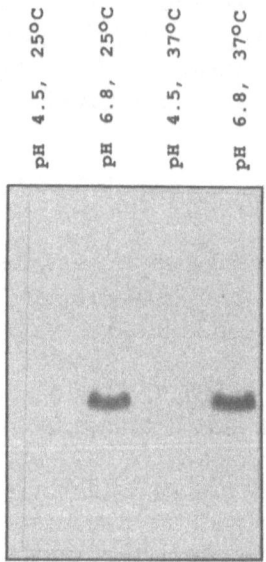

Fig. 1 Northern blot analysis of *PHR1* expression. Each lane was loaded with 20 µg RNA from strain SC5314 grown in Lee's medium, pH 4.5, 25 °C (lane 1), pH 6.8, 25 °C (lane 2), pH 4.5, 37 °C (lane 3), pH 6.8, 37 °C (lane 4). The blot was hybridized with the cDNA #73 clone.

A genomic copy of the #73 cDNA clone was obtianed by screening a l library containing DNA that was partially digested with *Sau*3A. Subsequently a 2.1 kb subcloned *Eco*R1 fragment which hybridized to the cDNA clone was isolated. We designated this gene *PHR1* for pH resposive.

The *PHR1* gene was sequenced and the deduced amino acid sequence was compared to sequences in the Genbank database. The predicted 60 kDa protein encoded by the *PHR1* gene showed 56% amino acid identity with a *Saccharomyces cerevisiae* membrane glycoprotein, gp115.[24,25] The amino acid sequence comparison is shown in Fig. 2. In addition to the overall 70% amino acid similiarity, N- and C-terminal hydrophobic domains and N- and O-linked glycosylation sites are also conserved between the two proteins.

The *S. cerevisiae* glycoprotein, gp115 is encoded by the *GGP1* gene. A *GGP1* gene desruption mutant exhibits a low growth phenotype, approximately 20-30% of the normal rate, and an abnormal cell morphology. The function of gp115, however, has yet to be determined. The protein, gp115, is covalently linked to the

Fig. 2 Comparison of the deduced amino acid sequence of the *C. albicans PHR1* gene and the *S. cerevisiae* gp115 protein. Dots present in the protein sequence represent amino acid gaps introduced to give maximal alignment. Hydrophobic N- and C-terminal domains are boxed. Conserved N-glycosylation sites are circled. The GPI anchor attachment site determined for gp115 is designated by an *.

lipid bilayer by glycosylphosphatidyl inositol (GPI) and its synthesis is cell cycle regulated.

This particular mode of attachment, referred to as GPI anchor, occurs in a wide variety of eukaryotic organisms. These modified proteins fall into diverse functional groups which include hydrolytic enzymes such as acetylcholinesterase,[26] cell surface antigens such as *Trypanosoma* variant surface glycoprotein[27] and cell adhesion proteins such as *S. cerevisiae* a-agglutinin.[28] GPI anchors share a common core structure consisting of a phosphatidylinositol molecule linked to a glycan exhibiting a non-acetylated glucosamine at its reducing end. The glycan is attached to ethanolamine via a phosphodester bond. The ethanolamine is linked via an amide bond to the C-terminal amino acid of the mature protein. Rather than being entirely embedded in the membrane, GPI linkage allows the entire protein to be exposed at the cell surface. GPI-linked proteins usually contain a cleavable N-terminal signal sequence. The C-terminal domain is composed predominantly of hydrophobic amino acids, and addition of the anchor involves removal of 17-31 amino acids at the C-terminal end. The evidence suggests that the GPI moeity is preassembled and transferred as a unit to the protein in the endoplasmic reticulum. The overall similarity and conserved hydrophobic N- and C-terminal domains between the *S. cerevisiae* gp115 and the *C. albicans PHR1* protein suggests that the *PHR1* protein may also be GPI linked. Experiments to confirm GPI-linkage to the membrane are under way.

Role of *PHR1* in Mutagenesis

In order to learn if the *PHR1* protein plays an essential role in hyphal production, we set out to delete the *PHR1* gene. Since *C. albicans* is a diploid organism, it was necessary to create a cell deficient in both allele copies of the *PHR1* gene. For this, we employed a gene replacement strategy to create double mutants, using the sequential gene disruption technique developed for *S. cerevisiae* by Alani *et al.*[29] This method uses a plasmid construct consisting of the URA3 genes flanked by direct repeats of the *E. coli hisG* genes. This construct is cloned into the gene to be disrupted and transformed into a ura⁻ strain, selecting for ura⁺ cells. Subsequent homologous recombination between the *hisG* sequences results in the spontaneous excision of the URA3 gene, producing now a ura⁻ derivate which can be transformed again to disrupt the second allele of the target genes.

A *phr1* null mutant was constructed by this method. Northern blot analysis was used to confirm that the strain, CAS-8, no longer expressed the 1.95 kb pH-responsive transcript. Total RNA was extracted from the single *PHR1* disruptant, CAS-6, the double disruptant, CAS-8, and the parent strain, CAS-3, grown in Lee's medium at pH 4.5 and pH 6.8. The entire *PHR1* gene was used as a hybridization probe. This filter was subsequently hybridized with an actin gene probe. Actin gene expression is constitutive in *C. albicans* and can be used to determine if equivalent amounts of RNA were loaded in each lane. The results of the Northern blot analysis (Fig. 3) show that the 1.95 kb pH-responsive transcript is completely absent in the *PHR1* mutant strain, CAS-8, thus confirming that we have constructed a cell deficient in *PHR1* protein.

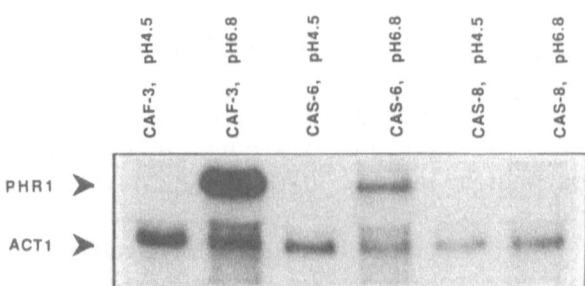

Fig. 3 Expression of *PHR1*. Each lane was loaded with 20 µg total RNA from the indicated strain grown at the indicated pH. The Northern blot was hybridized with genomic *PHR1* sequences and subsequently rehybridized with the ACT1 gene.

We were interested in determining the phenotypes associated with the *PHR1* double deletion. Since the *S. cerevisiae GGP1* mutant exhibited a slower growth rate compared to wild-type cells, we looked at the growth rate of the *PHR1* mutant. Cells were grown in YNB at pH 4.5 and pH 7.0 at 30 °C or in YPD at 25 °C and 37 °C. No differences in growth rate were observed.

We examined the ability of a *PHR1* mutant to form hyphae. Cells were grown to stationary phase at room temperature on YPD plates, then inoculated at a cell density of 8×10^6 cells in TCM199[30] at 37 °C for 4 h. Hypha formation was observed in the parent strain CAF-3. However, the *PHR1* double mutant strain CAS-8 was unable to form true hyphae under these conditions, but formed abnormal cells instead (Fig. 4). These results clearly implicate the *PHR1* gene as

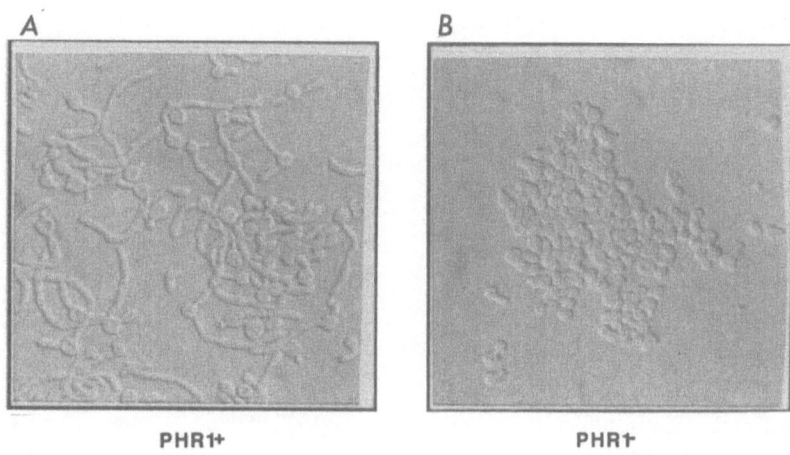

Fig. 4 Induction of hypha formation. Strains CAF-3 (panel A) and CAS-8 (panel B were induced to form hyphae in TCM199, pH 7.0 at 37 °C.

controlling a function essential to the production of hyphae. The nature of the gene product, inferred to be a surface, GPI-anchored protein, and the role which the protein plays in the process is currently under investigation. The role of *PHR1* protein in the process of hypha formation may be either as a structural protein necessary for germ-tube formation or may be involved in transducing the signal to form hyphae from the external environment. The resolution of the role played by the PHR protein will involve determining the cellular location; that is, membrane or cell wall. In the latter case, the role of a GPI anchor in the movement of the a-agglutinin of *S. cerevisiae* may provide an important paradigm. The GPI anchor is believed to be involved in transporting the a-agglutinin across the membrane, where the anchor is cleaved, allowing the a-agglutinin to be incorporated into the cell wall matrix. Our next studies on the PHR protein will be directed toward understanding the nature of the protein and whether or not it is GPI-linked, along with the cellular location, and studies to determine if the double deletion of the gene affects adherence or pathogenicity.

Isolation of a Moderately-Repetitive Element

In the course of our work on genes involved in dimorphism we isolated a genomic clone which hybridized with multiple DNA fragments in Southern blots of *C. albicans* genomic DNA digested with *EcoRI*. The repetitive sequences were

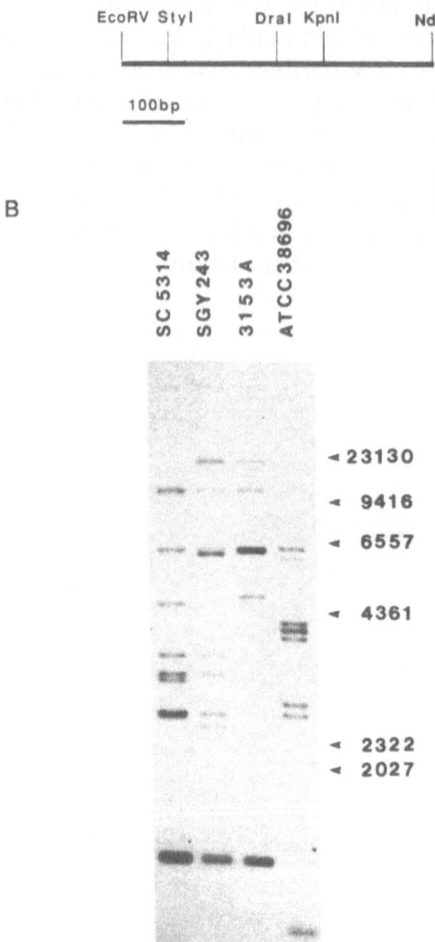

Fig. 5 Panel A. Restriction endonuclease sites present in the a element. Panel B. Southern blot of *Eco*RI digested genomic DNA hybridized with the alpha element. The C. albicans strain used as a source of the DNA is indicated at the top of each lane. The position and length in base pairs of λ DNA *Hind*III fragments is indicated on the right.

delimited to a region of approximately 500 bp defined by *Eco*RV and *Nde*I restriction sites (Fig. 5). Since short repetitive elements in *S. cerevisiae* are designated with Greek names, we chose to designate this *C. albicans* repeat with the alpha. The hybridization patterns obtained with genomic DNA isolated from four different *C. albicans* strains is shown in Fig. 5. The alpha element hybridized with multiple *Eco*RI fragments generated from the DNA of each strain, but the sizes and number of hybridizing fragments varied with the source of the DNA. The number of hybridization bands ranged from eight to ten and the size of the bands ranged from approximately 0.8 kb to 15 kb. Although a unique pattern was obtained for each strain, DNA from three of the four strains tested contained a number of hybridizing *Eco* RI fragments of equivalent size. For instance, DNA from strain SGY243 contained a total of ten hybridizing fragments, of which seven were identical in size to hybridization bands seen for DNA from strain SC5314. The fourth strain tested, ATTCC38696, exhibited an entirely unique pattern.

The alpha element was not part of a larger tandem repeating structure, thus giving rise to the similar hybridization patterns. In addition to the observation that the sequences flanking the alpha element were unique (data not shown), the element hybridized to multiple chromosomes (Fig. 6). The chromosomes of three

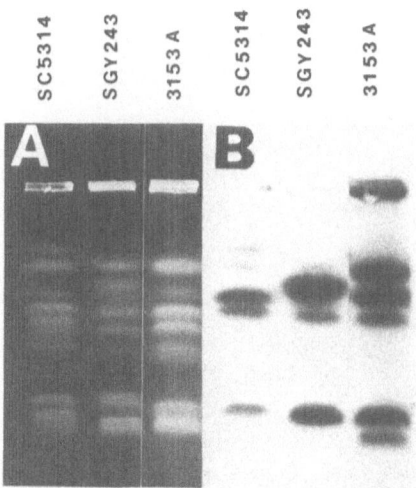

Fig. 6 CHEF gel analysis of α element distribution. Panel A shows the ethidium bromide staining pattern of chromosomes from the indicated strains separated by CHEF gel electrophoresis. Panel B shows the results of Southern blot analysis of the gel in Panel A when hybridized with the alpha element.

different strains were separated by CHEF gel electrophoresis. Southern blot analysis of the gel demonstrated that the alpha element hybridized to five of the ten chromosomes resolved from strain SC5314. Nine chromosomes were resolved from each of the other two strains tested and four or five of those hybridized with the alpha element. Thus it appeared that the alpha element was dispersed in the genome; however, the number of genetic loci involved and the heterozygous or homozygous nature of those loci were not clear.

Identification of a Retrotransposon-Like Element

To address these and other issues, additional copies of the alpha element were cloned from a λ phage library of *C. albicans* genomic DNA. The library was constructed from partial *Sau3A* digested genomic DNA of strain SC5314 cloned into lGEM 12 (Promega) and was screened by plaque hybridization with the 500 bp alpha element. Interestingly, a single phage isolate, λCJY-3, contained two copies of the element. Southern blot analysis of λCJY-3 DNA demonstrated two *Eco*RI fragments which hybridized with the alpha element (Fig. 7). The lenght of these fragments, 3.4 kb and 3.7 kb, was identical to hybridizable fragments observed in *Eco*RI digests of genomic DNA. Subsequent restriction endonuclease

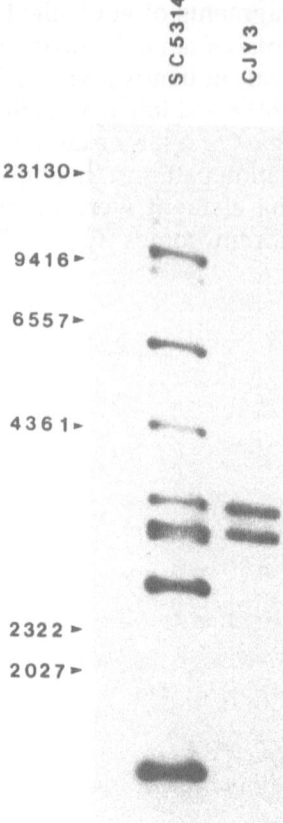

Fig. 7 Southern blot analysis of *Eco*RI digested DNA isolated from strain SC5314 or from the λ clone CJY-3 hybridized with the alpha element. The position and length in base pairs of λDNA *Hin*dIII fragments is indicated on the left.

analysis demonstrated that the two alpha elements in λCJY-3 were in the same orientation and were separated by a distance of approximately 5.5. kb (Fig. 8).

Two aspects of these data were suggestive of a retroviral-like element: the small size of the alpha element and the organization of these elements in λCJY-3 as direct repeats flanking a region of 5.5. kb. The long terminal repeats of retroviruses and retrotransposons are similar in size and organization.[31,32] This suggestion was further supported by the results of nucleotide sequence analysis of the cloned alpha elements. The two alpha repeats of λCJY-3 were found to contain a region of identical sequence 388 nucleotides in length (Fig. 8). The alpha element initially isolated was nearly identical in sequence, but contained a guanine insertion at position 132 and a T to A transversion at position 19. Each of the alpha elements was delimited by a six-base pair inverted repeat, 5'-TGTTCG....CGAACA-3'. The long terminal repeats of retrotransposons and retroviruses are similarly bounded by inverted repeats and the boundary nucleotides, 5'-TG...CA-3', are conserved.[31,32]

Replication of retroviruses and retrotransposons requires a plus-strand and a minus-strand primer binding site. A polypurine tract immediately 5' of the 3'-long terminal repeat serves as the plus-strand primer binding site during reverse transcription.[31,32] In an analogous manner, the nucleotide sequence or DNA located between the two alpha elements of λCJY-3 contained the purine rich sequence 5'-GAATCAGGGAG-3' adjacent to one of the alpha elements. The minus-strand primer-binding site, adjacent to the 5'-long terminal repeat, is complementary to various tRNAs which serve as primers for reverse transcription.[31,32] Adjacent to the other alpha element in λCJY-3, the position analogous to the minus-strand primer binding site, was the sequence 5'-GATTAGAAG-3'. This sequence is complementary to nucleotides 31 to 39 of *S. cerevisiae* tRNA^{Arg3}.[33] Given the high degree of homology between the

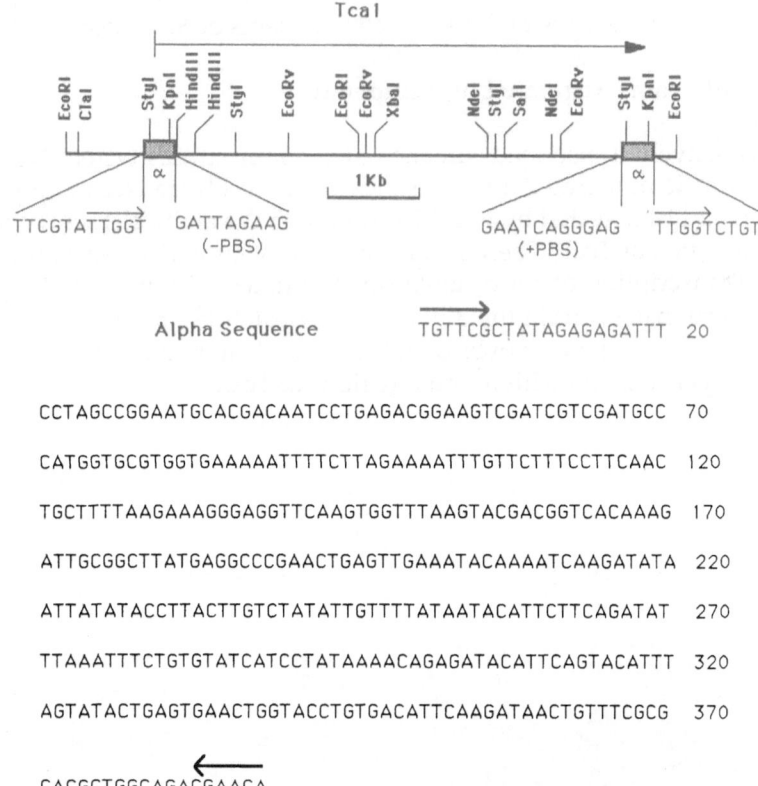

Fig. 8 Location of restriction endonuclease sites within the insert of λCJY-3 and the nucleotide sequence of selected regions. The location of the* elements within λCJY-3 is indicated by the boxed regions. The element is transcribed from left to right as depicted. The arrows above the sequence of the alpha elements indicate the position of the inverted repeats forming the borders of the alpha element. The direct repeats flanking Tca1, as well as the potential plus-strand primer binding site (+PBS) and minus-strand primer binding site (-PBS) are also indicated.

translational apparatuses of *C. albicans* and *S. cerevisiae*,[34,35] the analogous tRNA of *C. albicans* may serve as the primer for reverse transcription. It should be noted that although a full length tRNA is generally employed to reverse transcription, *copia* elements of *Drosophila spp.* employ a cleavage product containing nucleotides 1 to 39 of tRNA^{Met}.[36]

The presence of these many conserved structural features led us to conclude that the clone λCJY-3 contained a retrotransposon-like element consisting of direct repeats of the element enclosing a 5.5 kb internal region. We have designated this composite element Tca1 (transposon *Candida albicans*). The suggestion that Tca1 represents a mobile element was further supported by the presence of direct repeats of nucleotide sequence flanking the element. Integration of retrotransposons results in a four- to six-base pair duplication of the integration site such that the element is flanked by this short direct repeat. Examination of the nucleotide sequence surrounding Tcal revealed that the alpha elements were externally bounded by a direct repeat of the sequence surrounding Tcal revealed that the alpha elements were externally bounded by a direct repeat of the sequence 5'-TTGGT-3'. The first alpha element isolated was bordered by a different five-base-pair repeat, 5'-GATTA-3'. These data are consistent with independent integration events and target site duplications. They also imply that the first alpha element isolated resulted from intragenic recombination between the alpha repeats of a Tcal element analogous to the solo delta elements of *S. cerevisiae*.[37]

Expression of Tcal is Regulated by Temperature

Tcal is actively transcribed. In each of the four strains tested, Northern blot hybridization demonstrated the presence of a 6.9 kb transcript (Fig. 9). This transcript could be detected by hybridization with the alpha element alone or with the internal region of Tcal. These results are consistent with a retrotransposon-like element. Transcription of retrotransposons initiates within the 5'-long terminal repeat and terminates within the 3'-terminal repeat.[31,32] In addition to the 6.9 kb transcript, the alpha element hybridized to several shorter transcripts (Fig. 9) not detected by hybridization with internal regions of Tcal.

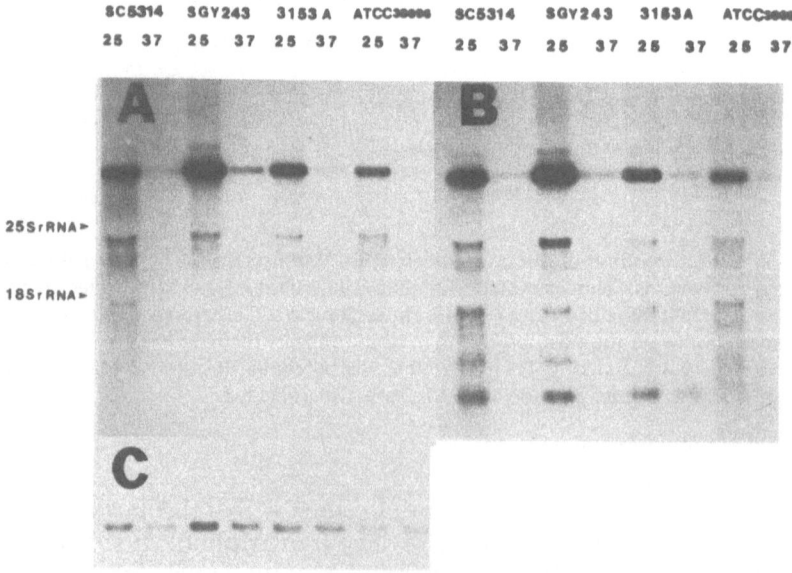

Fig. 9 Expression of Tca1. Each panel was prepared from the same Northern blot hybridized sequentially with a 1.8 kb *HindIII/EcoRI* internal fragment of Tca1 (Panel A), the element (Panel B), and the *C. albicans* actin gene (Panel C). Each lane contained 20 μg of total RNA isolated from the indicated strain grown at the indicated temperature. The electrophoretic position of the rRNA bands is indicated to the left of the figure.

The regulation of Tcal expression was most interesting. The amount of Tcal transcript was reduced 20- to 30-fold in cells grown at 37 °C relative to cells grown at 25 °C (Fig. 9). This was observed for each of the four strains tested. The decreased expression of Tcal did not appear to be related to the dimorphic transition of *C. albicans*. Strain SGY243 (lanes 3 and 4 of Fig. 9) did not undergo germ-tube formation under the conditions employed, yet exhibited the same decrease in Tcal transcript levels as did those strains which formed hyphae. Down-regulation at 37 °C was also evident for the alpha-specific transcript.

The strong response of Tcal expression to temperature may be of particular interest with regard to the virulence of *C. albicans*. Temperature is known to regulate expression of virulence determinants in a number of pathogenic bacteria.[38] Similarly, temperature may also regulate the expression of virulence determinants in *C. albicans* since cells grown at 25°C are reportedly more virulent than cells grown at 37 °C.[39] Consequently, understanding the regulation of Tcal expression may be directly relevant to the expression of *C. albicans* virulence genes. Furthermore, transposition of Tcal may be responsible, in part, for the variability in strain virulence seen in natural isolates. Just as Ty elements of *S. cerevisiae* can regulate adjacent gene transcription,[31] Tcal elements may place adjacent genes under temperature-dependent control. This could allow for increased expression of genes which improve survival of *C. albicans* outside of the mammalian host, or increased expression of genes required to establish host infection. On the other hand, the persistent down-regulation of virulence genes at 37 °C could be a well developed system for the organism to establish a <u>commensal</u> relationship with the host rather than a parasitic one. In this way loss of a Tcal element controlling virulence genes would result in their expression under their own promotor, active at 37 °C, with the ultimate effect being an active, invasive infection. Thus, Tcal may provide an efficient means for the organism to rapidly alter its virulence and evolutionary success.

REFERENCES

1. A.M. Gillum, E.Y.H. Tsay and D.R. Kirsch, Isolation of the *Candida albicans* gene for orotidine-5′-phosphate decarboxylase by complementation of *S. cerevisiae ura3* and *E. coli pyrF* mutations, *Mol. Gen. Genet.* 198: 179 (1984).
2. R. Kelly, S.M. Miller, M.B. Kurtz and D.R.Kirsch, Directed mutagenesis in *Candida albicans*: one-step gene disruption to isolate *ura3* mutants, *Mol. Cell. Biol.* 7: 199 (1987).
3. F.C. Odds, "*Candida and Candidosis*", Ballière Tindall, London (1988).
4. W.L. Whelan, R.M. Parridge and P.T. Magee, Heterozygosity and segregation in *Candida albicans*, *Mol. Gen. Genet.* 180: 107 (1980).
5. G.W. Bedell and D.R. Soll, Effects of low concentrations of zinc on the growth and dimorphism of *Candida albicans*: evidence for zinc-resistant and -sensitive pathways for mycelium formation, *Infect. Immun.* 26: 348 (1979).
6. K.L. Lee, H.R. Buckley and C.C. Campbell, An amino acid liquid synthetic medium for the development of mycelial and yeast forms of *Candida albicans*, *Sabouraudia* 13: 148 (1975).
7. Sambrook, E.F. Fritsch and T. Maniatis, "*Molecular Cloning: A laboratory Manual*", Cold Spring Harbor Laboratory Press, Cold Spring Harbor, N.Y. (1989).
8. L.R. Carlock, Analyzing lambda libraries, *Focus* 8: 6 (1986).
9. F. Sanger and A.R. Coulson, A rapid method for determining sequences in DNA by primed synthesis with DNA polymerase, *J. Mol. Biol.* 94: 441 (1975).
10. J. Devereux, P. Haeberli and O. Smithies, A comprehensive set of sequence analysis programs for the VAX, *Nucl. Acids Res.* 12: 387 (1984).
11. W.R. Pearson and D.J. Lipman, Improved tools for biological sequence comparison, *Proc. Natl. Acad. Sci. USA* 85: 2444 (1988).
12. S. Scherer and D.S. Stevens, A *Candida albicans* dispersed, repeated gene family and its epidemiological applications, *Proc. Natl, Acad. Sci. USA* 85: 1452 (1988).

13. C.J. Langford and D. Gallwitz, Evidence for an intron-containing sequence required for the splicing of yeast RNA polymerase II transcripts, *Cell* 33: 519 (1983).

14. B.B. Magee, Y. Koltin, J.A. Gorman and P.T. Magee, Assignment of cloned genes to the seven electrophoretically separated *Candida albicans* chromosomes, *Mol. Cell. Biol.* 8: 4721 (1988).

15. J.D. Boeke, F. LaCroute and G.R. Fink, A positive selection for mutants lacking orotidine-5'-phosphate decarboxylase activity in yeast: 5-fluoroorotic acid resistance. *Mol. Gen. Genet.* 197: 345 (1984).

16. F. Winston, F. Chumley and G.R. Fink, Eviction and transplacement of mutant genes in yeast, *Methods Enzymol.* 101: 211 (1983).

17. Bingham, K.S. Hall and J.L. Slonczewski, Alkaline induction of a novel gene locus, *alx*, in *Escherichia coli*, *J. Bacteriol.* 172: 2184 (1990).

18. M. Heyde and R. Portalier, Regulation of major outer membrane porin proteins of *Escherichia coli* K-12 by pH, *Mol. Gen. Genet.* 208: 511 (1987).

19. V.L. Headley and S.M. Payne, Differential protein expression by *Shigella flexneri* in intracellular and extracellular environments, *Proc. Natl. Acad. Sci. USA* 87: 4179 (1990).

20. Z. Aliabadi, Y.K. Park, J.L. Slonczewski and J.W. Foster, Novel regulatory loci: oxygen- and pH-regulated gene expression in *Salmonella typhimurium*, *J. Bacteriol.* 170: 842 (1988).

21. C. Parsot and J.J. Mekalanos, Expression of the *Vibrio cholerae* gene encoding aldehyde dehydrogenase is under the control of *ToxR*, the cholera toxin transcriptional activator, *J. Bacteriol.* 173: 2842 (1991).

22. R.D. King, J.C. Lee and A.L. Morris, Adherence of *Candida albicans* and other *Candida* species to mucosal epithelial cells, *Infect. Immun.* 28: 667 (1980).

23. J.D. Sobel, P.G. Myers, D. Kaye and M.E. Levison, Adherence of *Candida albicans* to human vaginal and buccal epithelial cells, *J. Infect. Dis.* 143: 76 (1982).

24. M. Vai, E. Gatti, E. Lacana, L. Popolo and L. Alberghina, Isolation and deduced amino acid sequence of the gene encoding gp115, a yeast glycophospholipid-anchored protein containing a serine rich region, *J. Biol. Chem.* 266: 12242 (1991).

25. C. Nuoffer, P. Jeno, A. Conzelmann and H. Reizman, Determinants for glycophospholipid anchoring of the *Saccharomyces cerevisiae GAS1* protein to the plasma membrane, *Mol. Cell. Biol.* 11: 27 (1991).

26. A.H. Futerman, M.G. Low, K.E. Ackermann, W.R. Sherman and I. Silman, Identification of covalently bound inositol in the hydrophobic membrane-anchoring domain of *Torpedo* acetyl-cholinesterase, *Biochem. Biophys. Res. Commun.* 129: 312 (1985).

27. M.A.J. Ferguson, K. Hadler and G.A.M. Cross, *Trypanosoma brucei* variant surface glycoprotein has a sn-1,2-dimyristyl glycerol membrane anchor at its COOH-terminus, *J. Biol. Chem.* 260: 4963 (1985).

28. A. Roy, C.F. Lu, D.L. Marykwas, P.N. Lipke and J. Kurjan, The *AGA1* product is involved in cell surface attachment of the *Saccharomyces cerevisiae* cell adhesion glycoprotein a-agglutinin, *Mol. Cell. Biol.* 11: 4196 (1991).

29. E. Alani, L. Cao and N. Kleckner, A method for gene disruption that allows repeated use of the URA3 selection in the construction of multiple disruption yeast strains, *Genetics* 116: 541 (1987).

30. J.W. Landau, N. Dabrowa and V.D. Newcomer, The rapid formation in serum of filaments by *Candida albicans*, *J. Investig. Dermatol.* 44: 171 (1965).

31. J. Boeke, Transposable elements in *Saccharomyces cerevisiae*, in: "Mobile DNA", D.E. Berg and M.M. Howe, eds., American Society for Microbiology, Washington, D.C. (1989).

32. H. Varmus and P. Brown, Retroviruses, in: "Mobile DNA", D.E. Berg and M.M. Howe, eds., American Society for Microbiology, Washington, D.C. (1989).

33. G. Keith and G. Dirheimer, Reinvestigation of the primary structure of brewer's yeast Arg-tRNA-3, *Biochem. Biophys. Res. Commun.* 92: 116 (1980).

34. P. Sundstrom, D. Smith and P.S. Sypherd, Sequence analysis and expression of the two genes for elongation factor-1a from the dimorphic yeast *Candida albicans*, *J. Bacteriol.* 172: 2036 (1990).

35. K.K. Myers, W.A. Fonzi and P.S. Sypherd, Isolation and sequence analysis of the gene for translation elongation factor 3 from *Candida albicans*, *Nucleic Acids Res.* 20: 1705 (1992).

36. Y. Kikuchi, Y. Ando and T. Shiba, Unusual priming mechanism of RNA-directed DNA synthesis in *copia* retrovirus-like particles of *Drosophila*, *Nature* 323: 824 (1986).

37. G.S. Roeder and G.R. Fink, DNA rearrangements associated with a transposable element in yeast, *Cell* 21: 239 (1980).

38. A.T. Maurelli, Temperature regulation of virulence genes in pathogenic bacteria: a general strategy for human pathogens?, *Microb. Path.* 7: 1 (1989).

39. P.P. Antley and K.C. Hazen, Role of yeast cell growth temperature on *Candida albicans* virulence in mice, *Infect. Immun.* 56: 2884 (1988).

THE MULTIPLE CHITIN SYNTHASE GENES OF *CANDIDA ALBICANS* AND OTHER PATHOGENIC FUNGI - A REVIEW [*]

P.W. Robbins[1], A.R. Bowen[2], J.L. Chen-Wu[3], M. Momany[4], P.J. Szaniszlo,[4] and J. Zwicker[5]

[1] Massachusetts Institute of Technology, Center for Cancer Research, Room E17-233A, 40 Ames Street, Cambridge, MA 02139;
[2] University of Michigan, Ann Arbor, MI;
[3] Vicam, 29 Mystic Avenue, Somerville, MA;
[4] Department of Microbiology, University of Texas, Austin, TX 78712;
[5] Massachusetts Institute of Technology, Center for Cancer Research, Cambridge, MA 02139

ABSTRACT

Chitin, the $\beta1,4$-linked polymer of N acetyl glucosamine, is a fibrous polysaccharide that in many yeasts helps to maintain the structure of the mother-bud junction and in filamentous fungi is often the major supporting component of the cell wall. We have isolated and sequenced two chitin synthase genes from *Candida albicans* and studied their expression by Northern blot analysis. The two genes show strikingly different levels of transcription during the yeast and hyphal growth stages of the organism.

The very similar DNA sequences of the *Saccharomyces cerevisiae* and *Candida albicans* chitin synthase genes. Fragments homologous to chitin synthase (~600 base pairs) were amplified from the genomic DNA of 14 fungal species. These fragments were sequenced, and their deduced amino acid sequences were aligned. With the exception of *S. cerevisiae CHS1*, the sequences fell into three distinct classes, which could represent separate functional groups. Within each class phylogenetic analysis was performed. Although not the major purpose of the investigation, this analysis tends to confirm some relationships consistent with current taxonomic groupings.

---

[*] This review is adapted from our publications: A.R. Bowen *et al.*, Classification of fungal chitin synthases, *Prod. Nat. Acad. Sci., USA* 89: 519-523 (1992), and J. Chen-Wu *et al.*, Expression of chitin synthase genes during yeast and hyphal growth phases of *Candida albicans*, *Mol. Microbiology* 6: 497-502 (1992).

INTRODUCTION

Candida albicans is an opportunistic dimorphic fungal pathogen of considerable medical importance. Much recent evidence suggests that of the two prominent morphological forms of the organism, yeast and hyphae, the hyphal form plays a major role in pathogenicity and is the form involved in tissue invasion. Among the changes that have been observed in the yeast-hyphal trasition in *Candida* is a three- to five-fold increase in cell wall chitin levels that takes place immediately following germ tube formation.[1] This sudden transition in levels of chitin synthesis is of interest from several points of view. The basic molecular mechanisms involved, the structural role played by chitin in the hyphal wall, and the potential therapeutic value of inhibiting this transition are all of interest.

Two types of chitin synthase enzymes have recently been found in yeast. One type, typified by *Saccharomyces CHS1* and *CHS2*, is a zymogen that requires proteolytic activation before enzymatic activity can be detected.[2] The second type of enzyme does not require proteolytic activation.[3] Previous enzymatic studies of *Candida* by Orlean and Kerridge[4] and Orlean[5] have shown that while levels of chitin syntase activity following trypsin activation are about the same in extracts from yeast and hyphal forms of *Candida*, activity in the absence of trypsin activation is 3.5 times higher in the hyphal than in the yeast preparations.

More recent work by Au-Young and Robbins[6] has confirmed the finding of Orlean and Kerridge. In addition, detailed analysis of the pH activity profiles of activated and non-activated membrane preparations suggests that *Candida* has two or more chitin synthases. One of these enzymes, a zymogen with an acidic pH optimum, was cloned by heterologous expression in *Saccharomyces cerevisiae*, while another was cloned with degenerate PCR primers (see below). Probes from the two genes were used to study differential expression during yeast and hyphal growth of the organism.

In extension of these studies we decided to use degenerate PCR primers that encoded short, completely conserved sequences within the three genes to probe genomic DNA from a variety of fungi. PCR-derived fragments were cloned inte M13, and single nucleotide sequencing runs were used to classify the clones. Representative clones were then completely sequenced, and the deduced amino acid sequences were put into groups by the CLUSTAL program. The aligned DNA sequences within these groups or classes were analyzed further with the FITCH program.

RESULTS AND DISCUSSION

Differential Chitin Synthase Gene Expression During Yeast and Hyphal Growth of *Candida*

A number of methods are available for the demonstration of germ tube formation and hyphal outgrowth in *Candida*.[7] We have chosen a modification of the method of Marriott.[8] Stationary-phase cells were inoculated with fresh medium at high (4×10^7 cells/ml) or low (4×10^6 cells/ml) density at 30 °C or 40 °C. Hyphal outgrowth occurred well in only the low-density culture grown at 40 °C. The high-density culture at 40 °C serves as a control for heat-shock effects. The procedure is shown in diagrammatic form in Fig. 1. Northern blot analyses of cells from four cultures were performed a number of times. Typical results are shown in Fig. 2. Lanes 2 and 3 show the level of message in low-density cells after

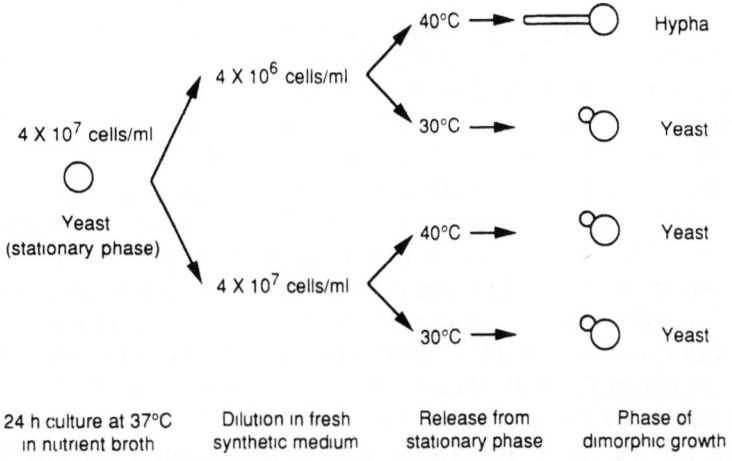

Fig. 1 Diagram of the Marriott[8] procedure for studying hyphal outgrowth in *C. albicans*. The high-density cells at 40 °C serve as a control for heat-shock effects (see text.).

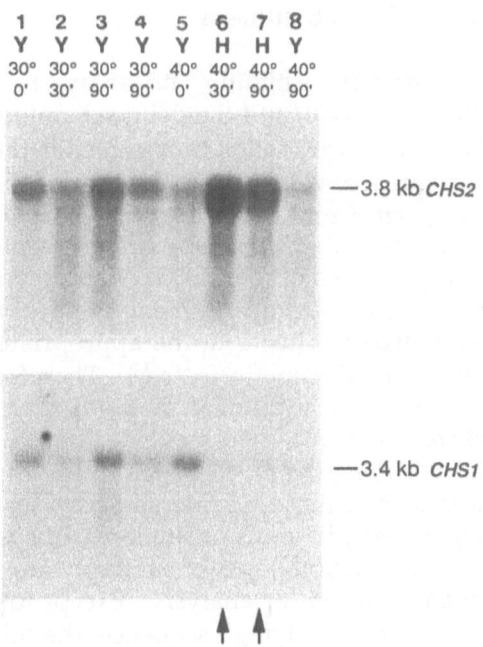

Fig. 2 Northern blot analysis of yeast and hyphal-phase *C. albicans*. The times and temperatures of growth as well as the cell densities are indicated. The procedure is presented in diagrammatic form in Fig. 1. Y indicates yeast and H hyphal growth phases. As explained in greater detail in the text, lanes 4 and 8 show the message level in control high-density yeast cultures at 30 °C and 40 °C, respectively.

30 and 90 min at 30 °C, while lane 4 indicates the message level in high-density cells after 90 min at 30 °C. As can be seen, the message for both CHS1 and CHS2 is present after 90 min at 30 °C. In contrast, cells undergoing hyphal outgrowth at 40 °C (lanes 6 and 7) displayed very high levels of CHS2 message but little CHS1 message after 30 and 90 min at 40 °C. The control high-density cells at 40 °C (lane 8) show only very low levels of either message.

As discussed below, fungal chitin synthase genes fall into three major sequence classes. We assume that the corresponding genes serve different functions, but studies of these functions are only beginning. The multiple chitin synthase genes of *Saccharomyces* have, however, been analyzed to some extent. *Saccharomyces CHS1* may serve as a 'repair enzyme'[9] or have another more obscure function. *Saccharomyces CHS2* is required for maintaining normal cell morphology, but gene-disruption studies suggest that it is not responsible for the major part of cellular chitin synthesis. A third chitin synthase, which is not a zymogen, appears to be responsible for the bulk of chitin made in *Saccharomyces*.[10]

The *Candida CHS1* and 2 enzymes are probably blot zymogens similar to *Saccharomyces CHS1* and 2. Results available to date suggest that *Candida CHS1* has a low pH optimum, as does *Saccharomyces CHS1*. However, the sequence of *Candida CHS1* is somewhat more similar to *Saccharomyces CHS2* than *Saccharomyces CHS1*. S. Silverman (unpublished) has shown that *Candida CHS1*, although expressed, did not correct the growth defects associated with *Saccharomyces CHS2* deletion. Gene deletion studies are now underway to study the functions of *C. albicans* CHS1 and CHS2 in detail.

Classification of Fungal Chitin Synthases

The list of fungi used for isolation and analysis of chitin synthase gene fragments is presented in Table 1, and the deduced amino acid sequences for the fragments obtained by PCR amplification is presented in our recent paper.[11]

The amino acid sequences were aligned by the program CLUSTAL,[12] which first derives a dendrogram from a matrix of all pairwise sequences similarity scores and then progressively aligns the most similar sequences. The dendrogram (Fig. 3) produced in the first step of the CLUSTAL program was calculated by the unweighted pair group method using arithmetic averages (UPGMA).[13] Although UPGMA dendrograms usually cluster in the appropriate manner, they are not intended to be used as phylogenetic trees.[12] Therefore, after the sequence alignment has been obtained, additional methods were consulted to deduce possible phylogenetic relationships.

As pointed out by Valencia *et al.*[14] in their study of ras protein sequences, closeness in sequences such as those considered here "can be interpreted in terms of similarity of function *and/of* in terms of similarity of species." As is clear from inspection of the tree in Fig. 3, both types of similarity are suggested for the chitin synthase gene fragments chosen for analysis. Except for the *S. cerevisiae CHS1* fragment, which is left as an "outlying" sequence, the other gene fragments fall into three classes, which could represent three separate functional groups. Although the classification was done by computer, the groups can be found on inspection by characteristic gaps and by residues, such as the proline that occurs after the first gap in class III. Within each CLUSTAL class some expected close relationships are seen. For example, similarities are apparent between *Aspergillus niger* and *A. nidulans*, which are known or suspected Ascomycetes of the genus *Emericella*, between *Histoplasma capsulatum* and *Blastomyces dermatitidis*, which are both Ascomycetes of the genus *Ajellomyces*, and among the opportunistic pathogens *Exophiala jeanselmei*, *Wangiella dermatitidis*, *Phaeococcomyces exophialea*, and *Xylohypha bantiana*, which are all members of the same form family (Dematiaciae) of the Fungi Imperfecti, but most likely represent loculomycetous Ascomycetes.[13,15,17]

To gain a more detailed picture of possible evolutionary relationships, the DNA sequences within each class were compared by programs available in Joseph Felsenstein's phylogeny inference package.

Table 1. Taxonomic affinities of fungal species and chitin synthase gene designations

Species	Affinity	Gene designations
Saccharomyces cerevisiae	Ascomycete	CHS1, CHS2
Schizosaccharomyces pombe		SpCHS1
Emericella (Aspergillus) nidulans		AdCHS1, AdCHS2
Aspergillus niger		AnCHS1, AnCHS2
Ajellomyces (Blastomyces) dermatitidis		BdCHS1, BdCHS2
Ajellomyces (Histoplasma) capsulatum		HcCHS1, HcCHS2, HcCHS3
Neurospora crassa		NcCHS1, NcCHS2, NcCHS3
Ustilago maydis	Basidiomycetes	UmCHS1, UmCHS2
Schizophyllum commune		ScCHS1
*Candida albicans**	Fungi Imperfecti	CaCHS1, Ca CHS2
Exophiala jeanselmei¥		EjCHS1, EjCHS2, EjCHS3
Phaeococcomyces exophialae		PeCHS1, PeCHS2
Rhinocladiella atrovirens		RaCHS1, RaCHS2
Wangiella dermatitidis WdCHS3		WdCHS1, WdCHS2,
Xylohypha bantiana		XbCHS1, XbCHS2

Taxonomic affinities are modified from Dixon and Frontling.[27]

*Although asexual, has many counterparts in the Hemiascomycetidae.[15,19]

¥ These species are considered to be asexual species of Loculomycetidae, or less often the Pyrenomycetidae.[16]

The distance matrix programs DNADIST and FITCH were used to produce the class I trees (Fig. 4). Similarity scores were obtained for all pairwise comparisons of the aligned DNA sequences and were transformed into a distance matrix by the program DNADIST. This program allows for different substitution rates between transitions and transversions, according to the 2-parameter model of Chimera. In turn, the distance matrix was used as input to the program FITCH, which calculated branching order and length. *S. cerevisiae CHS1*, shown by CLUSTAL to be an outlying species, was used to root the trees. In addition to confirm the close relationships suggested by the CLUSTAL program, the FITCH tree for class II places the sequences for the two Basidiomycetes *Schizophyllum commune* and *Ustilago maydis* together and separates the single *Schizosaccharomyces pombe* gene fragment into a class by itself (CLUSTAL had grouped *Ss. pombe* and *Sp. commune* together, a result totally inconsistent with modern fungal taxonomic concepts). The most striking aspect of the FITCH analysis, however, is the large evolutionary separation suggested between *S. cerevisiae* (and possibly *C. albicans* and *Ss. pombe*) and the other fungi. On morphological grounds, both *S. cerevisiae* and *Ss. pombe* are traditionally classified in the same Ascomycete order, Endomycetales.[18] Based on a variety of results, *C. albicans* is also generally thought

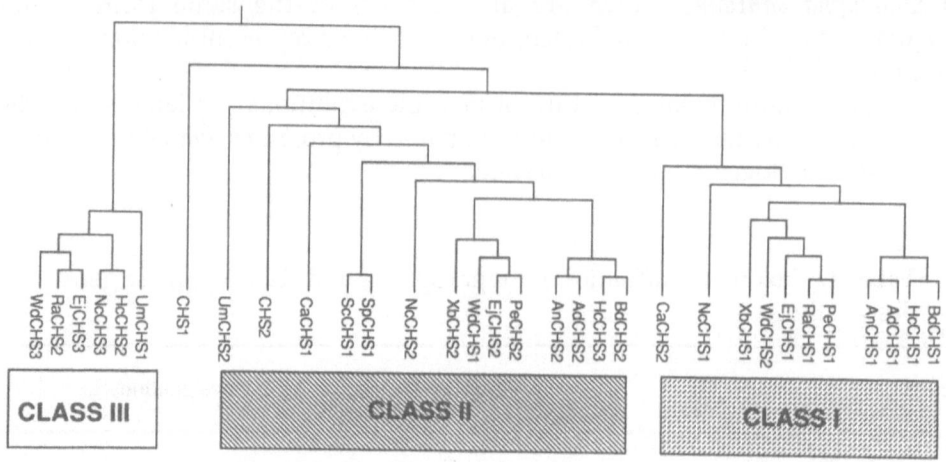

Fig. 3 Unweighted pair group method using arithmetic averages (UPGMA) dendogram showing three distinct chitin synthase classes. The tree was calculated by the program CLUSTAL from deduced amino acid sequences. Except for *S. cerevisiae CHS1*, the gene fragments fall into three groups. Branch lengths do not indicate a rigorous calculation of evolutionary distances, nor can phylogenetic relationships be inferred with confidence from this tree. Except for *CHS1*, the first two letters represent the species (see Table 1).

to be a member of the same order.[15,19] Should *C. albicans*, in fact, be an ascosporic yeast like *S. cerevisiae* and *Ss. pombe*, then it might not be surprising that these three fungi exhibit large separations from the remaining fungi investigated. This point, at least as related to *S. cerevisiae*, is of interest in the light of the analysis of glyceraldehyde-3-phosphate genes by Smith,[20] who postulates enormous evolutionary separation of *Saccharomyces* and related organisms from the filamentous fungi. Numerous other studies with *S. cerevisiae* and *Ss. pombe* suggest considerable evolutionary separation between these two hemiascomycetous species, which may be as great as that between *S. cerevisiae* and animals.[21]

The amplification of a chitin synthase fragment from *Ss. pombe* DNA was unexpected because members of the genus *Schizosaccharomyces* are generally thought to have no chitin in their cell walls. However, a recent paper by Sietsma and Wessels[22] reports the presence of glucosaminoglycan in *Ss. pombe*.

A second aspect of the FITCH analysis, which is quite remarkable is the clustering of fungi, in both the class I and class II analysis, which are traditionally recognized as related. For the class I tree, all the hyphomycetous and melanized Fungi Imperfecti are included in one branch, reflective of their possible loculoascomycetous affinities, whereas the remaining ascomycetous, ascocarpic fungi are included in another branch consisting of only the cleistothecial and the one perithecial species.[16,22] While the exclusive clustering of the two ascocarpic ascomycetous groups is not apparent in the class II analysis, the clustering together of these same fungi with the hyphomycetous, melanized organisms in a single main branch that diverges from a second branch with the two Basidiomycetes and a third branch that encompasses the known or suspected ascosporic yeast is still compatible with studies by Walker,[24] who proposes similar groupings of the

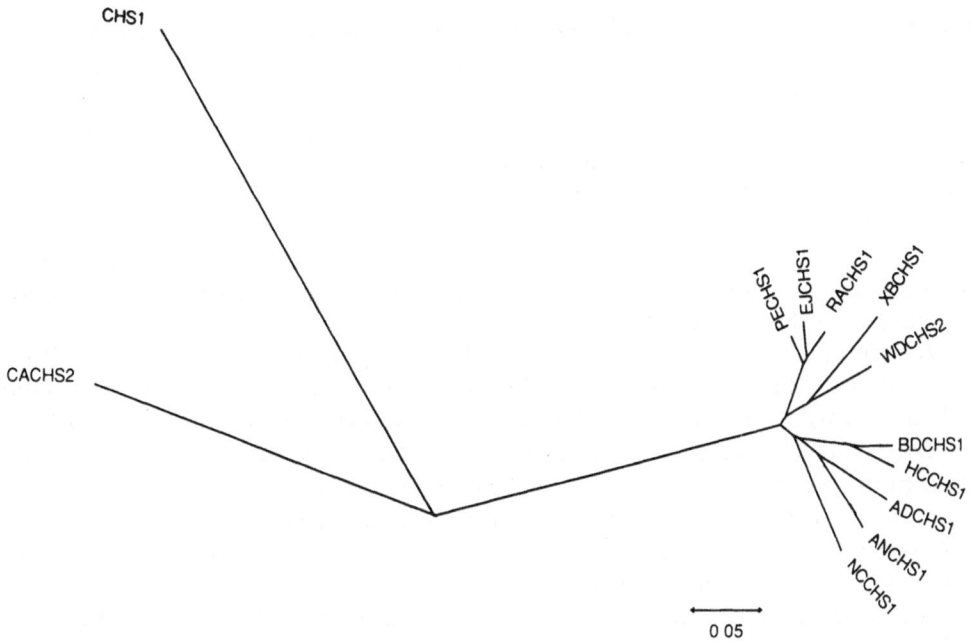

Fig. 4 Tree showing phylogenetic relationships of class I chitin synthase fragments. The DNA sequences were compared by distance matrix methods using the FITCH program. *S. cerevisiae* *CHS1* was included in the analysis as an outgroup to root the tree. Branch lengths reflect relative evolutionary distance and are defined by Felsenstein.[26] Except for *CHS1*, the first two letters represent the species (see Table 1).

Ascomycete species based on 5S ribosomal RNA sequences, class II chitin synthase phylogenetic tree contradicts at least one current sequence analysis of small subunit ribosomal RNAs, which did not suggest that Basidiomycetes are derived from Ascomycetes.[25] Possibly this contradiction relates only to the *CHS2* tree being rooted to *CHS1* of *S. cerevisiae*. Future analyses using different roots or no roots and additional sequences from other fungi with better established phylogenies should clarify this situation.

Bootstrap analysis was performed by the DNABOOT program[26] as a further means of validating interrelationships inferred by the FITCH analysis. Bootstrapping is a statistical method used to evaluate the confidence level of the phylogenetic estimate by random resampling of the data. One hundred bootstrap replicates were performed on each class, and the resulting consensus confirmed our taxonomic interpretation of the CLUSTAL and FITCH analyses. *E. jeanselmei*, *P. exophialae*, and *Rhinocladiella atrovirens* were grouped together in 95 of the 100 trials, giving them a 95% bootstrap confidence limit as a phylogenetic group. *H. capsulatum* and *B. dermatitidis* appear with a >99% confidence limit as a group. *C. albicans* and *S. cerevisiae* appeared as outgroup from the other fungal species in 100% of the bootstrap replicates. Although not showing >95% confidence limits, other close relationships between *U. maydis* and *Sp. commune*, and between two *Aspergillus* species were preducted by the DNABOOT program, which constructs trees using parsimony methods rather than distance matrix as FITCH does.

Although the sequence analysis suggests the presence of three classes of "zymogen type" chitin synthases in fungi, the results are considered only suggestive and are represented only to serve as a guide for further investigations. Several factors limit the value of our conclusions. In the first place, the analysis was limited to only one very highly conserved region of the genes in question. It will be necessary to extend the analysis to the complete sequences of as many of the genes as possible to determine whether the class and evolutionary relationships suggested here are maintained in the light of complete sequence information. A related point concerns the general function of the protein domain being analyzed. If, as expected, this domain is part of the catalytic region of the enzyme, different classes might differ somewhat in catalytic mechanism, pH optimum, etc. However, this type of variation may or may not be correlated with the biological functions of the enzymes -i.e., catalytic mechanism variants may or may not have different functions in different species.

A second limitation of the data is that all fragments were recovered with a single set of PCR primers. An incomplete set of gene fragments may well bias the analysis toward a subgroup of genes, although we do know that *S. cerevisiae CHS and CHS2*, which can be recovered from the genome with our primers, have quite different functions *in vivo*.[10] We know that we would almost certainly not recover DNA fragments for the "non-zymogen" chitin synthase III class of enzyme because the *S. cerevisiae CSD2 (CAL1)* gene lacks the sequences used to design our primers (C. Bulawa, personal communication).

A final limitation of our analysis is in the interpretation that can be made of evolutionary relationships. Becaus, probably for functional reasons, the sequences are all very similar, the apparent evolutionary distance and relationships may be different from those derived by other methods. On the other hand, all chitin synthases may catalyze the same reaction, using the same key residues. The variation seen in the segment examined may, in fact, reflect the accumulation of neutral amino acid changes; therefore, the greater the evolutionary separation, the greater the number of such changes.

In spite of these limitations, we feel that the analysis presented here will be valuable in planning and interpreting gene disruption experiments designed to unravel the functions of the multiple chitin synthase genes in fungi. They should also be useful in conjunction with more extensive sequence data in the analysis of evolutionary relationships among fungi.

ACKNOWLEDGMENTS

We thank the following Massachusetts Institute of Technology undergraduates for PCR fragment isolation and preliminary sequencing results: Monica McConnell, Susan Pauwels, Elly Bulboaca, Banu Ramachandran, Doug Jeffery, Jason Salter, Linda Sun, Wendy Wai, Daniel Wambold, Jork Zwicker, Richard Cheng, Cindy Hummel, Harry Hwang, Rachel McCarthy and Tracy Kindaid. This work was supported by Grants GM31318 (to P.W.R.) and CA14051 (to P. Sharp) from the National Institutes of Health. M.M. was supported by grants to P.J.S. from the Texas Applied Technology Program (TATP-4493) and University of Texas Research Institute Program Grant RR07091.

REFERENCES

1. A. Cassone, Cell wall of pathogenic yeasts and implications for antimycotic therapy, *Drugs Exptl. Clin. Res.* 12: 635 (1986).
2. A. Duran and E. Cabib, Solubilization and partial purification of yeast chitin synthase. Confirmation of the zymogenic nature of the enzyme, *J. Biol. Chem.* 243: 4419 (1978).
3. P. Orlean, Two chitin synthases in *Saccharomyces cerevisiae*, *J. Biol. Chem.* 262: 5732 (1987).

4. P.A.B. Orlean and D. Kerridge, Increase chitin synthase activity associated with the initiation of filamentous growth by *Candida albicans, VIIth International Specialized Symposium on Yeast,* Valencia, Spain, p. 93 (1981).

5. P. Orlean, Polysaccharide synthesis in the dimorphic fungus *Candida albicans.* Ph. D. Thesis, University of Cambridge, Cambridge, UK (1982).

6. J. Au-Young and P.W. Robbins, Isolation of a chitin synthase gene (*CHS1*) from *Candida albicans* by expression is *Saccharomyces cerevisiae, Mol. Microbiol.* 4: 197 (1990).

7. F.C. Odds, "*Candida* and candidosis: a review and bibliography", 2nd edn., Baillière and Tindall, London (1988).

8. M.S. Marriott, Isolation and chemical characterization of plasma membranes from the yeast and mycelial forms of *Candida albicans, J. Gen. Microbiol.* 86: 115 (1975).

9. E. Cabib, A. Sburlati, B. Bowers and S.J. Silverman, Chitin synthase I, an auxiliary enzyme for chitin synthesis, *J. Cell. Biol.* 108: 1665 (1989).

10. C.E. Bulawa and B.C. Osmonds, Chitin synthase I and chitin synthase II are not required for chitin synthesis *in vivo* in *Saccharomyces cerevisiae, Proc. Nat. Acad. Sci. USA* 87: 7424 (1990).

11. A.R. Bowen, J.L. Chen-Wu, M. Momany, R. Young, P.J. Szaniszlo and P.W. Robbins, Classification of fungal chitin synthases, *Proc. Nat. Acad. Sci. USA* 89: 519 (1992).

12. D.G. Higgins and P.M. Sharp, CLUSTAL: a package for performing multiple sequence alignment on a microcomputer, *Gene* 73: 237 (1988).

13. P.H.A. Sneath and R.R. Sokal, "Numerical Taxonomy", Freeman, San Francisco (1973).

14. A. Valencia, P. Chardin, A. Wittinghofer and C. Sander, The *ras* protein family: evolutionary tree and role of conserved amino acids, *Biochemistry* 30: 4637 (1991).

15. J.W. Rippon, "Medical Mycology: The Pathogenic Fungi and the Pathogenic Actinomycetes", Saunders, Philadelphia, 3rd Ed. (1988).

16. G.S. deHoog and M.R. McGinnis, Ascomycetous black yeasts, in: "Proceedings of an International Symposium on the perspectives of Taxonomy, Ecology and Phylogeny of Yeasts and Yeast-Like Fungi", G.S. deHoog, M.T.H. Smith and A.C.M. Weijman, eds., Elsevier, Amsterdam (1987).

17. K.B. Raper and D.J. Fennel, "The Genus *Aspergillus*", Krieger, Huntington, NY (1973).

18. J. Lodder, ed. "The Yeast: A Taxonomic Study", North-Holland, Amsterdam (1970).

19. S.M. Barns, D.J. Lane, M.L. Sogin, C. Bibeaw and W.G. Weisburg, Evolutionary relationships among pathogenic *Candida* species and relatives, *J. Bacteriol.* 173: 2250 (1991).

20. T. Smith, Disparate evolution of yeasts and filamentous fungi indicated by phylogenic analysis of glyceraldehyde-3-phosphate dehydrogenase genes, *Proc. Nat. Acad. Sci. USA* 86: 7063 (1989).

21. P. Nurse, Some of what you need to know about the molecular control of blood cell formation, *Cell* 61: 756 (1990).

22. J.H. Sietsma and J.G.H. Wessels, The occurrence of glucosaminoglycan in the wall of *Schizosaccharomyces pombe, J. Gen. Microbiol.* 136: 2261 (1990).

23. E. Moore-Landecker, "Fundamentals of the Fungi", Prentice Hall, Engelwood Cliffs, NJ, 3rd Ed. (1990).

24. W.F. Walker, *Sest. Appl. Microbiol.* 6: 48 (1985).

25. C.A. Illingworth, J.H. Andrews, C. Bibeau and M.L. Sogin, Phylogenetic placement of *Athelia bombacina, Aureobasidium pullulans,* and *Collectotrichum gloeosporioides* iferred from sequence comparisons of small-subunit ribosomal RNAs, *Exp. Mycol.* 15: 65 (1991).

26. J. Felsenstein, PHYLIP3.2 Manual," University of California Herbarium", Berkeley, CA (1989).

27. D.M. Dixon and R.A. Fromtling, Morphology, taxonomy and classification of the fungi, in:"Manual of Clinical Microbiology", A. Balows, W.J. Hawsler Jr., K.L. Hermann, H.D. Isenberg and H.J. Shadomy, eds., American Society Microbiology, Washington, 5th Ed. (1991).

KEY GENES IN THE REGULATION OF DIMORPHISM OF *CANDIDA ALBICANS*

Neil A.R. Gow, Rolf Swoboda, Gwyneth Bertram, Graham W. Gooday, and Alistair J.P. Brown

Department of Molecular & Cell Biology
Marischal College
University of Aberdeen
Aberdeen AB9 1AS, UK

ABSTRACT

The search for genes that regulate dimorphism is now a major focal point of *Candida* research. In this paper we make some general comments about approaches and methodologies related to this work and describe two contrasting strategies that we are using in the analysis of genes related to dimorphism. The disruption of a hypha-specific chitin synthase gene is described using a method that is well suited to work with this genetically recalcitrant organism. Secondly, we present the results of a screen of a cDNA library in a γ ZAPII expression system with antibodies from patients with candidiasis. The sera were enriched for antibodies recognizing proteins from the hyphal form, yet the screen yielded mainly sequences that are expressed in both the yeast and hyphal forms of the organism. Three of the ten sequences that were analyzed encode genes for glycolytic enzymes. Detailed Northern blot analysis revealed changes in the levels of these mRNA's that could be correlated with changes in growth that accompany dimorphism.

INTRODUCTION

Dimorphism and Molecular Genetics

Less than ten years ago, the notion of an understanding of dimorphism of *Candida albicans* at a genetic level seemed a distant prospect. However, practical and fundamental questions concerning the organism's pathogenicity and growth have created the impetus that has led to the design and application of molecular biological and genetic technologies that now make the analysis of genes in this

Dimorphic Fungi in Biology and Medicine, Edited by
H Vanden Bossche *et al*, Plenum Press, New York, 1993

organism a rapidly growing area of research. The search for genes that regulate, or are regulated by, the dimorphic switch of *Candida albicans* has employed two general approaches. Firstly, specific genes with products that are predicted to have regulatory roles (for example calmodulin, translation factors, etc.), or which may be important in the determination of cell shape (for example actin, tubulins, chitin synthases, etc.) have been cloned and analyzed. Secondly, screening for genes that have different mRNA or protein levels in the yeast and hyphal forms has been carried out. We have employed each of these in our studies of dimorphism in *Candida*.

Both these approaches are ultimately problematic when considering the regulation of dimorphism in *C. albicans*. The first depends on guess-work that is limited by inadequate knowledge of the physiological and biochemical basis of the control of dimorphism. Moreover, the relationship between dimorphism and pathogenicity is still not clear. Strategies targeted to specific genes have and will provide valuable information about the roles of the genes of choice, but they may fail to uncover many important regulatory elements (such as protein kinases, phosphatases, proteases, DNA binding proteins) that may be involved in temporal and spatial control of gene expression, metabolism and pattern formation. For investigation of the regulation of dimorphism, differential screening of cDNA libraries from yeast and hyphal forms has the benefit of avoiding preconceptions of the underlying mechanisms, but suffers from the inability of screens to discriminate between gene products that are the cause or the consequence of morphogenesis. Many changes uncovered by Northern or protein analysis are destined to be proven inconsequential as far as *regulation* of dimorphism is concerned. Also, controls must be designed with care to distinguish changes in transcription and translation that are due to changes in pH, temperature or medium composition, for example, rather than changes in cell shape. Another problem is that this method is unlikely to reveal key regulatory steps that operate at the post-translational level. Finally the function of many genes isolated via differential screening may not be identified by sequence comparison with those of known genes in the data bases, despite the rapid progress being made on the analysis of eukaryotic genes and the *Saccharomyces cerevisiae* genome sequencing project.

The analysis of genes in *C. albicans* has been severely limited by the genetic intransigence of the organism. The inability to perform genetic crosses, the scarcity of genetic markers and difficulties in creating stable mutants in an isogenic background of this diploid organism remain significant obstacles to the molecular analysis of the dimorphic transition. However, methodologies for the cloning of *C. albicans* genes by screening of gene libraries with homologous gene probes from *S. cerevisiae* and other organisms and/or by the functional complementation of mutations in *S. cerevisiae* are now well established.[1] Once a gene has been cloned, a logical step in the analysis of its function in the biology, morphogenesis or pathogenicity of *C. albicans* is to create null mutants and to characterize the resulting phenotype. A fundamental stumbling block in the functional analysis of genes identified by either approach is in the availability of convenient methods for the sequential disruption of several genes to investigate their function. This paper discusses the application of one recently developed method for the sequential gene disruption of *CHS2*, a gene encoding a chitin synthase that is expressed preferentially in the hyphal form, and describes the mutant phenotype. This technique requires only a single genetic marker to achieve multiple gene disruptions and is readily applicable to the creation of mutants of other *Candida* genes.[2]

Chitin Synthesis and Dimorphism

Chitin is the principle load-bearing and shape-giving component of fungal cell walls.[3] The hyphal cell wall contains 3-5 times the amount of chitin than the yeast cell wall[4] and has a structure that is unlike the microfibrillar forms of chitin found in hyphae of obligately filamentous fungi.[5] The specific chitin synthase activity of the hyphal form measured *in vitro* has been reported to be up to ten times that in the yeast form.[6] Therefore the regulation of chitin synthesis may yield valuable information about the dimorphic transition in *C. albicans*.

In *S. cerevisiae*, at least three chitin synthase genes (CHS1,CHS2,CSD2 - also called CHS3) have been cloned and characterized.[7-9] *C. albicans* homologues of the genes CHS1 and CHS2 (that encode genes for zymogenic chitin synthase enzymes) have been cloned and sequenced by Robbins and his co-workers.[10,11] The CSD2 homologue in *C. albicans* has been identified on low-stringency Southern blots (N.A.R. Gow, unpublished results) but has not yet been cloned. The *Candida* CHS1 and CHS2 genes are expressed differentially during yeast and hyphal growth. Using culture density as a variable to control cell morphology Chen-Wu *et al.*[11] showed that CHS1 was expressed preferentially in the yeast form whereas CHS2 mRNA occurred at high levels in the hyphal form.

MATERIALS AND METHODS

Growth Conditions

The yeast form of *C. albicans* was grown in the non defined yeast extract-peptone-dextrose medium (YPD)[12] or the synthetic minimal medium (SD).[12] Germ tubes were induced by inoculation into media containing 20% (v/v) newborn calf serum,[13] N-acetyl glucosamine[14] or by pH- and temperature-stimulated dimorphism[15] as indicated.

For kinetic experiments involving Northern analysis, hyphae were induced under two conditions, either by inoculating yeast cells into YPD plus serum at 37 °C, or using the regime of pH and temperature-stimulated dimorphism.[15] In addition, parallel cultures were set up to control for the effects of temperature, pH and serum. Cell numbers, and the percentage of cells in the yeast or hyphal forms were monitored throughout each experiment. During the course of the experiment, growth at 25 °C was exclusively by budding. RNA was prepared for Northern analysis at various times following the inoculation into fresh medium.

Gene Disruption of CHS2

Disruption of the *Candida* CHS2 gene has been achieved using a technique first described by Alani *et al.*[16] for gene disruption of *S. cerevisiae*. Accordingly a cassette containing a functional *Candida* URA3 gene flanked by 1.1 kb direct repeats of a bacterial hisG sequence [kindly provided by W. Fonzi & P. Sypherd; see ref (2)] was exploited. The plasmid for the gene disruption was obtained by introducing the hisG::URA3::hisG cassette into cloned CHS2 DNA to generate a disruption cassette which was flanked by 2.1 and 0.8 kb sequences of the CHS2 gene (Fig. 1). Gene disruption was achieved by integrative transformation of the *C. albicans ura3* strain SGY243, by use of linearized plasmid and selection on the basis of uridine prototrophy. Spheroplasts were prepared with β-glucuronidase (Sigma, type H-2S). Homologous recombination between the tandemly repeated hisG

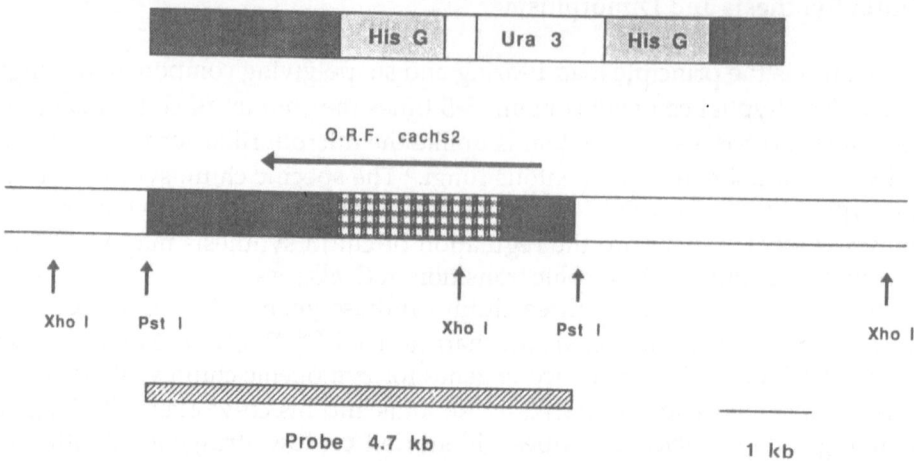

Fig. 1 *chs2::hisG::URA3::hisG* cassette used for the disruption of the *Candida albicans CHS2* gene. Below the disruption cassette is a map of the target *Candida* DNA with restriction sites used in the Southern analysis. The checked area indicates the portion of the open reading frame of the *CHS2* gene that was deleted in the disruptants. The region of the *CHS2* gene used as a probe is shown by the hatched bar.

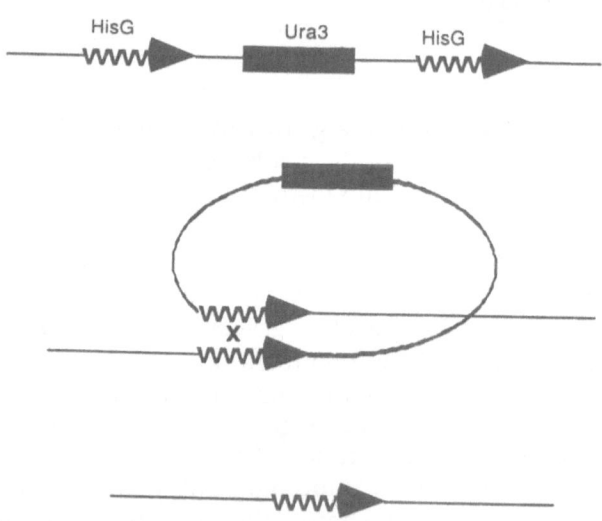

Fig. 2 Homologous recombination between direct repeats of the *hisG::URA3::hisG* cassette to remove the *URA3* selectable marker.

copies results in the excision of *URA3* (and one copy of the *hisG*) (Fig. 2) and the selection of *ura3* cells is facilitated by plating in the presence of 5-fluoroorotic acid (5-FOA) which kills Ura+ prototrophs.[17] The result is that the *ura3* auxotrophy is regenerated and the same disruption cassette can be used to disrupt the second copy of the gene (Fig. 3). Since the *ura3* phenotype is regenerated after each transformation step, multiple selectable markers are not required for sequential disruptions. This procedure could therefore be used to knock out further genes and thereby create multiple mutants.

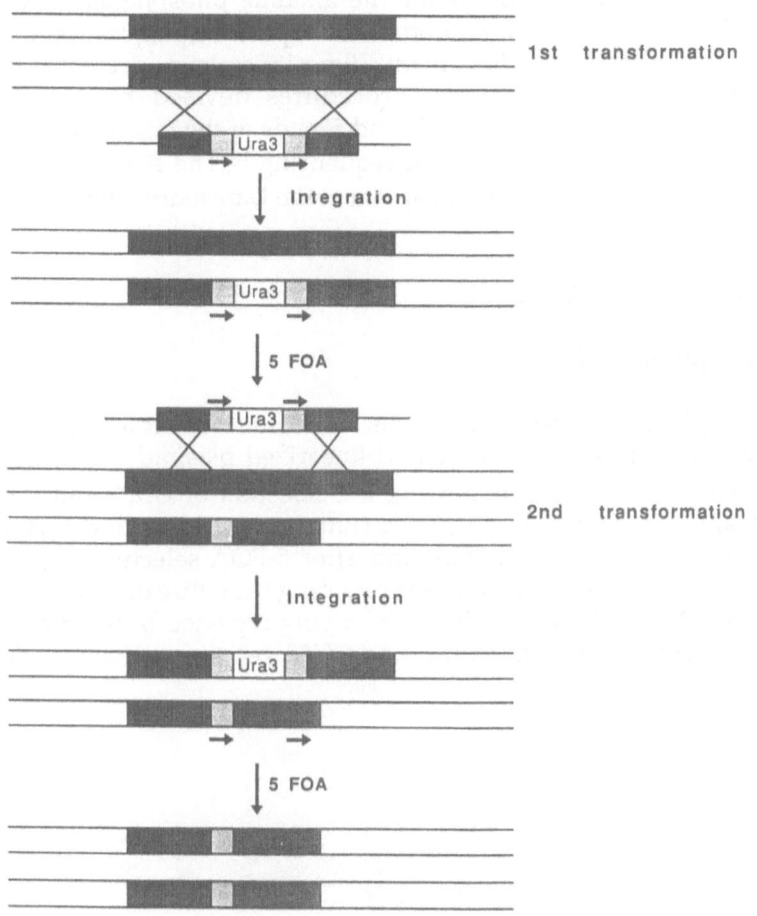

1st   transformation

Integration

5 FOA

2nd   transformation

Integration

5 FOA

Fig. 3 Strategy for sequential gene disruption of *Candida* gene with the "ura-blaster" cassette and the *ura3 C. albicans* strain SGY243. The first transformation leads to integration of the cassette by homologous recombination and disruption of the target allele to create the heterozygous mutant. Recombination between the *hisG* tandem repeats (Fig. 2) excises the *URA3* gene. Recombinants are selected on the basis of resistance to 5-FOA. Disruption of the second copy of the gene can then be achieved in the second transformation by recombination with the same disruption cassette. Growth in the presence of 5-FOA is used again to recover the *ura3* auxotrophy.

Construction and Screening of cDNA Library

RNA was prepared from *C. albicans* strain 3153 after 105 min growth in the hyphal form in serum-containing medium at 37 °C. Poly(A)-containing RNA was isolated by affinity chromatography with oligo(dT)-cellulose and this was used to prepare a cDNA library γ ZAPII kit according to the manufacturers instructions (Stratagene, Cambridge, UK). The cDNA expression library was screened with antibodies in pooled sera from five patients with candidiasis (most of whom were HIV positive), and five asymptomatic patients. The pooled antiserum was first preadsorbed with protein extracts made from *Escherichia coli* and the yeast form of *C. albicans* to reduce background and to enrich for hypha-specific antibodies. Positive clones were identified with alkaline-phosphatase-conjugated IgM and IgA human second antibodies and nitro blue tetrazolium and 5-bromo-4-chloro-3-

indolyl-phosphate as substrates for the alkaline phosphatase-mediated colour reaction. Ten clones picked at random from 83 positive plaques were then screened and purified a second time. Finally the 10 positive phages were converted to their Bluescript phagemids following procedures devised by the manufacturer (Stratagene, Cambridge, UK). The 3' and 5' ends of the inserts were sequenced by means of standard dideoxy plasmid sequencing.[18] The sequences obtained were then analyzed with the UWGCG program at the Daresbury laboratory.[19]

RESULTS AND DISCUSSION

Gene Disruption of CHS2

The *hisG::URA3::hisG* "ura-blaster" cassette was cloned between flanking regions of the *Candida CHS2* gene and linearized plasmid was transformed into spheroplasts of *C. albicans*. Following the selection of Ura⁺ transformants, *ura3* recombinants were isolated on plates containing uridine and 5-FOA. Genomic DNA of the transformants, before and after 5-FOA selection, was analyzed by Southern blotting using a 4.7 kb probe encoding the entire open reading frame and flanking regions as shown in Fig. 1. Following repeated transformation with the disruption cassette, a null mutant of the *CHS2* gene was obtained (Fig. 4). Three

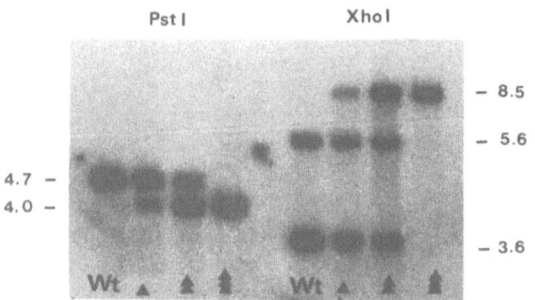

Fig. 4 Southern analysis of the sequential gene disruption of the *CHS2* gene of *C. albicans* by means of repeated *URA3* selection. The triangles indicate the number of transformations performed to generate the strain. For *PstI* digestions, the disruption results in the loss of the wild type 4.7 kb fragment to generate a new 4.0 kb band. For *XhoI* digests, the mutant allele generates a single band at 8.5 kb compared with the two wild type bands at 3.6 and 5.6 kb. Note the change in banding intensities reflecting the numbers of alleles present. After the first and second rounds of transformation the relative intensities of bands due to wild type and disrupted alleles are reversed. This can be interpreted by suggesting the transformant isolated after the first transformation had two wild type alleles and one disrupted allele. After the second transformation there are two mutant alleles and one wild type allele. The additional allele may have been created during the first round of transformation by a tandem duplication event as described in the text.

sequential rounds of transformation and 5-FOA selection were necessary to obtain a null mutant, suggesting initially that the original wild-type strain carries three copies of the *CHS2* locus. Southern analysis of the third round transformants prior to 5-FOA treatment revealed four bands, the unequivocal result of three *CHS2* alleles (Fig. 5). The largest predicted band in this Southern blot (13.2 kb) corresponded to an allele incorporating the full *hisG::URA3::HisG* cassette. While most transformants had a band at this position, some had higher molecular weight bands indicating that multiple copies of the cassette were integrated at a chs2 locus (Fig. 5). These multiple tandem copies of the *chs2::hisG::URA3::hisG* were always resolved by homologous recombination in 5-FOA-resistant progeny to a single copy lacking a wild type *URA3* gene. The formation of these multiple, tandem arrays of the *chs2::hisG::URA3::hisG* implies the formation *in vivo* of a circularized intermediate from the linear cassette used in the transformation. This raised the possibility that the apparent triploidy was an artefact due to the integration of circular forms of the linear cassette used in the disruption. Southern analysis of genomic DNA digested with *EcoRI* and *XhoI* revealed a new band of 5.0 kb which confirmed this was the case. An *EcoRI/ XhoI* band of 5.9 kb was expected for the disruption of a third *CHS2* allele, whereas a 5.0 kb band was expected for a tandem integration event.

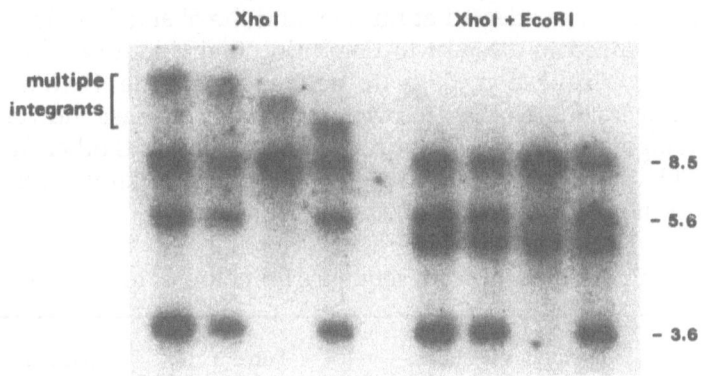

Fig. 5 Southern analysis of third round transformants prior to selection with 5-FOA. DNA from a null mutant, with no wild type *XhoI* bands at the 3.6 and 5.6 kb positions, is in the 3rd and 7th lanes (numbered from the left hand side). The bands at 8.5 kb are due to disrupted alleles. Lanes 1-4 also have higher molecular weight bands that are the result of multiple integrations of the disruption cassette. After double digestion with *XhoI* and *EcoRI* (lanes 5-8) a band at 5.0 kb is resolved indicating tandem duplication resulted in the apparent third *CHS2* allele, as described in the text.

Further indirect evidence for this tandem duplication was suggested by the finding that 75% of second round transformants were found to have the wild type pattern in *PstI* and *XhoI* DNA digests for Southern blots (not shown). This could have been due to recombination between homologous regions of the tandem copies of *CHS2* gene (created by integration and tandem duplication in the first round of transformation), resulting in the reconstruction of one wild type copy of the gene. Analogous results have been obtained by Fonzi *et al*, using the same "ura blaster" method for the sequential gene disruption of the *Candida ECE1* gene (W.A. Fonzi, personal communication).

Despite these complications, Southern analysis confirmed that the strategy employed was ultimately successful in producing a null mutant which was then analyzed with respect to phenotype.

Phenotype of the *chs2::HisG* Mutant

The homozygous *chs2::hisG* mutant was viable and therefore the *CHS2* gene product is not essential for growth. In addition the mutant was fully competent to form germ tubes upon stimulation of hyphal growth using 20% (v/v) calf serum, although the onset of germ tube formation was slightly delayed in the null mutant. The doubling time of the yeast form was slightly slower in YPD but not in the defined medium SD. Further characterization of the phenotype of this mutant in terms of its chitin content and chitin synthase activity is now in progress and the construction of a *CHS1* mutant is now planned employing the same disruption protocol. Therefore although the *CHS2* mRNA appears to be regulated in response to morphogenesis the gene product does not appear to be essential for dimorphic outgrowth.

Screening for Hypha-Specific Gene Sequences

We used antibodies to screen for sequences that encode proteins that are recognized during infections in humans, and which may be specific for the hyphal form of *C. albicans*. A total of 40,000 clones were screened with pooled sera that had been preadsorbed with proteins from the yeast form of *C. albicans*. Of 83 positive clones, 10 were selected at random and the 5' and 3' ends of each cDNA insert were sequenced to attempt to determine their identity. Of the ten, four cDNA's had significant homology to sequences in the GenBank or EMBL databases. Two were homologous to pyruvate kinases of yeast and vertebrates and one was homologous to alcohol dehydrogenases of yeast and other fungi (Table 1). The fourth cDNA showed strong homology to a gene of unknown function located downstream of the *SIR3* locus of *S. cerevisiae* (Table 1).

Table 1. Summary of significant sequence homologies to cDNAs 2, 3 and 4.

cDNA	Homologous Sequence	Percent Homology	Length of Overlap	References
cDNAs 2 & 3:				
5'-end	*S. cerevisiae* PYK1	74.9%	287bp	24,25
	Rat liver L-type Pyk	58.7%	230bp	26
	Chicken muscle Pyk	56.0%	250bp	27
3'-end	*S. cerevisiae* PYK1	59.9%	142bp	24,25,
cDNA4:				
5'-end	*S. cerevisiae* ADH1	76.6%	291bp	28
	S. cerevisiae ADH3	72.2%	259bp	29
	S. pombe adh	69.4%	170bp	30
	A. nidulans alcA	58.7%	254bp	31
3'-end	*S. cerevisiae* ADH1	73.2%	198bp	32
	S. cerevisiae ADH3	70.1%	201bp	29
cDNA10:				
5'-end	*S. cerevisiae* ORF 3' to *SIR3*	80.8%	313bp	32

The results show that some glycolytic enzymes can be strong antigens in *Candida* infections. However, further screening of the reactivity of antiserum from individual patients showed that immunoreactivity to *Candida* pyruvate kinase and alcohol dehydrogenase was not universal. Therefore these reactivities may not be exploited as diagnostic indicators of candidiasis. The finding that *Candida* glycolytic enzymes are common antigens has been reported elsewhere.[20-22] We can add pyruvate kinase to the list that includes alcohol dehydrogenase, aldolase, enolase and phosphoglycerate kinase.[20-22]

Northern Analysis of the Cloned cDNA's

Initially, Northern blots were performed on RNA isolated at a single time point following the induction of hyphal growth. These showed that the concentration of the pyruvate kinase and alcohol dehydrogenase mRNA's was different in yeast and hyphal cells and that the level of a specific mRNA depended on the protocol for stimulating germ tube growth. These findings initially encouraged us to think that transcriptional control of glycolytic enzymes may play a role in regulation of morphogenesis. However, a more detailed kinetic analysis of the transition from yeast to hyphal growth showed that the situation was more complex. The levels of the pyruvate kinase and alcohol dehydrogenase mRNA's were high in the overnight culture of yeast cells that was used as an inoculum. Upon dilution into fresh media, the levels of both mRNA's decreased markedly only to increase again during the onset of growth in either the yeast or the hyphal form. The levels of the pyruvate kinase and alcohol dehydrogenase mRNA's increased slowly in cultures at 25 °C, where growth was in the yeast form. In comparison, the increase in the levels of these mRNA's was significantly faster in cells growing at 37 °C, *prior* to the emergence of hyphae. However, the rate of induction of these mRNAs did not parallel the kinetics of hyphal induction. For example, the pyruvate kinase and alcohol dehydrogenase mRNA's increased rapidly in YPD at 37 °C in the absence of serum, although hyphae emerged slowly under these conditions. Thus the pyruvate kinase and alcohol dehydrogenase mRNA levels are regulated during growth, but the relationship with morphogenesis may be indirect. These experiments illustrate the large number of parameters and possible pitfalls that must be considered when investigating the "morphogenetic regulation" of a gene.

In conclusion, three general points emerge from this study. [1] The products of our differential screening protocol identified proteins that are common to both yeast and hyphal cell types. [2] Highly expressed genes for housekeeping activities may be responsive both to fluctuations in growth rate and environmental changes. [3] The data underline our general point that screening of cDNA libraries, using such strategies, cannot distinguish changes that may be the consequence and not the cause of dimorphism. The construction of pyruvate kinase and alcohol dehydrogenase mutants may help us to distinguish between these possibilities.

CONCLUSIONS

The *CHS2* gene of *C. albicans* is not essential for growth. Gene disruption of the *CHS2*, which is apparently expressed predominantly in the hyphal form, did not result in a significant perturbation of dimorphism or the inhibition of hyphal growth. Interestingly, disruption of *CHS1* in *S. cerevisiae* (the closest homologous gene to the *Candida CHS2* in brewer's yeast) was also non-lethal and had little effect on chitin content of the cell wall.[7] At least two other chitin synthase genes have been identified in *C. albicans* and the analysis of these genes is now underway. Despite complications arising through the formation of tandem arrays of transforming DNA, the method of gene disruption described here is likely to be widely applicable in the analysis of *Candida* genes.

Pyruvate kinase and alcohol dehydrogenase are dominant, but not ubiquitous immunogens during candidiasis. Northern analysis confirmed that an increase in the pyruvate kinase and alcohol dehydrogenase mRNA's accompany the dimorphic transition. It is possible that glycolytic gene regulation responds to changes in growth state that are a prerequisite for dimorphism. However, at this stage we cannot say whether this is cause or consequence of the dimorphic transition.

ACKNOWLEDGEMENTS

The gene disruption work described here was carried out while NARG was on sabbatical leave as an Aberdeen University Fellow, at Phil Robbins' laboratory. Thanks are due to Phil Robbins, Christine Bulawa, Alex Haüsler and Bill Fonzi for many useful discussions and to Bill Fonzi for providing the *hisG::URA3::hisG* construct. The cDNA screening project is supported by the Wellcome Trust.

REFERENCES

1. S. Scherer and P.T. Magee, Genetics of *Candida albicans*, *Microbiol. Rev.* 54 : 226 (1990).
2. C. Birse, W.A. Fonzi, M. Saporito, M. Irwin and P.S. Sypherd, Molecular Genetics of Dimorphism in *Candida albicans*, in: "*New Strategies in Fungal Disease*", J.E. Bennett, R.H. Hay, Peterson, P.K, eds., Churchill Livingstone, Edinburgh (1992).
3. N.A.R. Gow, Control of the extension of the hyphal apex, *Curr. Top. Med. Mycol.* 3 : 109 (1989).
4. F.W. Chattaway, M.R. Holmes and A.J.E. Barlow, Cell wall composition of the mycelial and blastospore forms of *Candida albicans*, *J. Gen. Microbiol.* 51: 367 (1968).
5. N.A.R. Gow and G.W. Gooday, Ultrastructure of the chitin in hyphae of *Candida albicans* and other dimorphic and mycelial fungi, *Protoplasma* 155: 52 (1983).
6. P.C. Braun and R.A. Calderone, Chitin synthesis in *Candida albicans*: comparison of yeast and hyphal forms, *J. Bacteriol.* 135 : 1472 (1978).
7. C.E. Bulawa, M. Slater, E. Cabib, J. Au-Young, A. Sburlati, L.W. Adair and P.W. Robbins, The *S. cerevisiae* structural gene for chitin synthase is not required for chitin synthesis *in vivo*, Cell 46 : 213 (1986).
8. A. Sburlati and E. Cabib, Chitin synthetase 2, a presumptive participant in septum formation in *Saccharomyces cerevisiae*, *J. Biol. Chem.* 261 : 15147 (1986).
9. C.E. Bulawa, *CSD2*, *CSD3* and *CSD4*, genes required for chitin synthesis in yeast: the *CSD2* gene product is related to chitin synthases and to developmentally regulated proteins in *Rhizobium* and *Xenopus*, *Mol. Cell. Biol.* 12: 1764 (1992).
10. J. Au-Young and P.W. Robbins, Isolation of a chitin synthase gene (*CHS1*) from *Candida albicans* by expression in *Saccharomyces cerevisiae*, *Mol. Microbiol.* 4 : 197 (1990).
11. J.L. Chen-Wu, J. Zwicker, A.R. Bowen and P.W. Robbins, Expression of chitin synthase genes during yeast and hyphal growth phases of *Candida albicans*, *Mol. Microbiol.* 6 : 497 (1992).
12. F. Sherman, Getting started with yeast, *Methods Enzymol.* 194: 3 (1991).
13. N.A.R. Gow and G.W. Gooday, Growth kinetics and morphology of colonies of the filamentous form of *Candida albicans*, *J. Gen. Microbiol.* 128 : 2195 (1982).
14. P. Gopal, P.A. Sullivan, P.A. and M.G. Shepherd, Enzymes of *N*- acetyl glucosamine metabolism during germ tube formation in *Candida albicans*, *J. Gen. Microbiol.* 128 : 2319 (1982).
15. J. Buffo, M.A. Herman and D.R. Soll, A characterization of pH-regulated dimorphism in *Candida albicans*, *Mycopathologia* 85 : 21 (1984).
16. E. Alani, L. Cao, N. Kleckner, A method that allows repeated use of *URA3* selection in the construction of multiply disrupted yeast strains, *Genetics* 116 : 541 (1987).
17. J.D. Boeke, F. La Croute and G.R. Fink, A positive selection for mutants lacking orotidine-5'-phosphate decarboxylase activity in yeast: 5-fluo-orotic acid resistance, *Gen. Genet.* 197 : 345 (1984).
18. F. Sanger, S. Nicklen and A.R. Coulsen, DNA sequencing with chain terminating inhibitors, *Proc. Natl. Acad. Sci. USA.* 74 : 5463 (1977).

19. J. Devereux, P. Haeberli and O. Smithies, A comprehensive set of sequence analysis programs for the VAX, *Nucleic Acids Res.* 12 : 387 (1984).

20. K.M. Franklyn, J.R. Warmington, A.K. Ott and R.B. Ashman, An immunodominant antigen of *Candida albicans* shows homology to the enzyme enolase, *Immunol. Cell Biol.* 68 : 173 (1990).

21. H.-D. Shen, K.-B. Choo, H.-H. Lee, J.-C. Hsieh, W.-L. Lin, W.-R Lee and S.-H. Han The 40-kilodalton allergen of *Candida albicans* is an alcohol dehydrogenase: molecular cloning and immunological analysis using monoclonal antibodies, *Clin. Exper. Allergy* 21 : 675 (1991).

22. A. Ishiguro, M. Homma, S. Torii and K. Tanaka Identification of *Candida albicans* antigens reactive with immunoglobulin E antibody of human sera, *Infect. Immun.* 60 : 1550 (1992).

23. C.E. Bulawa and B.C. Osmond, Chitin synthase I and chitin synthase II are not required for chitin synthesis *in vivo* in *Saccharomyces cerevisiae*, *Proc. Natl. Acad. Sci.* 87 ; 7424 (1990).

24. R.L. Burke, P. Tekamp-Olson and R.C. Najarian, The isolation, characterisation, and sequence of the pyruvate kinase gene of *Saccharomyces cerevisiae*, *J. Biol. Chem.* 258 : 2193 (1983).

25. T. McNally, I.J. Purvis, L. Fothergill-Gilmore and A.J.P. Brown, The yeast pyruvate kinase gene does not contain a string of non-preferred codons: revised nucleotide sequence, *FEBS Letts.* 247 : 312 (1989).

26. H. Inoue, T. Noguchi and T. Tanaka, Complete amino acid sequence of rat L-type pyruvate kinase deduced from the cDNA sequence, *Eur. J. Biochem.* 154 : 465 (1986).

27. N. Lonberg and W. Gilbert, Primary structure of chicken muscle pyruvate kinase mRNA, *Proc. Natl. Acad. Sci. USA* 80 : 3661 (1983).

28. J.L. Bennetzen and B.D. Hall, The primary structure of the *Saccharomyces cerevisiae* gene for alcohol dehydrogenase, *J. Biol. Chem.* 257 : 3018 (1982).

29. D. Pilgrim and M. Ciriacy, *S.cerevisiae* gene ADH3 encoding alcohol dehydrogenase III, complete coding sequence. *EMBL* K03292 (1986).

30. P.R. Russell and B.D. Hall, The primary structure of the alcohol dehydrogenase gene from the fission yeast *Schizosaccharomyces cerevisiae*, *J. Biol. Chem.* 258 : 143 (1983).

31. D.I. Gwynne, F.P. Buxton, S. Sibley, R.W. Davies, R.A. Lockington, C. Scazzocchio and H.M. Sealy-Lewis, Comparison of the cis-acting control regions of two coordinately controlled genes involved in ethanol utilization in *Aspergillus nidulans*, *Gene* 51 : 205 (1987).

32. D. Shore, M. Squire and K.A. Nasmyth, Characterisation of two genes required for the position-effect control of yeast mating-type genes, *EMBO J.* 3 : 2817 (1984).

SWITCHING AND THE REGULATION OF GENE TRANSCRIPTION IN CANDIDA ALBICANS

David R. Soll

Department of Biology
University of Iowa,
Iowa City, Iowa 52242, USA

ABSTRACT

Candida albicans strain WO-1 switches between a white and an opaque colony-forming unit in the white-opaque transition. We have cloned and sequenced two opaque-specific cDNAs, one white-specific cDNA and their respective genes, and we have examined how they are regulated during mass conversion from opaque to white. Two of the three genes have been identified by homology to other known genes. One encodes a pepsinogen and the other a potential lipid carrier involved in cell shape. Coordinate regulation, positions on different chromosomes and the absence of DNA reorganization suggest that these genes are not the primary sites of the switch event and that they are regulated by transacting factors, regulated by the switch mechanism.

INTRODUCTION

Most strains of *Candida albicans* switch spontaneously, reversibly and sometimes at extremely high frequencies between a number of general phenotypes distinguishable by colony morphology.[1-4] Switching can occur at sites of commensal carriage or infection[5,6] and can affect a number of putative virulence traits.[3,8-10] However, the molecular mechanisms involved in the basic switch event remain obscure in *C. albicans*,[3,11-13] and only recently information has begun to accumulate regarding genes involved in or regulated by the switching process.[8,14,15] The latter results have been collected primarily for the white-opaque phase transition in *C. albicans* strain WO-1, and will be the focus of this minireview.

Dimorphic Fungi in Biology and Medicine, Edited by
H. Vanden Bossche *et al.*, Plenum Press, New York, 1993

THE WHITE-OPAQUE TRANSITION

Although most strains of *C. albicans* exhibit reversible high frequency switching on agar limiting for zinc,[3,11] or can be induced to enter a high frequency mode of switching by treatment with a low dose of ultraviolet irradiation which kills less than 10% of the cell population,[1,2,16] only a few exhibit the "white-opaque transition".[4] Strain WO-1 was first isolated in 1985 from the blood and lungs of a bone marrow transplant patient at the Univerity of Iowa Hospitals and Clinics,[4] and was demonstrated to represent a bona fide strain of *C. albicans* both by DNA fingerprinting with the moderately repetitive probe Ca3,[5] which is species-specific,[17,18] by karyotyping,[3,14,19,20] by sugar assimilation pattern,[20] and by induced hypha formation.[21] In the white phase, cells of strain WO-1 form creamy white, hemispherical colonies on agar (Fig. 1A), representative of most strains of *C. albicans*, and budding cells are relatively round (Fig. 1C) forming round daughter cells with a budding pattern[4] similar to that of other strains of *C. albicans* and diploid *Saccharomyces cerevisiae*.[22] When induced to form hyphae, white cells do so readily, forming standard elongate hyphae compartmentalized and with little or no contriction at septa.[21] However, cells of strain WO-1 switch spontaneously, frequently and reversibly to an "opaque" colony forming unit.[4] Opaque colonies

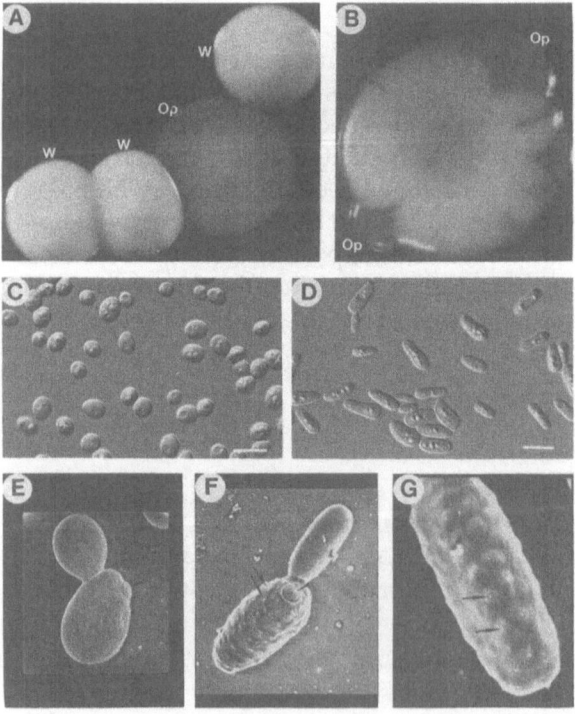

Fig. 1 Differences in colony and cellular phenotypes of *C. albicans* strain WO-1 in the white and opaque phases of the white-opaque transition. A, white (W) and opaque (Op) colonies; B, a white colony with opaque (Op) sectors; C, white cells from a white colony dome; D, opaque cells from an opaque colony dome; E, scanning electron micrograph of a budding cell (arrow head points to bud scar); F, G, scanning electron micrographs of opaque cells (arrows point to wall pimples).

are less creamy in color, and flatter than white colonies on difined nutrient agar (Fig. 1A). Opaque colonies or sectors (Fig. 1B) also stain preferentially with vital dyes like phloxine-B. Budding opaque cells are elongate, asymmetrical, sometimes with one end flatter and wider than the other (Fig. 1D), exhibit twice the average volume and mass of the white budding cells (Fig. 1E), and are distinguished by the presence of unique cell wall pimples (Fig. 1F en G) and a large intracellular vacuole.[21,23,24] In spite of these dramatic differences in cellular morphology, opaque cells contain roughly the same amount of DNA as white cells,[4] and possess the same karyotypes when assessed either by CHEF[19] or TAFE[3,8,14] chromosome separation.

Although estimates of the frequency of switching vary between studies,[4,19,23,24] it is safe to say that spontaneous switching in the opaque to white direction usually occurs at a higher frequency than in the white to opaque direction.[20] Differences in estimates are due to the methods employed and the fact that environmental parameters can drastically affect switching frequencies in both directions.[4,9,16,19,25] Therefore, rather than worrying about absolute switching frequencies, it suffices to keep in mind that switching frequencies are several orders of magnitude greater than point mutation frequencies, that switching is reversible at high frequency and that an average colony emanating from a single white cell on nutrient agar possesses between 0.1 and 5% opaque colony-forming cells, and vica versa.[4,23]

FIRST INDICATIONS OF GENE REGULATION IN THE WHITE-OPAQUE TRANSITION

The dramatic differences in the cellular morphology of white budding and opaque budding cells suggested that the phase transition involved differential gene expression, and this was supported by differences reported in the basic physiology of the cell (Table 1). White cells were shown to differ from opaque cells in the capacity to assimilate ribotol, xylitol, methyl-D-glucoside, and trehalose.[26] White and opaque cells also differed in sterol and lipid content,[27] adhesion and cohesion,[10] sensitivity to neutrophils and *in vitro* oxidants,[9] dye uptake,[19,23] sensitivity to common antifungal drugs,[19] and secretion of acid protease.[8] In addition, an antiserum generated against opaque budding cells identified 3 opaque-specific antigens, not present in white budding cells, of 14.5, 21 and 31 kDa.[24] Finally, 2 opaque-specific and 1 white-specific polypeptides were identified by a 2D-PAGE analysis of proteins pulse-labelled with [^{35}S]-methiodine during the mid-log phase of growth in liquid medium,[20] and opaque-specific polypeptides were also identified by a 2D-PAGE analysis of *in vitro* translation products utilizing white and opaque mRNA (Fig. 2).

GENETIC DEMONSTRATION OF GENES INVOLVED IN SWITCHING

Genetic evidence has also accumulated which suggests that specific genes are involved in, or are essential for, the switch to the opaque phenotype. McEachern *et al.*[14] demonstrated that the loss of the minichromosome from the chromosomal repertoire of strain WO-1 resulted both in the reduction of hypha formation and the frequency of opaque colony formation.

Table 1. Phenotypic traits affected by switching between the white and opaque phenotypes of strain WO-1.

Trait	References*
1. Colony morphology	Slutsky et al., 1987 (4)
2. Lipid and sterol content	Ghannoum et al., 1990 (27)
3. Susceptibility to antifungal agents	Soll et al., 1991 (20); Ghannoum et al., 1990 (27)
4. Cell morphology and wall archtecture	Slutsky et al., 1987 (4); Anderson et al., 1987 (23), 1989 (21), 1990 (24)
5. F-actin localization	Anderson and Soll, 1987 (23)
6. Dynamics of wall expansion observation	Staebell and Soll, unpublished
7. Adhesion to buccal epithelium and cohesion	Kennedy et al., 1988 (10)
8. Sensitivity to PMN's and oxidants	Kolotila and Diamond, 1990 (9)
9. Accessibility of vital dyes	Rikkerink et al., 1988 (19); Anderson and Soll, 1987 (23)
10. Antigenicity	Anderson et al., 1987 (23), 1989 (21), 1990 (24)
11. Acid protease secretion	Morrow et al., 1989 (8)
12. Sugar assimilation patterns	Soll, 1990 (26)
13. Environmental induction of switch to opposite phenotype	Slutsky et al., 1987 (4); Rikkerink et al., 1988 (19); Morrow et al., 1989 (16)

* References are numbered for complete listing in bibliography

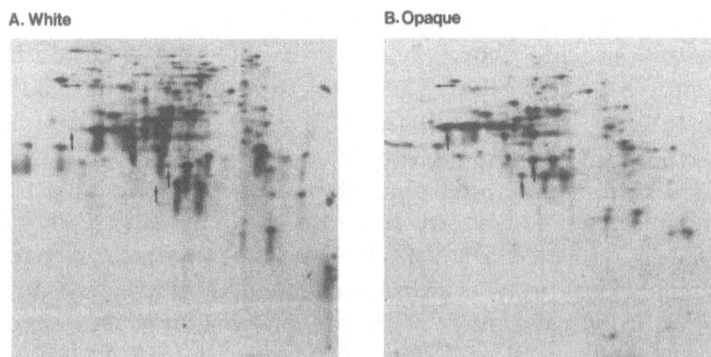

Fig. 2 *In vitro* translation products of mRNA isolated from white (A) and opaque (B) budding cell cultures. The mRNA preparations were translated in a cell-free reticulocyte translation system, and the [^{35}S]-methionine labeled proteins were separated by 2D-polyacrylamide gel electrophoresis. Arrows point to the positions of opaque-specific polypeptides.

Although the loss of the minichromosome did not result in the complete abolition of switching, the drastic reduction in frequency suggests that one or more genes on the minichromosome are necessary for the normally high frequency of switching.[14] Chu et al.[15] provided further genetic evidence for genes involved in the white-opaque transition through fusion experiments. They fused auxotrophic derivatives of WO-1 with unrelated strains that did not generate morphological variants on zinc-rich YPD agar. These fusants did not generate opaque colonies, but when chromosome loss was stimultated by heat shock, opaque colony formation as reestablished. Fusants of WO-1 were capable of opaque colony formation, demonstrating that increased ploidy or the act of fusion did not suppress switching. Chu et al.[15] concluded from their results that either the switching event or the capacity to generate the opaque phenotype was recessive, and that genes necessary for the white to opaque transition are carried on chromosome 3, and a repressor lies on chromosome 2, 5 or 6.

CLONING OF THE FIRST GENE REGULATED BY SWITCHING: AN OPAQUE-SPECIFIC PEPSINOGEN

To test directly whether genes are differentially transcribed in the alternative phases of the white-opaque transition, we constructed a cDNA library in λ gt 10 from opaque cell RNA, and hybridized duplicate blots with white or opaque cDNA probes, respectively.[8] Clones exhibiting positive signals exclusively with opaque cDNA were submitted to second and third screens. The first opaque-specific clone characterized was Op1a. By Northern blot hybridization , Op1a was demonstrated to be transcibed in cultures of opaque cells, but not in cultures of white cells grown in the same liquid nutrient medium at the same temperature of 25 °C (Fig. 3). White budding and hypha-forming cell populations, generated under the regime of pH-regulated dimorphism, were also dovoid of Op1a transcript (Fig. 3). When the Op1a cDNA was sequenced, it was found to be over 99% homologous to an aspartyl protease recently cloned and sequenced by Hube et al.[28] It contains the entire hydrophobic signal sequence and the N- terminal and C-terminal catalytic aspartyl domains characteristic of all pepsinogens.[29] We therefore have referred to the Op1 gene as *PEP1*. The *PEP1* gene was also identified in strain 3153A, but it was found to be unexpressed in the predominant o-smooth budding phenotype.[8] Although serum stimulated the secretion of acid protease in the white phase of strain WO-1 as well as other clinical isolates, there was no concomitant stimulation of *PEP1* transcription,[8] leading to the conclusion that a gene other than *PEP1* is stimulated by serum. Therefore, the evidence so far suggests that *PEP1* may also be under the regulation of switching in strains which exhibit switching repertoires other than the white-opaque transition.

Op1 cDNA hybridizes under conditions of high stringency to 2 chromosomal bands of strain WO-1 separated by either TAFE or CHEF.[8] When genomic DNA was digested with any of 14 different endonucleases, no differences in Southern blot hybridization patterns were evident between white and opaque cell DNA, suggesting that no gross reorganization occurs in or around the *PEP1* gene during the white-opaque transition.[8] Two of these endonucleases, *MspI* and *HpaII*, were methylation-sensitive isoschizomers.

CLONING OF THE SECOND OPAQUE-SPECIFIC GENE: Op4

The *PEP1* transcript is abudant, and many of the first cDNAs isolated from independently constructed opaque-specific libraries represented this gene. However, a second opaque-specific cDNA, Op4, was identified which represented an independent gene. Op4 is also expressed exclusively in opaque budding cell cultures (Morrow and Soll, unpublished observations) (Fig. 3). It is located on a

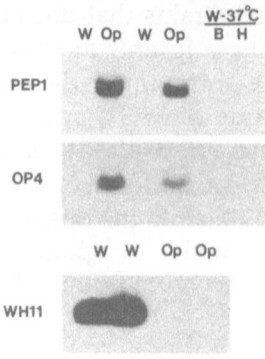

Fig. 3 Differential transcription of opaque-specific and white-specific genes. Northern blots were hybridized with the phase-specific cDNA probes noted. *PEP1*, a probe for a pepsinogen gene selectively expressed in opaque (Op) but not in white (W) cell cultures, and not in white cell cultures forming buds (B) or hyphae (H) under the regime of pH-regulated dimorphism at 37 °C; Op4, a probe for a second gene selectively expressed in opaque budding cells but not in white budding cells or white cell cultures forming buds or hyphae under the regime of pH-regulated dimorphism; Wh11, a probe for a gene selectively expressed in white budding cell cultures.

single chromosome in strain WO-1, different from the 2 chromosomes carrying the coregulated *PEP1* gene. Again, Op4 is not transcibed in semisynchronously budding or hypha-forming populations developed under the regime of pH-regulated dimorphism (Fig. 3), and there is apparently no reorganization of the gene as a result of switching. The sequence of Op4 revealed an open reading frame for 402 amino acids, but a search of both DNA and protein data bases revealed no significant homology with any previously identified genes. The gene is acidic, as is *PEP1*, and contains several alternating serine-rich and alanine-rich stretches.

CLONING OF THE FIRST WHITE-SPECIFIC GENE: Wh11

Since opaque-specific genes exist, we entertained the additional possibility that white-specific genes exist. However, several screens of white-specific cDNA libraries did not identify a white-specific cDNA. Therefore, we generated a white-specific cDNA library and submitted it to subtraction hybridization with an opaque DNA. The subtracted library was then differentially screened with white and opaque cDNA probes, and clones exhibiting positive signals only with white cDNA were submitted to second and third screens. The first white-specific cDNA isolated from the subtracted library was Wh11 (Srikantha and Soll, submitted for publication). Wh11 is transcribed in white budding cells but not in opaque budding cells (Fig. 3), the reverse of *PEP1* or Op4 (8, Morrow and Soll, unpublished observation). However, the Wh11 gene is also deactivated in the transition from bud to hypha in the white phase. Therefore, Wh11 is both white-specific and bud specific. The gene resides on chromosomes different from those for either *PEP1* or Op4, and does not appear to undergo reorganization during the white-opaque phase transition (Srikantha and Soll, submitted for publication). It is not only expressed in the white budding phase of strain WO-1, but also in the o-smooth budding phase of strain 3153A.

Wh11 is 63% homologous, with conservative changes, to a 59 amino acid sequence of the GLP1 gene of *S. cerevisiae*,[30] which is down-regulated by glucose and up-regulated by lipid. The 5-prime region upstream from the transcribed sequence of the Wh11 gene contains a number of putative regulatory sequences, including a TATA box, a CAAT sequence, a repressor-activator (RAP) recognition motif, heat shock consensus sequences, and CpG sequences, which are potential methylation sites. The initiation codon ATG of the thymidylate synthase gene is only 233bp downstream from the Wh11 gene.

Perhaps the most interesting aspect of the Wh11 gene is that it is transcribed only when the cell morphology is round. It is deactivated when the cell begins to grow as an elongate hypha or as an elongate opaque cell. Therefore, if the Wh11 gene product proves to be involved in lipid movement, it may represent a key factor in generating a round morphology through membrane development.

MASS CONVERSION, CELL DOUBLING, AND THE ACTIVATION AND DEACTIVATION OF PHASE-SPECIFIC GENES

It is clear from the preceding results that there exists both white-specific and opaque-specific genes, and that these genes must be activated and deactivated at the time of transition in each direction. Because an opaque cell culture can be induced to convert en masse to the white phenotype by an increase in temperature,[4,19,25] one can correlate population dynamics with regulatory events, at least in one direction of the switching system. In Fig. 4A, the dynamics of cell number and conversion from opaque to white colony forming units are presented for a population of opaque cells which had grown to mid-log phase at 25 °C, and were then diluted into fresh medium at 42 °C. After a lag period of 1.5 h, three semisynchronous cell doublings occurred by 3, 5 and 7 h. However, the transition from opaque to white occurred in 70% of the population between 3 and 7 h, concomitant with the second cell doubling, reaching 83% by the third cell doubling. This process will be referred to as "phenotypic" commitment. In three individual experiments, the transition occurred in 50 to 70% of the population concomitantly with the second cell doubling. When cells were inhibited from undergoing cell division either by resuspending them at 42 °C in distilled water,[19] or by not rotating the culture (Fig. 4A), cells remained viable but did not undergo phenotypic commitment.

Activation of Wh11 transcription occurs at approximately 3.5 h, coincident with the second cell doubling and the onset of commitment to white colony formation (Srikantha and Soll, manuscript submitted) (Fig. 4B). Inhibitions of cell multiplication and phenotypic commitment by the absence of culture rotation results in the inhibition of Wh11 activation (Fig. 4B) The deactivation of both *PEP1* and Op4 transcription occurs within the first hour after the increase in temperature from 25 °C to 42 °C, at least 2.5 h prior to the activation of Wh11 transcription (Fig. 4C). Since no signal of either *PEP1* or Op4 can be detected on a Northern blot 1 h after temperature is raised, the half-lives of both must be very short, at least after induction.

Since an increase in temperature results in a rapid loss in *PEP1* and Op4 mRNA several hours before the point of phenotypic commitment, the possibility was entertained that loss was due to temperature sensitivity and not to the induced phenotypic switch. To test this possibility, opaque cells were transferred from 25 °C to 42 °C, then returned to 25 °C at time intervals and tested for the induction of *PEP1* and Op4 mRNA accumulation. *PEP1* and Op4 MRNA was induced by the down shift up to the time of the commitment point, but not afterwards (Fig. 4D). This result underscores the pivotal role of the commitment event in transcriptional regulation of opaque-specific genes, which are turned off, and white-specific genes, which are turned on.

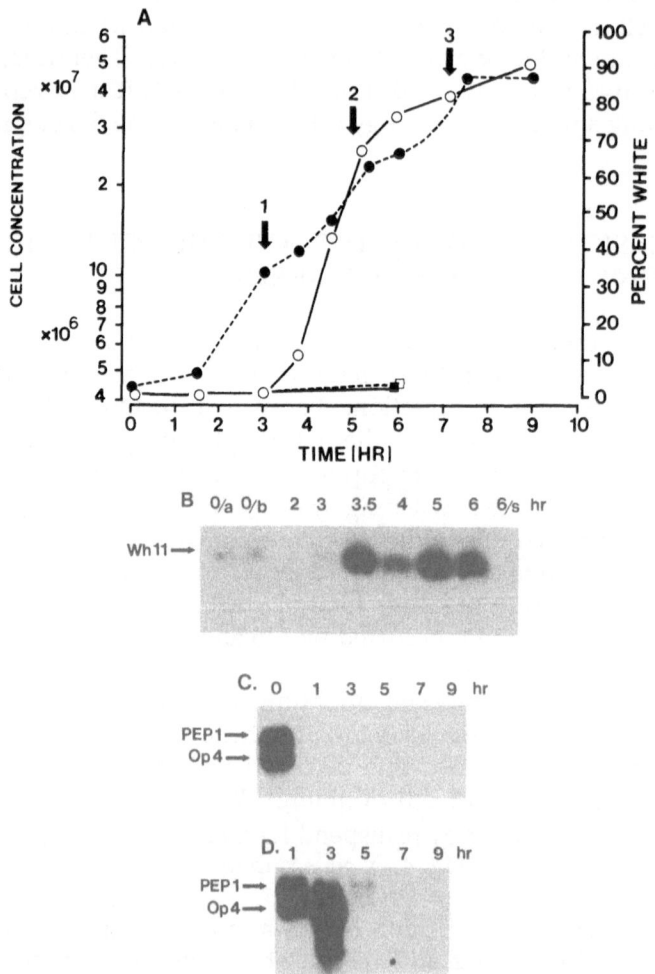

Fig. 4 The activation of Whll and deactivation of *PEP1* and Op4 during mass conversion from opaque to white induced by a shift in temperature from 25 °C to 42 °C. A, the dynamics of cell number (●) and phenotypic commitment (o, measured as the % population switching from opaque to white colony forming units) after a shift from 25 to 42 °C for an originally opaque cell culture. A parallel culture was treated as above, but the cells were not rotated at 42 °C (■, cell number; phenotypic commitment, □). B, a Northern blot hybridized with a Whll cDNA probe for mRNA from o time (prior to shift, o/a), o time (after shift, o/b), 2 h, 3 h, 3.5 h, 4 h, 5 h, and 6 h after shift, and 6 h after shift in absence of culture rotation (6/s). C, a Northern blot hybridized with a combination of *PEP1* and Op4 cDNA probes for RNA from 0 to 9 h after shift. D, a Northern blot hybridized with a combination of *PEP1* and Op4 cDNA probes for RNA from cultures shifted from 42 to 25 °C after 1, 3, 5, 7 and 9 h incubation at 42 °C.

CONCLUSION AND PERSPECTUS

We have cloned the first genes regulated by a switching system in *C. albicans*. Two of the three genes exhibit homology to genes cloned in other systems. However, although we have demonstrated that these genes are regulated by switching at the level of transcription, we have not demonstrated that expression of any of these genes is essential for the genesis of either the white or opaque phenotype. Gene knock-out experiments are now in progress which may provide us with this information. The role of these genes in switching may either be in the basic switch event or the expression of phenotypes. In the latter case, the switching genes which we have cloned would be regulated by the switch event, and the coordinate regulation of *PEP1* and Op4, genes residing on different chromosomes, suggests that these genes are down line from the switch event and may be regulated by trans-acting factors, a possibility now under investigation. We are now in the process of dissecting the promoters of both the white and opaque-specific genes for regions which are necessary for transcriptional regulation in the switching phases. In particular, we are interested in the processes of activation and silencing of transcription in both directions

ACKNOWLEDGEMENTS

The recent work performed in our laboratory and reviewed here was supported by PHS grant AI 23922 from the National Institutes of Health.

REFERENCES

1 B Slutsky, J Buffo and D R Soll, High frequency switching of colony morphology in *Candida albicans*, Science 230 666 (1985)

2 R Pomes, C Gil and C Nombela, Genetic analysis of *Candida albicans* morphological mutants, J Gen Microbiol 131 2107 (1985)

3 D R Soll, High-frequency switching in *Candida albicans*, Clin Microbiol Rev 5 183 (1992)

4 B Slutsky, M Staebell, J Anderson, L Risen, M Pfaller and D R Soll, "White-opaque transition" a second high-frequency switching system in *Candida albicans*, J Bacteriol 169 189 (1987)

5 D R Soll, C J Langtimm, J McDowell, J Hicks and R Galask, High-frequency switching in *Candida strains* isolated from vaginitis patients, J Clin Microbiol 25 1611 (1987)

6 D R Soll, R Galask, S Isley T V G Rao, D Stone, J Hicks, J Schmid, K Mac and C Hanna, "Switching" of *Candida albicans* during successive episodes of recurrent vaginitis, J Clin Microbiol 27 681 (1989)

7 D R Soll, R Galask, J Schmid, C Hanna, K Mac and B Morrow, Genetic dissimilarity of commensal strains of *Candida* spp carried in different anatomical locations of the same healthy women, J Clin Miocrobiol 29 1702 (1991)

8 B Morrow, T Srikantha and D R Soll, Transcription of the gene for a pepsinogen, *PEP1*, is regulated by white-opaque switching in *Candida albicans*, Mol Cell Biol 12 2997 (1992)

9 M P Kolotila and R D Diamond, Effects of neutrophils and *in vitro* oxidants on survival and phenotypic switching of *Candida albicans* WO-1, Infect Immun 58 1174 (1990)

10 M J Kennedy, A L Rogers, L R Hanselman, D R Soll and R J Yancey, Variation in adhesion and cell surface hydrophobicity in *Candida albicans* white and opaque phenotypes, Mycopathol 102 149 (1988)

11 D R Soll, High-frequency switching in *Candida albicans*, in "Mobile DNA", D E Berg and M M Howe, eds, American Society for Microbiology, Washington, D C (1989)

12. D.R. Soll, B. Morrow and T. Srikantha, High-frequency phenotypic switching in *Candida albicans*, *Trends in Genetics* in press.

13. S. Scherer and P.T. Magee, Genetics of *Candida albicans*, *Microbiol. Rev.* 54: 226 (1990).

14. M.J. McEachern and J.B. Hicks, Dosage of the smallest chromosome affects both the yeast-hyphal transition and the white-opaque transition of *Candida albicans* WO-1, *J. Bacteriol.* 173: 7436 (1991).

15. W.S. Chu, E.H.A. Rikkerink and P.T. Magee, Genetics of the white-opaque transition in *Candida albicans*: demonstration of switching recessivity and mapping of switching genes, *J. Bacteriol.* 174: 2951 (1992).

16. B. Morrow, J. Anderson, E. Wilson and D.R. Soll, Bidirectional stimulation of the white-opaque transition of *Candida albicans* by ultraviolet irradiation, *J. Gen. Microbiol.* 135: 1201 (1989).

17. D.R. Soll, M. Staebell, C. Langtimm, M. Pfaller, J. Hicks and T.V.G. Rao, Multiple *Candida* strains in the course of a single systemic infection, *J. Clin. Microbiol.* 26: 1448 (1988).

18. C. Sadhu, M.J. McEachern, E.P. Rustchenko-Bulgac, J. Schmid, D.R. Soll and J.B. Hicks, Telomeric and dispersed repeat sequences in *Candida* yeasts and their use in strain identification, *J. Bacteriol.* 173: 842 (1991).

19. E.H.A. Rikkerink, B.B. Magee and P.T. Magee, Opaque-white phenotype transition: a programmed morphological transition in *Candida albicans*, *J. Bacteriol.* 170: 895 (1988).

20. D.R. Soll, J. Anderson and M. Bergen, The developmental biology of the white-opaque transition in *Candida albicans*, in: "*Candida albicans*, Cellular and Molecular Biology", R. Prasad, ed., Springer-Verslag, Berlin (1991).

21. J.M. Anderson, L. Cundiff, B. Schnars, M. Gao, I. Mackenzie and D.R. Soll, Hypha formation in the white-opaque transition of *Candida albicans*, *Infect. Immun.* 57: 458 (1989).

22. D. Freifelder, Bud position in *Saccharomyces cerevisiae*, *J. Bacteriol.* 80: 567 (1960).

23. J.M. Anderson and D.R. Soll, Unique phenotype of opaque cells in the white-opaque transition of *Candida albicans*, *J. Bacteriol.* 169: 5579 (1987).

24. J.M. Anderson, R. Mihalik and D.R. Soll, Ultrastructure and antigenicity of the unique cell and pimple of the *Candida* opaque phenotype, *J. Bacteriol.* 172: 224 (1990).

25. M. Bergen, E. Voss and D.R. Soll, Switching at the cellular level in the white-opaque transition of *Candida albicans*, *J. Gen. Microbiol.* 136: 1925 (1990).

26. D.R. Soll, Dimorphism and high frequency switching in *Candida albicans*, in: "Genetics of *Candida albicans*", D.R. Kirsch, R. Kelly and M.B. Kurtz, eds., CRC Press, Boca Raton, Fla. (1990).

27. M.A. Ghannoum, I. Swairjo and D.R. Soll, Variation in lipid and sterol contents in *Candida albicans* white and opaque phenotypes, *J. Med. Vet. Mycol.* 28: 103 (1990).

28. B. Hube, C.J. Turner, F.C. Odds, H. Eiffert, G.J. Boulnois, H. Kochel and R. Ruchel, Sequence of the *Candida albicans* gene encoding the secretory aspartate proteinase, *J. Med. Vet. Mycol.* 29: 129 (1991).

29. D.R. Davies, The structure and function of the aspartic proteinases, *Annu. Rev. Biophys. Chem.* 19: 189 (1990).

30. R.L. Stone, V. Matarese, B.B. Magee, P.T. Magee and D.A. Bernlohr, Clonig, sequencing and chromosomal assignment of a gene from *Saccharomyces cerevisiae* which is negatively regulated by glucose and positively by lipids. *Gene* 96: 171 (1990).

CHARACTERIZATION OF *SACCHAROMYCES CEREVISIAE* PSEUDO-HYPHAL GROWTH

Carlos J. Gimeno, Per O. Ljungdahl, Cora A. Styles, and Gerald R. Fink

Whitehead Institute for Biomedical Research and
Department of Biology
Massachusetts Institute of Technology
Cambridge, Massachusetts 02142, USA

ABSTRACT

Diploid *Saccharomyces cerevisiae* strains undergo a dimorphic transition that involves changes in cell shape and the pattern of cell division and results in invasive filamentous growth in response to nitrogen starvation. Cells become long and thin and form pseudohyphae that grow away from the colony and invade the agar medium. Our data strongly suggest that pseudohyphae are initiated when yeast cells bud pseudohyphal cells in an asymmetric cell division. As pseudohyphae elongate, they become covered with yeast cells. Pseudohyphal cells may be vectors to deliver assimilative yeast cells to new substrates thereby allowing *S. cerevisiae* to forage for nutrients. Pseudohyphal growth requires the polar budding pattern of a/α diploid cells; haploid axially budding cells of identical genotype cannot undergo this dimorphic transition. Mutation of *SHR3*, a gene required for amino acid uptake, enhances the pseudohyphal phenotype.

INTRODUCTION

Dimorphic fungi are important animal and plant pathogens. For some of these fungi the ability to switch morphology has been linked to the ability to cause disease (see end of Discussion), suggesting that genes with roles in dimorphism could be considered as possible targets for antifungal drugs. The baker's yeast, *Saccharomyces cerevisiae*, undergoes a dimorphic switch from growth in a yeast morphology to growth in a filamentous pseudohyphal form.[1-4] Studies of *S. cerevisiae* dimorphism aided by the power of genetics in this organism may reveal principles of dimorphism common to all fungi and therefore applicable to dimorphic pathogens and antifungal therapy.

In *S. cerevisiae*, polarized cell division is required for pseudohyphal growth (abbreviated PHG).[1] In *S. cerevisiae*, mitotic cell divisions occur by budding.[5] The polarity of cell division is defined with respect to the position on the cell surface of previous budding events. Polarized cell division is manifested as two genetically programmed spatial patterns of cell division, axial for *MATa* or *MATα* cells and polar for *MATa/α* cells.[6-8] In the axial pattern the mother and daughter cells bud adjacent to their cell pole that defined the previous mother-daughter junction (see Table 1 for illustration).

Table 1. Budding Pattern of Pseudohyphal and Sated **a**/α Shr3⁻ Cells.

	First Buds		Second Buds	
	Pseudohyphal	Sated	Pseudohyphal	Sated
Cell Divisions	90	69	90	69
Free End	100%	100%	90%	73%
Birth End	0%	0%	10%	27%

Time lapse photography was used to determine bud site selection in both pseudohyphal and sated cells as described in the Experimental Procedures.

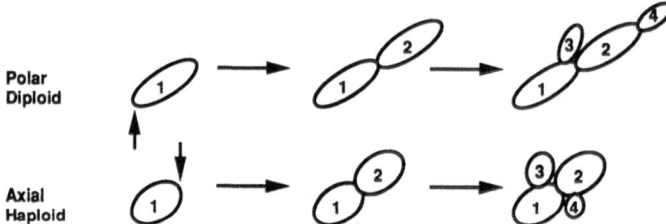

The polar budding pattern most often observed in diploid virgin pseudohyphal and sated cells as well as the axial budding pattern of haploid cells is shown in the drawing. In the 159 cell divisions reported axial haploid budding was never observed. In all cases cell 1 is a virgin cell. The vertical arrows indicate the birth end of cell 1. Cell 2 and cell 3 are the first and second daughters, respectively, of cell 1. Cell 4 is the first daughter of cell 2.

In the polar pattern a virgin mother's first several buds emerge at the pole opposite the one that defined the junction to its mother (we refer to this initial pattern as unipolar budding); subsequent buds emerge at either this or the opposite pole[6,7] (we refer to this latter pattern as bipolar budding). The biological function of axial haploid budding for mating has been discussed,[9] and as discussed below, the diploid polar budding pattern probably exists to allow PHG.[2]

Polar cell division is controlled genetically in *S. cerevisiae*.[10] The mating type locus programs cell type specific budding patterns probably by regulating budding pattern genes and consequently regulates PHG because PHG requires the diploid budding pattern. The current model proposes that budding pattern genes represented by *RSR1/BUD1* and *BUD2-BUD5* [8,11-13] are required for selection of the proper bud site and consequently for establishing the proper axis of cell division. *RSR1/BUD1*, *BUD2*, and *BUD5* convert the default random budding pattern to bipolar and subsequent action of *BUD3* and *BUD4* convert bipolar to axial. To explain the observed cell type specificity (diploids are bipolar, haploids axial), an elegant model was proposed[8] that either or both *BUD3* and *BUD4* are repressed by the mating type locus encoded repressor a1α2 found only in *MATa/α* cells. This model also suggests the molecular basis of mating type locus regulation of PHG.[1] Our finding that a random budding pattern inhibits PHG suggested that the *BUD* genes might exist to act in the pseudohyphal pathway and allow foraging.[1]

In this report we characterize the dimorphic transition to PHG in *S. cerevisiae*. The term dimorphic has been used to define fungi that can grow vegetatively in either a yeast or filamentous form.[14] A pseudohypha is defined as a "fragile chain of cells (usually yeasts, which have arisen by budding and have elongated without detaching from adjacent cells), with morphological characteristics intermediate between a chain of yeast cells and a hypha".[15] This dimorphic transition is induced by starvation for a nitrogen source and is controlled directly or indirectly by the *RAS* signal transduction pathway.[1] PHG in *S. cerevisiae* is a unique type of polarized cell division that requires unipolar budding, and a change in cellular morphology that results in the formation of macroscopic filamentous structures emanating away from the colony into unpopulated substrate. Pseudohyphal cells appear to originate from yeast cells by budding. As pseudohyphae elongate they become covered by vegetative yeast cells, suggesting that pseudohyphal cells may be vectors to deliver assimilative yeast cells to new substrates. Interesting similarities exist between the dimorphism of *S. cerevisiae* and that of certain human and plant fungal pathogens.

MATERIALS AND METHODS

Saccharomyces cerevisiae Strains and Plasmids

Construction of yeast strains used in this work, with the exception of CGX66, CGX68, CG143, and CG144 has been described elsewhere.[1] Yeast strains and plasmids used in this work are listed in Table 2. MB1000 and MB758-5B were crossed to make diploid strain CGX66. CGX66 was sporulated, tetrads were dissected, and a *MATa* segregant (CG143) and a *MATα* segregant (CG144) were isolated. CG143 and CG144 were crossed to make diploid CGX68 congenic to the Σ1278b genetic backround with the genotype *MATa/α*.

Media and Microbiological Techniques

Standard *S. cerevisiae* media were prepared and yeast genetic manipulations were performed as described previously.[16] YPD and YPAD are standard rich media. Other media and microbiological techniques have been described in detail elsewhere.[1] In nonstandard media name abbreviations S represents synthetic, H represents histidine, and D represents dextrose. In the following nonstandard media, sole nitrogen source(s) name abbreviations are SLAHD (low ammonia), SPHD (proline), SAHD (ammonia), SRHD (arginine), SPRHD (proline and arginine), and SPAHD (proline and ammonia).

Table 2

Saccharomyces cerevisiae Strains[a]

Strain	Genotype
F35[b]	*MATa/α HO/HO shr3-101/shr3-101*
MB1000	*MATα*
MB758-5B	*MATa ura3-52*
CG25	*MATa ura3-52 shr3-102*
CG41	*MATα ura3-52 shr3-102*
CG62	*MATa/α ura3-52/ura3-52 shr3-102/shr3-102* (pPL210)
CG64	*MATa/α ura3-52/ura3-52 shr3-102/shr3-102* (pRS316)
CG67	*MATα/α ura3-52/ura3-52 shr3-102/shr3-102*
CG85	*MATa/a ura3-52/ura3-52 shr3-102/shr3-102*
CG143	*MATa*
CG144	*MATα*
CGX19	*MATa/α ura3-52/ura3-52 shr3-102/shr3-102*
CGX31	*MATa/α ura3-52/ura3-52*
CGX66	*MATa/α ura3-52/URA3*
CGX68	*MATa/α*

Plasmids

Name	Description
pRS316	*URA3* marked centromere vector[51]
pPL210	1.4 kb fragment containing *SHR3* in pRS316[17]

[a] All *S. cerevisiae* strains are congenic to the Σ1278b genetic background.[19]

[b] Construction of F35 is described in reference 17.

Amino Acid Transport Assays

Proline transport assays were performed as previously described.[17]

Bud Site Selection Assays

Our assay to determine bud site selection of both yeast and pseudohyphal cells has been described elsewhere.[1]

Light and Scanning Electron Microscopy and Quantitation of Yeast Cell Dimensions

Microscopy and cell measurement methods have been described elsewhere.[1]

RESULTS

A *S. cerevisiae* cell grown on standard media multiplies until it forms a visible structure, an approximately hemispherical colony with a smooth circular outline. This morphology is strikingly homogeneous, with little variation from colony to colony. *S. cerevisiae* has a second distinct mode of proliferation, PHG. PHG results from a reiterated pattern of unipolar cell division to form the chain of cells that constitute the pseudohypha. In colonies formed by cells undergoing PHG, which resemble colonies formed by filamentous fungi, the pseudohyphae radiate outward in all directions (Fig. 1).

The Dimorphic Switch to Pseudohyphal Growth Is Induced by Nitrogen Starvation

The transition from unpolarized colonial growth to PHG occurs on agar based synthetic growth medium deficient in nitrogen. Wild-type cells form pseudohyphae on standard minimal medium containing low levels of ammonia (SLAHD) or proline as sole nitrogen source (SPHD). On the low ammonia medium all of the wild-type colonies form pseudohyphae (CGX31, Fig. 2A and 2E), whereas on proline medium (SPHD) small regions of PHG are apparent in about a quarter of the colonies (CGX31, Fig. 2B and 2F). CGX31 does not form pseudohyphae when grown on standard ammonia based medium (SD) or media with the same composition as SPHD but containing as sole nitrogen source(s) standard levels of ammonia (SAHD), arginine (SRHD), proline and ammonium sulfate (SPAHD), or proline and arginine (SPRHD) (data not shown).

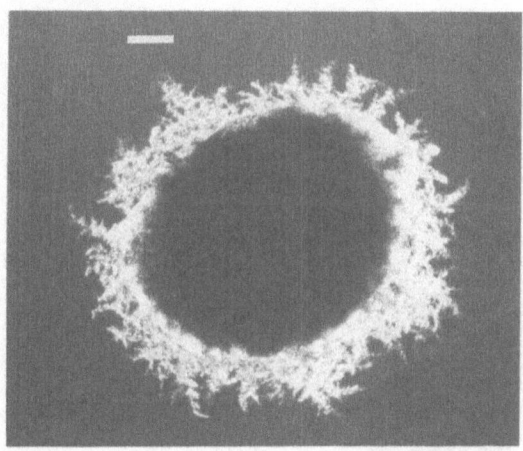

Fig. 1 Morphology of a polarized colony produced by pseudohyphally growing *S. cerevisiae*. A wild-type strain, CGX68 (*MATa/α*), was pregrown on a YPAD plate (a rich medium) overnight at 30° C and then streaked for single cells on SLAHD medium. After 4.75 days of growth at 30° C a representative colony from the part of the streakout with low colony density was photographed. The scale bar represents 0.1 mm.

Of all strains tested, those with the Σ1278b background undergo the most uniform and easily controlled transition from unpolarized to PHG on both low ammonia and proline medium. Many laboratories commonly use strains derived from this background[18-20] because they are extremely sensitive to the ammonia repression of nitrogen assimilation pathways.[21,22] Σ1278b and its derivatives cross well with other standard laboratory strains such as S288C[23] and comprise part of the set of interbreeding laboratory isolates known collectively as *S. cerevisiae*.

Mutations in the *SHR3* Gene Enhance Pseudohyphal Growth

MATa/α Shr3⁻ strains are important tools for studying pseudohyphal development because they undergo more rapid and uniform dimorphic transitions to PHG than wild-type strains when grown on SPHD medium, where 8.7 mM proline is the sole source of nitrogen (compare Fig. 2B and 2F with Fig. 2C and 2G). Shr3⁻ strains have enhanced PHG on SLAHD medium also (data not shown), an observation that can be explained if ammonia uptake, like amino acid uptake,[17] is impaired by *shr3* mutations. To prove it is loss of function of *SHR3* that is responsible for enhanced PHG, we transformed CGX19 (*MATa/α shr3-102/shr3-102 ura3-52/ura3-52*) with a centromere based plasmid containing either no insert

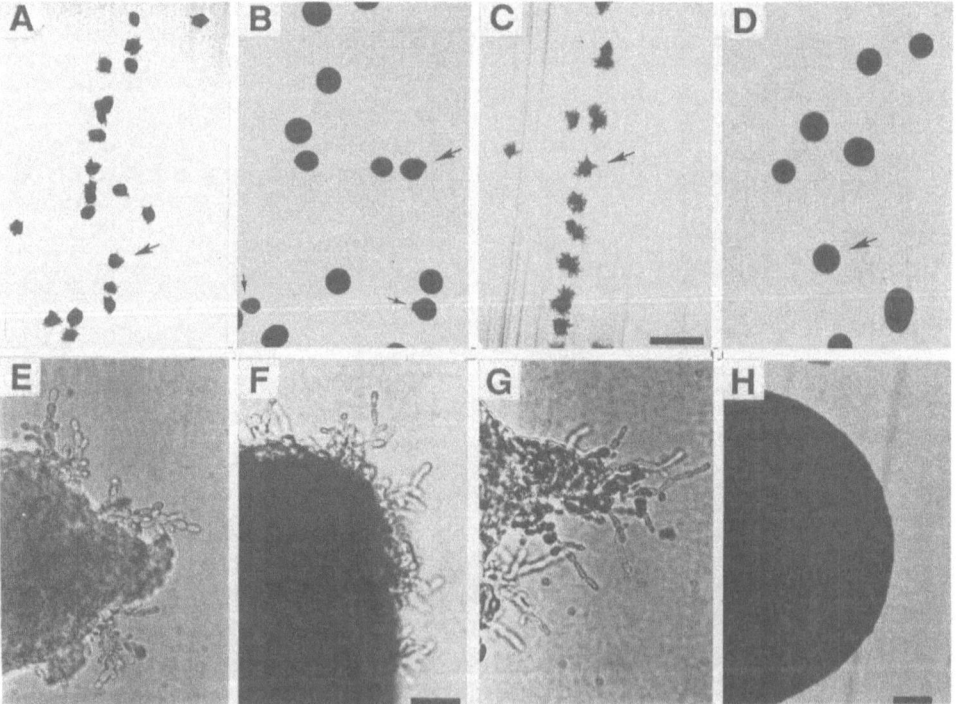

Fig. 2 Genetic and physiological characterization of *S. cerevisiae* pseudohyphal growth. CGX31 (*MATa/α ura3-52/ura3-52*) and CGX19 (*MATa/α ura3-52/ura3-52 shr3-102/shr3-102*) were streaked for single cells on SLAHD plus uracil, SPHD plus uracil, or SAHD plus uracil plates, incubated at 30° C for 48 hours, and the resulting colonies were photographed. Panels A - D show low magnification views of colonies of (A) strain CGX31 growing on SLAHD plus uracil, (B) CGX31 growing on SPHD plus uracil, (C) CGX19 growing on SPHD plus uracil, and (D) CGX19 growing on SAHD plus uracil. In Panel B the three colonies with pseudohyphae are designated with arrows. Panels E - H show high magnification views of the colonies marked by large arrows in Panels A - D. Panels A - D have the same scale with the scale bar in Panel C representing 0.5 mm. Panels E - G have the same scale with the scale bar in Panel F representing 30 μm. The scale bar in Panel F represents 30 μm.

(pRS316) or the *SHR3* (pPL210) gene. Transformants containing pRS316 (CG64) showed PHG identical to that exhibited by CGX19 whereas the pPL210 transformants (CG62) did not. Only a minority of colonies of diploid cells homozygous for *shr3* in a S288C backround have pseudohyphae and the number of pseudohyphae per colony is much lower than that observed in a comparable Shr3⁻ Σ1278b strain. Diploid cells derived from one *shr3* S288C parent and one *shr3* Σ1278b parent show the PHG characteristic of pure Σ1278b Shr3⁻ diploids.

To better understand the mechanism by which *shr3* mutations enhance PHG on SPHD medium, we measured proline transport rates at both high and low proline concentrations (respectively 10 mM and 0.004 mM) for **a**/α *shr3-101/shr3-101* and *shr3-102/shr3-102* strains used in this paper for PHG studies and an **a**/α wild-type strain. Transport rates at 10 mM proline primarily reflect the activity of multiple low affinity transporters while transport rates at 0.004 mM measure mainly the activity of the high affinity proline permease PUT4.[24] Proline transport rates at the 10 mM concentration give an idea of how efficiently proline is taken up by yeast cells on SPHD plates.

Our results summarized in Fig. 3 reveal that both of the Shr3⁻ strains tested take up proline 7-10 times more slowly than Shr3⁺ strains at both the high and low proline concentrations. This result suggests that multiple independent proline transport systems are impaired by the *shr3* mutation. Consistent with these data, Shr3⁻ cells grow slower than Shr3⁺ cells on SPHD medium where proline is their sole source of nitrogen (compare Fig. 2B and 2C). Shr3⁻ strains grow normally on SAHD medium (Fig. 2D) where ammonium sulfate is the sole nitrogen source and on SPAHD medium where proline and ammonium sulfate are the sole nitrogen sources (data not shown) indicating that it is the inability of Shr3⁻ strains to efficiently take up proline and not some other nutrient that causes the slow growth on SPHD medium. These results suggest that for Shr3⁻ cells growing on SPHD medium the rate limiting nutrient for growth is proline, the sole nitrogen source, and that Shr3⁻ cells growing on SPHD medium are starved for nitrogen.

On medium containing standard levels of ammonium sulfate as sole nitrogen source Shr3⁺ (CGX31) and Shr3⁻ (CGX19) cells grow at similar rates, and neither strain forms pseudohyphae (data not shown and Fig. 2D and 2H). The fact that CGX19 fails to form pseudohyphae when proline medium (SPHD) contains ammonia (SPAHD) supports the contention that it is nitrogen starvation that induces PHG.

Pseudohyphal Growth Is a Diploid Specific Pathway

Diploid but not haploid *S. cerevisiae* strains give rise to pseudohyphae. To study the effect of ploidy and the genotype at the mating type locus on PHG, we have constructed a congenic set of yeast strains carrying a mutant allele of *SHR3*. Fig. 4 compares the morphology of the diploid strain CGX19 (Fig. 4A) and with its two haploid parents (Fig. 4B and 4C) carrying the *shr3-102* mutation. No *shr3-102* haploids we have analyzed manifest PHG; all form typical hemispherical unpolarized colonies on SPHD. *MATa/a shr3-102/shr3-102* (CG85) and *MATα/α shr3-102/shr3-102* (CG67) isogenic derivatives of CGX19 (*MATa/α shr3-102/shr3-102*) also do not form pseudohyphae on SPHD; instead they form hemispherical colonies identical to those shown in Fig. 2D. The cell type specificity of pseudohypal growth is controlled in part by the alleles of the mating type locus.

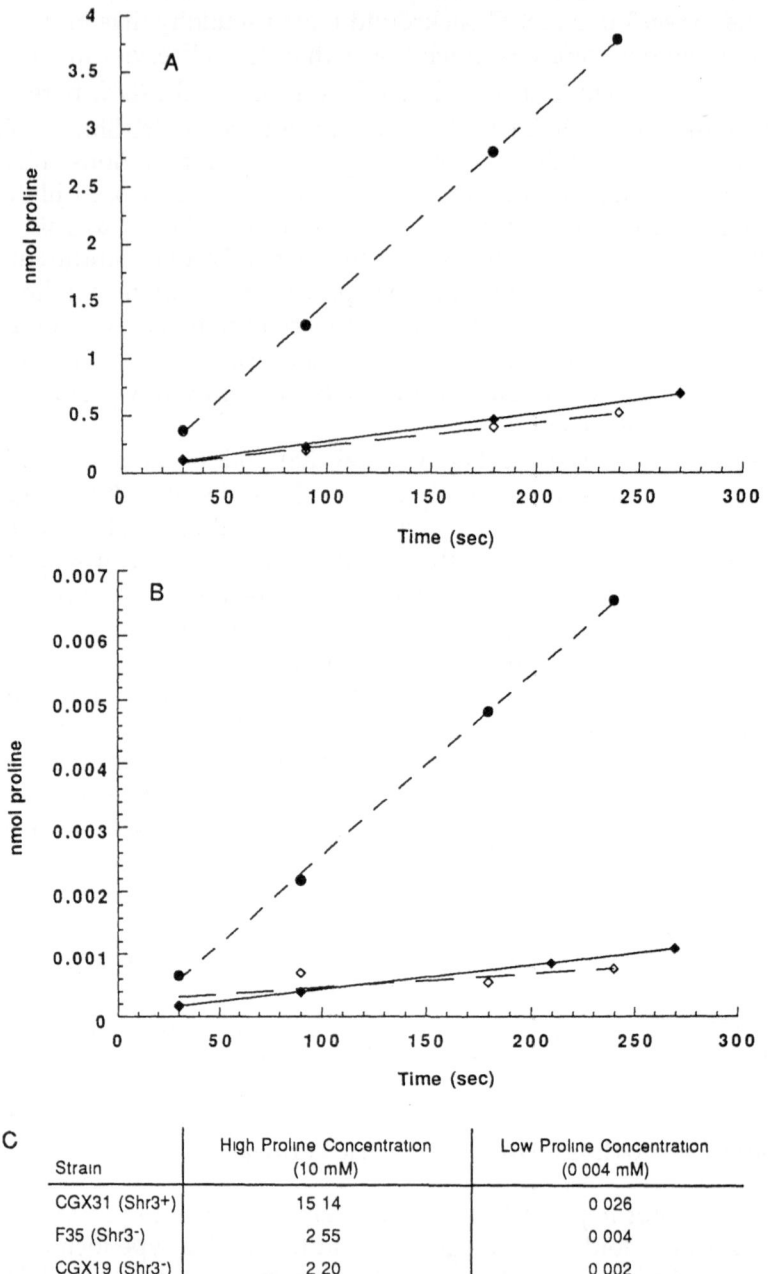

C		
Strain	High Proline Concentration (10 mM)	Low Proline Concentration (0 004 mM)
CGX31 (Shr3+)	15 14	0 026
F35 (Shr3-)	2 55	0 004
CGX19 (Shr3-)	2 20	0 002

Fig. 3 Proline uptake into Shr3+ and Shr3- Cells.
CGX31 (Shr3+), F35 (Shr3-), and CGX19 (Shr3-) cells were grown in SUD medium. Proline uptake was assayed as previously described.[17] Uptake rates were determined at two proline concentrations, 10 mM (A) and 0.004 mM (B). Subsamples were withdrawn and filtered at the times indicated. Symbols: CGX31 (•); F35 (♦), and CGX19 (◇). The proline uptake rates (nmol min^{-1} mg^{-1} protein) are tabulated in (C).

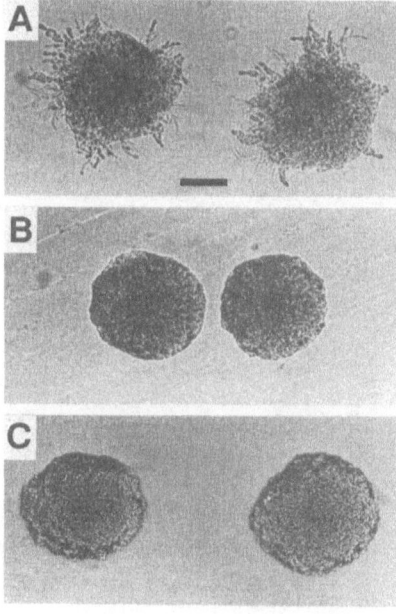

Fig. 4 Effect of ploidy on pseudohyphal growth

(A) CGX19 (*MATa/α ura3-52/ura3-52 shr3-102/shr3-102*) and its two haploid parents, (B) CG41 (*MATα ura3-52 shr3-102*) and (C) CG25 (*MATa ura3-52 shr3-102*) were streaked for single cells on the same SPHD plus uracil plate, incubated at 30° C for 49 hours, and the resulting colonies were photographed The scale bar represents 60 μm

Pseudohyphal Growth Results from Unipolar Cell Division

The unipolar cell divisions that characterize polar diploid budding are critical for the elaboration of PHG. We define virgin cells as those that have had no daughters and sated cells as those growing vegetatively on rich medium. We observed the budding pattern of virgin sated CGX19 cells or of virgin CGX19 cells growing in pseudohyphae by time lapse photomicroscopy Fig. 5 shows the results of a time lapse experiment where the development of a pseudohypha was monitored for 6 hours with interpretative drawings summarizing the results.

We assayed budding pattern quantitatively by determining the site of emergence of the first and second buds of virgin pseudohyphal and sated cells by time lapse observation (Table 1) Following the conventions of Freifelder,[6] the pole of a bud which contacts its mother cell is called the birth end and the opposite pole the free end. The first bud of 90 virgin terminal pseudohyphal cells and 69 virgin sated cells of strain CGX19 emerged without exception on the free end of its mother cell. The first bud of a diploid is therefore a good marker for the free end of this cell. The second bud of 90 virgin terminal pseudohyphal cells emerged in 90% of the cases again on the free end of its mother cell after two doubling times had elapsed. The shape of these cells together with their immobility in the agar matrix permitted easy identification of a cell's poles The proliferation of ancestral cells prevented us from scoring events at the birth end in the other 10% of the cases where no bud was present at the free end after two doubling times. The second bud of each of 69 sated cells emerged from the mother cell's free end, which we identify in this case as the same cell pole from which the first bud

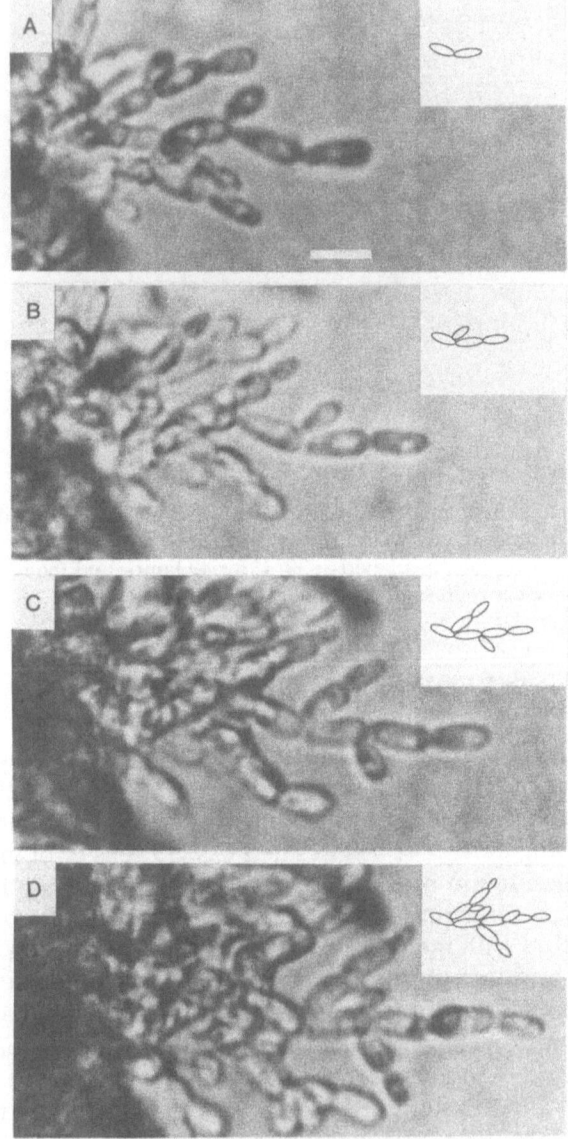

Fig. 5 Time lapse photographic analysis of pseudohyphal growth

Strain CGX19 (*MATa/α ura3 52/ura3 52 shr3 102/shr3 102*) was streaked for single cells on SPHD plus uracil medium   After 3 days of growth at 30° C a visible pseudohypha was chosen and photographed at 2-hour intervals while growing at 30° C   Panel A is the initial time point   An interpretative drawing of the elongating pseudohypha appears in the upper right hand corner of each panel   The scale bar represents 10 µm emerged, 73% of the time and from the birth end, defined as the opposite pole, 27% of the time. Clearly, the first bud of virgin CGX19 cells emerges in a unipolar manner from the free end regardless of the cell's growth mode. The second bud also emerges unipolarly in the majority of cell divisions. From this sequence it can be seen that serial reiteration of unipolar budding by terminal pseudohyphal cells results in polarized chain elongation. It can be seen also that the second bud of a virgin terminal cell can initiate a new lateral chain oriented at an angle from the main lineage.

It is important to note that in some lineages lateral budding was completely absent whereas apical growth continued (data not shown). In other words, daughters divided for several divisions while the mother cells did not, suggesting that in these lineages cell division may be repressed after a cell gives birth to its first daughter. In the fungal literature this phenomenon is known as apical dominance.[25]

The Cells of the Pseudohypha Are a Morphologically Distinct Cell Type

We compared the dimensions of pseudohyphal and sated cells of the same genotype. In the first experiment we grew Shr3⁻ (CGX19) cells on YPD or SPHD plus uracil media. We then prepared cells taken from the surface of the agar for scanning electron microscopy (SEM). Fig. 6 shows scanning electron micrographs of a typical ellipsoidal CGX19 cell from the YPD plate as well as a CGX19 pseudohyphal cell from the SPHD plus uracil plate. To be certain that the surface

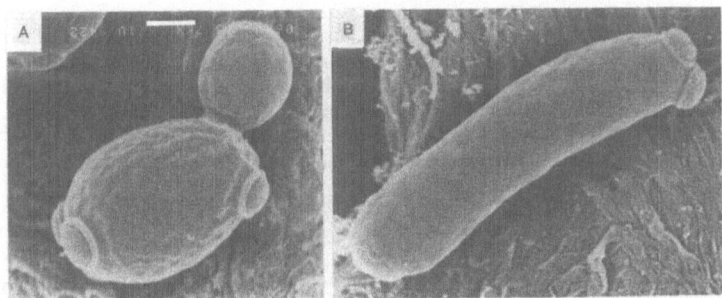

Fig. 6. SEM analysis of starvation induced cell morphology changes.
CGX19 (*MATa/α ura3-52/ura3-52 shr3-102/shr3-102*) growing vegetatively at 30° C was streaked for single colonies on prewarmed (A) YPD and (B) SPHD plus uracil plates. These plates were incubated at 30° C for 31 hours and then prepared for SEM as described in the Materials and Methods. Panel A shows a representative YPD grown sated yeast cell that has been budding in a bipolar manner. The budding pattern of the cell can be deduced from the positions of the bud scars, the protrusions visible on the surface of the cell. Panel B shows a pseudohyphal cell with two bud scars visible at one pole, a conformation predicted by polar budding. Panels A and B have the same scale with the scale bar in Panel A representing 1 μm.

grown cells in SEM micrographs were representative of cells in invasive pseudohyphae, we also measured the dimensions of the latter by light photomicroscopy (Table 3). Given the difference in imaging methods, the two sets of measurements agree well and give similar axial ratios.

Pseudohyphal Cells Appear to Originate by Budding from Yeast Cells

The first pseudohyphal cell in a pseudohypha could originate in at least two ways. A yeast cell could intiate the growth of a pseudohypha by undergoing morphogenesis and changing its shape to the pseudohyphal morphology or it could retain its yeast morphology and produce a pseudohyphal cell by budding, or stated more precisely, by assymmetric cell division. To distinguish between these two models we pregrew a Shr3⁻ strain on rich medium, which supports the yeast morphology, transferred it to SPHD medium, which rapidly activates the PHG program in Shr3⁻ strains (data not shown), and after a short period of time analyzed 4-cell microcolonies microscopically (Fig. 7A). In these microcolonies, cell morphology can be unambiguously assigned only to the two cells with daughters because it is unknown what part of the cell cycle unbudded cells are in.

Table 3

Strain	Cell Type	Medium	Cell Length (μm)	Cell Width (μm)	Axial Ratio Length/Width
CGX19	Pseudohyphal (SEM)	SPHD+U	6.7±1.0 (3)	1.9±0.1 (3)	3.5
CGX19	Sated (SEM)	YPD	4.2±0.4 (7)	3.0±0.2 (7)	1.4
CGX19	Pseudohyphal (LM)	SPHD+U	9.2±1.7 (11)	2.7±0.3 (11)	3.4
CGX19	Sated (LM)	YPD	5.7±0.8 (19)	3.9±0.4 (19)	1.5
CGX19	Blastospore-Like (LM)	SPHD+U	5.6±0.5 (10)	4.4±0.4 (10)	1.3

Dimensions of Pseudohyphal Cells, Sated Cells, and Blastospore-Like Cells: CGX19 cells (*MATa/α ura3-52/ura3-52 shr3-102/shr3-102*) were measured in all cases. Cell dimensions are based on scanning electron (SEM) and light (LM) photomicrographs as described in the experimental section. Cell length is the length of the longest axis of the cell. Cell width is the width of the cell at the midpoint of its longest axis. The axial ratio is the average cell length divided by the average cell width. The tabulated values are averages with standard deviations listed. The number of cells measured for each table entry appears in parentheses after the standard deviation.

The diploid polar budding pattern allows one to infer the mother-daughter relationships of the four cells in the microcolony (see illustration in Table 1). The cell with yeast morphology indicated by the arrow is the progenitor of the microcolony and has clearly retained its yeast shape. Its first daughter is clearly a pseudohyphal cell. Given that pseudohyphal cells form by budding in elongating pseudohyphae (Fig. 5), this observation suggests that incipient pseudohyphae are initiated by yeast cells that bud to form pseudohyphal cells by assymmetric cell division. Incipient pseudohyphae initiated by pseudohyphal cells were not observed.

As Pseudohyphae Elongate they Become Covered with Yeast Cells

The elongated pseudohyphal cells have been observed to give rise to either of two cell types. Elongated cells may divide to produce an elongated daughter with roughly the same final dimensions as the mother cell or alternatively a blastospore-like cell with roughly the dimensions of a sated yeast cell (Table 3). Blastospores are defined as round or oval budding yeast cells arising from pseudohyphae.[26] Both the elongated pseudohyphal cell and the blastospore-like cell can be produced either apically or laterally. The blastospore-like cells, which we also call yeast cells, produced by the pseudohyphal cell may be a new cell type or they may be identical to vegetative yeast cells. These yeast cells are clearly proliferating because they can be observed to bud (visible in Fig. 7B indicated by a large arrow and in ref. 1). Pseudohyphae are often observed to invade the agar

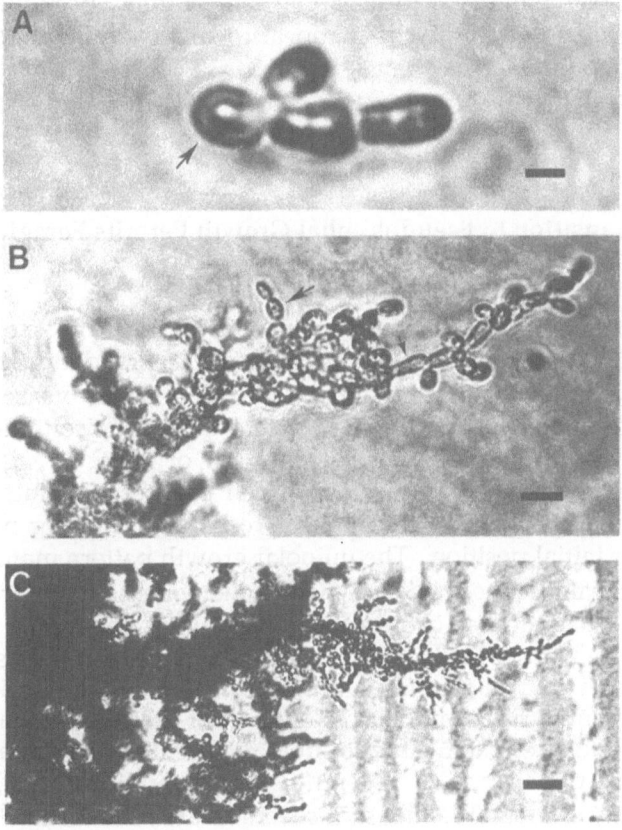

Fig. 7. Origin and development of *S cerevisiae* pseudohyphae

(A) A Shr3⁻ strain, CGX19 (*MATa/α shr3-102/shr3-102 ura3-52/ura3-52*), was pregrown overnight on YPAD medium and then streaked for single cells on a SPHD plus uracil plate. An incipient pseudohypha after 7 hours (A) and an invasive pseudohypha after 4.75 days (B) are shown. In (A), the arrow indicates the cell with the yeast morphology that initiated the pseudohypha. In (B) the large arrow indicates a mitotic yeast cell and the small arrow indicates a pseudohyphal cell. A *Shr3*⁺ wild-type strain, CGX68 (*MATa/α*), was pregrown overnight on YPAD medium and then streaked for single cells on a SLAHD plate. A macroscopic invasive pseudohypha after 10.6 days is shown (C). The scale bars in Panels A, B and C represent respectively 3, 10 and 30 μm.

and subsequently begin budding blastospore-like cells at the base of the pseudohypha (Fig. 7B and 7C). As long as the pseudohypha continues to grow, the first few cells at the growing tip are pseudohyphal cells. Pseudohyphae have a backbone of pseudohyphal cells which becomes covered with vegetative yeast cells that comprise the majority of the biomass of the pseudohypha.

Pseudohyphal Cells Invade the Semisolid Agar Growth Medium

Pseudohyphal cells penetrate the surface of the agar plate and grow down into the medium. Shr3⁻ diploids as well as other standard strains growing in the sated mode on rich medium grow by spreading out on the surface of the agar. Even on SPHD medium, most strains grow on the surface. By contrast, Shr3⁻ diploids on SPHD medium are invasive and grow into the agar, presumably in search of food (Fig. 7B). Wild-type diploids grown on SLAHD also produce robust invasive pseudohyphae (Fig. 7C). The invasive growth is easily observed in a dissecting microscope and is further demonstrated by the observation that a microneedle must pierce the agar to reach the cells of many pseudohyphae. The mothers and daughters within the chain appear to be physically attached because they often can be manipulated as a unit.

DISCUSSION

The Dimorphic Transition to Pseudohyphal Growth Permits Foraging for Nutrients

We have described PHG, a dimorphic transition in the life cycle of *S. cerevisiae*. The pseudohypha in *S. cerevisiae* consists of a lineage of first daughters associated in a chain. There have been references to PHG for this yeast[3,4,26-29] (and references in these sources) but no detailed description of the conditions required for its induction. Fig. 8 diagrams our current view of the *S. cerevisiae* life cycle. The radial pattern and invasive character of cell proliferation into the growth substrate clearly is a mechanism that permits cells to forage for nutrients at a distance from their initial position. The unipolar growth pattern manifest by yeast pseudohyphae is the major mechanism by which filamentous fungi proliferate.[25]

Some simple calculations illustrate how PHG allows yeast cells to forage for nutrients. As can be seen in Fig. 1, colonies formed by cells in the pseudohyphal mode have two domains, an interior symmetrical domain (the dark part of the colony) and an assymmetric corona (the bright part of the colony) that defines the outer boundary of the colony. The interior domain resembles a normal colony and is composed mainly of yeast cells. The radius of the interior domain of the 4.75 day old Fig. 1 colony is about 0.27 mm so it covers an area of 0.23 mm². The radius of the entire colony including the corona is on average about 0.35 mm so the whole colony covers an area of 0.38 mm². It can be calculated then that the corona covers 0.15 mm², 40% of the of the entire colony area. PHG may have helped this colony explore and exploit a 1.7 fold greater area than would have been possible with yeast growth alone. Furthermore, the pseudohyphae in the colony continue to elongate for as long as they have sufficient nutrients so large areas can be explored with PHG. By contrast, the symmetrical colony core eventually runs out of nutrients and ceases to enlarge. This simple example has underestimated the new territory explored by the pseudomycelium because it ignores the fact that the pseudohyphae grow in three dimensions (they invade the substrate).

In another example, this time at the level of the individual pseudohypha, the tip of the pseudohypha in Fig. 7C (from a 10.6 day old colony) is 0.59 mm from the colony center and the interior domain of this colony ends 0.21 mm from this point (data not shown). This pseudohypha has extended away from the initial point of colonization a 2.8 fold greater linear distance than the edge of the symmetrical colony core. Invasive PHG allows *S. cerevisiae* to reach and exploit substrates that

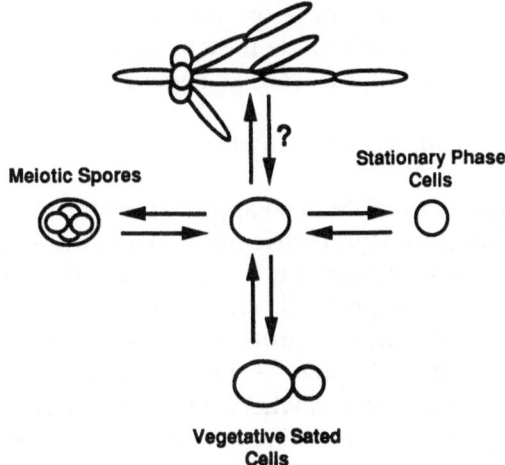

Foraging Pseudohyphae and Blastospore-Like Cells

Meiotic Spores

Stationary Phase Cells

Vegetative Sated Cells

Fig. 8. Developmental pathways of diploid yeast cells

Under favorable environmental conditions yeast cells grow vegetatively by budding. Environmental stresses can cause diploid yeast to enter the pseudohyphal pathway, differentiate into a growth arrested stationary phase cell, or undergo meiosis and sporulation with the resulting production of an ascus with its four dormant spores. We presume that either pseudohyphal cells or their associated yeast cells can resume vegetative growth. Stationary phase cells and diploid cells formed after germination and mating of ascospores both resume vegetative growth under favorable environmental conditions.

are distant from the colony or inside the growth substrate and for these reasons inaccessible to cells growing in the yeast morphology.

Pseudohyphal Cells May Deliver Yeast Cells to New Substrates

The growing tip of a mature elongating pseudohypha is composed of pseudohyphal cells. Pseudohyphal cells behind the growing tip often bud vegetative yeast cells (also referred to as blastospore-like cells) (Fig. 7B). These yeast cells divide and comprise the majority of the biomass of the pseudohypha (Fig. 7C). So, a pseudohypha is composed of a backbone of pseudohyphal cells that become covered with vegetative yeast cells as the pseudohypha grows. This phenomenon suggests that pseudohyphal cells are specialized to rapidly invade new substrates, like the interior of a grape, and deliver assimilative yeast cells to them. In this model it is the yeast cells, and not the pseudohyphal cells, that colonize new substrates and assimilate the majority of the nutrients present in them. The phenomenon of apical dominance discussed in a subsequent section is consistent with this idea. Moreover, the elongated shape of pseudohyphal cells allows them to extend 1.5 times faster than if they had the yeast cell morphology, giving further support to the idea that these are foraging cells.

Interestingly, apical growth predominates over lateral growth in pseudohyphae, that is to say pseudohyphae do not branch as much as they could if each cell along the chain budded a pseudohyphal cell once per doubling time. A nutrient gradient along the length of the pseudohypha could preclude lateral growth. Alternatively, the inhibited lateral growth could reflect a genetically programmed phenomenon. That pseudohyphal cells often laterally bud yeast cells contributes to apical dominance because these yeast cells appear not to form

97

branching pseudohyphae (see Fig. 7B and 7C). Apical dominance may be advantageous because it prevents clonal pseudohyphae from competing with each other for nutrients and favors the exploration of new substrates. These observations suggest that the PHG mode may be thought of as: yeast cell at one position → pseudohyphal cells → yeast cells at a new position.

The Requirements for Pseudohyphal Growth

Cell Shape Changes. Pseudohyphae appear to be initiated by a cell with the yeast morphology budding by asymmetric cell division a pseudohyphal cell (Fig. 7A) and not by yeast cells undergoing a shape change to the pseudohyphal morphology. The shape change in *S. cerevisiae* is similar to the dimorphic transition from yeast growth to hyphal growth in *Candida albicans* where yeast cells form germ tubes which initiate hyphae.[30] Interestingly, it is also similar to the white-opaque transition of *C. albicans*, where white or opaque cells do not switch morphologies themselves but give rise to progeny or descendents of the alternative cell type.[31]

Pseudohyphal cells are longer and thinner than sated cells growing on rich medium. As cells become longer and thinner, the tip of the cell becomes more defined. The ellipsoidal shape of a sated diploid yeast cell confers upon it a low surface area. Pseudohyphal cells have a shape with an increased surface area and consequently an increased absorptive surface. A consequence of the elongated shape of pseudohyphal cells is that cell growth is polarized along the same axis as cell division. This dual polarity of cell growth and cell division enhances the ability of the growing chain of cells to escape the colony because the growth of each new individual unit of the chain of cells along the axis of cell division incrementally moves the column along. The velocity of pseudohyphal elongation is proportional to the length of the cells which comprise the chain. Thus, a pseudohypha composed of pseudohyphal cells elongates 1.5 times faster than a pseudohypha composed of yeast cells, assuming the same doubling time for both of these cell types.

Starvation for Nitrogen. Growth on low ammonia or proline as sole nitrogen source (SPHD) induces PHG. Growth on high amounts of ammonium sulfate or a mixture of proline and ammonium sulfate suppresses PHG. Proline is known to be a poor nitrogen source for wild-type cells,[32] and is even a poorer source for *shr3* cells because of their impaired proline uptake ability. Further evidence for the starved state of *shr3* cells is that even when growing in the presence of abundant nutrients, they activate the *GCN4* amino acid starvation pathway.[17]

Diploidy and the *BUD* Genes. Only a/α diploids and not **a** or α haploids or **a**/**a** or α/α diploids show PHG, indicating that the mating type locus controls this dimorphic transition. Cells expressing both *MATa* and *MATα* bud in a polar manner whereas those expressing only *MATa* or *MATα* bud in the axial pattern.[6-8] The simplest explanation for the control of PHG by the mating type locus is that the polar budding pattern of *MATa/α* diploid cells permits linear chains of cells to form; the axial pattern leads to budding at the junction of two cells and cannot extend the column.[6]

The budding pattern of diploids is controlled by five *BUD* genes and is programmed by the mating type locus.[6-8,11-13] *MATa/α* cells that bud randomly because of the *rsr1asn16* (Bud1⁻)[33] mutation are unable to form pseudohyphae.[1] On the basis of these experiments we have proposed that one role of the *BUD* genes in yeast biology is to enable cells in the diploid phase to forage for nutrients under conditions of nitrogen starvation.[1]

S. cerevisiae strains isolated from nature typically are diploid, homothallic (having an intact mating type switching system), and competent to form pseudohyphae. Proline medium (SPHD) resembles grape juice, a natural substrate of S. cerevisiae that elicits PHG.[1] In nature, haploids have a transient existence functioning as gametes specialized for mating. Since in nature diploids are specialized for assimilating nutrients, they manifest the mechanisms S. cerevisiae has evolved for coping with nutrient stress such as sporulation and meiosis and PHG. This difference in roles of haploids and diploids in the S. cerevisiae life cycle probably explains the logic of ploidy regulation of PHG.[2]

Unipolar Cell Division. Cell division in the pseudohypha is polarized in one direction, the direction away from the mass of cells in the colony and out into the substrate. This polarization is achieved by four constraints on cell division. First, a terminal pseudohyphal cell always buds at its free end, the one opposite the junction with its mother. Second, the site of bud emergence on the daughter is close to 180° from that junction. Third, daughters stay juxtaposed to their mothers exactly along the axis of cell division, either because they are physically connected or because they are constrained by the agar. Fourth, the first daughter of the founding mother cell (the cell that gives rise to the column) divides in a direction away from the mass of cells in the colony. This initial division coupled with the three other constraints leads to the polarized extension of pseudohyphae away from the colony into unpopulated substrate.

Sated cells like pseudohyphal cells initially bud in a unipolar fashion. The first bud of a virgin sated cell (one that has no prior daughters) as well as the majority of the second buds from these cells, emerge from the pole opposite the junction of the virgin with its mother. If sated diploid cells can bud in a unipolar pattern then why don't they form pseudohyphae? The answer is that subsequent steps in the budding and growth of sated cells create asymmetries that preclude the formation of a pseudohypha.[1]

Invasiveness. Pseudohyphal cells are invasive and grow into the agar, presumably foraging for nutrients. The invasiveness of pseudohyphal cells can be explained in several ways. One might imagine that the force of unipolar cell division by thin pseudohyphal cells is sufficient to propel a column through the agar. In the pseudohyphal cycle, cell separation, a late step in cell division, may be delayed leaving the daughters attached to the mother. This linked structure might be able to generate more force than a single cell because the previous generations could act as an anchor for the cell at the apex. Although this mechanical model may be correct, we know that the connection between the apical cell in the pseudohyphal column and its mother can sometimes be broken by mechanical agitation with a microneedle.

The secretion of hydrolytic enzymes is likely to be an important factor permitting invasive growth. Lytic enzymes capable of hydrolyzing polysaccharides may be secreted by strains capable of PHG. The secretion of proteases is common to many invading pathogens such as Candida albicans.[34] Hydrolytic enzymes in C. albicans are important in creating a pathway for penetration into the host tissue. By analogy, the invasive habit of pseudohyphal S. cerevisiae cells may be a growth pattern used in nature to penetrate natural substrates such as grapes.

Proline Regulates the Dimorphic Switches of Several Fungi

We have shown that proline activates the PHG program in *S. cerevisiae*. The ascomycete *Ophiostoma ulmi*, the dimorphic yeast that causes Dutch elm disease, also regulates its dimorphic transition in response to proline.[35] In this fungus, by contrast with *S. cerevisiae*, proline supports the yeast phase and ammonium sulfate supports the invasive mycelial form. Nutritional regulation by proline may be related to disease because elms resistant to infection contain high proline levels in their xylem sap while susceptible elms do not.[36] Proline levels also regulate the development of the human pathogen *C. albicans*.[37-39] In *C. albicans*, as in *S. cerevisiae*, proline supports filamentous growth while high ammonium sulfate supports yeast phase growth.

Implications of Dimorphic Growth and Diploidy for Pathogenesis

The interconversion of a yeast form and a filamentous form is typical of many pathogenic fungi.[14] In *C. albicans*, the ability to undergo a dimorphic transition may be important for pathogenesis.[40] In tissue infected with *C. albicans*, yeast cells, pseudohyphae and true hyphae are found.[30] Diploid *S. cerevisiae*[6] and *C. albicans*[41] cells possess a unipolar budding pattern which is important for PHG. The requirement of unipolar diploid budding for PHG could explain why *C. albicans* is found only as a diploid (no haploid form or sexual cycle has been observed[30]). Perhaps *C. albicans* once had a haploid phase, but with time and selection, genes required for meiosis and therefore the generation of haploid cells were lost. In analogy to *S. cerevisiae* budding pattern regulation, *C. albicans* may have a cryptic mating type locus required for programming polarized cell division patterns.

Ustilago maydis, the causitive agent of corn smut, is pathogenic only in its dikaryotic filamentous form (reviewed in ref. 42). The haploid phase of this fungus grows exclusively in a yeast form and is nonpathogenic. Although *S. cerevisiae*, an ascomycete, and *U. maydis*, a basidiomycete, are quite distant on a phylogenetic scale, the major morphogenetic event in each species, conversion of the yeast to a filamentous form, has similar physiological and genetic control. Both *S. cerevisiae* and *U. maydis* haploids grow as yeast cells unable to develop into their filamentous form. *S. cerevisiae* diploids and *U. maydis* dikaryons heterozygous for mating type loci undergo an environmentally triggered dimorphic transition from a yeast form to a filamentous form. As described in this paper, *S. cerevisiae* MATa/α diploids undergo a dimorphic transition from yeast growth to filamentous growth when transferred to medium where they are starved for nitrogen. Similarly, *U. maydis* diploids heterozygous for both the *a* and *b* mating type loci grow in a yeast form on most media but when transferred to charcoal medium or to their corn host undergo a dimorphic transition to filamentous growth.[42]

In *MATa/α S. cerevisiae* diploids the *MATa/α* encoded heterodimeric repressor a1α2[43,44] must be required for the conversion because isogenic *MATα/α* or *MATa/a* strains do not undergo the dimorphic transition. We surmise that a1α2 repression of either or both *BUD3* and *BUD4*, two genes required for the diploid budding pattern, is required for conversion to the PHG mode. It is also possible that a1α2 regulates other processes required for PHG in addition to budding pattern and we are currently investigating this possibility. In *U. maydis* the *b* mating type locus is also thought to encode transcription factors[45-49] and diploids homozygous for *b* neither grow filamentously nor cause disease.[50] Heterozygosity at the *b* locus may be required to repress haploid specific cell division patterns. Mating type locus control of dimorphism probably occurs in fungi pathogenic to humans and mating type locus encoded genes should be considered as antifungal drug targets.

ACKNOWLEDGEMENTS

The authors thank Sheldon Penman for helping with the SEM experiment, and Marjorie Brandriss for providing yeast strains. We thank Linda Bisson, Neil Gow, Regine Kahmann, Boris Magasanik, Donald Pfister, David Soll, and members of the Fink lab for helpful conversations. We thank Ruth Hammer and David Pellman for comments on the manuscript. Fig. 2, 4, 5, 6, and 8 and Tables 1, 2, and 3 and the text related to these tables and figures are adapted from reference 1 with kind permission from Cell Press, the copyright holder. This work was supported by a Howard Hughes Medical Institute Predoctoral Fellowship to C.J.G., NIH Research Fellowship GM12038-01 to P.O.L., and NIH Research Grants GM40266 and GM35010 to G.R.F. G.R.F. is an American Cancer Society Professor of Genetics.

REFERENCES

1. C.J. Gimeno, P.O. Ljungdahl, C.A. Styles, and G.R. Fink, Unipolar cell divisions in the yeast *S. cerevisiae* lead to filamentous growth: regulation by starvation and *RAS*, *Cell* 68 : 1077 (1992).
2. C.J. Gimeno and G.R. Fink, The logic of cell division in the life cycle of yeast, *Science* 257 : 626 (1992).
3. E.C. Hansen, Recherches sur la physiologie et la morphologie des ferments alcooliques. V. Méthodes pour obtenir des cultures pures de *Saccharomyces* et de microorganismes analogues, *Compt. rend. trav. lab. Carlsberg, Sér. physiol.* 2: 92 (1886).
4. G.H. Scherr and R.H. Weaver, The dimorphism phenomenon in yeasts, *Bacteriol. Rev.* 17: 51 (1953).
5. J.R. Pringle, and L.H. Hartwell, The *Saccharomyces cerevisiae* cell cycle, in : "The Molecular Biology of the Yeast *Saccharomyces*: Life Cycle and inheritance", J.N. Strathern, E.W. Jones, and J.R. Broach, eds., Cold Spring Harbor Laboratory Press, Cold Spring Harbor (1981).
6. D. Freifelder, Bud position in *Saccharomyces cerevisiae*, *J. Bacteriol.* 80: 567 (1960).
7. J.B. Hicks, J.N. Strathern, and I. Herskowitz, Interconversion of yeast mating types III. Action of the homothallism (*HO*) gene in cells homozygous for the mating type locus, *Genetics* 85: 395 (1977).
8. J. Chant and I. Herskowitz, Genetic control of bud site selection in yeast by a set of gene products that constitute a morphogenetic pathway, *Cell* 65: 1203 (1991).
9. K.A. Nasmyth, Molecular genetics of yeast mating type, *Ann. Rev. Genet.* 16: 439 (1982).
10. D.G. Drubin, Development of cell polarity in budding yeast, *Cell* 65: 1093 (1991).
11. A. Bender and J.R. Pringle, Multicopy suppression of the *cdc24* budding defect in yeast by *CDC42* and three newly identified genes including the *ras*-related gene *RSR1*, *Proc. Natl. Acad. Sci. USA* 86 : 9976 (1989).
12. J. Chant, K. Corrado, J.R. Pringle, and I. Herskowitz, Yeast *BUD5*, encoding a putative GDP-GTP exchange factor, is necessary for bud site selection and interacts with bud formation gene *BEM1*, *Cell* 65: 1213 (1991).
13. S. Powers, E. Gonzales, T. Christensen, J. Cubert, and D. Broek, Functional cloning of *BUD5*, a *CDC25*-related gene from *S. cerevisiae* that can suppress a dominant-negative *RAS2* mutant, *Cell* 65: 1225 (1991).
14. M.G. Shepherd, Morphogenetic transformation of fungi, *Curr. Top. Med. Mycol.* 2: 278 (1988).
15. E.G.V. Evans and M.D. Richardson, "Medical Mycology: a Practical Approach", Information Press Ltd., Oxford (1989).
16. F. Sherman, G.R. Fink, and J.B. Hicks, "Methods in Yeast Genetics", Cold Spring Harbor Press, Cold Spring Harbor (1986).

17. P.O. Ljungdahl, C.J. Gimeno, C.A. Styles, and G.R. Fink, SHR3, a novel component of the secretory pathway specifically required for localization of amino acid permeases in yeast, *Cell* 71: 463 (1992).

18. R.K. Mortimer and J.R. Johnston, Genealogy of principal strains of the yeast genetic stock center, *Genetics* 113: 35 (1986).

19. M. Grenson, M. Mousset, J.M. Wiame, and J. Bechet, Multiplicity of the amino acid permeases in *S. cerevisiae*. I. Evidence for a specific arginine transporting system, *Biochim. Biophys. Acta* 127: 325 (1966).

20. M.C. Brandriss and B. Magasanik, Genetics and physiology of proline utilization in *Saccharomyces cerevisiae*: enzyme induction by proline, *J. Bacteriol.* 140: 498 (1979).

21. J. Rytka, Positive selection of general amino acid permease mutants in *Saccharomyces cerevisiae*, *J. Bacteriol.* 121: 562 (1975).

22. J.-M. Wiame, M. Grenson, and H.N. Arst Jr., Nitrogen catabolite repression in yeasts and filamentous fungi, *Adv. Microb. Physiol.* 26: 1 (1985).

23. A.H. Siddiqui, and M.C. Brandriss, A regulatory region responsible for proline-specific induction of the yeast *PUT2* gene is adjacent to its TATA box, *Mol. Cell. Biol.* 8: 4634 (1988).

24. P.F. Lasko and M.C. Brandriss, Proline transport in *Saccharomyces cerevisiae*, *J. Bacteriol.* 148: 241 (1981).

25. A.D.M. Rayner, The challenge of the individualistic mycelium, *Mycologia* 83: 48 (1991).

26. J. Lodder, "The Yeasts: a Taxonomic Study", North-Holland Publishing Co., Amsterdam (1970).

27. A. Guilliermond, "The Yeasts", John Wiley and Sons, Inc., New York (1920).

28. C.M. Brown and J.S. Hough, Elongation of yeast cells in continuous culture. *Nature* 206: 676 (1965).

29. V.L. Eubanks and L.R. Beuchat, Effects of antioxidants on growth, sporulation, and pseudomycelium production by *Saccharomyces cerevisiae*, *J. Food Sci.* 47: 1717 (1982).

30. F.C. Odds, "*Candida* and Candidosis", Baillière Tindall, London (1988).

31. M.S. Bergen, E. Voss, and D.R. Soll, Switching at the cellular level in the white-opaque transition of *Candida albicans*, *J. Gen. Microbiol.* 136: 1925 (1990).

32. T.G. Cooper, Nitrogen metabolism in *Saccharomyces cerevisiae*., in: "The Molecular Biology of the Yeast *Saccharomyces*: Metabolism and Gene Expression", J.N. Strathern, E.W. Jones, and J.R. Broach, eds., Cold Spring Harbor Laboratory Press, Cold Spring Harbor (1982).

33. R. Ruggieri, A. Bender, Y. Matsui, S. Powers, Y. Takai, J.R. Pringle, and K. Matsumoto, *RSR1*, a *ras*-like gene homologous to *Krev-1/smg21A/rap1A*: role in the development of cell polarity and interactions with the Ras pathway in *Saccharomyces cerevisiae*, *Mol. Cell Biol.* 12: 758 (1992).

34. F. Macdonald, and F.C. Odds, Virulence for mice of a proteinase-secreting strain of *Candida albicans* and a proteinase deficient mutant, *J. Gen. Microbiol.* 129: 431 (1983).

35. R.K. Kulkarni, and K.W. Nickerson, Nutritional control of dimorphism in *Ceratocystis ulmi*, *Exp. Mycol.* 5: 148 (1981).

36. D. Singh, and E.B. Smalley, Nitrogenous compounds in the xylem sap of American elms with Dutch elm disease, *Canad. J. Bot.* 47: 1061 (1969).

37. G.A. Land, W.C. McDonald, R.L. Stjernholm, and L. Friedman, Factors affecting filamentation in *Candida albicans*: relationship of the uptake and distribution of proline to morphogenesis, *Infect. Immun.* 11: 1014 (1975).

38. N. Dabrowa, S.S.S. Taxer, and D.H. Howard, Germination of *Candida albicans* Induced by proline, *Infect. Immun.* 13: 830 (1976).

39. A.R. Holmes, and M.G. Shepherd, Proline-induced germ-tube formation in *Candida albicans*: role of proline uptake and nitrogen metabolism, *J. Gen. Microbiol.* 133: 3219 (1987).

40. D.R. Soll, Current status of the molecular basis of *Candida* pathogenicity, in: "The Fungal Spore and Disease Initiation in Plants and Animals", G.T. Cole and H.C. Hoch, eds., Plenum Press, New York (1991).

41. W.L. Chaffin, Site selection for bud and germ tube emergence in *Candida albicans*, *J. Gen. Microbiol.* 130: 431 (1984).

42. F. Banuett, *Ustilago maydis*, the delightful blight, *Trends Genet.* 8: 174 (1992).

43. C. Goutte and A.D. Johnson, a1 protein alters the DNA binding Specificity of α2 repressor, *Cell* 52: 875 (1988).

44. I. Herskowitz, A regulatory hierarchy for cell specialization in yeast, *Nature* 342: 749 (1989).

45. J.W. Kronstad and S.A. Leong, Isolation of two alleles of the *b* locus of *Ustilago maydis*, *Proc. Natl. Acad. Sci. USA* 86: 1384 (1989).

46. B. Schulz, F. Banuett, M. Dahl, R. Schlesinger, W. Schäfer, T. Martin, I. Herskowitz, and R. Kahmann, The *b* alleles of *U. maydis*, whose combinations program pathogenic development, code for polypeptides containing a homeodomain-related motif, *Cell* 60: 295 (1990).

47. J.W. Kronstad and S.A. Leong, The *b* mating-type locus of *Ustilago maydis* contains variable and constant regions. *Genes Dev.* 4: 1384 (1990).

48. B. Gillissen, J. Bergemann, C. Sandmann, B. Schroeer, M. Bölker, and R. Kahmann, A two-component regulatory system for self/non-self recognition in *Ustilago maydis*, *Cell* 68: 647 (1992).

49. I. Herskowitz, Yeast branches out, *Nature* 357: 190 (1992).

50. F. Banuett and I. Herskowitz, Different *a* alleles of *Ustilago maydis* are necessary for maintenance of filamentous growth but not for meiosis, *Proc. Natl. Acad. Sci. USA* 86: 5878 (1989).

51. R.S. Sikorski and P. Hieter, A system of shuttle vectors and yeast host strains designed for efficient manipulation of DNA in *Saccharomyces cerevisiae*, *Genetics* 122: 19 (1989).

PHASE TRANSITION IN *WANGIELLA DERMATITIDIS*: IDENTIFICATION OF CELL-DIVISION-CYCLE GENES INVOLVED IN YEAST BUD EMERGENCE

Chester R. Cooper, Jr.

Molecular Genetics Program and Laboratories for Mycology
Wadsworth Center for Laboratories and Research
New York State Department of Health
P. O. Box 509
Albany, New York 12201-0509, USA

ABSTRACT

Yeast-to-multicellular-form (Y→Mc) conversion in the darkly-pigmented (dematiaceous), pathogenic fungus *Wangiella dermatitidis* represents an excellent model to investigate similar phase transitions exhibited by other dematiaceous fungi upon invasion of host tissue. The yeast-phase cell cycle of *W. dermatitidis* and its role in Y→Mc conversion was studied by previous investigators using temperature-sensitive (ts) morphological mutants. Both the mutants, designated multicellular (Mc) strains, and the parental wild type grow as virtually indistinguishable budding yeasts at the permissive temperature (25 °C). However, at the restrictive temperature (37 °C), the Mc strains quantitatively undergo Y→Mc conversion whereas the wild type continues to grow as a yeast. In two Mc strains, Mc2 and Mc3, the ts lesions reflect distinct cell-cycle execution points for yeast bud emergence. This suggests that the ts lesions delimit separate mutations in one or more genes.

Because *W. dermatitidis* is an asexual fungus, a parasexual system of genetic analysis was established to investigate the relationship of the ts lesions in strains Mc2 and Mc3. Specifically, spheroplasts of melanin-deficient (albino) auxotrophs independently derived from strains Mc2 and Mc3 were fused and regenerated at 25 °C in minimal medium. The resulting fusion products were dematiaceous and prototrophic. At 37°C, all fusion products exhibited polarized growth predominantly as uninucleate yeasts and less frequently as moniliform hyphae. These results indicated that the mutation in each Mc parental strain was complemented by a wild-type allele present in its fusion partner. Subsequent analysis of albino ts segregants derived from one fusion product demonstrated that the ts lesions in strains Mc2 and Mc3 were non-allelic. Hence, the genes defined by these mutations have been designated *CDC1* and *CDC2*, respectively. These are the first cell-division-cycle genes identified in a dematiaceous fungus.

Their discovery provides the basis for additional studies into the molecular mechanisms involved in fungal morphogenesis and pathogenesis.

INTRODUCTION

Mycotic diseases of humans and animals appear to be routinely caused by less than 300 of the approximately 100,000 recognized fungal species.[1-3] Most zoopathogenic fungi are considered to be opportunistic etiologic agents in that they are typically isolated from debilitated or immunosuppressed individuals.[2-5] By comparison, some 50 different fungi are encountered as pathogens with relative frequency among individuals having no readily-discernable predisposing conditions prior to infection.

Critical to the establishment of disease by this latter group of fungi is how successful they are at evading the host's immune system and adapting to the harshness of the tissue environment. These requisites are often associated with unique phase transitions concomitant with tissue invasion. Such transitions, variously termed dimorphism, polymorphism, or pleomorphism, had been defined by medical mycologists to mean phenotypic duality (or plurality) of form in which a fungus exhibits distinct saprophytic and parasitic morphologies.[6-9] Because some non-pathogenic fungi morphologically respond to environmental conditions in a similar fashion,[10,11] dimorphism (or polymorphism) is now defined as the ability of a fungus to grow in at least two different vegetative morphologies and to express these forms through distinct, vegetative-phase transitions.[12]

Conceivably, phase transition of a medically-important fungus upon tissue invasion may contribute to its survival and reproduction *in vivo*. If so, then those specific molecular mechanisms that induce the morphological change could be considered potential virulence factors. By ascertaining the molecular basis for this alteration of form, potential insights into the control of fungal diseases as well as a better understanding of eukaryotic cellular development may be provided. Therefore, an ideal transition in which to investigate these mechanisms would be one that is readily inducible under laboratory conditions and is characteristic or diagnostic of a specific type of infection. These criteria are fulfilled by the polymorphic nature of certain pathogenic members of the form-family Dematiaceae.

PHASE TRANSITIONS AMONG THE DEMATIACEAE

The form-family Dematiaceae consists of anamorphic (asexual) fungi that are darkly-pigmented (dematiaceous) due to the presence of melanin.[1-3] The latter substance, which has been associated with virulence, is synthesized by a pathway common to most, if not all, fungi belonging to this taxon.[13-15] As a group, the Dematiaceae exhibit pronounced polymorphism including morphs such as true hyphae bearing conidia (e.g., phialoconidia, blastoconidia, annelloconidia, etc.), thin- and thick-walled yeasts, pseudohyphae, moniliform hyphae, and swollen, septate forms.[16-18] Colonies of a particular form-species may contain a variety of these morphologies depending upon their age, environment, and nutritional status. These parameters can also be used to manipulate readily certain dematiaceous fungi to undergo phase transitions *in vitro*. Some of these in-vitro transitions presumably simulate the natural pathological process through the formation of cellular entities that strongly resemble characteristic and diagnostic *in-vivo* morphologies.

Chromoblastomycosis and phaeohyphomycosis are two types of subcutaneous infections caused by dematiaceous fungi.[2,3,16] It is believed that both types of infection are initiated when either spores or hyphal elements of the etiological agents are traumatically implanted in tissue. In cases of chromoblastomycosis, the invading propagules subsequently grow isotropically and form characteristic "sclerotic bodies". The latter are defined as thick-walled and darkly-pigmented muriform structures, i.e., containing intersecting vertical and horizontal septa. In contrast, the etiologic agents of phaeohyphomycosis grow in the host tissue predominantly as dematiaceous yeasts, hyphae, pseudohyphae, swollen entities, or any combination thereof. By definition, the sclerotic bodies associated with chromoblastomycosis are not found in cases of phaeohyphomycosis. However, at least two fungi usually considered to cause only phaeohyphomycosis have the genetic capability to form multicellular entities both *in vitro* and *in vivo* that are indistinguishable from sclerotic cells (e.g., see Fig. 12 in reference 16 and Fig. 12 in reference 19; see also references 20 and 21) For purposes of this treatise, the terms "sclerotic", "muriform", and "multicellular" are considered to be synonymous.

Historically, medical mycologists studying the dematiaceous fungi have focused much attention upon the conditions which influence the formation of the sclerotic body. Early attempts to cultivate muriform cells *in vitro* and *in vivo* or to induce their formation were only marginally successful (see reference 18 for a historical review). The critical discovery in the production of the multicellular morphology was made by Szaniszlo and co-workers.[17,21-23] These investigators were able to induce efficiently the production of sclerotic cells in strains of *Fonsecaea pedrosoi*, *Phialophora verrucosa*, *Cladosporium carrionii*, and *Wangiella dermatitidis* by incubating cultures of these dematiaceous fungi under acidic conditions. Thick-walled, multiply-septate bodies were formed that were virtually indistinguishable from the characteristic muriform cells of chromoblastomycosis. In the case of *W. dermatitidis*, yeast cells quantitatively converted to the sclerotic morphology within 2-5 days after inoculation into media acidified to pH 2.5.

It is interesting to note that of the four form-species induced to form sclerotic bodies, only *W. dermatitidis* is not a known agent of chromoblastomycosis, but does cause phaeohyphomycosis. When other agents of phaeohyphomycosis, specifically *Exophiala jeanselmei* and *Xylohypha bantiana*, were similarly treated, no sclerotic bodies were observed.[17,21-23] These results suggested that *W. dermatitidis* could be used as a unique model in which the molecular mechanisms that influence both morphogenesis as well as the development of chromoblastomycosis and phaeohyphomycosis could be studied.

MULTICELLULAR-FORM DEVELOPMENT IN *WANGIELLA DERMATITIDIS*

The polymorphic nature of *W. dermatitidis* is expressed in three well-defined modes of vegetative growth (Fig. 1). Blastic, apical, and isotropic modes of development are primarily associated with growth in the yeast, hyphal, and sclerotic morphologies, respectively. Various aspects of the readily-controllable transitions that occur among these various morphologies have been extensively reviewed.[17,24,25] The best characterized transition is the conversion of yeasts to sclerotic forms, i.e., the change from blastic to isotropic development. The following discussion describes this morphogenic process in more detail.

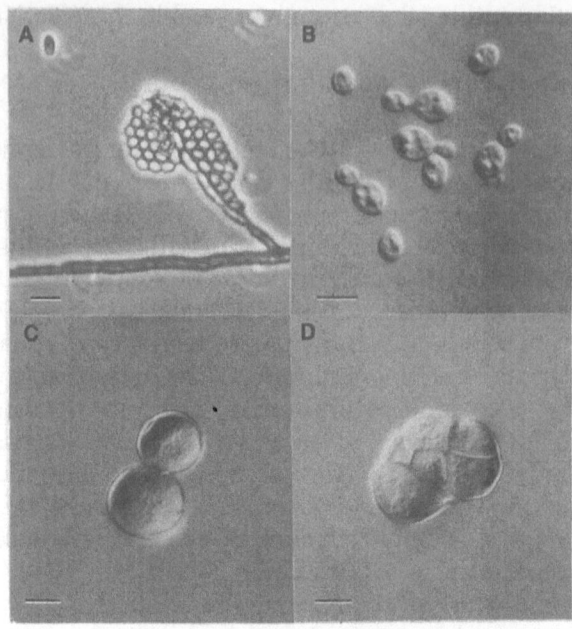

Fig. 1 The various morphs associated with the different modes of vegetative development expressed by *W. dermatitidis*. Apical development is best represented by the mould phase which typically forms conidiophores bearing phialoconidia (A; phase contrast optics). Budding yeasts characterize the blastic mode of development (B; Nomarski optics). Isotropic development initially causes the formation of swollen, non-septated (Stage I) cells (C; Nomarski optics) that subsequently produce internal septations characteristic of the multicellular (Stage II) cell (D; Nomarski optics). The latter two morphs are often termed "sclerotic bodies". The scale bar in each figure equals 5 μm.

Early studies with *W. dermatitidis* showed that acid-induced formation of the sclerotic morphology results from the cessation of yeast bud emergence without the inhibition of cellular growth, nuclear division, or cytokinesis.[21] The process by which this entity is induced can be arbitrarily divided into two stages (Fig. 1C and 1D). Stage I is marked by the formation of swollen, unbudded cells having multiple nuclei and thickened cell walls. In Stage II, the cellular growth that characterizes Stage I development continues and one or more transverse septa form, thereby generating the multicellular phenotype. An analogous transition from blastic to isotropic development is also exhibited by certain temperature-sensitive (ts), cell-cycle mutants of *Saccharomyces cerevisiae*, e.g., strains possessing mutations in the *CDC24*, *CDC42*, or *CDC43* genes.[26,27] A search for similar ts, morphological mutants of *W. dermatitidis* produced several that were designated as multicellular (Mc) strains.[28,29] The Mc strains, like the parental wild type, grow as budding yeasts at the permissive temperature (25 °C). At the restrictive temperature (37 °C), however, the Mc strains quantitatively undergo yeast-to-multicellular-form (Y→Mc) conversion. In contrast, the wild type continues to grow as a yeast at 37 °C.

Conversion of yeasts to multicellular forms in these ts Mc strains follows the same two stage process noted above for acid-induced cells. The process is initiated when the cell-cycle event of yeast bud emergence is inhibited while cellular growth, nuclear division, and cytokinesis continue.[28,29] The resulting terminal phenotypes are indistinguishable from the thick-walled and multinucleate Stage I and II cells induced by acidic culture conditions (Fig. 2). In addition, Y→Mc conversion is accompanied by quantitative changes in the major cell-wall polymers chitin, β-glucan, and mannoprotein, as well as qualitative changes in wall architecture.[24,30-34] The relative amount of chitin in the cell wall increases ten fold, whereas β-glucans and mannoproteins decrease by one half. Chitin deposition becomes delocalized, shifting from the bud/birth scar regions to an inner wall layer and the transverse septa of the multicellular phenotype. The incorporation of β-glucans and mannoproteins also becomes delocalized with most of the latter eventually being sloughed off as the chitinous inner wall layer forms. Inhibition of the biosynthesis of any of these polymers during Y→Mc conversion causes increased cellular death due to lysis. Interestingly, incubation of strain Mc3 at 25 °C in the presence of aculeacin A, an inhibitor of β-glucan biosynthesis, induces the formation of the multicellular morphology. By comparison, when the wild type is incubated at 25 °C or 37 °C in the presence of chitin or β-glucan biosynthesis inhibitors, chains of yeast cells form that often possess aberrant, incurvate septa or a newly-developing cell inside another.

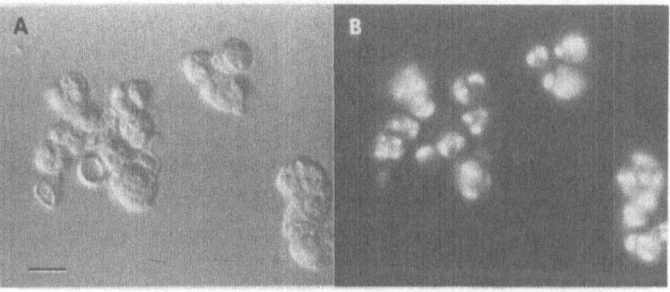

Fig. 2 Stage I and II cells of strain Mc3 formed during Y→Mc conversion at 37 °C (A; Nomarski optics) and corresponding fluorescent photomicrograph showing individual cells with multiple nuclei (B). The nuclei visualized by staining with mithramycin.[28,29] The scale bar equals 10 μm and is applicable to both figures.

The role of the yeast-phase cell cycle of *W. dermatitidis* in Y→Mc conversion has been extensively studied.[24,28,29,34-36] The cell cycle consists of at least three interdependent control pathways in which no less than twelve landmark events have been temporally mapped (Fig. 3). Among these are the execution points of the multicellular lesions from two different Mc strains, Mc2 and Mc3. These execution points occur at different times in the yeast bud emergence pathway of the cell cycle, suggesting that the ts lesions in these strains, previously designated as *mcm2* and *mcm3* (see reference 34), respectively, occupy different genetic loci, i.e., they are separate mutations in one or more genes. However, because *W. dermatitidis* is an asexual fungus, confirmation of this conjecture by classical methods of genetic complementation is not possible. This difficulty has blocked past attempts to understand the genetic basis of Y→Mc conversion.

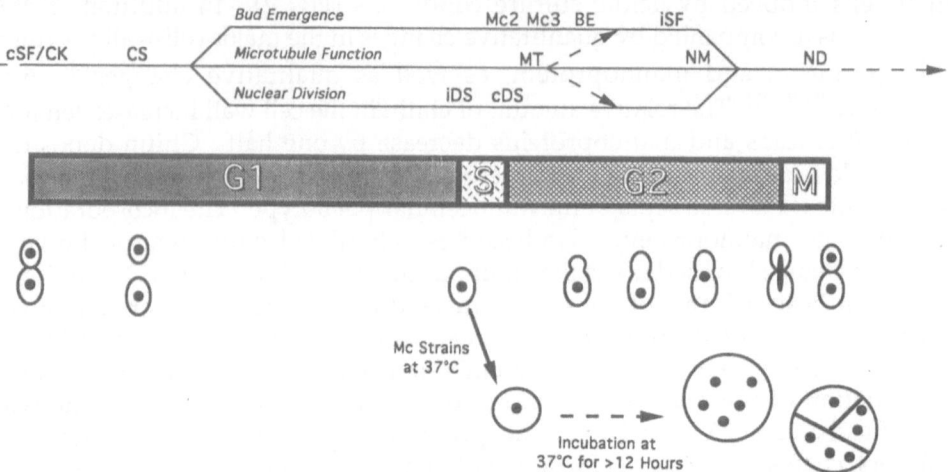

Fig. 3 Diagrammatic representation of the yeast-phase cell cycle of *W. dermatitidis* and corresponding cellular morphologies. The top of the diagram depicts the interdependent control pathways governing bud emergence, microtubule function, and nuclear division. The relative position of particular execution points with respect to each other and the phases of the cell cycle (middle bar) are shown. The bottom portion of the figure relates the relative cellular and nuclear morphologies to the cell cycle. For comparison purposes only, an alternative mode of morphogenesis is included that is representative of the Y→Mc conversion at 37 °C by Mc strains. This pathway is not meant to correspond in scale to the cell cycle as depicted in this figure. Abbreviations: cSF, completion of septum formation; CK, cytokinesis; CS, cell separation; Mc2 and Mc3, execution points for these respective mutants; BE, bud emergence; iSF, initiation of septum formation; MT, microtubule polymerization; NM, nuclear migration; ND, nuclear division; iDS, initiation of DNA synthesis; cDS, completion of DNA synthesis; G1, "gap 1" of the cell cycle; S, DNA synthesis portion of the cell cycle; G2, "gap 2" of the cell cycle; M, mitosis portion of the cell cycle. Adapted in part from reference 24.

PARASEXUAL GENETIC ANALYSIS IN *WANGIELLA DERMATITIDIS*

General Design and Methodology

The impediment of genetic analysis due to asexuality is not unique to *W. dermatitidis*. Most zoopathogenic fungi are asexual. Of the few species that are sexual, their reproductive cycles are not readily amenable to classical methods of genetic analysis or have yet to be extensively exploited.[37,38] Moreover, despite significant progress in the development of modern molecular techniques to study pathogenic fungi, the inherent difficulties associated with these organisms (e.g., complex life cycles, diploidy, etc.) has hampered the general applicability of these methodologies. Hence, genetic studies of the medically-important fungi have been effectively limited to those investigations involving parasexual methods of analysis.

Perhaps the most notable and successful system for parasexual genetic analysis of a pathogenic fungus is the one developed in *Candida albicans*.[39] In this system, the genetic information needed to establish complementation groups and allelic relationships is obtained primarily from the phenotypic analysis of spheroplast-fusion products and segregants derived from them. Using this system as a model, similar methodology was applied in studying the ts lesions, *mcm2* and *mcm3*, of *W. dermatitidis*.[40] For this purpose, melanin-deficient (albino), auxotrophs of the Mc strains were derived. Each possessed a genetic defect for melanin biosynthesis (*mel3* or *mel4*), as well as an auxotrophic requirement for either adenine (*ade*), arginine (*arg*), methionine (*met*), or uracil (*ura*). Spheroplasts of each strain were independently produced in a high-salt buffer. Pairs of appropriate mutants were then mixed together in an osmotic buffer containing Ca^{2+} ions and polyethylene glycol. Following centrifugation, the resulting spheroplast pellet was suspended in osmotically-stabilized agar and overlayed on to minimal medium. Within 2 weeks of incubation at 25 °C, darkly-pigmented colonies appeared and were phenotypically characterized. Occasionally, a non-pigmented colony would arise. However, in these experiments, fusion products were defined and selected as dematiaceous colonies growing on minimal medium (Fig. 4A).

Additional genetic data were obtained from fusion product segregants. In particular, albino sectors arising from the darkly-pigmented colonies were isolated because they could be easily discerned from fusion products and are indicative of a genetic segregation event (Fig. 4B). These segregants were phenotypically analyzed and those capable of Y→Mc conversion were further characterized as described below.

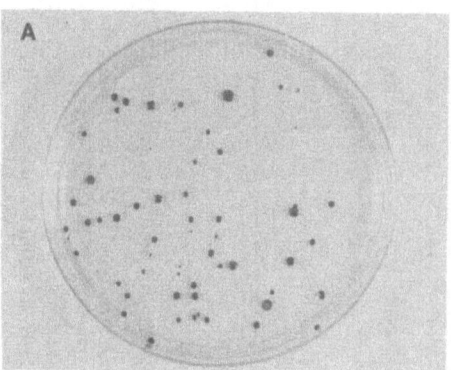 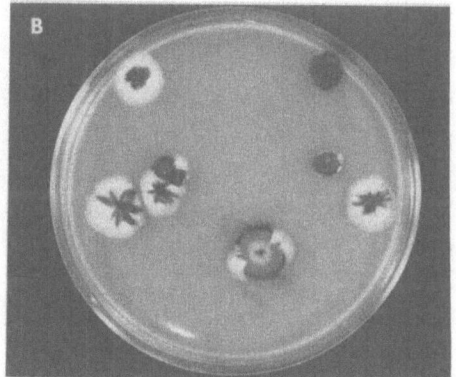

Fig. 4 A) Colonies of black fusion products growing on minimal medium following spheroplast fusion of albino, auxotrophic Mc strains of *W. dermatitidis*. B) Colonies of MBC-treated fusion products growing on complete medium and giving rise to sectors of albino segregants.

Analysis of Fusion Products

Numerous fusion products were isolated and phenotypically characterized.[40] All were quite distinct from their albino, auxotrophic parental strains (Table 1). At 25 °C and 37 °C, all fusion products exhibited polarized growth mainly as budding yeasts and to a lesser extent by apical extension of moniliform hyphae. Fusion products were greater in size than either parental strain and typically contained a single nucleus which also appeared to be larger than those of their parental strains (Fig. 5). Rarely were two nuclei were observed in a single cell. These observations suggested that karyogamy occurred shortly after plasmogamy, thereby giving rise to an extensively heterozygotic isolate. This was further supported by quantitative DNA analyses that showed fusion products contained 1.7 to 3.5 more DNA than

Table 1. Distinguishing phenotypes of selected groups of parental strains, fusion products, and segregants

Group	Colony color	Auxotrophic phenotype	Mode of development at 37°C	Range of DNA content (fg/cell)	Cell type at 39°C	Relevant *mcm* genotype
Parental strains:						
Mc2-derived	White	Met⁻ or Arg⁻	Isotropic	43.7[a]	Stage I	*mcm2*
Mc3-derived	White	Ade⁻ or Ura⁻	Isotropic	36.0–42.3	Stage II	*mcm3*
Fusion products	Black	None	Blastic[b]	73.8–133.6	Yeast[b]	*MCM2/mcm2*; *MCM3/mcm3*
Segregants of strain 3u2m-428:[c]						
Class A	White	Met⁻ or Ura⁻	Isotropic	67.1[a]	Stage I	*mcm2*
Class B	White	Met⁻ or Ura⁻	Isotropic	40.9–42.2	Stage II	*mcm3*
Class C	White	None	Isotropic	55.7–62.6	Stage I, Stage II	*mcm2, mcm3*
Class D	White	None	Blastic[b]	n.d.[d]	Yeast[b]	*MCM2/mcm2*; *MCM3/mcm3*[e]

[a]The DNA content of only one strain was determined.

[b]Occasionally, apically-developing moniliform hyphae were present.

[c]Segregants were isolated following MBC-treatment of fusion product strain 3u2m-428. The parental strains of this fusion product possessed the following genetic markers: strain Mc2W-1, *mcm2, mel4, met*; strain Mc3W-15, *mcm3, mel3, ura*.

[d]n.d., not determined.

[e]Other genotypes resulting in yeast growth at the restrictive temperature are possible.

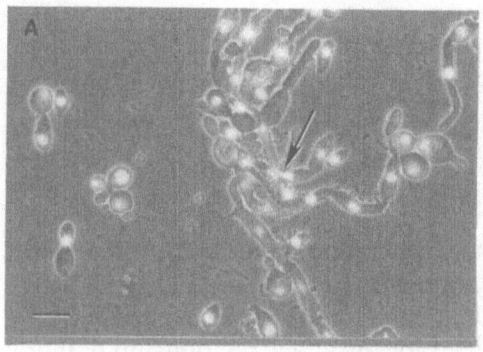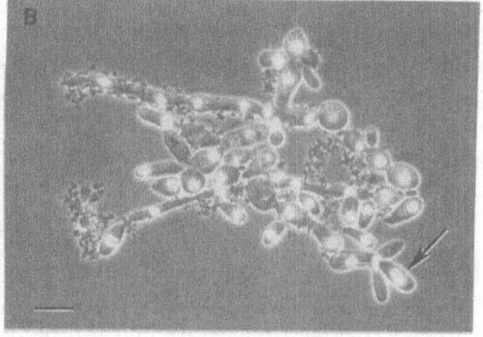

Fig. 5 Yeast and moniliform hyphae of fusion products 3u2m-1113 (A) and 3u2m-1115 (A). In most cells, a single, large nucleus is recognizable following staining with mithramycin. Also, a mitotically-dividing nucleus is visible (A; arrow) as is a cell containing two nuclei (B; arrow). The scale bar in each figure equals 10 mm. (Combined fluorescent and phase contrast optics)

either parental strain (Table 1). Furthermore, the ability of fusion products to grow as yeasts at 37 °C indicated that the *mcm2*-bearing strains carried a wild-type allele of *mcm3*, and vice versa. However, the presence of moniliform hyphae is not easily explained. Perhaps the controls that govern hyphal and yeast growth in this polymorphic fungus are closely related because either mode of polarized development eventually gives rise to the other when culture conditions are appropriately manipulated.[17,25] Alternatively, the moniliform hyphae might have resulted from gene dosage effects, the presence of previously unexpressed mutations in the parental strains, or poorly understood intergenic relationships among the multiple morphogenetic loci present in the fusion product. Nonetheless, the presence of these morphologies indicated that genetic complementation did occur because the various cell-division-cycle events involved in polarized cellular development were again predominantly associated in one well-regulated cell cycle, the result of which was usually yeast vegetative reproduction.

Induction of Fusion Product Segregants

All spontaneous albino segregants of the dematiaceous fusion products were non-ts and prototrophic.[40] However, the rate of spontaneous segregation was low, thereby making segregant analysis laborious because large numbers of colonies had to be plated and screened. To facilitate these studies, greater frequencies of albino segregation were induced by treating fusion products with the mitotic inhibitor methyl benzimidazole-2-yl carbamate (MBC). This drug reportedly causes chromosomal loss and has been used to map genetic loci in *S. cerevisiae*.[41,42] Exposure of fusion products to MBC significantly increased the number of albino segregants by an average of 21%.

The MBC-induced albino segregants from selected fusion product strains exhibited a variety of phenotypes.[40] The majority, like the spontaneously occurring segregants, were non-ts and prototrophic. Yet, unlike untreated fusion products, exposure to MBC also produced segregants having ts or auxotrophic phenotypes. Table 1 describes the general phenotypes of four classes of segregants derived from fusion product strain 3u2m-428. These classes are based upon mode of cellular development and the type of Y→Mc conversion kinetics expressed at restrictive temperatures (see below). Consistent with the presumed function of MBC, selected ts segregants contained less DNA than strain 3u2m-428 (85.9 fg/cell), but 1.0 to 1.6 the amount of DNA found in the parental Mc strains.

Allelic Relationship of the ts Lesions

The production of ts albino segregants from pigmented and prototrophic fusion products not only completes the requirements for a parasexual cycle,[43] but also permits a more detailed analysis of the genetic markers employed in such studies. For example, evidence can be obtained for the allelic relationship of those mutations expressing similar phenotypes. With respect to the similar modes of isotropic development expressed at 37 °C by ts mutants bearing either *mcm2* or *mcm3*, restoration of blastic development in fusion products of *W. dermatitidis* could have resulted from either intra- or intergenic complementation. The particular type of complementation is a consequence of the allelic relationship of *mcm2* and *mcm3*, i.e., the latter are either mutations in a single gene or lesions in separate genes. As described below, the phenotypic characterization of ts segregants provided evidence that the genes represented by these mutations are non-allelic.

Of the four classes of albino segregants derived from fusion product strain 3u2m-428, three were ts and exhibited the same terminal phenotype (i.e., Stage II cells) at 37 °C. Although it is not possible to distinguish between these classes of segregants at this temperature, previous studies have shown that strains Mc2 and Mc3 express different Y→Mc conversion kinetics when incubated at 39 °C.[28,29] Therefore, the ts segregants of 3u2m-428 and their respective parental strains were phenotypically scored following incubation at the higher temperature. At 39 °C, albino auxotrophs derived from strain Mc2 interrupted Y→Mc conversion and formed only Stage I cells. About 75% of these forms were judged non-viable due to changes in the integrity of the cell wall (Table 1; Fig. 6A). In contrast, albino

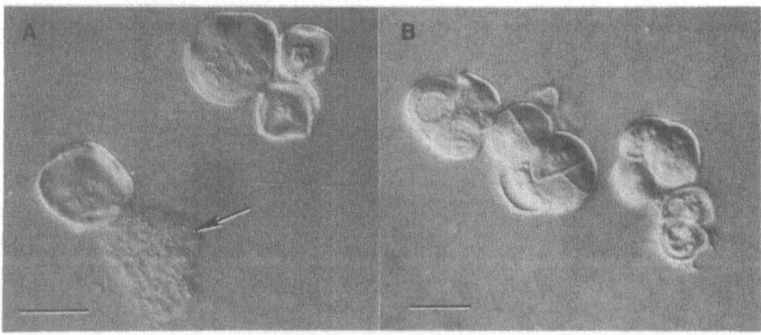

Fig. 6 Cellular development of Mc2-derived (A) and Mc3-derived (B) strains at 39°C. Note that the Mc2-derived cells arrested as Stage I cells. One cell has ruptured and released its cytoplasmic contents (arrow). Mc3-derived cells completed Y→Mc conversion and formed Stage II cells. The scale bars equal 10 μm. (Nomarski optics)

auxotrophs derived from strain Mc3 tended to complete Y→Mc conversion with comparatively higher levels of viability (Fig. 6B). When selected albino ts segregants were similarly treated, three distinct types of conversion kinetics and development were noted. Two were equivalent to those expressed by the parental strains, whereas a third type was somewhat "intermediate". Such a situation would most likely only exist if both *mcm2* and *mcm3* were present in each uninucleate cell of an asynchronously-growing culture of some fusion-product strains. At the time of a shift to the restrictive temperature, some cells of such a yeast culture would be at a cell-cycle stage prior to the Mc2 execution point, whereas others would be at a cell-cycle stage between the Mc2 and Mc3 execution points (Fig. 7). Consequently, some cells then would arrest at Stage I of the Y→Mc conversion process and others would arrest at Stage II of this process, thereby giving the entire culture a mixed appearance of the two parental phenotypes. Such results are consistent with the hypothesis that *mcm2* and *mcm3* are indeed representative of mutations in wild-type genes belonging to separate complementation groups.

Intragenic complementation between *mcm2* and *mcm3* might have restored polarized development in fusion products if these lesions are alleles of the same wild-type gene. There are precedents for intragenic complementation among fusion products (see reference 39). However, the possibility of intragenic complementation between *mcm2* and *mcm3* appeared unlikely. Expression of a ts phenotype in an albino segregant indicated that a recessive *mcm* lesion had been "unmasked" by loss of the dominant wild-type allele. If intragenic complementation accounted for the blastic mode of development among the fusion products, then only Mcm2⁻ or Mcm3⁻ segregants should have occurred.

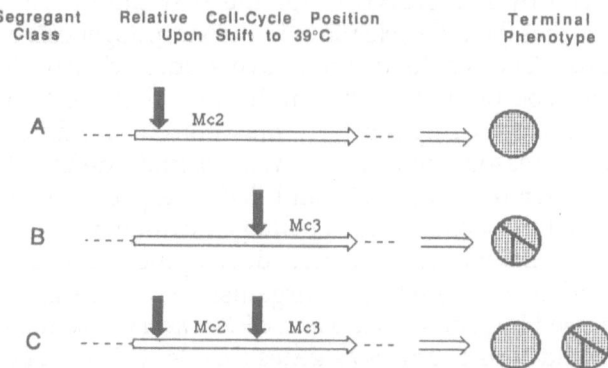

Fig. 7 Diagrammatic representation depicting the effect of the shift to the restrictive temperature of 39 °C upon the terminal phenotypes of different classes of albino ts segregants from fusion product strain 3u2m-428. These classes of segregants presumably possess the lesions (A) *mcm2*, (B) *mcm3*, or (C) both *mcm2* and *mcm3*. Upon shifting an asynchronously-growing culture of class C segregants to the non-permissive temperature of 39 °C (dark arrows), some would be at a cell-cycle position before the Mc2 execution point, whereas some would be at a position between the Mc2 and Mc3 execution points. The former would arrest as Stage I cells and the latter would form Stage II cells, thereby giving the entire culture a "mixed" phenotype. In the other classes of segregants (A and B), all cells would arrest in a single terminal phenotype as either Stage I or Stage II cells, respectively.

Simultaneous expression of both lesions would not be possible because phenotypic expression of one lesion was due to the loss of the other. Thus, if any of the albino ts segregants exhibited a third phenotype different from those expressed by either parental strain, this would likely suggest that restoration of blastic development in the fusion products was not by virtue of intragenic complementation between the *mcm2* and *mcm3* lesions, but rather by complementation of the lesions by wild-type alleles existing at different genetic loci in the opposing fusion partner. Therefore, the observation of a third ts phenotype among the segregants of strain 3u2m-428 does suggest that *mcm2* and *mcm3* represent mutations in non-allelic, wild-type genes governing bud emergence.

Attempts were also made to ascertain the allelic nature of the *mcm2* and *mcm3* lesions by performing genetic backcrosses.[44] Spheroplasts were made from the ts albino segregants of strain 3u2m-428 and fused with spheroplasts derived from the parental albino, auxotrophic Mc strains. Fused spheroplasts from these crosses failed to regenerate under standard selective experimental conditions (minimal medium at 37 °C) as well as conditions devised to encourage the growth of fusion products (e.g., nutritionally supplemented media, incubation at 25 °C, etc.). However, controls consisting of mixtures of spheroplasts derived from the parental albino, auxotrophic Mc strains did fuse and regenerated as dematiaceous colonies growing on minimal medium incubated either under permissive or restrictive temperatures. In addition, other controls consisting of spheroplasts independently produced from individual parental strains and ts albino segregants formed white colonies on appropriately supplemented media. These data suggested that the backcrosses were incompatible perhaps due to the unstable nature of anueploid fusion products, toxic gene dosage effects, or the expression of lethal genes.

IMPLICATIONS FOR FUTURE STUDIES

The establishment of a parasexual cycle in *W. dermatitidis*, in accord with the principles described by Pontecorvo,[43] for the first time permitted the genetic analysis of the role that the cell cycle plays in the morphogenesis of a polymorphic, pathogenic fungus. The results of the above studies clearly demonstrated the usefulness of this type of analysis by implicating that the two morphogenetic lesions in *W. dermatitidis* actually represent mutations in two different cell-division-cycle genes. For this reason, the genes in *W. dermatitidis* defined by the *mcm2* and *mcm3* lesions have been renamed *CDC1* and *CDC2*, respectively. These are the first *CDC* genes identified in a dematiaceous pathogenic fungus.

In addition, parallel modes of cellular development appear to exist between *W. dermatitidis* and the better-studied organism *S. cerevisiae*. Specifically, the phenotypes expressed by *cdc1(=mcm2*) and *cdc2(=mcm3)* resemble those produced by lesions in the wild-type *S. cerevisiae* genes *CDC24*, *CDC42*, and *CDC43*.[26,27] In this fungus, these three genes are responsible for bud site assembly and initiation.[27,46] The gene product of *CDC24* is probably a Ca^{2+}-binding protein bound to the plasma membrane via its interaction with the product of *CDC42*. The latter encodes a GTP-binding protein and is most likely prenylated by the *CDC43* gene product. This prenylation permits the attachment of the *CDC24* gene product to the plasma membrane. At the site of attachment, the cytoskeletal components necessary for the assembly of a new bud are organized in a manner to allow blastic development. Presumably, the dysfunction of any of these three genes does not allow the cell to organize the cytoskeletal matrix essential for bud emergence. The resulting phenotype is a large, swollen and unbudded cell having multiple nuclei, delocalized incorporation of cell-wall polymers, and a dramatically altered wall architecture. The resemblance between these cells and those formed by the Mc strains of *W. dermatitidis* is remarkable. Therefore, future studies directed toward understanding the molecular function of *CDC1* and *CDC2* in *W. dermatitidis* may reveal that the mechanisms responsible for yeast morphogenesis in both these fungi are very similar.

At present, the exact relationship of the *CDC1* and *CDC2* genes to pathogenesis is unclear. Recent studies suggest that the transition of yeasts to the multicellular phenotype has no role in the initiation of infection, but rather permits

W. dermatitidis to survive and persist in the harsh host environment.[45] In these studies, mice infected with either the pigmented or albino (*mel3*) strain of Mc3 responded in a generally similar histopathological manner. Both Mc strains formed the multicellular phenotype in addition to the morphs typically associated with phaeohyphomycosis. However, infection with the pigmented Mc3 strain was always fatal. The pigmented Mc strain was also shown to be as virulent as the wild type strain. In contrast, it took nearly twice as long for the Mel3⁻ derivative of Mc3 to cause death, and then only at the highest inoculum used. In those mice that survived, the fungus continued to persist in various morphologies within infected tissue. The collective results are consistent with previous reports associating melanin with virulence.[14,15] Given the above findings, it is more likely that development of muriform cells *in vivo* is more suited to a role which perpetuates the characteristic chronic and insidious nature of infections caused by dematiaceous fungi. Long-term studies using animal models of infection may help determine if the multicellular phenotype does indeed have a role in pathogenicity.

Finally, similar cell-division-cycle genes must certainly exist in other dematiaceous fungi of greater clinical significance. Experimental protocols like those described above could be readily applied to these pathogens. Such investigations could provide the basis for a general understanding of cellular development in the Dematiaceae. In turn, this may have broad implications in the development of methods for the control and treatment of mycotic diseases caused by dematiaceous fungi.

ACKNOWLEDGEMENTS

The continuing encouragement, support, and patience of Paul J. Szansizlo is deeply appreciated. The author developed the parasexual system for genetic analysis of *W. dermatitidis* in his laboratory at the University of Texas at Austin. Financial support for this work was provided by grants to P.J.S. from the University Research Institute of the University of Texas at Austin and from the Advanced Technology Program of the State of Texas. Also, the effort of Ira F. Salkin in reviewing this manuscript and providing helpful criticism is gratefully acknowledged.

REFERENCES

1. M. R. McGinnis, "Laboratory Handbook of Medical Mycology", Academic Press, New York (1980).
2. J. W. Rippon, "Medical Mycology: The Pathogenic Fungi and the Pathogenic Actinomycetes", 3rd ed., W. B. Saunders, Philadelphia (1988).
3. K. J. Kwon-Chung and J. E. Bennett, "Medical Mycology", Lea and Febiger, Philadelphia (1992).
4. J. M. B. Smith, The pathogenesis of opportunistic mycoses in man, *Microbiol. Sci.* 3: 122 (1986).
5. T. J. Walsh and P. A. Pizzo, Nosocomial fungal infections: a classification for hospital-acquired fungal infections and mycoses arising from endogenous flora or reactivation, *Ann. Rev. Microbiol.* 42: 517 (1988).
6. G. C. Ainsworth, Pathogenicity of fungi in man and animals, in : "Microbial Pathogenicity", J. W. Howie and A. J. O'Hea, eds., Cambridge University Press, Cambridge (1955).
7. D. H. Howard, The morphogenesis of the parasitic form of the dimorphic fungi, *Mycopathol. Mycol. Appl.* 18: 127 (1962).
8. F. Mariat, Saprophytic and parasitic morphology of pathogenic fungi, in : "Microbial Behavior, In Vivo and In Vitro", H. Smith, ed., Cambridge University Press, Cambridge (1964).
9. J. W. Rippon, Dimorphism in pathogenic fungi, *Crit. Rev. Microbiol.* 8:49 (1980).

10. A. H. Romano, Dimorphism, in : "The Fungi: An Advanced Treatise", Vol. 2, G. C. Ainsworth and A. S. Sussman, eds., Academic Press, New York (1966).

11. P. R. Stewart and P. J. Rogers, Fungal dimorphism: a particular expression of cell wall morphogenesis, in : "The Filamentous Fungi", vol. 3, J. E. Smith and D. R. Berry , eds., John Wiley and Sons, New York (1978).

12. P. J. Szaniszlo, An introduction to dimorphism among zoopathogenic fungi, in : "Fungal Dimorphism: With Emphasis on Fungi Pathogenic for Humans", P. J. Szaniszlo, ed., Plenum Press, New York (1985).

13. B. E. Taylor, M. H. Wheeler and P. J. Szaniszlo, Evidence for pentaketide melanin biosynthesis in dematiaceous human pathogenic fungi, *Mycologia* 79: 320 (1987).

14. M. H. Wheeler and A. A. Bell, Melanins and their importance in pathogenic fungi, *Cur. Topics Med. Mycol.* 2: 338 (1987).

15. D. M. Dixon, P. J. Szaniszlo and A. Polak, Dihydroxynaphthalene (DHN) melanin and its relationship with virulence in the early stages of phaeohyphomycosis, in : "The Fungal Spore and Disease Initiation in Plants and Animals", G. T. Cole and H. C. Hoch, eds., Plenum Press, New York (1991).

16. M. R. McGinnis, Chromoblastomycosis and phaeohyphomycosis: New concepts, diagnosis, and mycology, *J. Amer. Acad. Dermatol.* 8: 1 (1983).

17. P. J. Szaniszlo, C. W. Jacobs, and P. A. Geis, Dimorphism: morphological and biochemical aspects, in : "Fungi Pathogenic for Humans and Animals", part A (biology), D. Howard, ed., Dekker, New York (1983).

18. B. H. Cooper, *Phialophora verrucosa* and other chromoblastomycotic fungi, in : "Fungal Dimorphism: With Emphasis on Fungi Pathogenic for Humans", P. J. Szaniszlo, ed., Plenum Press, New York (1985).

19. T. Matsumoto, T., A. A. Padhye, L. Ajello and P. G. Standard, Critical review of human isolates of *Wangiella dermatitidis, Mycologia* 76: 232 (1984).

20. W. Naka, T. Harada, T. Nishikawa and R. Fukushiro, A case of chromoblastomycosis: with special reference to the mycology of the isolated *Exophiala jeanselmei, Mykosen* 29: 445 (1986).

21. P. J. Szaniszlo, P. H. Hsieh and J. D. Marlowe, Induction and ultrastructure of the multicellular (sclerotic) morphology in *Phialophora dermatitidis, Mycologia* 68: 117 (1976).

22. P. H. Hsieh, "Growth and Developmental Characteristics in *Phialophora dermatitidis* and Other Chromoblastomycotic Fungi", M. A. Thesis, Univ. of Texas, Austin (1974).

23. J. D. Marlowe, "The Development of the Sclerotic Cell in Agents of Chromoblastomycosis: An Ultrastructural Study", M. A. Thesis, Univ. of Texas, Austin (1977).

24. P. J. Szaniszlo, P. A. Geis, C. W. Jacobs, C. R. Cooper, Jr. and J. L. Harris, Cell-wall changes associated with yeast-to-multicellular-form conversion in *Wangiella dermatitidis*, in : "Microbiology-1983", D. Schlessinger, ed., American Society for Microbiology, Washington, D.C. (1983).

25. P. A. Geis and C. W. Jacobs, Polymorphism of *Wangiella dermatitidis*, in : "Fungal Dimorphism: With Emphasis on Fungi Pathogenic for Humans", P. J. Szaniszlo, ed., Plenum Press, New York (1985).

26. J. R. Pringle and L. H. Hartwell, The *Saccharomyces cerevisiae* cell cycle, in : "The Molecular Biology of the Yeast *Saccharomyces*: Life Cycle and Inheritance", J. N. Strathern, E. W. Jones and J. R. Broach, eds., Cold Spring Harbor Monographs, Cold Spring Harbor, New York (1981).

27. K. Madden, C. Costigan and M. Snyder, Cell polarity and morphogenesis in *Saccharomyces cerevisiae, Trends Cell Biol.* 2: 22 (1992).

28. R. L. Roberts and P. J. Szaniszlo, Temperature-sensitive multicellular mutants of *Wangiella dermatitidis, J. Bacteriol.* 135: 622 (1978).

29. R. L. Roberts, R. J. Lo and P. J. Szaniszlo, Nuclear division in temperature-sensitive multicellular mutants of *Wangiella dermatitidis, J. Bacteriol.* 137: 1456 (1979).

30. P. A. Geis, "Chemical Composition of the Yeast and Sclerotic Cell Walls of *Wangiella dermatitidis*", Ph. D. Dissertation, Univ. of Texas, Austin (1981).

31. J. L. Harris and P. J. Szaniszlo, Localization of chitin in walls of *Wangiella dermatitidis* using colloidal gold-labeled chitinase, *Mycologia* 78: 853 (1986).

32. C. R. Cooper, Jr., J. L. Harris, C. W. Jacobs and P. J. Szaniszlo, Effects of polyoxin AL on cellular development in *Wangiella dermatitidis, Exper. Mycol.* 8: 349 (1984).

33. C. R. Cooper, Jr., "Localization of Cell-Wall Components During Vegetative Morphogenesis in *Wangiella dermatitidis*", M. A. Thesis, Univ. of Texas, Austin (1983).

34. C. W. Jacobs, R. L. Roberts, and P. J. Szaniszlo, Reversal of multicellular-form development in a conditional morphological mutant of the fungus *Wangiella dermatitidis, J. Gen. Microbiol.* 131: 1719 (1985).

35. R. L. Roberts and P. J. Szaniszlo, Yeast-phase cell cycle of the polymorphic fungus *Wangiella dermatitidis, J. Bacteriol.* 144: 721 (1980).

36. C. W. Jacobs and P. J. Szaniszlo, Microtubule function and its relation to cellular development and the yeast cell cycle in *Wangiella dermatitidis, Arch. Microbiol.* 133: 155 (1982).

37. K. J. Kwon-Chung, Genetics of fungi pathogenic for man, *CRC Crit. Rev. Microbiol.* 3: 115 (1974).

38. W. L. Whelan, The genetics of medically important fungi, *CRC Crit. Rev. Microbiol.* 14: 99 (1987).

39. R. T. M. Poulter, Parasexual genetics of *Candida albicans*, in : "*Candida albicans*: Cellular and Molecular Biology", R. Prasad, ed., Springer-Verlag, Berlin (1991).

40. C. R. Cooper, Jr., "Induced Parasexual Genetic Analysis of the Imperfect, Pathogenic Fungus *Wangiella dermatitidis*: Identification of Selected Morphogenic and Biosynthetic Complementation Groups", Ph. D. Dissertation, Univ. of Texas, Austin (1989).

41. J. S. Wood, Genetic effects of methyl benzimidazole-2-yl-carbamate on *Saccharomyces cerevisiae, Mol. Cell. Biol.* 2: 1064 (1982).

42. J. S. Wood, Mitotic chromosome loss induced by methyl benzimidazole-2-yl-carbamate as a rapid mapping method in *Saccharomyces cerevisiae, Mol. Cell. Biol.* 2: 1080 (1982).

43. G. Pontecorvo, The parasexual cycle in fungi, *Ann. Rev. Microbiol.* 10: 393 (1956).

44. C. R. Cooper, Jr., unpublished data.

45. D. M. Dixon, J. Migliozzi, C. R. Cooper, Jr., O. Solis, B. Breslin, and P. J. Szaniszlo, Melanized and non-melanized multicellular form mutants of *Wangiella dermatitidis* in mice: mortality and histopathology studies, *Mycoses* 35: 17 (1992).

46. D. G. Drubin, Development of cell polarity in budding yeast, *Cell* 65: 1093 (1991).

Cell Biology and Biochemistry

MORPHOLOGICAL ASPECTS OF FUNGAL DIMORPHISM

Marcel Borgers

Dept of Morphology, Life Sciences
Janssen Research Foundation
2340 Beerse, Belgium

ABSTRACT

From a purely structural point of view, the term dimorphism is very restrictive for the various fungi that are dealt with in this symposium. Polymorphism would be more accurate. Ultrastructural features of the following fungal organisms are presented : *Candida albicans, Pityrosporum ovale, Histoplasma capsulatum, Paracoccidioides brasiliensis , Sporothrix schenckii* and *Coccidioides immitis*. The detailed ultrastructure of these fungi is documented by transmission/scanning electron microscopy, with standard preparation procedures and auxiliary techniques, such as enzyme cytochemistry and immunocytochemistry.

This multidisciplinary approach to morphology aims at describing differences in the number and composition of subcellular organelles between the various fungi and between different morphogenetic forms of the same organism. Particular attention is given to several constituents, namely the cell wall, plasmalemma and its derived structures, nucleus, central vacuolar system, mitochondria, peroxisomes, and cytosol. Environmental *(in vivo)* and culture *(in vitro)* conditions as determinants for morphology and morphogenetic transition include growth inhibitory interventions, interactions with host defence cells, composition of nutrients, temperature, pH, etc.

The relation between pathogenicity and morphologic adaptation of fungi is intriguing. The structural changes that accompany the expression of invasiveness or the inhibition thereof are seen primarily at the level of the cell periphery and consist of alterations in the density and thickness of the cell wall, altered patterns of the plasmalemma-cytoskeletal complex, changes in subplasmalemmal membranes, and differences in the directional movements of subcellular organelles.

Although an important number of morphological aspects of polymorphism have been identified, many remain enigmatic. With the advent of new morphologically oriented, molecular biological markers, these aspects can be tackled more adequately in future research.

INTRODUCTION

Descriptive fungal cytology has, for many years, struggled with the problem of adequate permeation of chemical fixatives through the cell wall and plasmalemma. This is especially true for species such as *Candida albicans* and *Pityrosporum ovale*.

The slow permeation of chemical fixatives into the cytoplasm has posed a serious problem for the morphological identification of internal organelles of these species. Potassium permanganate, a commonly employed fixative, revealed the membranous components fairly well but failed to display ribosomes and the various nuclear and nucleolar substructures.[1,2] Moreover, this fixative cannot be used for the preservation of cells for enzyme cytochemistry. With a modification of the conventional preparation procedures,[3] this has been largely solved by glutaraldehyde fixation, which makes all subcellular organelles fairly visible and yet enzymes retain their activity.[4,5] The detailed substructure of the cell walls are also well displayed with this methodology. A more recently developed approach, microwave-assisted fixation of fungi, has proved to have enormous advantages over classical fixation because of its rapid immobilization of structures and enhanced permeation of fixatives. The changes in distribution, size and shape of subcellular organelles that accompany morphologic transition in several dimorphic species have been described. In addition, marker enzymes for the various subcellular organelles have been identified, which may aid the understanding of inter-organelle relationships that are not obvious from morphological examination alone.

The species and their different morphological forms that are dealt with in this paper are *C. albicans*, *P. ovale*, *Coccidioides immitis*, *Paracoccidioides brasiliensis* and *Sporothrix schenckii*.

CANDIDA ALBICANS

C. albicans is polymorphic and grows as a multipolarly budding yeast, as a pseudohypha of elongated yeast-like cells, and as a true septate hypha (Fig. 1). Both the true hypha and pseudohypha may regenerate the yeast phase by producing clusters of yeast-like blastoconidia. Morphogenetic and ultrastructural aspects have been reported extensively.[1,2,6-10]

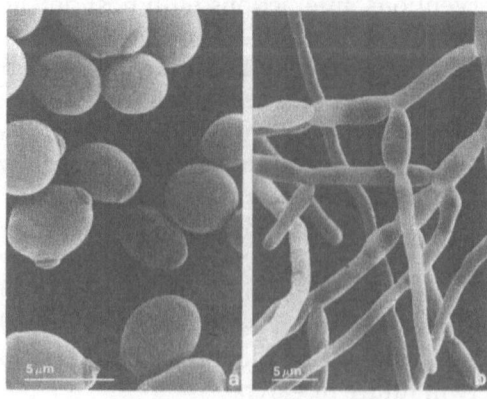

Fig. 1 *C. albicans.* SEM pictures of surface morphology of yeast (a) and branching mycelial (b) phase cells.

Yeast cells of *C.albicans* are round to oval, measure 2-5 µm in diameter, and are capable of multipolar budding. The yeast cell wall is a laminated structure from 100 to 300 nm thick consisting of 5-8 layers of different density, depending on the culture conditions and the fixation procedure used.[1,9] The innermost portion of the wall contains predominantly chitin and protein. Mannans and glucans span throughout most of the wall structure, and the outermost fibrillar layers appear to contain mannans and proteins.

Yeast cells have ovoid nuclei, round to elongate mitochondria containing linearly arranged cristae, ribosomes, sparse profiles of endoplasmic reticulum often connected with the nuclear membrane, and a vacuolar system (Figure 2a).

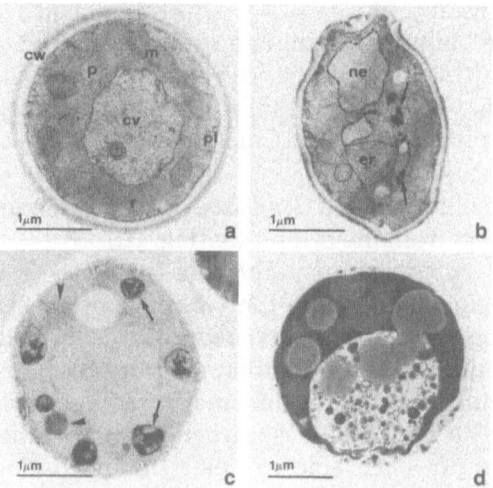

Fig. 2 *C albicans*. TEM of yeast phase cells.

a. Ultrastructural detail of subcellular organelles (m : mitochondria; p : peroxisomes; cw: multilayered cell wall; pl undilated plasma membrane: r: ribosomes; cv: central vacuole, positive for acid phosphatase activity).
b Intracellular membranous compartments such as nuclear envelope (n e), endoplasmic reticulum (e r) and collapsed parts of the vacuolar apparatus (arrows) are stained for alkaline phosphatase activity
c. Mitochondria (arrows) are reactive for cytochrome peroxidase, and peroxisomes (arrowheads) for catalase.
d Transition of cytoplasmic lipid droplets into the central vacuolar system

The cytoplasm is surrounded by an undulating cell membrane, which possesses tubular invaginations called lomasomes. The latter may perform secretory functions in the formation or incorporation of cell wall materials such as mannoproteins. Since an organized Golgi apparatus is not observed in *C. albicans*, it is possible that lomasomes in association with the endoplasmic reticulum might in some manner be used by these organisms as a functional equivalent for the Golgi apparatus.[1] When membrane structures are assayed for marker enzymes such as alkaline phosphatase, not only the endoplasmic reticulum and the nuclear envelope display the enzyme activity but also the vacuolar apparatus possesses this enzyme This fact possibly points to a relationship between these organelles not previously considered (Fig 2b).

Vacuoles are intracytoplasmic structures found in all pathogenic fungi. They can vary in size and shape and are common in the subapical regions of most of the pathogenic fungi. They are especially numerous in older cells. In C. *albicans* yeast cells, there is generally a large vacuolar apparatus, although several smaller units can occur. Their enzymic contents comprise acid-, neutral- and alkaline monophosphatases (Fig. 2a), neutral di- and polyphosphatases, and other proteolytic and lipolytic enzymes. Hence, the vacuolar system may be considered as a lysosomal apparatus.

Mitochondria appear as ubiquitous intracytoplasmic organelles throughout the pathogenic fungi. They can vary considerably in their size and shape. They may be spherical to rod-shaped or thread-like in outline and measure about 0.3-1.0 μm in diameter and up to 3.0 μm in length. The relative numbers and morphology of fungal mitochondria can vary with the age of the fungal cell or with the stage of its life cycle. Unlike mammalian cell mitochondria, C. *albicans* mitochondria contain, besides cytochrome oxidase also cytochrome peroxidase and NADH oxidase activities[5] (Fig. 2c).

Peroxisomes are bound by a single membrane and are approximately 0.5-1.2 μm in diameter. They may show spherical, elongate, or irregularly shaped profiles in thin section, and have a granular to fibrillar matrix of variable electron opacity. Nucleoids may occur in their matrix. Catalase can be used as a marker enzyme for C. *albicans* peroxisomes.[5]

Lipid bodies are ubiquitous in C. *albicans*, and are frequently seen entering the central vacuolar system from the cytoplasm (Fig. 2d).

The remaining cytosol is usually filled with free ribosomes (Fig. 2a). Techniques that specifically enhance the visualisation of glycogen reveal it to be prominent. Most often, glycogen is clustered towards the cell periphery.

Although microtubules and microfilaments are recognised as components of virtually all eukaryotic cells, little information is available on the occurrence, characteristics, and distribution of these structures among the pathogenic fungi. In C. *albicans*, we have not been able to detect microtubules. The sensitivity of microtubules to the preparatory procedures to preserve the yeasts might be responsible for the absence of microtubules in electronmicroscopic sections.

In budding cells, new cell material evaginates from a polar site and enlarges to a diameter less than that of the parent cell. The nucleus migrates into the neck region between parent and bud, divides, and a septum is formed. Mitochondria and membranous strands are transported from the parent to the daughter compartment. The septum is laid down centripetally in two stages: an electron-transparent, primary septum is followed by secondary deposition of material of higher electron density on each side of the primary septum. After separation, the primary and secondary septa remain as part of the mother cell wall and form a bud scar, and the secondary septum alone forms the birth scar of the daughter cell. For details of the morphology of phase transitions, excellent reviews are given by Garrison,[1] Szanislo *et al.*[2] and Odds.[9]

Mycelium

During hyphal development, new cell material grows in an elongated form, the nucleus migrates into the neck of the germ tube, and a septum is formed in two stages as in the yeast form. A transit of all organelles into the germ tube can be observed. Mitochondria elongate and sometimes reach a length of 10 μm (Fig. 3a).

Germ tube formation from yeast cells is accompanied by substantial rearrangement of the cell wall layers. The common finding is that the cell wall of germ tubes is thinner than that of parent yeast cells. This is also obvious under *in vivo* circumstances, and thus a yeast can be distinguished from a transverse-cut mycelium in an electronmicroscopic section.

The pseudomycelial morphology of *C. albicans* consists of elongated yeasts possessing complete septa (Fig. 3b). Individual cells are from 2 to 5 μm in diameter and 26 to 30 μm in length.

True mycelial cells of *C. albicans* are usually longer and thinner than yeast cells and measure from 0.8 to 1.3 μm in diameter. A perforated septum 0.17-0.22 μm wide, with no evidence of Woronin bodies or septal plugs, separates the hyphal cells. The hyphal cell plasma membrane is significantly different from the yeast cell membrane and exhibits invaginations that are deeper and much more frequent. In older cells, the vacuolar system is highly developed and may comprise more than 50% of the cell volume.

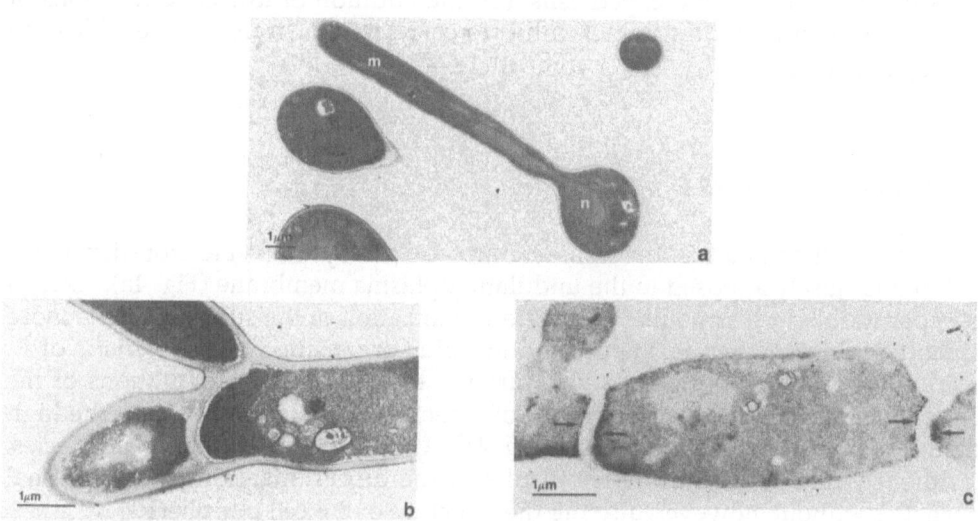

Fig. 3 *C. albicans*. TEM of yeast-mycelium transitions.
a. Germ tube formation showing the difference in cell wall thickness between the yeast and the emerging germ tube. Nucleus (n) is transiting to the germ tube and an elongated mitochondria (m) of several micrometers long is seen in the germ tube.
b. Detail of a branching mycelium.
c. Adenosine triphosphatase activity is localized exclusively at the septa (arrows).

Under some conditions the hyphae or pseudohyphae of *C. albicans* produce chlamydospores. These are thick-walled, nondeciduous, intercalary or terminal, asexual spores. Chlamydospores have a double- or triple-layered cell wall, which is much thicker than that of the yeast or mycelial cell and which is continuous, in part, with that of the "suspensor cell".

A number of macromolecular and enzymic differences exist among the various morphologies of *C. albicans* . Adenosine tri-, di-, and monophosphatases differ in concentrations between yeast and mycelial forms and redistribution or focal activation of these enzymes at sites of bud and/or germ tube formation takes place (Fig. 3c).

Acid and alkaline phosphatase are demonstrable cytochemically in both the yeast and mycelial morphologies. Phospholipase is an intracellular enzyme whose activity *in vivo* is suggested to be associated with pathogenicity and which, together with acid phosphatase and glucosaminidase, is concentrated at the bud site of yeasts and the apical tips of germ tubes and pseudomycelia, but is found randomly distributed at the cell membrane of nondividing yeasts.

In general, the enzymic differences between the yeast and mycelial cells are more quantitative rather than qualitative.

Influencing yeast-mycelium transition may have the following important therapeutic consequences. Since invasion and often symptomatology depend critically on the development of the hyphal form, the inhibition of further outgrowth of this particular form by antifungal treatment leads to eradication of the infection.[11] Moreover, the cooperative action of systemically absorbed antifungals with the natural host defense system (polymorphonuclear leukocytes) may be a relevant factor leading to clearance of infection. This is due to the fact that: *(1)* leukocytes avidly engulf small yeast-form cells but are limited in their capacity to destroy them; *(2)* yeast cells that are engulfed but not immediately killed produce germ tubes, thereby escaping destruction *(3)* leukocytes are rather defenceless against these escaped cells; *(4)* the addition of low concentrations of antifungals such as azoles, which inhibit yeast-mycelial transition, enables the leukocytes to eradicate *C. albicans* completely.[12]

PITYROSPORUM OVALE

P.ovale cells in culture are characterized by a very thick, electron-dense cell wall that is closely apposed to the undulating plasma membrane (Fig. 4a). Due to its impermeable wall structure, adequate visualization of the substructure is more difficult in *P. ovale* than in the other dimorphic organisms. Characteristic of *P. ovale's* ultrastructure, besides its wall, are the highly regular indentations of the plasma membrane, the virtual absence of a central vacuolar system and the high amount of lipid bodies in the cytosol (Fig. 4a). Unlike the other dimorphic species, treatment with azole- antifungals results in the disorganization of the internal organelles without, however, altering the structure of the cell periphery.

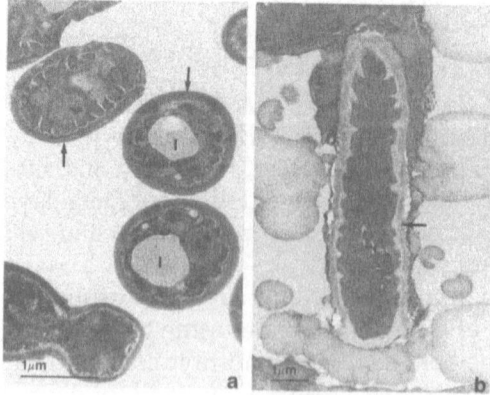

Fig. 4 *P. ovale.* TEM pictures of yeast form cells in culture and b) a hyphal form penetrating a stratum corneum cell of a patient with pityriasis versicolor. Note the thick wall (arrows) and typical lipid droplets (l) in the cytoplasm.

Faergemann[13,14] introduced culture conditions that induced yeast-hyphae phase transition in *P. ovale*. These cells resembled the short hyphae found in patients with pityriasis versicolor. Again, as in candidosis, invasion of squamous cells in pityriasis versicolor was seen to be by hyphae, not by yeast-form cells.[15,16] A characteristic cell wall, very similar to that of the yeasts as far as thickness and density is concerned, is present in the invading hyphae (Fig. 4c). To my knowledge, marker enzymes for different subcellular organelles have not been characterized at the ultrastructural level for this species.

COCCIDIOIDES IMMITIS

Coccidiodes immitis exhibits a unique dimorphism in the host, characterized by the transition of arthroconidia to spherules. The arthroconidia are derived by disarticulation of a saprophytic mycelial phase consisting of septate hyphae. The arthroconidia are usually barrel-shaped, measuring 2.5-4 by 3-6 µm. The spherules, which become as large as 80 µm or more in diameter, develop by the isotropic enlargement of the arthroconidia and give rise to a self-perpetuating spherule-endospore phase. Reproductive growth of the spherule-endospore phase in the host or *in vitro* is via liberation of endospores followed by their development into new endosporulationg spherules.

The cell wall composition of the morphological forms of *Co. immitis* indicates the same qualititative composition for spherules as for mycelial walls. Chitin has been identified in and is critical to the structural integrity of the spherule wall. Detailed ultrastructure is given in the work of Sun et al.[17,18]

Most arthroconidia contain two or three nuclei, but may have as few as one or as many as eight nuclei. At first the cells of the spherule are multinucleate, but as cleavage continues, the number of nuclei per cell becomes diminished until each endospore has one or, rarely, two nuclei. Upon maturation the spherule walls begin to rupture and endospores are released.

Electron microscope studies demonstrate that the spherule wall is approximately 500 nm or more in thickness and is composed of at least two layers. Upon transition of spherules to mycelium, the outer electron-dense layer of the wall ruptures, and the inner, less electron-dense portion becomes continuous with the wall of the germ tube. The new germ tube wall is thin approximately 150 nm in thickness. Simple septa, septal pores, Woronin bodies, and septal pore plugs are characteristic of *Co. immitis* mycelium. As in *C. albicans*, ultrastructural changes after azole treatment concern primarily the cell wall, cell membranes and the central vacuolar system.[19]

PARACOCCIDIOIDES BRASILIENSIS

This dimorphic fungus has both a saprophytic morphology, which is filamentous, and a parasitic morphology, which is that of a multiple-budding yeast.

Blastoconidia have a diameter of 2 to 3 µm and are directly associated with hyphae or borne singly on the tips of short conidiophores. Chlamydospores, ranging in size from 6.5 to 30 µm in diameter, are found on the hyphae.[1,2]

Hyphae have cell dimensions of 0.8-2.4 µm by 6.4-29 µm. These cells are multinucleate and possess mesosomes, lomasomes and numerous mitochondria whose long axes are parallel to those of the hyphal unit. Very often, the cytosol is filled with lipid globules and concentric lamellae or myelin-like structures (Fig. 5a). The cell walls of the hyphae are from 80 to 150 nm in width and are composed of from one to three discernible layers. The outermost layer is electron dense and displays finely fibrillar material.

The parasitic morphology is that of a budding yeast, and can be of two types. The first involves the production of narrow-necked buds, 2-10 μm in diameter whereas the second involves "cryptosporulation". In cryptosporulation, the chromatin is divided into chromidia, which function as potential spores. The chromidia become associated with the cell membrane and form numerous, small buds, which encircle the mother cell (Fig. 5b). The cell wall is 200 - 600 nm thick and comprises from one to three layers.

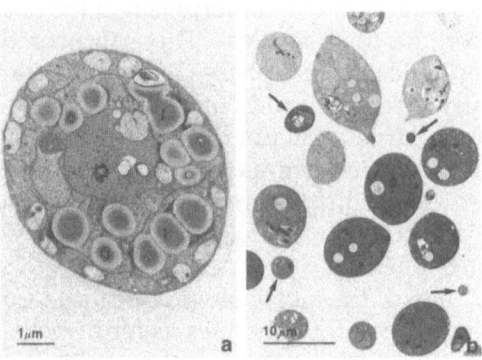

Fig. 5 *P. brasiliensis.* TEM pictures of cultured yeast cells showing the polymorphism of cells in culture. a) A large yeast cell containing numerous lipid inclusions (l) and b) survey of large and small elements (arrows) probably representing cross-sectioned buds.

The production of elongated buds with diameters greater than those of the hyphae indicates a yeast-mycelium transition. These multinucleate, transitional cells have thinner cell walls than do mother yeast cells, and also have perforated septa similar to those found in hyphae.

In *P. brasiliensis* the structural changes elicited during exposure to antifungal treatment differ greatly according to the morphogenetic phase of the organism used.[20] Most sensitive to azoles were mycelium-to-yeast cultures, in which, for example, 10^{-10} M itraconazole completely prevented the development of the yeast form and induced complete necrosis of the mycelial cells. In yeast-to-yeast cultures, the majority of cells became fully necrotic after one day of exposure to 10^{-7} M itraconazole. In yeast-to-mycelium cultures, itraconazole inhibited transformation into the mycelial form at a concentration of 10^{-6} M.[21]

SPOROTHRIX SCHENCKII

S. schenckii, the agent of sporotrichosis, exhibits yeast-hyphal dimorphism. Hyphae of *S. schenckii* are 1.5-3 μm in diameter, septate, and granular in appearance. Hyphal cell walls are thin (80-140 nm thick) and under the electron microscope appear to be bilayered. Pyriform to ovoid conidia arise laterally on the hyphae or from the conidiophore regions.

The yeast phase of *S. schenckii* ranges in shape from fusiform, to spherical, but no cytoplasmic differences correlate with the different cell shapes. The yeast wall is bilayered and consists of an inner, electron-transparent layer and an outer, electron-dense layer.

Osmiophilic microfibrillar material is apparent at the outermost cell wall surface of yeast-like cells of *S. schenckii*, but is much less apparent or even lacking at the outer cell walls of hyphal cells of the corresponding mycelial phase.[1]

Subcellular organelles of the cytoplasm are similar to those in other dimorphic organisms and unlike in *C. albicans*, there have been no attempts to study enzyme distributions in this species.

REFERENCES

1. R.G. Garrison, Ultrastructural cytology of pathogenic fungi, in :" Fungi Pathogenic for Humans and Animals", D.H. Howard, ed., Marcel Dekker Inc., New York (1983).
2. P.J. Szaniszlo, C.W. Jacobs and P.A. Geis, Dimorphism: morphological and biochemical aspects, in : "Fungi Pathogenic for Humans and Animals", D.H. Howard, ed., Marcel Dekker Inc., New York (1983).
3. M. Borgers and S. De Nollin, The preservation of subcellular organelles of *Candida albicans* using conventional fixatives, *J. Cell. Biol.* 62: 574 (1974).
4. S. De Nollin, F. Thoné and M. Borgers, Enzyme Cytochemistry of *Candida albicans*, *J. Histochem. Cytochem.* 23: 758 (1975).
5. S. De Nollin and M. Borgers, An ultrastructural and cytochemical study of *Candida albicans* after in vitro treatment with imidazoles, *Mykosen* 19: 317 (1976).
6. D. Poulain, V. Hopwood and A. Vernes, Antigenic variability of *Candida albicans*, *Crit. Rev. Microbiol.* 12: 223 (1985).
7. F.M. Harold, To shape a cell : an inquiry into the causes of morphogenesis of microorganisms, *Crit. Microbiol. Rev.* 54: 381 (1990).
8. M. Hrmovà and L. Drobnica, Induction of mycelial type of development in *Candida albicans* by low glucose concentration, *Mycopathol.* 76: 83 (1981).
9. F.C. Odds, Morphogenesis in *Candida albicans*, *Crit. Rev. Microbiol.* 12: 45 (1985).
10. M. Borgers, Ultrastructural correlates of antimycotic treatment, in : "Current Topics in Medical Mycology", M.R. McGinnis, ed., Springer Verlag, New York, (1988).
11. M. Borgers and M-A. Van de Ven, Degenerative changes in fungi after itraconazole treatment, *Rev. Infect. Dis.* 9: S33 (1987).
12. M. De Brabander, F. Aerts, J. Van Cutsem, H. Vanden Bossche and M. Borgers, The activity of ketoconazole in mixed cultures of leukocytes and *Candida albicans*, *Sabouraudia* 18: 197 (1980).
13. J. Faergemann, A new model for growth and filament production of *Pityrosporum ovale* (*orbiculare*) on human stratum corneum *in vitro*, *J. Invest. Dermatol.* 92: 117 (1989).
14. J. Faergemann and M. Borgers, The effect of ketoconazole and itraconazole on the filamentous form of *Pityrosporum ovale*, *Acta Derm Venereol* (Stockh.) 70: 172 (1990).
15. M. Borgers, G. Cauwenbergh, M-A. Van de Ven, A. del Palacio Hernanz and H. Degreef, Pityriasis versicolor and *Pitorosporum ovale*, Morphological and ultrastructural considerations, *Int. J. Dermatol.* 26: 586 (1987).
16. A. del Palacio-Hernanz, J. Guarro-Artigas, M.J. Figueras-Salvat, J. Esteban-Moreno and S. Lopez-Gomez, Changes in fungal ultrastructure after short-course ciclopiroxolamine therapy in pityriasis versicolor, *Clin. & Exp. Dermatol.* 15: 95 (1990).
17. S.H. Sun and M. Huppert, A cytological study of morphogenesis in *Coccidioides immitis*, *Sabouraudia* 14: 184 (1976).
18. S. Sun, S.S. Sekhon and M. Huppert, Electron microscopic studies of saprobic and parasitic forms of *Coccidioides immitis*, *Sabouraudia* 17: 265 (1979).
19. M. Borgers, H.B. Levine and J.M. Cobb, Ultrastructure of *Coccidioides immitis* after exposure to the imidazole antifungals miconazole and ketoconazole, *Sabouraudia* 19: 27 (1981).
20. M.B. Negroni de Bonvehi, M. Borgers and R. Negroni, Ultrastructural changes produced by ketoconazole in the yeast-like phase of *Paracoccidioides brasiliensis* and *Histoplasma capsulatum*, *Mycopathologia* 74: 113 (1981).
21. M. Borgers, Changes in fungal ultrastructure after itraconazole treatment, in : "Recent Trends in the Discovery, Development and Evaluation of Antifungal Drugs", R.A. Fromtling, ed., J.R. Prous Science Publ., Barcelona (1987).

MATHEMATICAL ANALYSIS OF THE CELLULAR BASIS OF FUNGAL DIMORPHISM

Salomon Bartnicki-Garcia,[1] and Gerhard Gierz[2]

[1] Department of Plant Pathology
[2] Department of Mathematics
   University of California
   Riverside, California, 92521, USA

ABSTRACT

The hyphoid equation $y = x \cot (xV/N)$ provides a mathematical foundation for the cellular basis for dimorphism. Its parameters, N and V, define two morphogenetically critical factors: the amount of wall-building vesicles produced per unit time and the rate of advance of the VSC (vesicle supply center). The V/N ratio describes the essence of vesicle dynamics of the cell. High values determine a hyphal morphology; low values, a yeast cell morphology. Manipulation of this ratio by computer graphics produces a full spectrum of morphologies seen in dimorphic fungi: from hyphal tubes to budding yeast cells plus other variations including pseudomycelium formation and triangular cell morphogenesis. In this manner, the specific dimorphic features exhibited by fungi of the genera *Mucor*, *Candida*, *Saccharomyces*, and *Trigonopsis* have been duplicated on the computer screen. The ultimate validity of the present mathematical analysis of dimorphism, which is centered mainly on the rate of displacement of the VSC, depends on the existence of a VSC in living fungal cells. There is good circumstantial evidence to believe that the Spitzenkörper found in the growing hyphal tips of higher fungi functions as a VSC, and to extrapolate that other similar vesicle accumulations seen in regions of wall formation, represent VSCs. Accordingly, one may predict that the key molecular event(s) regulating dimorphism would be those responsible for moving the VSC.

PROLOGUE

Louis Pasteur would have been delighted with this large international rendezvous of students of fungal dimorphism. The great French microbiologist was one of the very first scientists to be captivated by the morphological duality of fungi.[1] In the 1870's, Pasteur labored diligently to elucidate the physiological reasons for the dimorphism exhibited by *Mucor mucedo*. With much foresight, he saw in dimorphism an opportune phenomenon to correlate morphology and physiology, and used it as a supporting pillar for his theories on fermentation. Pasteur, who valued highly both basic and applied science, would have appreciated the bringing together of a fine selection of students of fungal dimorphism from the medical, industrial and basic science communities. We are today comparing our progress in the field of dimorphism, fully cognizant that the exact explanation of the phenomenon that intrigued Pasteur so much, so long ago, has yet to be found.

INTRODUCTION

The mycelial-yeast dimorphism exhibited by diverse fungi offers a classic example of the morphological plasticity in fungi.[2,3] Whether a fungus grows in the yeast form (produces spheroidal-ellipsoidal cells by budding) or whether it grows as a mycelium (produces a system of branched long tubular cells or hyphae) has been the subject of numerous studies. This article is an attempt to develop a unified view of dimorphism in fungi based on our mathematical model of morphogenesis which places the hyphoid equation as the mathematical foundation of the cellular basis of morphological development. This is also a platform for making predictions on the likely molecular events underlying morphogenesis, dimorphism in this case. This article is a further elaboration of views presented earlier.[4]

The Dual Simplicity and Complexity of Dimorphism.

Whereas other authors in this book have indicated their concern that fungal dimorphism appears more complicated than previously believed, we submit that dimorphism can be much simpler than previously conceived. Both standpoints are defensible. There is no question that our knowledge of dimorphic fungi has increased vastly, particularly in the fine details about the biochemistry, physiology, and cytology of these fungi; to this is added the recent impetus to unravel their genetics with the tools of molecular biology. In such an arena, the outlook is indeed one of formidable growing complexity. On the other hand, our recent computer-based reconstruction of fungal morphogenesis[5-7] has given us a new insight into the possible mechanism for generation of fungal shapes. The hyphoid model suggests that the underlying principle of dimorphism is much simpler than anticipated, at least conceptually.

Experimental Foundation

Our mathematical analysis of fungal dimorphism,[4] and for that matter fungal morphogenesis in general,[5-7] is based on the following experimental findings:
1) The shape of the fungal cell is determined by the shape of its wall. Dimorphism is therefore the ability of a fungus to construct cell walls with two different morphologies--an idea first advanced by the late Walter J. Nickerson.[8]
2) The shape of the wall is not primarily determined by the chemical composition of the wall but by the manner in which the wall is built.[9]

Accordingly, dimorphism may be reduced to being the expression of two different patterns of cell wall growth: a sharply (apically) polarized pattern of wall growth for the hypha, and a weakly or non-polarized pattern for yeast cells.[9,10]

3) Cell walls of fungi are made from the materials (enzymes and preformed polymers) discharged by cytoplasmic vesicles.[11-14]

It follows from these observations that:

1) The pattern . distribution of wall-building vesicles is the final determinant of the shape of the cell wall.[7,14]

2) The secret of dimorphism resides in the cellular events that govern the pattern of spatial distribution of wall-building vesicles.

RESULTS AND DISCUSSION

According to our computer-derived model,[4-7] a surprisingly simple mechanism can account for the pattern of spatial distribution of wall-building vesicles and, hence, for the morphological duality we call dimorphism.

The Hyphoid Equation

The model assumes that wall-destined vesicles arise from a certain point in the cell called the vesicle supply center or VSC. This idealized point source could actually be the geometric center of a complex vesicle generating apparatus present in fungal cells, and whose overall effect is to release vesicles in all directions. Vesicles need only be endowed with the ability to move towards the cell surface in any random direction. The movement of the VSC is the key to cell morphogenesis (Fig. 1). Thus, if the VSC is held stationary while vesicles are

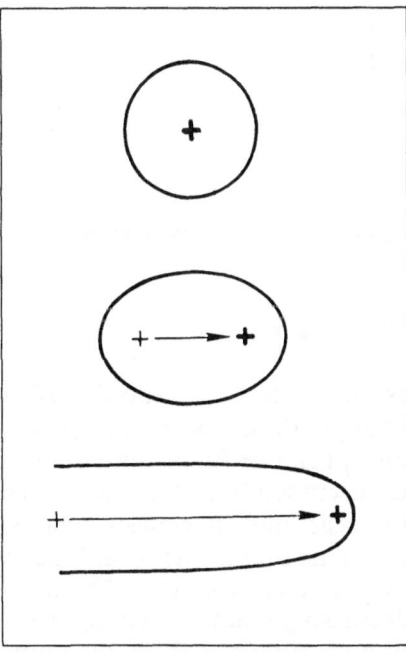

Fig. 1 Schematic representation of the relationship between the relative rate of movement of the vesicle supply center (VSC) and morphogenesis. Three relative rates of displacement, zero, slow and fast, generate spheroidal, ellipsoidal and hyphoidal shapes, respectively.

released, the resulting shape would be a spherical cell. If the VSC is advanced continuously in a linear fashion while vesicles are released, a hyphal tube would be produced. The model yielded the equation:

$$y = x \cot (xV/N)$$

This equation, named the hyphoid, defines the shape and size of a fungal cell by two physiological parameters: N, represents the amount of wall-destined vesicles released from the VSC per unit time; V, the rate of linear displacement of the VSC. A plot of the hyphoid equation produces a curve, the hyphoid (Fig. 2), that describes the ideal shape of a fungal hypha in median longitudinal section[5,6] and corresponds closely with the shape of actual hyphae (see examples of coincidence in previous publications[5-7]).

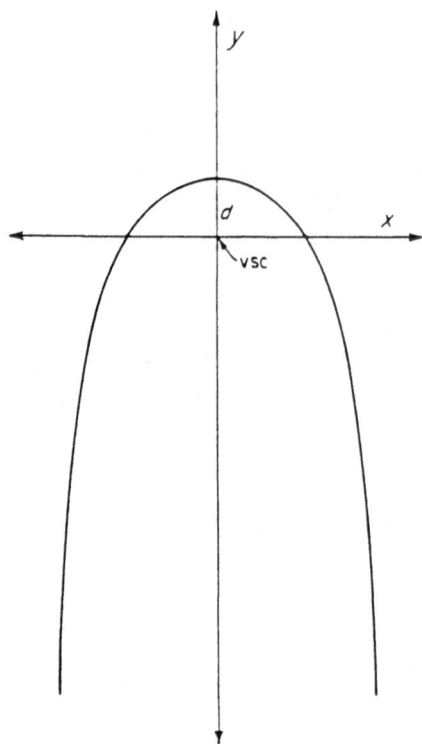

Fig. 2 Hyphoid curve plotted on an arbitrary scale from equation $y = x \cot (xV/N)$.

The V/N ratio Defines Dimorphism

According to our postulate, the V/N ratio describes the essence of vesicle dynamics of the cell. By manipulating this ratio, the wide spectrum of cell shapes seen in mycelial-yeast dimorphism of fungi can be generated. For the sake of simplicity, in the examples discussed in this article, we will maintain N constant while varying V. Basically, yeast morphogenesis obtains when the value of V is either zero (spherical cell) or small (ellipsoidal cell), i.e., when the VSC does not advance or advances slowly, respectively (Fig. 1). Increasing the values of V produces longer cells of decreasing diameter giving rise to distinct hyphal tube morphologies. It is important to stress that in obtaining the wide spectrum of morphologies, from spheres to near cylinders, all other cellular conditions are believed to remain equal, i.e. the cell continues to produce and discharge the same kind and quantity of wall-building vesicles, the only difference is whether the entire vesicle delivery apparatus, or VSC, moves or not, and if it does, how fast.

Reality of the VSC

The hyphoid model defines a mechanism to generate new surface. It describes the shape of new surface generated from precursors (vesicles) released in all directions by an advancing source (VSC) of these precursors. In the model, the VSC is a point source, an abstraction. In actual fungal cells, the VSC may represent the geometric center of the entire vesicle-producing mechanism of the cell, or the distribution site from which vesicles start on the final leg of their journey to the cell surface. The validity of the model in real life would depend largely on the existence of a cell entity akin to the VSC, namely an apparatus for the distribution of wall-building vesicles. Significantly, fungal cells do have accumulations of vesicles near wall-growing regions that may betray the existence of a VSC: the most conspicuous one is an organelle found in the actively growing hyphae of higher fungi--the Spitzenkörper.[15] The position, composition, and behavior of the Spitzenkörper make it a likely candidate to function as a VSC. Superimposed images of hyphoid curves on to sections of real cells show a remarkable coincidence between the position of the VSC and that of the Spitzenkörper.[5]

Do all Cells have VSCs or Spitzenkörper?

A classical Spitzenkörper can be seen by light microscopy, under appropriate optics, in the actively growing tips of higher fungi. Although in many other hyphae, a Spitzenkörper cannot be seen by light microscopy, by electron microscopy one always finds, in growing cells, vesicle accumulations that may function like a Spitzenkörper, e.g. in hyphal tips of lower fungi[12,13,16] or near the pole of apiculate yeast cells.[17] Vesicle concentrations are also apparent prior to, and during, bud initiation in Saccharomyces cerevisiae,[18] and also at the site of germ tube emergence in mycelial fungi.[19,20] We suggest that these conglomerates of vesicles represent functional VSCs.

Hyphal Morphogenesis

Fig. 3 simulates the onset of hyphal morphogenesis in two different fungi: *Mucor* and *Candida*. In *Mucor*, the mother cell is a sphere whose growth is simulated by a stationary VSC (Fig. 3A, frames a-d) sending out "vesicles" randomly in all directions. Hyphal development was simulated by programming the VSC to advance in a straight line at relatively fast rate while continuing to release "vesicles" in all directions (Fig. 3A, frames e-j). The sequence shown in Fig. 3A simulates the spore germination process in *Mucor*.[21] In *Candida*, the mother cell is usually an ellipsoidal cell; this shape was generated by programming the VSC to move at a slow speed (Fig. 3B, frames a-c); the production of a hyphal tube was simulated by a substantial increase (4.75-fold) in the speed of the VSC (frames d-j). The sequence shown in Fig. 3B is a simulation of the conversion of yeast morphology to hyphal morphology in *Candida albicans*.[22]

Budding Yeast Cells

The budding yeast morphology can be mimicked closely by the same computer program. To simulate the ellipsoidal budding cell morphogenesis typical of well known yeasts, e.g. *S. cerevisiae* or *C. albicans*, an ellipsoidal mother cell was first generated by a VSC programmed to advance slowly (Fig. 4A, frames a-c); when the cell reached a certain size, the VSC was accelerated for a brief period (frames d-e). This maneuver caused the appearance of a protuberance or bud on the mother cell; subsequently, the rate of displacement of the VSC was returned to its original value to create the daughter ellipsoidal bud. By appropriate modified manipulation of the VSC, the multipolar spherical budding manifested by species of *Mucor*[2,23,24] or by *Paracoccidioides brasiliensis*[25] was reproduced on the computer screen (Fig. 4B). The spherical mother cell was generated by a stationary VSC and the triple budding was achieved by splitting the original VSC into 3 separate fast-moving VSCs advancing away from each other to generate three buds; after the incipient buds emerged, the VSCs were rendered stationary to generate the spherical daughter cells.

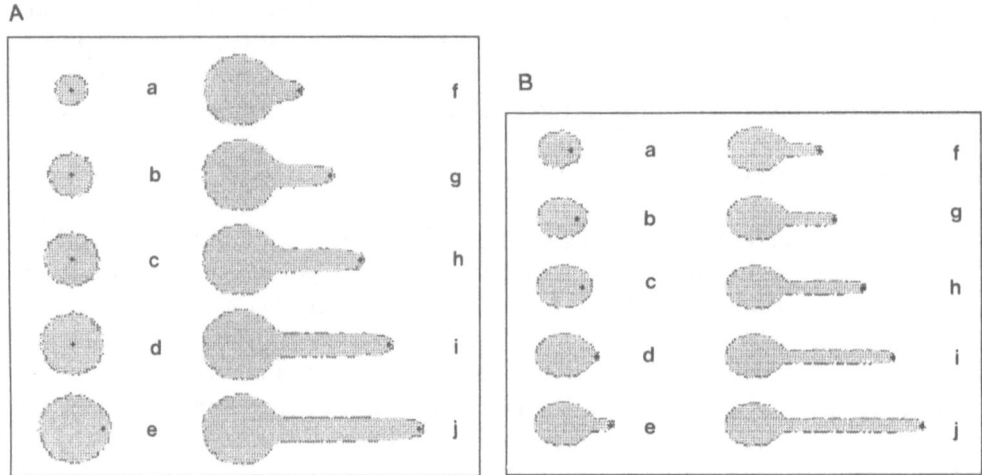

Fig. 3 Computer simulations of hyphal development Panel A simulates the emergence of a hyphal tube from a spherical cell, as seen in the germination of *Mucor* sporangiospores [21] See earlier article for explanation of program [5] In frames a-d the source of computer vesicles (VSC) (black cursor) remained stationary and therefore a growing circle was generated by the random increase in surface from a VSC instructed to send out "vesicles" randomly in all directions. Hyphal development (frames e-j) was simulated by programming the VSC to advance in a straight line ($V = 2$) while continuing to release "vesicles" in all directions Panel B simulates the production of a germ tube from an ellipsoidal cell, in the manner of *Candida albicans*. The elliptical mother yeast cell morphology was generated by programming the VSC to move at a slow speed ($V = 0.8$) (frames a-c), hyphal development (frames e-j) was simulated by programming the VSC to advance in a straight line at a fast rate ($V = 3.8$) while continuing to release "vesicles" in all directions.

Dimorphism

What we believe is the key difference in dimorphism is readily apparent in the parallel simulation of yeast cell development and hyphal tube formation (Fig. 5) where a simple change in the V/N ratio produced an imitation of the dimorphic behavior of *Candida*. At the beginning (frames a-e), both sequences were identical in the number of vesicles programmed and the slow rate of advancement of the VSC to give rise to the ellipsoidal mother cells; also, in both sequences bud

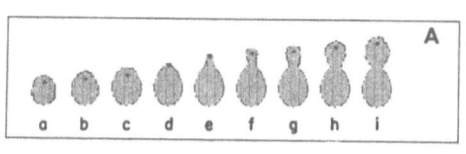 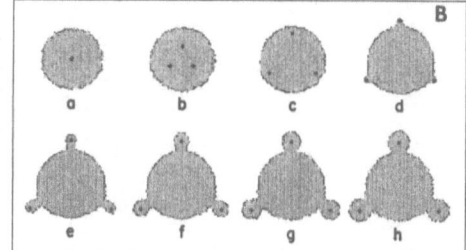

Fig. 4 Computer simulation of the budding process of yeast cells. Panel A is the typical case of an ellipsoidal mother cell producing another ellipsoidal cell, as in *Saccharomyces cerevisiae* or *Candida albicans*. The quasi-elliptical mother cell (frames a-c) was generated by a slow-moving VSC ($V =$ 0.8) (black cursor). Bud extrusion was achieved by accelerating briefly the VSC ($V =$ 3.8; frames d-e) and then reducing the rate to its original value ($V =$ 0.8) to grow a daughter cell (frames f-i). Panel B shows the multipolar budding of spherical yeast cells from a large mother cell. This type of yeast cell is typical of *Mucor* spp. and *Paracoccidioides brasiliensis*.[25] The spherical mother cell was generated by a stationary VSC (frame a). The triple budding was achieved by splitting the original VSC into 3 separate fast-moving VSCs ($V =$ 5) proceeding centrifugally at 120° from each other (frames b-e); after bud extrusion, the VSCs were rendered stationary (frames f-h) to generate the spherical daughter cells.

extrusion (or germ tube emergence) was simulated by accelerating the VSC 4.75-fold (frames d-e). The dimorphic response was generated in frames f-j by programming a single difference in VSC movement. In the yeast case, after bud extrusion, the rate of displacement of the VSC was returned to its initial slow value to generate another ellipsoidal cell. In the hyphal case, the rate of displacement of the VSC was kept at the same high pace and a slim tube was produced. In Fig. 5, the simulation of hyphal tube formation was made more realistic than those in Fig. 3 by introducing a wobbling motion to the VSC. The random oscillation produces the somewhat tortuous shape commonly seen in hyphae, rather than an unlikely perfectly straight tubular shape. A wobbling VSC simulates more closely Spitzenkörper behavior.

Pseudomycelium

Yeasts often produce elongated buds in succession that remain joined to give a mycelial-like appearance, the so-called pseudomycelium. This intermediate dimorphic state, manifested by many dimorphic fungi including *Candida*[22] and *Saccharomyces*,[26-28] can be readily duplicated under the computer hyphoid program (Fig. 6). The constricted hypha appearance, or "pseudohypha", was produced in a hypha whose VSC was accelerated periodically for a brief time to bud a new pseudohyphal segment (Fig. 6, frame b). The length of each segment was determined by the intervals between VSC accelerations (buddings). Periodically, new VSCs were programmed and positioned in the right place and direction to simulate branch formation. The long pseudohyphal segments in Fig. 6 were made to resemble those of *Candida* pseudomycelium. In the branches, the speed of the new VSCs was increased, and the length of the segments decreased, to simulate apical dominance. Pseudomycelium with shorter segments, as seen in *S. cerevisiae*,[27,28] may be programmed by simply increasing the frequency of budding, i. e. decreasing the length of each segment.

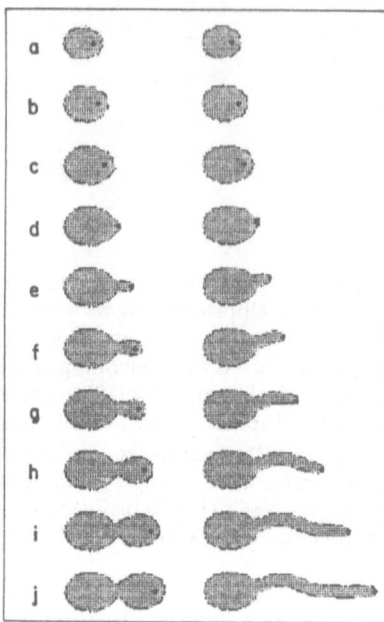

Fig. 5 Computer simulation of dimorphism in *Candida albicans*. Both yeast (left side) and hyphal (right side) sequences were programmed to grow at the same rate. The black cursor indicates the position of the VSC. The ellipsoidal cell morphology was programmed by keeping the VSC speed slow ($V = 0.8$) in the first frames (a-c). Budding was achieved by a brief 4.75 fold acceleration of the VSC (frames d-e). For yeast cell morphogenesis, the VSC speed was returned to 0.8 to allow the growth of a daughter ellipsoidal cell (frames f-j). For hyphal development, the VSC speed remained at 3.8 to generate the long tubular shape (frames f-j). Note that the same parameters used to program the sequence in Fig. 3B were used here except that the VSC was programmed to oscillate randomly with a certain frequency and amplitude. This oscillation mimics the motions of the Spitzenkörper and gives a more realistic computer simulation of the dimorphism of *Candida albicans*.

Dimorphic Index

Fungal dimorphism is often a graded response; within a culture, one often finds a gamut of cells with yeast, hyphal and intermediate morphologies. This morphological gradient, we believe, is a manifestation of variability in dimorphic ratios (V/N) among individual cells in the culture. It should be possible to calculate appropriately weighted average values for the dimorphic ratio of mixed-morphology cultures and correlate them with the morphology index values devised by Merson-Davies and Odds[29] to quantitate the relative abundance of yeast, hyphal and intermediate forms in a mixed-morphology culture.

Ellipsoidal-Triangular Cell Dimorphism

The power of the model to simulate morphogenesis extends beyond the sphere-cylinder duality of mycelial-yeast dimorphism to cover other special cases such as the development of triangular cells in the yeast *Trigonopsis variabilis*[30] (Fig. 7). In Fig. 8, a cell was programmed to grow from events generated by 3 separate

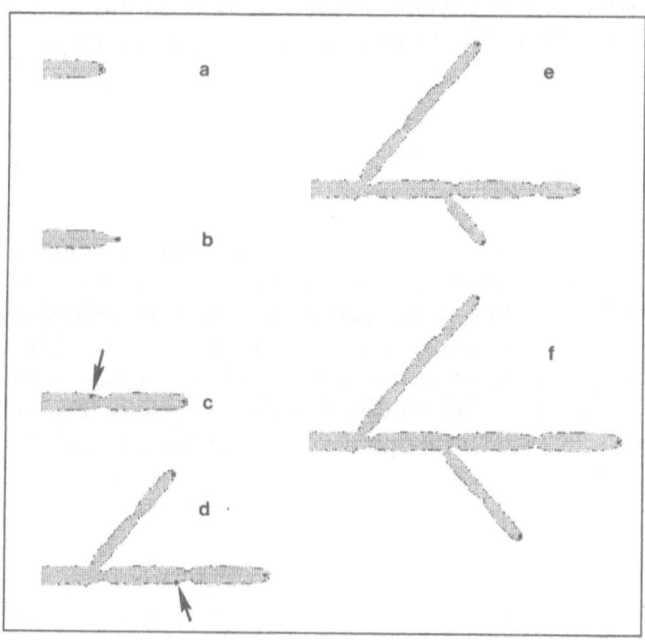

Fig. 6 Computer simulation of pseudomycelium formation. This sequence simulates the pseudohyphal development of *Candida albicans*. The pseudohypha appearance was achieved by growing a hyphal tube with a VSC moving at $V = 2$ (frame a); periodically, the VSC was accelerated briefly four-fold ($V = 8$) to produce an apical bud (frame b); the VSC was then allowed to return to its original value to generate another hyphal segment (frame c). Branches were created by placing new VSCs (arrows) at the appropriate places and programming them to grow ($V = 3$) at a 50° from the mother hypha.

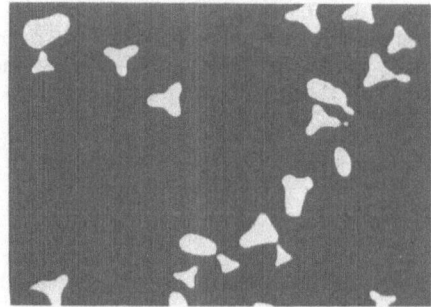

Fig. 7 Dimorphism of *Trigonopsis variabilis*. Microscopic appearance of the organism grown with methionine as nitrogen source (From Sentheshanmuganathan and W.J. Nickerson[30]).

VSCs moving away from one another at 120° angles. This produced a gradual morphological progression from triangular shapes (frames b-e) to three-pointed stars (frames f-i). These shapes are nearly identical to those seen in cultures of this yeast grown in the presence of methionine. The model suggests that the primary effect of methionine in this peculiar yeast would be to induce the formation of three polarity centers (VSCs) that travel radially away from one another.

CONCLUSIONS

Cellular Basis of Dimorphism

The success of the model in simulating dimorphism gives us reason to postulate that the same basic mechanism operates in real cells, namely that morphology would be regulated by vesicle dynamics as defined by the hyphoid equation (V/N ratio). The critical factor in dimorphism would be the rate (V) of displacement of the distribution center of wall-building vesicles (VSC). Basically, a stationary or slow-moving VSC would give rise to spherical or ellipsoidal yeast cells, respectively, while a fast-moving VSC would produce hyphal tubes.

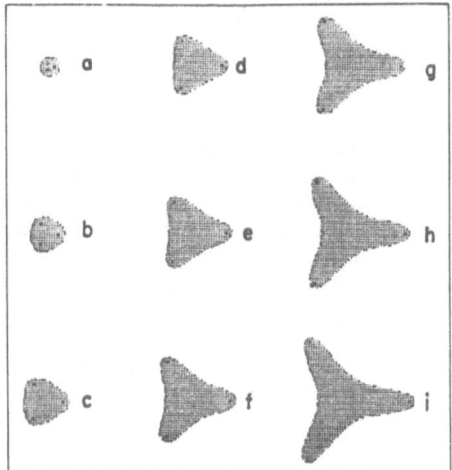

Fig. 8 Computer simulation of the triangular-cell morphology of *Trigonopsis variabilis*. A cell was programmed to grow from events generated by 3 separate VSCs moving at a speed of $V = 1$ away from one another at 120° angles. Note the gradual progression from triangular shapes (frames b-e) to three-pointed stars (frames f-i).

Molecular Basis of Dimorphism

We predict that the molecular basis of dimorphism resides in the macromolecules that comprise and/or regulate the mechanism for the displacement of the vesicle supply center(s) of the cell. From our current knowledge of fungal cell biology,[31-36] we have speculated that elements of the cytoskeleton are the most likely candidate(s) for the displacement of the biological equivalent of the VSC.[5]

Factors Regulating Dimorphism

Dimorphism provides classic examples of environmentally controlled morphogenesis.[37] The hyphoid model suggests that the factors affecting dimorphism are those that impact the vesicle dynamics of the cell; namely, the biochemical, physiological, cytological and genetic events involved in the production and discharge of wall-building vesicles and the mechanism of displacement of vesicles and the VSC. One potentially useful corollary of our hypothesis is that it focuses the search for factors governing dimorphism around the events necessary for the orderly production and distribution of vesicles, and their organizing elements. However, even with this narrow focus, a large number of biochemical events potentially related to vesicle physiology need to be considered. Consequently, it should not be surprising that numerous environmental factors have been reported to impact the morphology of dimorphic organisms. Similarly, the many genes that might affect vesicle dynamics, directly or indirectly, could also appear as regulators of dimorphism. The repeated frustration of investigators of dimorphism in trying to find a single biochemical or genetic explanation for dimorphism is quite understandable. According to the views presented here, the best hope for identifying a single, common molecular cause for fungal dimorphism might be found in the structure responsible for the displacement of the VSC.

ACKNOWLEDGMENTS

We are indebted to David Bartnicki for help with the computer graphs and programs. The experimental work that led to the mathematical model was supported in part by grants from the NIH (GM-33513) and NSF (INT-8413728).

REFERENCES

1. L. Pasteur, Études sur la bière. Gauthier-Villars, Paris (1876).
2. S. Bartnicki-Garcia, Symposium on the biochemical bases of morphogenesis in fungi III. Mold-yeast dimorphism of *Mucor*, *Bacteriol. Rev.* 27: 293 (1963).
3. P.J. Szaniszlo and J.L. Harris, "Fungal Dimorphism. With emphasis on Fungi Pathogenic for Humans", Plenum Press, New York (1985).
4. S. Bartnicki-Garcia and G. Gierz, Predicting the molecular basis of mycelial-yeast dimorphism with a new mathematical model of fungal morphogenesis, in: "More Gene Manipulations in Fungi", J.W. Bennett and L.L. Lasure, eds., Academic Press, San Diego (1991).
5. S. Bartnicki-Garcia, F. Hergert and G. Gierz, Computer simulation of fungal morphogenesis and the mathematical basis for hyphal (tip) growth, *Protoplasma* 153: 46 (1989).
6. S. Bartnicki-Garcia, F. Hergert and G. Gierz, A novel computer model for generating cell shape: application to fungal morphogenesis, in: "Biochemistry of Cell Walls and Membranes of Fungi", P.J. Kuhn, A.P.J. Trinci, M.J. Jung, M.W. Goosey and L.G. Copping, eds., Springer-Verlag, Berlin Heidelberg (1990).
7. S. Bartnicki-Garcia, Role of vesicles in apical growth and a new mathematical model of hyphal morphogenesis, in: "Tip Growth in Plant and Fungal Cells", I.B. Heath, ed., Academic Press, San Diego (1990).
8. W.J. Nickerson, Symposium on biochemical bases of morphogenesis in fungi. IV. Molecular bases of form in yeasts, *Bacteriol. Rev.* 27: 305 (1963).
9. S. Bartnicki-Garcia and E. Lippman, Fungal morphogenesis: cell wall construction in *Mucor rouxii*, *Science* 165: 302 (1969).
10. S. Bartnicki-Garcia and E. Lippman, Polarization of cell wall synthesis during spore germination of *Mucor rouxii*, *Exp. Mycol.* 1: 230 (1977).
11. M. Girbardt, Die Ultrastruktur der Apikalregion von Pilzhyphen, *Protoplasma* 67: 413 (1969).
12. S.N. Grove and C.E. Bracker, Protoplasmic organization of hyphal tips among fungi: vesicles and Spitzenkörper, *J. Bacteriol.* 104: 989 (1970).
13. I.B. Heath, J.L. Gay and A.D. Greenwood, Cell wall formation in the Saprolegniales: cytoplasmic vesicles underlying developing walls, *J. Gen. Microbiol.* 65: 225 (1971).

14. S. Bartnicki-Garcia, Fundamental aspects of hyphal morphogenesis in: "Microbial Differentiation", J.M. Ashworth, J.E. Smith, eds., Cambridge University Press, Cambridge, U.K., (1973).

15. M. Girbardt, Der Spitzenkorper von *Polystictus versicolor, Planta* 50: 47 (1957).

16. W.K. McClure, D. Park and P.M. Robinson, Apical organization in the somatic hyphae of fungi, *J. Gen. Microbiol.* 50: 177 (1968).

17. E.K. McCully and C.E. Bracker, Apical vesicles in growing bud cells of heterobasidiomycetous yeasts, *J. Bacteriol.* 109: 922 (1972).

18. H. Moor, Endoplasmic reticulum as the initiator of bud formation in yeast (*S. cerevisiae*), *Arch. Mikrobiol.* 57: 135 (1967).

19. C.E. Bracker, Cytoplasmic vesicles in germinating spores of *Gilbertella persicaria, Protoplasma* 72: 381 (1971).

20. S.N. Grove and C.E. Bracker, Protoplasmic changes during zoospore encystment and cyst germination in *Pythium aphanidermatum, Exp. Mycol.* 2: 51 (1978).

21. S. Bartnicki-Garcia, N. Nelson and E. Cota-Robles, Electron microscopy of spore germination and cell wall formation in *Mucor rouxii, Arch. Mikrobiol.* 63: 242 (1968).

22. N.A.R. Gow and G.W. Gooday, A model for the germ tube formation and mycelial growth form of *Candida albicans, Sabouraudia* 22: 137 (1984).

23. S.L. Lara and S. Bartnicki-Garcia, Cytology of budding in *Mucor rouxii*: wall ontogeny, *Arch. Microbiol.* 97: 1 (1974).

24. M. Orlowski, Mucor dimorphism, *Microbiol. Rev.* 55: 234 (1991).

25. L.M. Carbonell and J. Rodriguez, Transformation of mycelial and yeast forms of *Paracoccidioides brasiliensis* in cultures and in experimental inoculations, *J. Bacteriol.* 90: 504 (1965).

26. S.D. Steele and J.J. Miller, Pseudomycelium development and sporulation in *Saccharomyces fragilis, Can. J. Microbiol.* 20: 265 (1974).

27. P.W. Thompson and A.E. Wheals, Duplication cycle in filamentous forms of *Saccharomyces cerevisiae, J. Gen. Microbiol.* 122: 151 (1981).

28. C.J. Gimeno, P.O. Ljungdahl, C.A. Styles and G.A. Fink, Unipolar cell divisions in the yeast *S. cerevisiae* lead to filamentous growth: regulation by starvation and RAS, *Cell* 68: 1077 (1992).

29. L.A. Merson-Davies and F.C. Odds, A morphology index for characterization of cell shape in *Candida albicans, J. Gen. Microbiol.* 135: 3143 (1989).

30. S. Sentheshanmuganathan and W.J. Nickerson, Nutritional control of cellular form in *Trigonopsis variabilis, J. Gen. Microbiol.* 27: 437 (1962).

31. A.E.M. Adams and J.R. Pringle, Relationship of actin and tubulin distribution to bud growth in wild-type and morphogenetic-mutant *Saccharomyces cerevisiae, J. Cell. Biol.* 98: 934 (1984).

32. H.C. Hoch and R.C. Staples, The microtubule cytoskeleton in hyphae of *Uromyces phaseoli* germlings: its relationship to the region of nucleation and to the F-actin cytoskeleton, *Protoplasma* 124: 112 (1985).

33. J.M. Anderson and D.R. Soll, Differences in actin localization during bud and hypha formation in the yeast *Candida albicans, J. Gen. Microbiol.* 132: 2035 (1986).

34. P. Runeberg and M. Raudaskoski, Cytoskeletal elements in the hyphae of the homobasidiomycete *Schizophyllum commune* visualized with indirect immunofluorescence and NBD phallacidin, *Eur. J. Cell Biol.* 41: 25 (1986).

35. L.J. McKerracher and I.B. Heath, Cytoplasmic migration and intracellular organelle movements during tip growth of fungal hyphae, *Exp. Mycol.* 11: 79 (1987).

36. K. Yokoyama, H. Kaji, K. Nishimura and M. Miyaji, The role of microfilaments and microtubules in apical growth and dimorphism of *Candida albicans*, J. Gen. Microbiol. 136: 1067 (1990).

37. P.R. Stewart and P.J. Rogers, Fungal dimorphism in: "Fungal Differentiation", J.E. Smith, ed., Marcel Dekker, New York, (1983).

QUANTIFICATION OF *CANDIDA* MORPHOLOGY *IN VITRO* AND *IN VIVO*

Frank C. Odds

Department of Bacteriology and Mycology
Janssen Research Foundation
B-2340 Beerse, Belgium

ABSTRACT

Studies of morphogenesis in *Candida albicans* have until recently been hampered by the lack of a quantitative marker for shape phenotypes. The lack of such a marker means that experiments designed to study molecular processes possibly related to morphology depend on subjective descriptions of morphological form and are often biased to the use of growth environments that favor development of the fungus at the morphological extremes of true hyphae and of yeast forms. A "morphology index" (Mi), calculated from three cell measurements, was first described in 1989. This index reduces gross cell morphology to a number, usually between 1 and 4.5, in which high Mi indicates hyphal forms and low Mi indicates yeast forms, while Mi in the range 2–3 reflects intermediate, "pseudohyphal" morphologies.

Cellular chitin content and rates of cell envelope expansion both correlated linearly with Mi, while general cell envelope expansion appeared to be switched off at Mi>2 and expression of surface epitopes reactive with two monoclonal antibodies was "switched on" only at Mi>3. Determinations of Mi for *C. albicans* cell forms in vaginal smears failed to show any association between morphology and clinical symptomatology among the patients studied. It is hoped that future applications of Mi will include experiments in which aspects of gene expression are related to gross morphologic phenotype, and thus provide a means for directly determining morphology-specific regulatory processes at the molecular level.

INTRODUCTION

"Dimorphism" is an inappropriate term in the context of cell shape in *Candida albicans*. The name implies two distinct and, possibly, mutually exclusive morphological phases that are triggered to convert from one to the other by environmental stimuli. Application of the "dimorphic" principle to *C. albicans* leads to misconceptions such as the notion that one "phase" can be defined as "commensal" and the other "pathogenic". The reality of *C. albicans* morphology is that individual cells of this fungus exhibit a phenotype on a spectrum of morphologies, ranging from virtually spherical yeast forms at one extreme to elongate, cylindrical hyphae at the other.

Studies examining molecular and genetic factors regulating *C. albicans* cell shape *in vitro* are often designed around environmental conditions that favor development of *C. albicans* at the two extremes of its morphological phenotype. But this approach does not avoid the difficulty of distinguishing those molecular phenotypes that are strictly related to morphological development from those that merely reflect environmental differences, and it ignores the intermediate cell shapes that are so frequently enountered in *Candida*-infected tissues *in vivo*. Brawner *et al.*[1] have warned about the need to study expression of particular molecules in *C. albicans* in as great a diversity of strains and environments as possible before that expression is described as a "form-specific" event.

Descriptions of *C. albicans* phenotypes have for many years been based on subjective descriptions of cell morphologies, and this practice has sometimes led to confusion. The term "pseudohypha" in particular is open to wide subjective interpretation and has been used to describe many different cell types on the continuum between yeast forms and true hyphae.[2] Yokoyama & Takeo[3] proposed that the terms "short pseudohyphae" and "long pseudohyphae" should be used for the intermediate morphologies. However, these names still depend on subjective assessment of what are, often, complex mixtures of morphological forms.

An objective measure of *C. albicans* cell morphology would represent a step forward in the sense that it would facilitate inter-laboratory comparison of morphological data, it would remove subjective interpretations of microscopic images and it would allow quantitative associations to be drawn between molecular/genetic and morphological phenomena.

QUANTITATION OF CELL MORPHOLOGY IN *C. ALBICANS*

Morphology Index

A quantitative measure of *C. albicans* cell morphology was first described in 1989.[4] This "morphology index" (Mi) is a unit-free number calculated from measurements of the maximum cell length (h), maximum cell diameter (d) and septal diameter (s) of a *C. albicans* cell (Fig. 1). It can be seen from Figure 1 that the ratio d/s is 1 for parallel-sided hyphae, but increases with the roundness of the cell, approaching a theoretical maximum of infinity for a truly spherical cell with a septal diameter of 0. The ratio h/d is 1 for a spherical cell but becomes large for a cell unit within a hypha (Fig. 1). The ratio h/d alone gives some measure of the rotundity of a cell, but it was found to be too insensitive a measure in practice to discriminate reproducibly between subtle differences in the ovoid character of *C. albicans* yeast cells. Therefore the two ratios, h/d and d/s were combined in a single expression, viz.:

$$2 + 1.78 \times \log_{10}\left(\frac{hs}{d^2}\right)$$

Applied to real microscopic measurements of *C. albicans* cells, this expression results in a number usually in the range 1.0–4.5, where an Mi below 2 usually indicates a yeast form and an Mi greater than 3 usually indicates a true hypha. Mi between 2 and 3.5 usually indicates a pseudohyphal cell.[4] Figure 2 illustrates several different *C. albicans* morphologies together with the Mi calculated for representative cells.

Measurement of the three dimensions necessary to calculate Mi and the calculation itself can be automated by use of computerized image analysis techniques. The extent and speed of automatic measurement and calculation depends on the image analysis hardware and software available to the user. If large numbers of cells (100 or more) can be measured to determine their Mi, it becomes possible to estimate the mean Mi of a cell population and its variance.

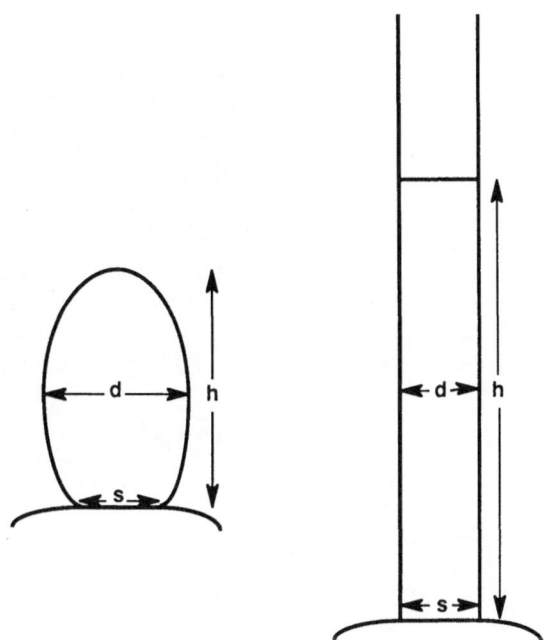

Fig. 1 Basis of Mi calculation. The dimensions h, d and s are determined for an ovoid yeast form (left) and a hyphal cell segment (right). These measurements are combined in the expression

$Mi=2+1.78 \times \log 10 \left(\frac{hs}{d^2}\right)$ to yield a number that ranges in practice from around 1 to around 4, which

represent the extremes of spherical yeast cells and long true hyphal cells. The Mi for the examples in the figure are 1.7 (left) and 3.3 (right).

Thus, Mi can be used both to associate population characteristics such as cell wall composition with average population or morphologies, and to associate parameters that can be measured in individual cells (e.g. binding of monoclonal antibodies, cell envelope expansion rate) with Mi in the same individual cells.

Mi of Different *C. albicans* Strains and in Different Growth Environments

Measurement of Mi distributions for *C. albicans* cells cultured in 12 different growth environments showed that considerable variation was possible in the nature of induced morphologies.[4] Figure 3 provides examples of mean Mi for a selection of growth environments. Cultures incubated at temperatures of 30 °C generally contain cells with a low mean Mi (1.3–1.4) and small standard deviations. Cultures incubated at temperatures at or above 37 °C often give mean Mi values >3, also with small standard deviations. However, out of 12 different growth environments tested, two, Eagle's minimum essential medium at 30 °C and yeast nitrogen base/glucose medium at 37 °C, gave mean Mi values between 2 and 3 with large standard deviations. These environments were therefore interpreted as supporting growth of a diversity of *C. albicans* morphologies.[4]

Experiments with various wild-type *C. albicans* strains and with morphological variants grown in five different environments confirmed the discriminatory power of the morphology index.[4] Three morphological phenotypes were recognized among 5 variants tested: two variants with a tendency to produce cell morphologies with Mi of approx. 2 under all conditions except in serum and

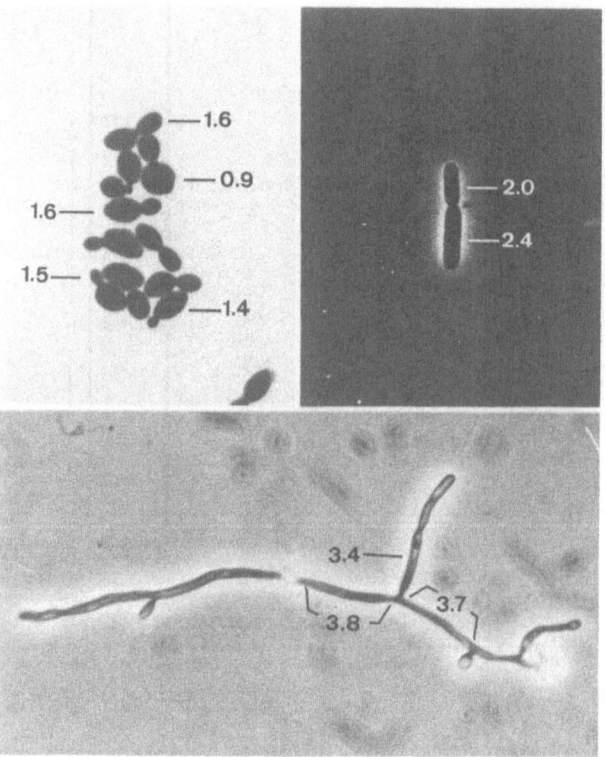

Fig. 2 Examples of *C. albicans* morphological forms shown with the Mi determined for representative cells. Round to ovoid yeast forms (top left) show Mi from 0.9 to 1.6; elongated yeast forms (short pseudohyphal cells? top right) give Mi of 2 0 and 2.4 and hyphal cell segments (below) show Mi =3 4 or higher.

EMEM at 37 °C, when a higher mean Mi was recorded; two yeast-form variants with mean Mi between 1.3 and 1.6 under all growth conditions, including incubation in serum; and one pseudohyphal variant, with a mean Mi of approx. 3 under all growth conditions.[4]

Chitin Content and Epitope Expression in *C. albicans* Cells of Different Mi

Chitin content in *C. albicans* has been associated with morphological change in the fungus for many years.[5-8] The previous studies have demonstrated only elevated cell wall chitin content in hyphae as compared with yeast forms, but expression of chitin content as a function of population Mi now suggests that there is a proportionality between the two parameters: cell wall chitin content and Mi both increase together.[4]

By contrast, determination of cell reactivity with two monoclonal antibodies in indirect fluorescence tests showed no gradual change in epitope expression of cells related to Mi; instead there was an abrupt increase in fluorescent antibody reactivity in cells with Mi>3.0 regardless of the growth environment used.[9] This observation suggests that expression of both epitopes was fully switched on only in cells with an essentially true hyphal morphology.

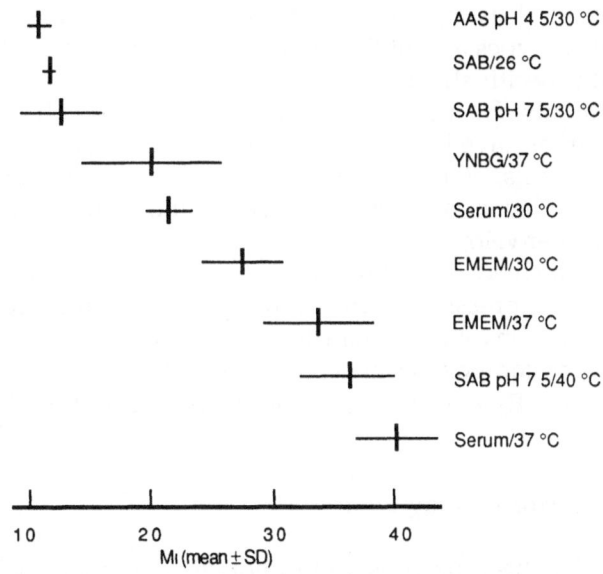

Fig. 3 Mean Mi ± SD illustrated for nine different growth environments. AAS, amino-acid salts medium of Lee *et al.*; SAB, Sabouraud glucose broth; EMEM, Eagle's minimal essential medium. (Data from L.A. Merson-Davies, PhD thesis, University of Leicester, 1991).

Nature of Cell Envelope Expansion in *C. albicans* Cells of Different Mi

Staebell & Soll[10] devised an ingenious method for estimation of sites of *C. albicans* cell envelope expansion in which the dimensions of growing cell evaginations were determined relative to microscopic latex beads adherent to the cell surface. By this means it was possible to estimate the relative contributions of apical expansion (expansion occurring at the tip of the developing cell) and general expansion (occurring all over the cell envelope) during the evagination and maturation processes of buds and hyphae. It was shown that apical expansion predominated in both morphological forms of *C. albicans*, but that yeast forms also underwent a phase of development in which general expansion predominated.

These observations were confirmed in experiments based on similar microscopic culture methodology in which rates of apical and general envelope expansion were related to the final Mi of a newly formed cell.[11] It was found that general envelope expansion was a function restricted to cells with Mi<2.0; only low levels of general expansion were measured in cells with final Mi greater than 2. By contrast, the gross overall rates of envelope expansion were found to be proportional to final Mi, showing an essentially linear correlation in which the rate of expansion was, on average, more than 10-fold greater in cells with Mi~4 as compared with Mi~1.[11] Since this figure is calculated from 2-dimensional measurements of the periphery of expanding cells, the difference between rates of hyphal and yeast-form cell wall deposition in terms of 3-dimensional cell surface area is likely to be even greater still.

C. albicans Morphogenesis In Vitro: an Integrated View

The number of experiments so far published in which Mi has been used as an objective marker of cell morphology is small, but they suggest that use of Mi to determine cell shape provides additional experimental information that might otherwise be missed. It is highly useful for thorough characterization of variant morphological phenotypes and in this context may help future efforts to associate specific genes with specific shapes.

The experiments in which Mi was used as a cell and population morphological marker have indicated at least three underlying components in C. albicans morphogenesis. One is the continuously variable component that is reflected in proportional associations between Mi and chitin content and between Mi and gross rate of envelope expansion. A second is the expression of particular epitopes almost exclusively in cells of the hyphal form, indicating a switching type of up-regulation of expression of these epitopes. The third component is the switch-off of general envelope expansion in cells even with slightly elongate morphologies. These components are compatible with known evidence for "extent of cell elongation" (ECE) genes[12] and for yeast-specific and hypha-specific genes and gene products.[12,13]

Mi and C. albicans Morphology In Vivo

Thus far, Mi has been used as a morphological marker for infecting fungal cells in vivo in only one study.[14] In this, the mean Mi of C. albicans cells in vaginal smears from 26 Candida-infected patients was compared with the clinical symptomatology of the patients. No association was found between the two parameters, and in most instances the mean Mi of cells in the smears was >2.0, indicating a predominant tendency towards pseudohyphal and hyphal forms in the smears, regardless of the patient's clinical condition. However, this study did not examine Candida morphologies in biopsies from vaginal lesions, so the morphologies measured in smears may not accurately reflect the morphologies of invading C. albicans cells, and there were too few instances of Candida-positive smears from totally asymptomatic patients to permit firm conclusions about relations between morphology and pathologic status.

Future experiments with Mi will be aimed at quantifying the morphologies of cells at different stages of tissue invasion in experimentally infected animals to establish which morphological forms are predominant as infection progresses in different sites. Up to now, evidence supporting a role for hypha formation as an invasive feature in Candida infection has relied entirely on subjective interpretation of histopathological sections in which it is impossible to judge accurately whether round fungal forms are truly yeast cells or transverse sections through hyphae. By trypsinization and gentle homogenization of infected tissues it should be possible to recover intact C. albicans cells whose distribution of morphologies can then be accurately assessed in terms of Mi.

CONCLUSION

The morphology index is an easily determined marker that can be used *in vitro* and *in vivo* to associate other parameters with overall *C. albicans* cell shapes. In its so far limited application to morphology experiments Mi has been shown to facilitate the precise demarcation of morphologic effects and possible cell biological and molecular associations. The Mi provides a simple and useful objective estimation of morphology phenotype that can be determined both *in vitro* and in infected tissues *in vivo*. It is hoped that its future application will include experiments in which aspects of gene expression are related to gross morphologic phenotype, and thus provide a means for directly determining morphology-specific regulatory processes at the molecular level.

REFERENCES

1. D.L. Brawner, J.E. Cutler and W.L. Beatty, Caveats in the investigation of form-specific molecules of *Candida albicans*, *Infect. Immun.* 58: 378 (1990).
2. F.C. Odds, "*Candida* and candidosis", Bailliere Tindall, London (1988).
3. K. Yokoyama and K. Takeo, Differences of asymmetrical division between the pseudohyphal and yeast forms of *Candida albicans* and their effect on multiplication, *Arch. Microbiol.* 134: 251 (1983).
4. L.A. Merson-Davies and F.C. Odds, A morphology index for characterization of cell shape in *Candida albicans*. *J. Gen. Microbiol.* 135: 3143 (1989).
5. F.W. Chattaway, M.R. Holmes and A.J.E. Barlow, Cell wall composition of the mycelial and blastospore forms of *Candida albicans*, *J. Gen. Microbiol.* 51: 367 (1968).
6. M. Hrmová and L. Drobnica, Induction of mycelial type of development in *Candida albicans* by the antibiotic monorden and *N*-acetyl-D-glucosamine, *Mycopathologia* 79: 55 (1982).
7. D.S. Schwartz and H.W. Larsh, An effective medium for the selective growth of yeast or mycelial forms of *Candida albicans*, *Mycopathologia* 70: 67 (1980).
8. P.A. Sullivan, C.Y. Yin, C. Molloy, M.D. Templeton and M.G. Shepherd, An analysis of the metabolism and cell wall composition of *Candida albicans* during germ tube formation, *Can. J. Microbiol.* 29: 1514 (1983).
9. L.A. Merson-Davies, V. Hopwood, R. Robert, A. Marot-Leblond, J.-M. Senet and F.C. Odds, Reaction of *Candida albicans* cells of different morphology index with monoclonal antibodies specific for the hyphal form, *J. Med. Microbiol.* 35:.321 (1991).
10. M. Staebell and D.R. Soll, Temporal and spatial differences in cell wall expansion during bud and mycelium formation in *Candida albicans*, *J. Gen. Microbiol.* 131: 1467 (1985).
11. L.A. Merson-Davies and F.C. Odds, Expansion of the *Candida albicans* cell envelope in different morphological forms of the fungus, *J. Gen. Microbiol.* 138: 461 (1992).
12. C. Birse, W.A. Fonzi, S. Saporio, M. Irwin and P.S. Sypherd, Molecular genetics of dimorphism in *Candida albicans*, in: "New Strategies in Fungal Disease", J.E. Bennett, R.J. Hay and P.K. Peterson, eds., Churchill-Livingstone, Edinburgh (1992).
13. R. Finney, C.J. Langtimm and D.R. Soll, The programs of protein synthesis accompanying the establishment of alternative phenotypes in *Candida albicans*, *Mycopathologia* 91: 3 (1985).
14. L.A. Merson-Davies, R. Malet, S. Young, V.C. Riley, P. Schober, P.G. Fisk and F.C. Odds, Quantification of *Candida albicans* morphology in vaginal smears, *Eur. J. Obstet. Gynecol. Reprod. Biol.* 42: 49 (1991).

NATURE AND CONTROL OF CELL WALL BIOSYNTHESIS

Maxwell G. Shepherd,[1] and Pramod K. Gopal[2]

[1] Experimental Oral Biology Unit
Division of Health Sciences
University of Otago
Dunedin, New Zealand
[2] Dairy Research Institute
Palmerston North, New Zealand

ABSTRACT

The structural integrity of the skeletal cell wall is critical for the survival of fungi in their natural environments. The wall and outer surface are important in cell adhesion, cell shape maintenance and as a barrier to metabolites and drugs. The outer fuzzy layer of the wall is composed of proteins and mannoproteins and is involved in pathogenicity, phagocytosis and adherence. The exoskeleton is composed of polymers producing a scaffold that imparts rigidity and determines the cell shape. Adhesins have been identified on the outer cell surface and they interact with the host proteins laminin, fibrinogen and C3d of human complement. The ß-glucans impart structural strength to the wall and at least three distinct glucans exist; a highly branched ß1,6 glucan, a highly branced ß1,3 glucan and an insoluble mixed ß1,3/ß1,6 glucan complexed to chitin. The chitin is linked to the mixed ß-1,3/ß-1,6 glucan through a 1,6 linkage of a glucose from the ß1,3 glucan to a GlcNAc residue of the chitin. Wall fractions partially degraded by zymolyase were subjected to two-dimensional [^{1}H] and [^{13}C] NMR and the heteronuclear spectrum confirmed i) a pure glucan of glucose and ii) a ß-linked polymer. A further zymolyase treatment produced a mixed ß1,3/ß1,6 glucan with mannoprotein bound to it. High resolution solid state [^{13}C] NMR has been performed on the insoluble glucan and these data showed a ß1,3 glucan with ß1,6 side chains and a degree of polymerisation >400. Chitin and glucan are synthesized by transmembrane enzymes catalyzing the vectorial synthesis of these polymers and the products are extruded through into the wall. When the products emerge from the cell membrane they undergo intussusception into the expanding wall. At this point the molecules are linear and questions that remain unanswered

Dimorphic Fungi in Biology and Medicine, Edited by
H. Vanden Bossche *et al.*, Plenum Press, New York, 1993

include how the glucans are branched, cross linked with other polymers such as chitin and complexed with mannoproteins. Glucanases have been implicated in the controlled hydrolytic modification of wall glucans for morphogenesis and development of the final wall structure. A secreted β-glucan branching enzyme from *C. albicans* has recently been described. A model for cell wall growth and development will be presented.

INTRODUCTION

The survival of fungi in their natural environment depends upon the structural integrity of the exoskeletal cell wall. The wall proper and the material on its outer surface are important in cell adhesion, cell shape maintenance and as a barrier to metabolites and drugs. In addition, the cell wall contains components that act as immunogenic determinants and immunomodulators as well as secreted enzymes. The outer fuzzy layer of the wall is composed of proteins and mannoproteins and is involved in pathogenicity, phagocytosis and adherence. This review will examine: the dynamic changes associated with the outer wall surface during morphogenesis, the cell wall composition and ultrastructure, the outer surface components of the cell wall, the structure of the wall polymers that form the exoskeleton, and the mechanism of biosynthesis for the cell wall polymers. Finally, a model for cell wall growth and development will be presented.

THE CELL AND ITS WALL

The exoskeleton of the fungal wall is composed of a network of interlaced polymers producing a scaffold that imparts rigidity and determines the cell shape. Although the wall is made up of a covalently interlocked network a mechanism clearly exists for the network to enlarge during cellular multiplication and growth. Conceptually this requires the cleavage of covalent bonds during expansion of the fungal cell wall. For this to occur there must be selective and transient cleavage followed by the introduction of new material. During this period the wall must not lose its ability either to maintain the cellular shape or resist damage to the cell. The enzymes responsible for "clipping" the structural glucans to allow the insertion of material have not been identified. However, a number of fungal autolysins and glucanases have been identified and, similarly to bacteria, once cell wall synthesis is inhibited wall degradation and lysis is initiated.

The wall acts as a protective barrier and is an obstacle that must be considered in the construction and design of antifungal drugs that need to enter the cell for their effective action. The ß-glucans and chitin of the pathogen are not present in the host and hence these compounds and particularly the enzymes involved in their biosynthesis and degradation are potentially safe targets for antifungal agents. Given the overall importance of the wall to the nature of its components it is not surprising that the architecture of the fungal cell wall and the mechanisms of wall synthesis and degradation are currently areas of intensive research effort. The cell wall comprises a mosaic of components and it is the discussion of these components and their mechanisms of biosynthesis that is the subject of this paper. The cell envelope and cell wall of *Candida albicans* has recently been reviewed by Shepherd *et al.*,[1] Reiss,[2] Shepherd,[3] Odds,[4] Shepherd & Gopal,[5] Shepherd.[6]

DYNAMIC NATURE OF THE CELL WALL

The major difficulty encountered in the investigation and description of the cell wall of *Candida albicans* is defining a typical wall. In addition to the qualitative and quantitative differences found between the yeast and mycelial cells[7,8] we know that the wall of yeast cells alters during the growth cycle and in different media.[9-11] Tronchin et al.[12] have described the dynamic changes that occur on the cell surface during germ-tube formation including the rapid synthesis and degradation of surface mannoproteins. We have recently studied the coaggregation of streptococci with *C. albicans*[13] and have found that starving the yeast cells for a period as short as 30 minutes promoted the bacterial/fungal attachment. The induction of the adhesin on the yeast cells was inhibited by the protein synthesis inhibitor trichodermin, implying that new protein synthesis was occurring at the surface. Collectively, these studies highlight the subtle and rapid changes that occur in the cell wall under different environmental and physiological conditions.

CELL WALL COMPOSITION AND ULTRASTRUCTURE

The cell wall makes up approximately 30% of the total weight of the cell and a number of studies have shown that the wall is composed of mannoproteins (20-30%); ß1,3-linked glucans (25-35%); ß1,6-linked glucans (35-45%) and small quantities of chitin (0.6-2.7%); protein (5-15%) and lipid (2-5%).[8,13-16] The glucan and mannoprotein components of yeast and hyphal cells are similar although hyphal cells contain at least 3 times as much chitin as yeast cells.[8,14,17] Transmission electron microscopy studies have shown the wall to be composed of a number of different layers and the thickness and number of these layers varies depending upon the morphology and the stage of growth.[18-22] Fig. 1 is a schematic diagram showing the architecture, composition and organisation of the *C. albicans* cell wall. The outermost layer is composed of mannoprotein although mannoprotein is also distributed throughout the entire wall. This has been shown with lectin binding studies and extraction of the cell wall with different solvents,[23,24] cytochemical staining[25] and cross reaction with ferritin-conjugated antibodies.[26] The remainder of the wall is made up of a network of the structural microfibrillar components, ß-glucans and chitin. These molecules form a scaffold which is enmeshed in a matrix of amorphous components including mannoproteins and proteins. Tronchin et al.[27] used wheat agglutinin to show that one layer of chitin was distributed in the inner wall layer near the plasma membrane but that there were smaller amounts found dispersed throughout the outer wall layers. In this organization the fibrils would be resistant to stretching while the matrix would resist compression and act to trap soluble components such as wall enzymes. It should be noted, however, that the interpretation to electron micrographs must be made with caution as artifacts can arise from differential staining, staining of minor components, drying and fixation procedures.

OUTER SURFACE OF *C. ALBICANS*

A number of workers have demonstrated an outer cell wall coat variously called slime layer, mucous layer or fuzzy layer[19,22,28,29] (Fig. 2). and this layer is believed to play an important role in pathogenicity, phagocytosis and adherence. In addition the mannoproteins on the outer surface are the main immunogenic determinants in a *Candida* cell. It is known that this outer layer undergoes considerable alteration depending on the mode of growth.[18,20,22] McCourtie & Douglas[9] have demonstrated that the formation of this outer fibrillar/floccular layers is promoted by growing yeast cells of *C. albicans* in the presence of high concentrations of galactose and sucrose. The study of Howlett and Squiers[30] demonstrated the importance of the floccular layer in mediating adhesion of *Candida* to oral mucosal cells. Horisberger and Clerc[31] have described an apical concentration of anionic sites on germ-tubes and developing hyphae. The anionic sites are associated with the outer fuzzy coat and are believed to be derived from the phosphate groups on mannoproteins. These authors suggest that the anionic sites play an important role in tissue colonization.

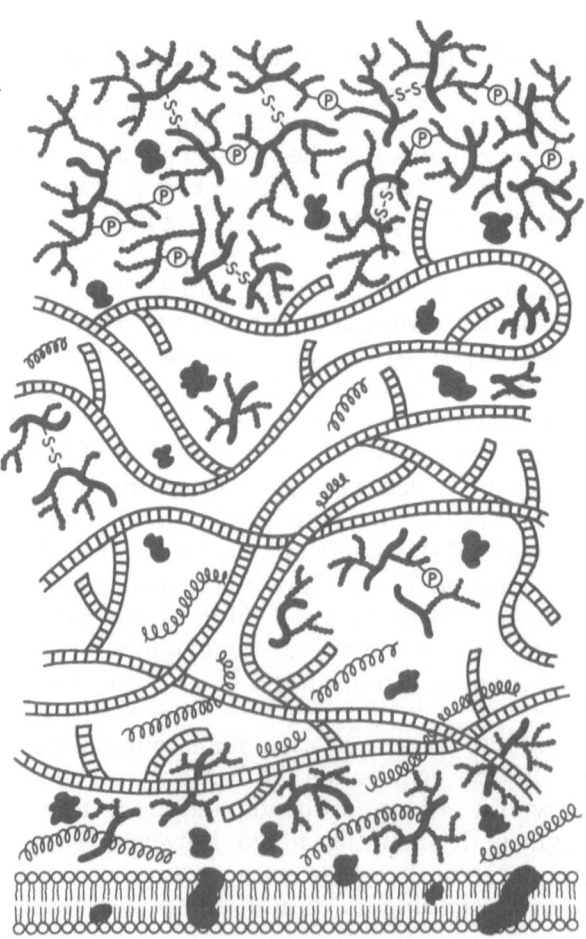

Fig. 1 Schematic diagram of the architecture of the *C. albicans* cell wall. The molecules depicted are: chitin (ꬊ); ß glucan microfibrils (⤳);mannoprotein (⤸); extracellular enzymes (●); disulphide bridges (- S - S -); phosphoester linkages (—P—).

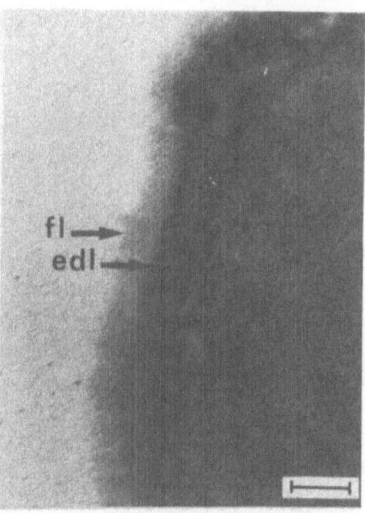

Fig. 2 Cell surface of *C. albicans* showing the fuzzy layer (fl) external to the outer electron-dense layer (edl). The fuzzy layer extends over the bud scar at the top of the field. Poststaining with potassium permanganate obscured most of the intracellular detail. Marker bar = 0.2 µm. (From Hubbard *et al.*,[22] with permission).

Finally, the outer surface of the *C. albicans* wall contains receptors. The receptor for the C3 fragments of human complement C3d and iC3b has been characterized by several groups.[32-34] In mammalian cells the iC3b interacts with the CR3 receptor on phagocytic cells promoting phagocytosis and stimulating the release of intracellular microbiocidal agents such as superoxide or myeloperoxidase. Therefore, the receptor may allow *Candida* cells to bind the iC3b non covalently and avert recognition by neutrophil CR3. In this way, the *C. albicans* surface receptor for the C3 fragment promotes pathogenicity by inhibiting phagocytosis.[34] Immunofluorescence studies have shown that plasma and matrix host proteins bind specifically to the surface of germ-tubes[35] and this adherence is mediated by adhesins with molecular weights of 60, 68 and 200 kDa. Tronchin *et al.*,[12] have shown with Western blotting experiments that these adhesins interact with the host proteins laminin, fibrinogen and C3d and speculate that the same fungal components posses a variety of biological functions. Laminin has been implicated in both adhesion and pathogenesis of bacterial infections and therefore the germ-tube-specific surface receptors for laminin could mediate attachment of these cells to basement membranes and so contribute to the establishment of candidiasis.

CELL WALL POLYMERS

Glucan. Two lines of evidence indicate that the ß-glucans impart the structural strength. Firstly, osmotically sensitive cells are generated after degradation of whole yeast cells with a purified ß1,3 glucanase.[36] Secondly, protoplasts of *C. albicans* regenerated in a simple medium comprising magnesium chloride and glucose give osmotically resistant cells that have not incorporated mannoprotein into the wall.[37] Fig. 3 shows the fibrils of glucan forming a mesh around the protoplast cell. Although mannoprotein was not incorporated into the walls of these cells it was secreted into the medium. The classical protocol for obtaining fungal wall glucans (reviewed in ref. 38) is fractionation on the basis of solubility in acid and alkali. Analysis of *C. albicans* glucans obtained in this manner by methylation, gas-liquid chromatography (glc) mass spectroscopy and [13C] NMR showed that ß1,6 glucan was the major polymer and that it was highly branched and soluble. There was also a branched ß1,3 glucan and mixed ß-1,3/ß-1,6-glucan linked to chitin. The ratio of ß-1,3-to β-1,6-glucan was similar in yeast and hyphal cells indicating that the morphology of *C. albicans* is not determined by the ratio of these linkages. However, the insoluble glucan from germ-tube forming

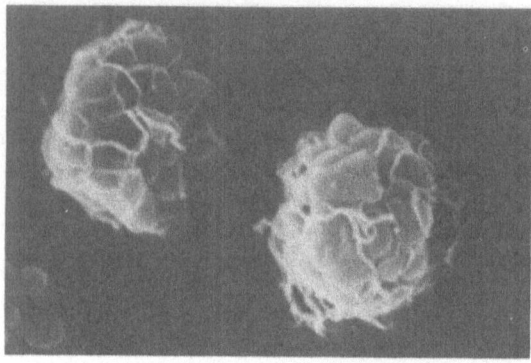

Fig. 3 Scanning electron microscopy of glucan nets forming on regenerating spheroplasts. The conditions for regeneration were 3 h at 30 °C in a medium of 0.6 M MgCl$_2$ and 50 mM glucose.

cells contained considerably more β-1,3 linkages than that found in yeast and hyphal cells.[7] If the ß-1,3: ß-1,6 ratio in the mother cell is taken into account it would appear that during the early stages of germ-tube formation there is almost exclusive synthesis of ß1,3-linked glucans and this is analogous to the situation observed with regenerating spheroplasts[37,39] which preferentially synthesize ß-1,3-glucans. Surarit *et al.*[40] isolated the insoluble glucan-chitin fraction from regenerating spheroplasts of *C. albicans* and mass spectrometry analysis confirmed a glycosidic linkage between the 6th position of *N*- acetylglucosamine on chitin and the 1 position of a ß-1,6-glucan. Chitin is dispersed around the entire wall[27] and its association with glucan would provide a structural scaffold to which the remaining components attach. There are also reports of glucan-mannoprotein complexes[41] but the nature of these linkages has not been established. A major difficulty in cell wall chemistry is obtaining molecules for analysis that are identical to those in the intact wall of the organism under study. Extremes of temperatures and pH and extraction with solvents cause degradation and rearrangement. Gopal and Shepherd (unpublished results) have solubilized glucans from the wall by partial zymolyase degradation. The first glucan fraction to be solubilized was analyzed by two-dimensional [1H] and [13C] NMR and the heteronuclear spectrum confirmed (i) a pure glucan of glucose and (ii) a ß-linked polymer. While the first zymolyase treatment released a ß1,3 glucan, the second zymolyase treatment gave a mixture of ß1,3-ß1,6 glucan and mannoproteins. High resolution solid state [13C] NMR has been performed on the insoluble glucan (Gopal & Shepherd, unpublished results) and the spectra analyzed according to the assignments given by Saito.[42] Solid state [13C] NMR gives information on the structure of both non crystalline and crystalline insoluble polymers. The spectra showed (Fig. 4) typical features of a ß-1,3 glucan with the anomeric peak of C1 at 103.7 ppm and the C2 and C5 peaks at 74.2 ppm. The most noteworthy feature of the spectrum was the doublet signal due to the C3; Ca at 85.7 ppm and Cb at 83.5 ppm. These two C3 signals correspond to the laminarin-type and paramylon-type conformations respectively.[42] Finally, from the relative proportions of the two C3 signals we conclude that the degree of polymerization is greater than 400. From these data and by analogy with other structural glucans such as cellulose and paramylon it is concluded that this molecule would indeed act as a major structural component of the wall. There is also evidence that these structural ß-glucans form helical structures.[43]

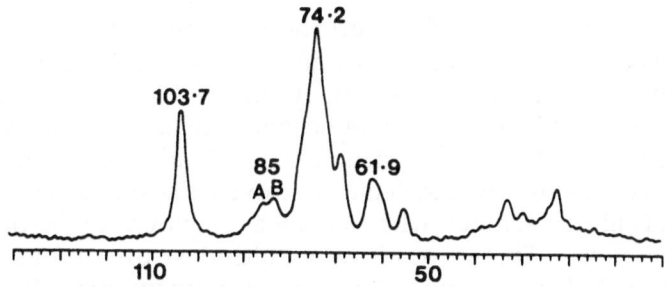

Fig. 4 The 75.43 MHz [13C]-CP-MAS NMR spectrum of *C. albicans* insoluble wall glucan. Contact time was 500 μsec.

Mannoproteins. Despite the large number of studies completed on *C. albicans* mannoprotein[44] and reviewed in refs. 2-4 detailed structures for neither the polysaccharide portion nor the protein portion are known. In summary, the mannoprotein fraction makes up approximately 20 to 30% of the cell wall, the mannoproteins are the major antigenic components of intact and mycelial cells and, as shown in Fig. 5, their structures are similar to that of the mannoproteins found in *S. cerevisiae*. The major portion of the mannoprotein is an α1,6-linked

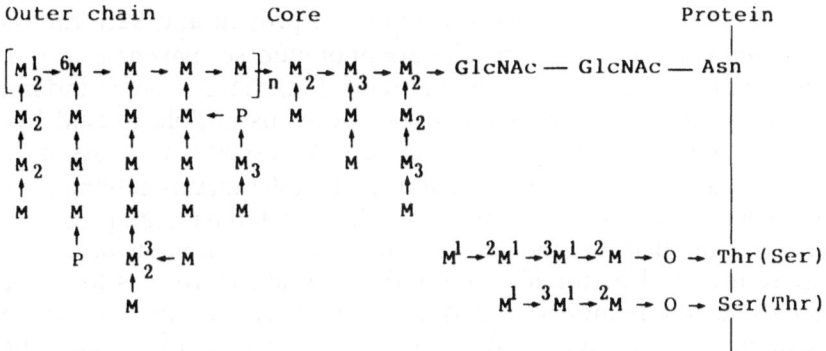

Fig. 5 Proposed structure of *C. albicans* ATCC10261 mannoprotein.

159

polymannose joined to the protein through a chitobiose bridge and asparagine. In the outer chain of *C. albicans* mannoproteins 60-70% of the mannose units are substituted at the 2 and 3 position and oligosaccharides of mannose are joined by 1,2 linkages and occasionally 1,3 linkages. It is the heterogeneity in these outer regions which give rise to antigenic determinants and the two serotypes of *C. albicans*. Elorza et al.,[45] have estimated that 12% of the mannoprotein of the wall is O-linked. The molecular weight of the isolated mannoprotein depends on the method of preparation. The O-linked mannoproteins are not as complex as the N-linked molecules. Gopal and Shepherd (unpublished) carried out controlled ß elimination on a highly purified cell wall mannoprotein fraction and fractionated the products on a Biogel P2 sizing column. A mannobiose, mannotriose and mannotetraose was isolated from the column. The proton NMR of the oligosaccharides gave the linkage analysis shown in Fig. 5. These linkages are different from those found in *S. cerevisiae* and may be important in the taxonomy of the *Candida* species. In the *C. albicans* wall mannoproteins the phosphate content is between 0.2 and 1% whereas in *S. cerevisiae* mannoproteins the phosphate content is less than 0.1%. The presence of this highly charged phosphate group on the outer regions of the wall would affect the way these cells are attracted to and indeed attached to other surfaces. The anionic sites on developing germ-tubes[31] are believed to be from the negatively charged phosphate groups on the mannoproteins. Details of the mannophosphate linkage are not available as most of the published studies on the structures of *C. albicans* mannoproteins used iolation procedures that would have destroyed phosphoester linkages. Early studies used Fehling's solution where acid was employed to dissociate the copper mannan complex. This procedure would destroy α1,3-mannobiosyl and phosphodiester linkages. The cetavalon-borate mannoprotein complexes are larger, but, again separation of the complex must be done in the absence of strong acid. Molloy et al.,[46] carried out vectorial iodination on proteins of the cell surface and showed that a zymolyase degradation of a wall preparation released material that was not contaminated with membrane and cytosolic proteins but did release 93% of the iodine with a specific activity 45-fold higher than the original SDS extract. The requirement for ß1,3 glucanase to release the [^{125}I] labelled material is consistent with either covalent attachment or tight entrapment of the labelled proteins in the cell wall glucans. The major iodinated material moved as a diffuse band at 260 kDa but it is not clear whether this is one mannoprotein or a mixture. By analogy with the study of Frevert and Ballou[47] on *S. cerevisiae* this 260 kDa mannoprotein may well be the major structural glycoprotein of the cell wall. The 260 kDa fraction was composed of 1.5% protein and 98.4% hexose and analysis of the peptide portion showed it contained 36-38% serine and threonine residues typical of mannoproteins enriched in O-linked oligosaccharides. Elorza et al.[45] found *C. albicans* mannoproteins to contain 7% protein and 92% carbohydrate, predominantly mannose with a small amount of glucose. Saxena et al.[48] analyzed mannans from two relatively avirulent mutant strains of *C. albicans* and found that the mannan on the a virulent strains was more susceptible to acid hydrolysis indicating differences in phosphoester linkages. Cell surface mutants of *C. albicans* have also been analyzed by Whelan et al.,[49] and these mutants exhibited a range of different phenotypes and chemotypes. The changes in the mannoprotein structure of the mutants implied that the immunodominant determinants of *C. albicans* are not directly involved in colonization ability. Again there was loss of a signal characteristic of the mannosyl a-phosphate linkage. The mannoproteins of *C. albicans* define two serotypes; serotype A and serotype B[50,51] and these two serotypes are distinguished by the side chain oligosaccharides attached to the backbone 1,6 mannose. The mannoprotein immunochemistry and immunology has been reviewed by Reiss.[2] Briefly, in serotype A mannan the immunodominant hapten is believed to be a linear α-1,2-linked mannohexaose with an α-1,3 linkage at the penultimate sugar from the reducing end. The B serotype appears to contain a branched molecule. It should be noted, however, that the structures postulated do not take into account the important role of phosphate residues in antigenic responses[52] and there is still some debate on the role of ß-linked mannose residues the presence of which are indicated from proton NMR spectra.

Proteins. The number and quantity of proteins in the cell wall remains a matter of considerable debate. Some reports show as many as 40 discrete bands on elecrophoresis. However, the study of Molloy *et al.*,[46] highlights the contamination that occurs in walls obtained by traditional extraction methods. In that study[46] there were only 12 wall proteins that could be resolved by electrophoresis and autoradiography that are accessible to vectorial iodination. The threadlike fibrils extending from *C. albicans* to epithelial cells are believed to be proteins. In addition

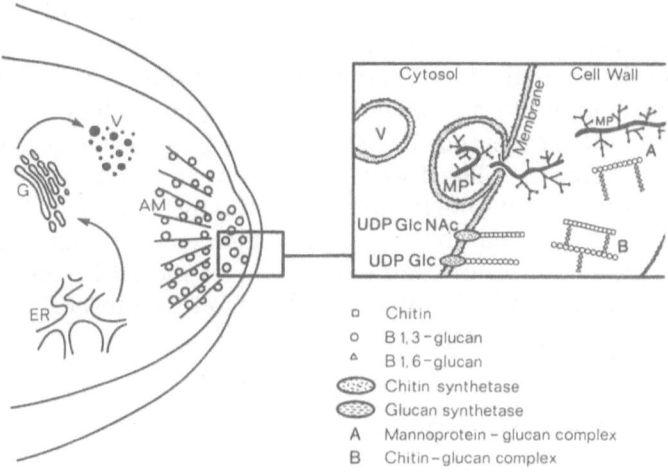

Fig. 6 Cellular and biochemical mechanisms included in *C. albicans* wall formation. AM: actin microfilaments; V: vesicles; ER: endoplasmic reticulum.

the self-aggregating clumps of germ-tubes can be dispersed by digestion with trypsin or sulphydryl reagents but not with α-mannosidase.[53,54] Finally, it is known that the wall contains a number of enzymes including *N*-acetyl glucosaminidase, acid phosphatase, proteinase, glucanase and chitinase (reviewed in ref. 4). There is evidence that the proteinase acts as an aggressin and is a determinant in the pathogenicity of *C. albicans*.[55]

CELL WALL BIOSYNTHESIS AND MODEL OF WALL FORMATION

The essential elements of wall biosynthesis are shown in Fig. 6. It is proposed that vesicles originate in the endoplasmic reticulum in a membraneous body equivalent to the Golgi apparatus. These vesicles contain new material for the plasma membrane as well as membrane bound enzymes, soluble enzymes, highly processed mannoproteins and perhaps a number of wall polymer primers. These vesicles proceed to the point of growth in the cell along actin fibrils and at the plasma membrane there is accretion of this new membrane from the vesicle into the plasma membrane. Yokoyama *et al.*[56] have demonstrated that actin microfilaments are important in apical cell elongation of *C. albicans* hyphae whereas the cytoplasmic microtubules are not essential. The plasma membrane can be recycled through the process of endocytosis. After fusion of the vesicle with the plasma membrane the mannoproteins are released into the cell wall (Fig. 6, inset).

Chitin and glucan are synthesized by transmembrane enzymes catalyzing the vectorial synthesis of these polymers with the precursors UDP *N*-acetylglucosamine and UDP glucose inside the cell and the product extruded through into the wall. We do not understand the fine controls of these enzymes nor the mechanisms that activate them at the points of growth.

The mechanism for the synthesis of the cell wall ß-1,6-glucan is still unknown. No *in vitro* synthesis of yeast ß-1,6-glucan has been described. The K1 killer toxin of *S. cerevisiae* kills sensitive strains by binding to linear ß-1,6-glucan and this provides the basis of a selection procedure for isolating ß-1,6 deficient mutants.[57]

Gene disruption of the KRE1 gene leads to resistance to killer toxin and a 40% reduction in cell wall ß-1,6-glucan: mutations in KRE5 and KRE6 also lead to defects in cell wall ß-1,6-glucan synthesis.[58] Boone *et al.*[58] propose a model for ß-1,6-glucan synthesis where the KRE5 and KRE6 gene products are required for the production of an acceptor glucan containing 1,3 and 1,6 branch points and that the KRE1 gene product adds extended chains of linear ß-1,6-glucan molecules onto the branched acceptor glucan. The KRE1 gene product has a functional amino terminal signal sequence that directs the protein into the yeast secretory pathway. Following glycosylation it may be localized at the yeast cell surface. There is, however, no direct evidence for the involvement of the KRE gene products in glucan synthesis and it is possible that they function by preserving rather than synthesizing the yeast ß-1,6-glucan.

SECONDARY WALL FORMATION

The development of the cross-linked and secondary wall structure is illustrated schematically in Fig. 7. The study of Sonnenberg *et al.*[59] on *Schizophyllum commune* provides evidence for cross-linking occuring in the subapical region of the developing hyphae. A definitive description of hyphal and yeast cell growth requires an understanding of the mechanisms by which the wall expands outside the cell. Staebell & Soll[60] have shown that the characteristic cell wall shape of the *Candida* yeast and hyphal cells is determined by the relative contribution of apical and general cell wall synthesis. Their data revealed that in yeast cells, apical growth accounted for 70% of the surface expansion during the first two-thirds of development. When a bud reaches two-thirds of its final size, apical growth shuts down and the final size is accomplished through general expansion. Hyphal development, however, involves continuous apical extension with less than 10% of growth attributable to general surface expansion. Evidence from studies on other organisms is based on microscopy[61], autoradiography[62-64] and fluorescent labelling[65,66] and supports these conclusions. The overall cell wall dynamics are dictated by both synthetases and hydrolases. For some time it has been postulated that the cross-linking of the wall and the intussusception of new material during general wall expansion involves (a) localised glucanases that clip the glucan molecules and allow new polymer to be inserted and (b) a branching

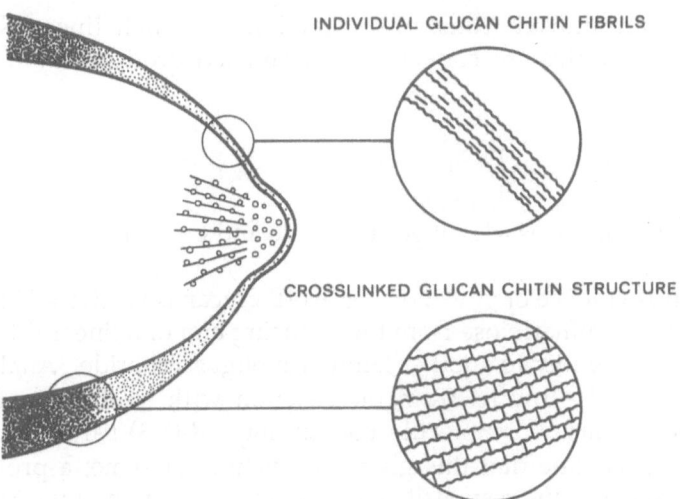

INDIVIDUAL GLUCAN CHITIN FIBRILS

CROSSLINKED GLUCAN CHITIN STRUCTURE

Fig. 7 Secondary wall development during growth of *Candida albicans*. It is proposed that individual ß1,3- and ß1,6-glucan and chitin microfibrils are formed at the point of apical growth. Subsequently, there is secondary wall synthesis which includes additional polymer synthesis and cross-linking of glucan and chitin.

enzyme located in the wall. The rate of growth and the final morphology are clearly controlled by both temporal and spatial control of enzymes and involves the delivery of polymers to the appropriate position in the wall. For some time a physiological role for the cell wall hydrolases ("autolysins") has been sought.

Fungal cell wall lytic enzymes and their regulation have been reviewed by Nombela *et al.*[67] who argue that the physiology and regulation of the ß-glucanases and chitinases point towards an involvement of these enzymes in cellular morphogenesis and development. However, gene disruption of the most abundant exo-1,3ß glucanase of *S. cerevisiae* showed this enzyme to be dispensible as there was no discernible phenotypic change in the deficient mutant. Nombela's group has extended these studies to the genes complementing *S. cerevisiae* lysis mutants. One gene cloned on the basis of its capacity to complement the lytic phenotype was shown to code for a putative serine/threonine protein kinase (M_r 55,666). This protein carries the domains specific for protein kinases and shares a significant degree of identity with yeast protein kinases implicated in the mitotic cell cycle.[68] The cloned protein kinase

gene is not the structural gene (Lyt2) but a suppressor of the lytic phenotype (SLT2 gene). Site-directed mutagenesis indicated that expression of the protein kinase SLT2 is required for proper morphogenesis of the cell wall; overexpression or blocking of the expression of the gene being lethal for the cells (Martin *et al.*, unpublished data). A similar critical role for protein kinases in yeast morphogenesis for *Schizosaccharomyces pombe* has been described by Levin & Bishop.[69] Whether the protein kinase genes that have a regulatory function are coupled to cell wall synthetases, cell wall autolysins or branching enzymes remains to be determined.

Recently, a secreted ß glucan-branching enzyme from *Candida albicans* has been described.[70] The 34,000 M_r protein was purified from the culture medium and catalyses a glucanosyl transferase reaction in which linear ß1,3-linked oligosaccharide substrates are converted to branched products by the reaction mechanism:

$$E + G_n \quad \rightarrow \quad E.G_{n-2} + G_2$$
$$E.G_{n-2} + G_n \quad \rightarrow \quad E + G_{2n-2}$$

where E = enzyme, G_n is ß1,3 oligosaccharide and G_2 is laminaribiose

There was no evidence of any endo- or exo-ß glucanase activity. The enzyme specifically releases laminaribiose from the reducing end of a linear ß-(1,3)-glucan and transfers the remainder to another laminarin oligosaccharide. Analysis of the G8 NMR (^{13}C and ^1H) product from the reaction with laminaripentaose (G5) showed it to be a branched molecule containing a ß-(1,3)-ß-1,6 branch point. Hartland et al.[70] speculate that this glucan branching enzyme, a protein of Mr 34,000 Da, is the key enzyme responsible for the transformation of the initial linear ß1,3 glucan into the branched cell wall ß1,3-ß1,6 glucan of C. albicans. It has been previously demonstrated that the M_r 34,000 protein from C. albicans is a major secreted wall protein.[71,45] Hartland et al. (unpublished data) have also found

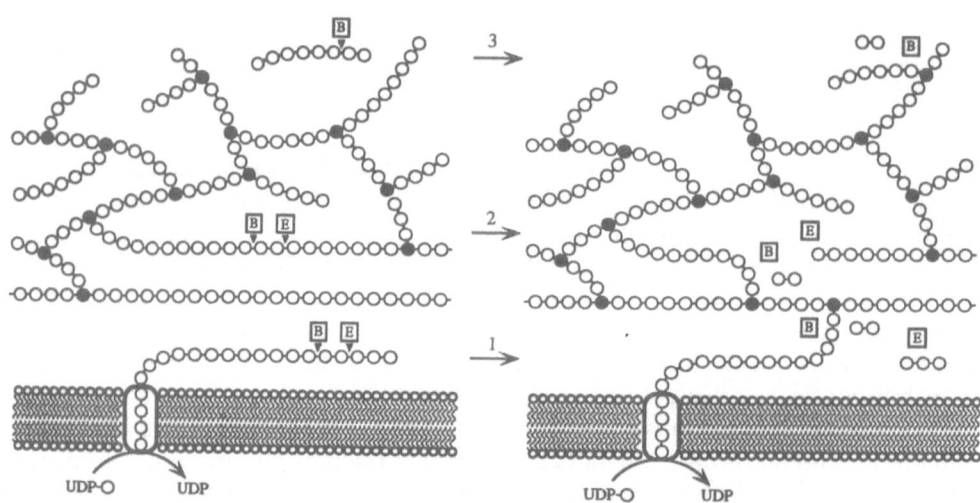

Fig. 8 Model for ß-glucan branching in the C. albicans wall: E: endoglucanase; B: glucanosyl transferase. 1, 2 and 3 show three different ways that the endoglucanase and glucanosyl transferase can combine to crosslink the wall glucans.

similar activity associated with the conserved cell wall protein M_r 31,500) of S. cerevisiae. A model for production of branched ß-(1,3) glucans from the combined action of endoglucanase (E) and glucanosyl transferases (B) is shown in Fig. 8 (R. Hartland, Ph.D. Thesis, 1992).

REFERENCES

1. M.G. Shepherd, R.T.M. Poulter and P.A. Sullivan, *Candida albicans*: Biology, genetics, and pathogenicity, *Ann. Rev. Microbiol.* 39: 579 (1985).

2 E Reiss, 'Molecular Immunology of Mycotic and Actinomycotic Infections', Elsevier, New York (1986)

3 M G Shepherd, Cell Envelope of *Candida albicans*, *Crit Rev Microbiol* 15 7 (1987)

4 F C Odds, '*Candida* and Candidosis , Balliere Tindall, London (1988)

5 M G Shepherd and P K Gopal, *Candida albicans* Cell Wall Physiology and Metabolism, in 'Candida and Candidamycosis , E Tumbay, H P R Seeliger and O Ang, eds , Plenum Press, New York (1991)

6 M G Shepherd, The structure and function of *Candida albicans* cell wall, *Jpn J Med Mycol* 32 (Suppl) 63 (1991)

7 P K Gopal, M G Shepherd and P A Sullivan, Analysis of wall glucans from yeast, hyphal and germ-tube forming cells of *Candida albicans*, *J Gen Microbiol* 130 3295 (1984)

8 P A Sullivan, Y Y Chiew, C Molloy, M D Templeton and M G Shepherd, An analysis of the metabolism and cell wall composition of *Candida albicans* during germ-tube formation, *Can J Microbiol* 29 1514 (1983)

9 J McCourtie and L J Douglas, Relationship between cell surface composition of *Candida albicans* and adherence to acrylic after growth on different carbon sources, *Infect Immun* 32 1234 (1981)

10 M J Kennedy and R L Sandin, Influence of growth conditions on *Candida albicans* adhesion, hydrophobicity and cell wall ultrastructure, *J Med Vet Mycol* 26 79 (1988)

11 B W Hazen and K C Hazen, Dynamic expression of cell surface hydrophobicity during initial yeast cell growth and before germ-tube formation of *C albicans*, *Infect Immun* 56 2521 (1988)

12 G Tronchin, J P Bouchara and R Robert, Dynamic changes of the cell wall surface of *Candida albicans* associated with germination and adherence, *Eur J Cell Biol* 50 285 (1989)

13 H F Jenkinson, H C Lala and M G Shepherd, Coaggregation of *Streptococcus sanguis* and other streptococci with *Candida albicans*, *Infect Immun* 58 1429 (1990)

14 F W Chattaway, M R Holmes and A J E Barlow, Cell wall composition of the mycelial and blastospore forms of *Candida albicans*, *J Gen Microbiol* 51 367 (1968)

15 H F Hasenclever and W O Mitchell, A study of yeast surface antigens by agglutination inhibition, *Sabouraudia* 3 288 (1964)

16 R J Yu, C T Bishop, F P Cooper, H F Hasenclever and F Blank, Structural studies of mannans from *Candida albicans* (serotype A and B) *Candida parapsilosis*, *Candida stellatoidea* and *Candida tropicalis*, *Can J Chem* 45 2205 (1967)

17 P C Braun and R A Calderone, Chitin synthesis in *Candida albicans* comparison of yeast and hyphal forms, *J Bacteriol* 133 1472-1477 (1978)

18 A Cassone, N Simonetti and V Stippoli, Ultrastructural changes in the wall during germ-tube formation from blastospores of *Candida albicans*, *J Gen Microbiol* 77 417 (1973)

19 W Djaczenko and A Cassone, Visualization of new ultrastructural components in the cell wall of *Candida albicans* with fixatives containing TAPO, *J Cell Biol* 52 186 (1971)

20 C Scherwitz, R Martin and H Ueberberg, Ultrastructural investigations of the formation of *Candida albicans* germ tubes and septa, *Sabouraudia* 16 115 (1978)

21 D Poulain, G Tronchin, J F Dubremetz and J Biguet, Ultrastructure of the cell wall of *Candida albicans* blastospores study of its constitutive layers by the use of a cytochemical technique revealing polysaccharides, *Ann Microbiol* 129A 141 (1978)

22 M J Hubbard, P A Sullivan and M G Shepherd, Morphological studies of N-acetylglucosamine induced germ tube formation by *Candida albicans*, *Can J Microbiol* 31 696 (1985)

23 A Cassone, E Mattia and L Boldrini, Agglutination of blastospores of *Candida albicans* by concanavalin A and its relationship with the distribution of mannan polymers and the ultrastructure of the cell wall, *J Gen Microbiol* 105 263 (1978)

24 G Tronchin, D Poulain and J Biguet, Cytochemical and ultrastructural studies of the cell wall of *Candida albicans* I Localization of mannan by means of concanavalin A on ultrathin sections *Arch Microbiol* 123 245 (1979)

25 A Cassone, D Kerridge and E F Gale, Ultrastructural changes in the cell wall of *Candida albicans* following cessation of growth and their possible relationship to the development of polyene resistance, *J Gen Microbiol* 110 339 (1979)

26 R A Venezia and R C Lachapelle, The use of ferritin-conjugated antibodies in the study of cell wall components of *Candida albicans*, *Can J Microbiol* 19 1445 (1973)

27 G Tronchin, D Poulain, J Herbaut and J Biguet, Localization of chitin in the cell wall of *Candida albicans* by means of wheat germ agglutinin, *Eur J Cell Biol* 26 121 (1981)

28. M. Pesti, E.K. Novak, L. Ferenczy and A. Svoboda, Freeze fracture electron microscopical investigation of *Candida albicans* cells sensitive and resistant to nystatin, *Sabouraudia* 19: 17 (1981).

29. G. Tronchin, D. Poulain, J. Herbaut and J. Biguet, Cytochemical and ultrastructural studies of *Candida albicans*. II. Evidence for a cell wall coat using Concanavalin A, *J. Ultrastruct. Res.* 75: 50 (1981).

30. J.A. Howlett and C.A. Squier, *Candida albicans* ultrastructure: Colonization and invasion of oral epithelium, *Infect. Immun.* 29: 252 (1980).

31. M. Horisberger and M.F. Clerc, Ultrastructural localization of anionic sites on the surface of yeasts, hyphal and germ-tube forming cells of *Candida albicans*, *Eur. J. Cell Biol.* 46: 444 (1988).

32. F. Heidenreich and M.P. Dierich, *Candida albicans* and *Candida stellatoidea*, in contrast to other *Candida* species, bind iC3b and C3d but not C3b, *Infect. Immun.* 50: 598 (1985).

33. R.A. Calderone, L. Lineman, E. Wadsworth and A.L. Sandberg, Identification of C3d receptors on *Candida albicans*, *Infect. Immun.* 56: 252 (1988).

34. B.J. Gilmore, E.M. Retsinas, J.S. Lorenz and M.K. Hostetter, An iC3b receptor on *Candida albicans*: Structure, function, and correlates for pathogenicity, *J. Infect. Dis.* 157: 38 (1988).

35. A. Bouali, R. Robert, G. Tronchin and J.M. Senet, Characterization of binding of human fibrinogen to the surface of germ-tubes and mycelium of *Candida albicans*, *J. Gen. Microbiol.* 133: 545 (1986).

36. P. Gopal, P.A. Sullivan and M.G. Shepherd, Metabolism of [^{14}C]glucose by regenerating spheroplasts of *Candida albicans*, *J. Gen. Microbiol.* 130: 325 (1984).

37. P.K. Gopal, P.A. Sullivan and M.G. Shepherd, Isolation and structure of glucan from regenerating spheroplasts of *Candida albicans*, *J. Gen. Microbiol.* 130: 1217 (1984).

38. G.H. Fleet, Composition and structure of yeast cell walls, in: "Current Topics in Medical Mycology", M.R. McGinnis, ed., Springer-Verlag, New York (1985).

39. D.R. Kreger and M. Kopecka, On the nature and formation of the fibrillar nets produced by protoplasts of *Saccharomyces cerevisiae* in liquid media: an electronmicroscopic, X-ray diffraction and chemical study, *J. Gen. Microbiol.* 92: 207 (1975).

40. R. Surarit, P.K. Gopal and M.G. Shepherd, Evidence for a glycosidic linkage between chitin and glucan in the cell wall of *Candida albicans*, *J. Gen. Microbiol.* 134: 1723 (1988).

41. J. Friis and P. Ottolenghi, The genetically determined binding of Alcian blue by a minor fraction of yeast cell walls, *Comp. Rend. Trav. Lab. Carlsberg.* 37: 327 (1970).

42. H. Saito, R. Tabeta, T. Sasaki and Y. Yoshioka, A high-resolution solid-state ^{13}C NMR study of (1-3)-ß-D-glucans from various sources. Conformational characterisation as viewed from the conformation-dependent 13C chemical shifts and its consequence to gelation property, *Bull. Chem. Soc. Jpn.* 59: 2093 (1986).

43. R.H. Marchessault and Y. Deslandes, Texture and crystal structure of fungal polysaccharides, in: "Fungal Polysaccharides", P.S. Sandford, K. Matsuda, eds., American Chemical Society Symposium series no. 126. American Chemical Society, Washington DC (1980).

44. G. Kogan, V. Pavliak and L. Masler, Structural studies of mannans from the cell walls of the pathogenic yeast *Candida albicans* serotypes A and B. and *Candida parapsilosis*, *Carbohydr. Res.* 172: 243 (1988).

45. M.V. Elorza, A. Marcilla and R. Sentandreu, Wall mannoproteins of the yeast and mycelial cells of *Candida albicans*: nature of the glycosidic bonds of polydispersity of their mannan moieties, *J. Gen. Microbiol.* 134: 2393 (1988).

46. C.E. Molloy, M.G. Shepherd and P.A. Sullivan, Identification of envelope proteins of *Candida albicans* by vectorial iodination, Microbios 57: 73 (1989).

47. J. Frevert and C.E. Ballou, *Saccharomyces cerevisiae* structural cell wall mannoprotein, *Biochemistry* 24: 753 (1985).

48. A. Saxena, C.F. Hammer and R.L. Cihlar, Analysis of mannans of two relatively avirulent mutant strains of *Candida albicans*, *Infect. Immun.* 57: 413 (1989).

49. W.L. Whelan, J.M. Delga, E. Wadsworth, T.J. Walsh, K.J. Kwon-Chung, R. Calderone and P.N. Lipke, Isolation and characterization of cell surface mutants of *Candida albicans*, *Infect. Immun.* 58: 1552 (1990).

50. D.F. Summers, A.P. Grollman and H.F. Hasenclever, Polysaccharide antigens of the *Candida* cell wall, *J. Immunol.* 92: 491 (1964).

51. H.F. Hasenclever and W.O. Mitchell, Antigenic studies of *Candida*. I. Observation of two antigenic groups in *Candida albicans*, *J. Bacteriol.* 82: 570 (1961).

52 Y Okubo, Y Honma and S Suzuki, Relationship between phosphate content and serological activities of the mannans of *Candida albicans* strains NIH A-207, NIH B-792, and J-1012, *J Bacteriol* 137 677 (1979)

53 L Rahary, R Bonaly, J Lematre and D Poulain, Aggregation and disaggregation of *Candida albicans* germ-tubes, *FEMS Microbiol Lett* 30 383 (1985)

54 A R Holmes, R D Cannon and M G Shepherd, Mechanisms of aggregation accompanying morphogenesis in *Candida albicans*, *Oral Microbiol Immunol* 7 32 (1992)

55 F Staib, Proteolysis and pathogenicity of *Candida albicans* strains, *Mycopathol Mycol Appl* 37 383 (1985)

56 K Yokoyama, H Kaji, K Nishimura and M Miyaji, The role of microfilaments and mictrotubules in apical growth and dimorphism of *Candida albicans*, *J Gen Microbiol* 136 1067 (1990)

57 H Bussey, D Saville, K Hutchins and R G E Palfree, Binding of yeast killer toxin to a cell wall receptor on sensitive *Saccharomyces cerevisiae*, *J Bacteriol* 140 888 (1979)

58 C Boone, S S Sommer, A Hensel and H Bussey, Yeast KRE genes provide evidence for a pathway of cell wall ß-glucan assembly, *J Cell Biol* 110 1833 (1990)

59 A S M Sonnenberg, J H Sietsma and J G H Wessels, Spatial and temporal differences in the synthesis of (1-3)-ß- and (1-6)-ß linkages in a wall glucan of *Schizophyllum commune*, *Exp Mycol* 9 141 (1985)

60 M Staebell and D R Soll, Temporal and spatial differences in cell wall expansion during bud and mycelium formation in *Candida albicans*, *J Gen Microbiol* 131 1467 (1985)

61 N F Robertson, The fungal hypha, *Trans Br Mycol Soc* 48 1 (1965)

62 S Bartnicki-Garcia and E Lippman, Fungal morphogenesis cell wall construction in *Mucor rouxii*, *Science* 165 302 (1969)

63 E Galun, Morphogenesis of *Trichoderma* autoradiography of intact colonies labelled by [^{3}H] N-acetylglucosamine as a marker of new cell wall biosynthesis, *Arch Microbiol* 86 305 (1972)

64 G W Gooday, An autoradiographic study of some fungi, *J Gen Microbiol* 67 125 (1971)

65 E Galun, A Braun, A Frensdorff and E Galun, Hyphal walls of isolated lichen fungi Autoradiographic localization of precursor incorporation and binding of fluorescein-conjugated lectins, *Arch Microbiol* 108 9 (1976)

66 D Hunsley and D Kay, Structure of the Neurospora hyphal apex immunofluorescent localization of wall surface antigens, J Gen Microbiol 95 233 (1976)

67 C Nombela, J Pla, E Herreros, G Gil, M Molina and M Sanchez, Novel targets for antifungal drugs in New Strategies in Fungal Disease , J E Bennett, R J Hay and P K Peterson, eds , Churchill Livingstone, Edinburgh (1992)

68 S I Reed, J A Hadwiger and A T Lorincz, Protein kinase activity associated with the product of the yeast cell division cycle gene CDC28, *Proc Nat Acad Sc of U S A* 82 4055 (1985)

69 D E Levin and J M Bishop, A putative protein kinase gene (kin1+) is important for growth polarity in *Schizosaccharomyces pombe*, *Proc Nat Acad Sc of USA* 87 8272 (1990)

70 R P Hartland, G W Emerson and P A Sullivan, A secreted ß-glucan-branching enzyme from *Candida albicans*, *Proc R Soc Lond* B 246 155 (1991)

71 E Herroro, P Sanz and R Sentandreu, Cell wall proteins liberated by Zymolyase from several asconycetous and imperfect yeasts, *J Gen Microbiol* 133 2895 (1987)

POSSIBLE ROLES OF MANNOPROTEINS IN THE CONSTRUCTION OF CANDIDA ALBICANS CELL WALL

Rafael Sentandreu, Maria Victoria Elorza, Salvador Mormeneo,
Raquel Sanjuan, and Maria Iranzo

Departament de Microbiologia
Facultat de Farmacia
Universitat de València
València, Spain

ABSTRACT

The shape of *Candida albicans* cells depends on their cell walls and some of their mannoproteins may act as modulators of the final molecular architecture. If that were the case, the wall mannoproteins might form part of what could be called a "morphogenetic code".

Experimental results obtained have shown that the wall mannoproteins of both yeast and mycelial cells may be grouped in two families: i)mannoproteins of high molecular weight, rich in mannan and ii) mannoproteins devoid of the mannan moiety. Both groups of proteins are *O*-glycosylated and secreted independently of each other, but released from cell walls by ß-glucanases forming part of the same supramolecular complexes, indicating that some type of connection between them is formed externally to the plasma membrane in the domains of the cell wall itself.

INTRODUCTION

The imperfect fungus *Candida albicans* is an opportunistic pathogen that grows with either a yeast or mycelial morphology, depending upon environmental conditions.[1] It is frequently associated with warm-blooded animals, including humans, as a yeast commensal or pathogen of various tissues and tracts. The incidence of both superficial and invasive candidosis has increased drastically, due probably to iatrogenic factors, immunological diseases, etc. *C.albicans* colonizes human tissues forming septate hyphae and/or pseudophyphae in addition to yeast cells.

The shape of *C. albicans* cells depends on the cell wall, but this structure also plays additional roles during the interaction of *C. albicans* with the mammalian host acting as a virulence factor, elicitor of antibodies, etc.[2] The cell walls of *C.albicans* have a complex molecular architecture formed by the structural polysaccharides (glucan and chitin) which in turn are embedded in an amorphous matrix composed mainly of mannoproteins. The composition of the yeast and mycelial cell walls is qualitatively similar although there are variations in the different polysaccharides contents.[3] The manner in which the cell wall components are synthesized and interconnected in the structure is largely unknown although covalent bonds between chitin-glucan,[4-6] glucan-mannoproteins[7,8] and chitin-mannoproteins,[9] have been reported for *C. albicans* as well as other fungal species.

Little information is available on the changes that take place in the cell wall molecular organization during the yeast-mycelium transition.[10] But "morphogenetic orders" carried out according to a pre-established "program" might direct the interaction and assembly of the wall polymers, once they are found externally to the plasma membrane, to produce the final wall architecture. The major questions concerning this working and provocative hypothesis are: i) which of the cell wall components are endowed with the information needed and as a consequence act as morphogenetic modulators? and ii) how are the catalytic systems present in the domains of the cell wall regulated?

The purpose of the present work is to give an answer to the first question using results from our own and other groups. The second question cannot be answered at this moment and will remain open for the time being, but it is kept in mind throughout this work as it may help in understanding some aspects of it.

ASSOCIATION OF PROTEINS WITH WALL STRUCTURE.

The cell wall proteins of *C. albicans* and other fungal species show a high diversity but can be divided into two major types depending upon their degree of interaction with the cell wall structure. One type interacts by means of non-covalent bonds, and is released in a soluble form by ionic detergents (e.g. sodium dodecyl sulphate)[11] and chaotropic agents (e.g. urea).[12] The second type of proteins ("intrinsic mannoproteins") seems to be covalently bound to other wall components, as they are solubilized only following degradation with hydrolytic enzymes of the networks formed by glucan and chitin[7,9] or by disrupting S-S bonds with mercaptoethanol or dithiothreitol.[9,13]

Significant amounts (>60%) of proteins are solubilized with SDS treatment of isolated walls of both morphologies.[11] Over 40 protein species with molecular weights ranging from 15 to 120 kDa, including a 34 kD a mannoprotein molecule (p34), were detected by sodium dodecyl sulphate-polyacrylamide gel electrophoresis (SDS-PAGE) and no important qualitative differences between yeast and mycelial walls were found.[11] Mannoproteins related to the p34 have been detected in other yeast species indicating that they are a highly conserved type of wall mannoproteins;[14] they have a molecular weight of 31,000 to 35,000 Da depending upon the fungal species, and are found associated with the structural polysaccharides, as well as with the intrinsic proteins.[7,9,15,16] These species, having affinity for all the cell wall components, might play a key role in organizing the architecture of this structure. Klebs and Tanner cloned the gene that codifies the *Saccharomyces cerevisiae* protein, describing it as having a sequence similar to those genes of some plant ß-glucanases.[16]

In the case of *C. albicans* a glucosyl transferase activity has also been assigned to this mannoprotein.[17] Such proteins that present different domains with affinity to glucan, chitin and other mannoproteins might interconnect them with non-covalent bonds by forming a network of relatively low strength. As a consequence this type of proteins could be functionally similar to those regulating formation of microtubules and actin filaments from different subunits.

The low-strength network could direct the formation of more stable linkages (covalent, ionogenic, etc.) producing a structure of higher strength (the final cell wall). Obviously these processes must take place in the domains of the wall itself and would therefore be the most characteristic feature of the wall construction.[10]

The wall mannoproteins that remain after treatment with detergents or chaotropic agents are solubilized following degradation of the skeleton formed by the structural polymers. Zymolyase (a ß-glucanase complex with proteolytic and other minor activities) releases between 50 and 60% of the mannoproteins present in the wall debris while the residual proteinaceous material is solubilized by chitinase in significant amounts.[7,9] Total solubilization of the intrinsic mannoprotein is gained only after some alternative digestions with both hydrolytic systems[7] suggesting that the components of some parts of the wall form interlocking structures.

Though the total amount of the intrinsic mannoproteins in yeast and mycelial cell walls is similar, morphology-specific molecules have been detected. Zymolyase releases from the walls a high molecular weight mannoprotein material which is resolved by SDS-PAGE in four bands in the case of yeast and in two bands in the mycelial cells.[5] Although the bands observed from the walls could be the result, in part, of the proteolytic activities present in the enzymic preparations, the detected differences in mobility suggest the existence of diverse molecular organizations. In confirmation of this suggestion the use of monoclonal antibodies has demonstrated the presence of antigens that are expressed exclusively by either the yeast or mycelial cells.[2,18]

RELATIONSHIP BETWEEN WALL COMPONENTS AND CELL SHAPE

Both chitin and glucans have been considered responsible for the cell wall morphology and morphogenesis in *C. albicans* [18] and in other fungal species,[19] but these polymers have the tendency to crystallize, and their degree of crystallization *in vivo* is apparently very low; moreover the fact that the polysaccharides are very rigid molecules hints that they may be responsible for the morphology of the structure but not for its morphogenesis. On the other hand proteins are very flexible molecules from the structural point of view, and may be endowed with catalytic activities, suggesting that they are responsible for cell wall morphogenesis.

Recently we found a monoclonal antibody (MAb 4C12), that recognizes an antigen specific to mycelial cell walls,[20] and detects the epitope in the materials solubilized by SDS, mercaptoethanol, Zymolyase and chitinase,[9] hinting that the mycelial antigen impregnates the entire wall structure. Furthermore, inhibition of the dimorphic transition by 1,4-diaminobutanone (an agent that deregulates polyamine biosynthesis), blocks formation of the antigen reacting with the MAb 4C12,[21] but does not interfere with yeast cell wall composition, in turn suggesting that the drug switches off the expression of a specific protein and, as a result, formation of mycelial cells can not take place.

In addition to *C.albicans*, changes in cell morphology related to mannoproteins have been detected in *S. cerevisiae* a cells. Supplementation of α-factor cells of mating type a results in the modification of their morphology, giving rise to pear-like cells called "shmoos", with a concomitant modification in the glycosylation pattern of the mannoproteins.[22]

As a working hypothesis it can be assumed that the wall mannoproteins may have critical roles in leading the wall organization. If so the mannoproteins should be located not only in the external surfaces of the wall, as has been previously described,[2,23] but also in the internal areas. Studies carried out after partial degradation of the structural skeleton with chitinase or Zymolyase, to open up the compact wall structure, demonstrated that mannoproteins are located randomly throughout the wall.[9]

INTERACTION AND ASSEMBLY OF WALL COMPONENTS

Wall construction is the result of different processes that must be coordinated to produce the final structure; some occur intracellularly (synthesis and secretion of mannoproteins and extrusion of glucan and chitin[19]) while others take place externally to the plasma membrane (interaction and assembly of all wall components[10]).

Once the cell wall components are found in the periplasmic space they probably maintain a dynamic equilibrium due to the formation and breakage of bonds of low affinity that help to initiate formation of other ones of high affinity (covalent bonds). These high-affinity bonds might be the result of reactions taking place between different molecules and under different conditions leading to the extension of the wall. Very few studies have been published about the mechanisms implicated in the construction of the walls. One possibility is to dissect them with drugs that interfere with different steps. The use of drugs that inhibit mannoproteins synthesis results in alterations in cell wall structure but other cellular activities are also disturbed. Another possibility is to interfere with the processes occurring externally to the plasma membrane that should alter specifically the interaction between molecules and, as a consequence, cell wall assembly and organization. Calcofluor white, a dipole substance that has the ability to form hydrogen-bonds with polysaccharides, interferes with the formation of their crystalline structures. In the case of *C. albicans* the stain produces abnormal cell walls[24] by intercalation of the dipole molecules between the nascent polysaccharide molecule,[25] demonstrating the existence of a temporal discontinuity between synthesis and crystallization. This gap allows the interaction on mannoproteins and the structural polysaccharides. If this hypothesis is true, then wall assembly may occur in two successive steps. Initially mannoproteins such as p34, exhibiting affinity to glucan and chitin as well as to other mannoproteins, would form temporal molecular networks, that during the second step will be stabilised by the formation of additional and high affinity linkages.

In order to follow the interaction and assembly of specific molecules we have used two monoclonal antibodies. Analytical as well as cytological determinations have demonstrated that the MAb 4C12 reacts specifically with an epitope which is found only in mycelial cells walls, whereas a second one (1B12), recognizes an epitope that is present in both yeast and mycelial cells.

Initially the presence of precursors of the wall mannoproteins was sought in the supernatant of regenerating protoplasts. When the molecules secreted by mycelial protoplasts are analyzed by Western blotting, the MAb 4C12 recognizes a band with an apparent mobility of 180 KDa (p180), which is not modified when it is secreted in the presence of tunicamycin.[20] These results suggest that the antigen detected by the MAb is devoid of N-linked sugar chains. On the other hand the antigen gives a positive reaction with concanavalin A, indicating that it is an O-glycosylated protein. When the materials examined with MAb 4C12 are obtained from cells following Zymolyase digestion, materials with low electrophoretic mobility are detected.[9,20] These materials suffer a significant reduction in their molecular weight after treatment with endo-N-acetylglucosaminidase (Endo-H). These results support the notion that the antigen is an O-glycosylated protein when it is secreted, but carries N-glycosydic chains when it is solubilized from the walls following degradation of the glucan network. These observations imply that the reactions responsible for the acquisition of the N-glycosydic chains occur as a result of their covalent incorporation to the wall structure. We know nothing about the processes involved in the incorporation of the antigen to the wall, but it appears that cross-linking with N-glycosylated mannoproteins does take place. To confirm that the behavior of the antigen detected with MAb 4C12 was not specific for this molecule, but rather represents a general one, another monoclonal antibody (MAb 1B12) that recognizes an epitope, present in both yeast and mycelial cell walls, has been used (results to be published elsewhere). The information obtained with both monoclonal antibodies indicates that various mannoproteinaceous antigens following secretion are incorporated into the wall by forming covalent connections with other materials, thus demonstrating that significant modifications take place as a result of this process.

CONCLUDING REMARKS

The wall of the fungal cell is one of its most striking supramolecular structures. Chemically it is very complex, so significant genetic information is probably needed to synthesize and coordinate all the products and processes involved to produce them. Moreover it is also apparent from the data reported above that the functional properties of the cell walls depend not only on their chemical composition, but mainly on how the individual constituents are spatially arranged.

From the functional point of view, and as a consequence for the viability of C. albicans, the reactions that occur in the domains of the wall, and which produce the final structure, are as important as those taking place within the cytoplasm.

For the regulation of the structure of the wall of both morphologies, processes that occur intracellularly (synthesis and secretion or extrusion of wall polysaccharides), as well as others that take place outside the plasma membrane (interaction and assembly of wall components), must be coordinated. From the results reported in this review, it seems that mannoproteins are able to produce low-affinity bonds with other wall components that may play a pivotal role in modulating their spatial distribution, facilitating formation of high affinity connections (covalent bonds) between cell wall components leading to the final three-dimensional network characteristic of each morphology.

The high insolubility of the intrinsic mannoproteins, due probably to their covalent attachment to the glucan-chitin framework, is the major obstacle to their study. Using monoclonal antibodies as probes for the corresponding antigens, we have been able to analyze the precursor-product relationship of two individual mannoproteins. A significant conclusion that can be drawn is that two types of mannoproteins seem to form part of the wall structure: one of them is *O*-glycosylated, whereas the other is *O*- and *N*-glycosylated. The second important conclusion is that the intrinsic mannoproteins undergo significant modifications during their incorporation into the wall due to different reactions occurring in the domain of the wall itself.

To understand the changes that are responsible for the yeast-mycelium transition, it is very important to determine not only the chemical nature of the wall polysaccharides but also how these polymers are distributed and interconnected. Analysis of the dynamic aspects involved in the cell wall construction will dominate future research. Which are the genetic and physiologic parameters that modulate the expression of morphology related genes; how the gene products are secreted and which catalytic systems are responsible of the final wall architecture? The combined research efforts of biochemists, microbiologists, molecular geneticists and other specialists would help to furnish answers to all these questions.

ACKNOWLEDGMENTS

This work was supported by grants from the Dirección General de Investigación Científca y Tecnica (PB90-0424), the Commission of the European Communities (EEC Group on *Candida albicans* Cell Biology and Pathogenicity and Cl1*.0631.M) and Glaxo, S.A.

M.V.E. is on leave from the Instituto de Microbiología Bioquímica , C.S.I.C. (Salamanca).

REFERENCES

1. F.C.Odds, "*Candida* and Candidosis", Balliere Tindall, London (1988).
2. A. Cassone, Cell wall of *Candida albicans*: its functions and its impact on the host, in :"Current topics in medical mycology", vol.3, M.P. McGinnis and M.Borgers, eds., Springer Verlag, Berlin, Heidelberg (1980).
3. M.G. Shepherd, R.T.M. Poulter and P.A. Sullivan, *Candida albicans*: biology, genetics, and pathogenicity, *Ann.Rev. Microbiol.* 39: 579 (1985).
4. R. Surarit, P.K. Gopal and M.G. Shepherd, Evidence for a glycosidic linkage between chitin and glucan in the cell wall of *Candida albicans, J. Gen. Microbiol.* 134: 1723 (1988).
5. J.G.H. Wessels, J.H. Sietsma and A.S.M. Sonnenberg, Wall synthesis and assembly during hyphae morphogenesis in *Schizophyllum commune, J. Gen. Microbiol.* 129: 1607 (1983).
6. P.C.Mol and J.G.H.Wessels, Linkages between glucosaminoglycan and glucan determine alkali-insolubility of the glucan in walls of *Saccharomyces cerevisiae, FEMS Microbiol. Lett.* 41: 95 (1987).
7. M.V. Elorza, A. Marcilla and R. Sentandreu, Wall mannoproteins of the yeast and mycelial cells of *Candida albicans*: nature of the glycosidic bonds and polydispersity of their mannan moieties, *J. Gen. Microbiol.* 134: 2393 (1988).
8. J. Van Rinsum, F.M. Klis and H.Van Den Ende, Cell wall glucomannoproteins of *Saccharomyces cerevisiae* mnn9, *Yeast* 7: 717 (1991).
9. A. Marcilla, M.V. Elorza, S. Mormeneo, H. Rico and R. Sentandreu, *Candida albicans* mycelial wall structure: supramolecular complexes released by Zymolyase, chitinase and ß-mercaptoetanol, *Arch. Microbiol.* 155: 312 (1991).
10. R. Sentandreu, J.P. Martinez, M.V. Elorza and S. Mormeneo, Relationships between dimorphism, cell wall structure and surface activities in *Candida albicans*, in "*Candida albicans*, cellular and molecular biology", R. Prasad, ed., Springer Verlag, Berlin, Heidelberg (1991).
11. M.V. Elorza, A. Murgui and R. Sentandreu, Dimorphism in *Candida albicans*: contribution of mannoproteins to the architecture of yeast and mycelial cell walls, *J. Gen. Microbiol.* 131: 2209 (1985).

12. W.L. Chaffin and D.M. Stocco, Cell wall proteins of *Candida albicans, Can. J. Microbiol.* 29: 1438 (1983).

13. F.W. Chattaway, S. Shenolikar and A.J.E. Barlow, The release of acid phosphatase and polysaccharide- and protein-containing components from the surface of the dimorphic forms of *Candida albicans* by treatment with dithiothreitol, *J. Gen. Microbiol.* 83: 423 (1974).

14. E. Herrero, P. Sanz and R. Sentandreu, Cell wall proteins liberated by Zymolyase from several ascomycetous and imperfect yeast, *J. Gen. Microbiol.* 133: 2895 (1987).

15. A. Murgui, M.V. Elorza and R. Sentandreu, Tunicamycin and papulacandin B inhibit incorporation of specific mannoproteins into the walls of *Candida albicans* regenerating protoplasts, *Biochim. Biophys. Acta* 884: 550 (1986).

16. F. Klebl and W. Tanner, Molecular cloning of a cell wall exo-ß-1,3-glucanase from *Saccharomyces cerevisiae, J. Bacteriol.* 171: 6259 (1989).

17. R.P. Hartland, G.W. Emerson and P.A. Sullivan, A secreted ß-glucan-branching enzyme from *Candida albicans, Proc.. Roy. Soc. London* 246: 115 (1991).

18. A. Cassone and A. Torosantucci, Immunological moieties of the cell wall, in "*Candida albicans,* cellular and molecular biology", R. Prasad, ed., Springer Verlag, Berlin, Heidelberg (1991).

19. E. Cabib , B. Bowers, A.Sburlati and S.J. Silverman, Fungal cell synthesis: the construction of a biological structure, *Microbiol. Sci.* 5: 370 (1988).

20. M.V. Elorza, S. Mormeneo, F. Garcia de la Cruz, C. Gimeno and R. Sentandreu, Evidence for the formation of covalent bonds between macromolecules in the domain of the wall of *Candida albicans, Biochem. Biophys. Res. Commun.* 162: 1118 (1989).

21. J.P. Martinez, J.L. Lopez-Ribot, M.L. Gil, R. Sentandreu and J. Ruiz-Herrera, Inhibition of the dimorphic transition of *Candida albicans* by the ornithine decarboxylase inhibitor 1,4-diaminobutanone: alteration in the glycoprotein composition of the cell wall, *J. Gen. Microbiol.* 136: 1937 (1990).

22. E. Herrero, E. Valentin and R. Sentandreu, Effect of α-factor on individual wall mannoproteins from *Saccharomyces cerevisiae* a cells, *FEMS Microbiol. Lett.* 27: 293 (1985)

23. J.G.H. Wessels and J.H. Sietsma, Fungal cell walls: a survey, in "Plant Carbohydrates II", vol. 13B, W. Tanner and F.A. Loewus, eds., Springer Verlag, Berlin, Heidelberg, New York (1981).

24. M.V. Elorza, H. Rico and R. Sentandreu, Calcofluor white alters the assembly of chitin fibrils in *Saccharomyces cerevisiae* and *Candida albicans* cells, *J. Gen. Microbiol.* 128: 1577 (1983).

25. C.A. Vermeulen and J.G.H. Wessels, Chitin synthesis by a fungal membrane preparation. Evidence for transient non- crystalline state of chitin, *Eur.J. Biochem.* 158: 411 (1986).

IS THERE A ROLE FOR STEROLS AND STEROIDS IN FUNGAL GROWTH AND TRANSITION FROM YEAST TO HYPHAL-FORM AND VICE VERSA? AN OVERVIEW

Hugo Vanden Bossche, and Patrick Marichal

Department of Comparative Biochemistry
Janssen Research Foundation
B-2340 Beerse, Belgium

ABSTRACT

Ergosterol seems to be not only the best-suited sterol to maintain fungal membrane integrity and activity; there is evidence that this 24-alkylated sterol is also involved in critical non-membrane associated functions. A number of fungal species contain steroid-binding proteins and vertebrate steroid hormones seem to affect fungal morphogenesis. In this article it is speculated that ergosterol may also play a role in the yeast-to-mycelium transformation. The possible role of steroid hormones and endogenous ligands in fungal growth and morphogenesis is also discussed.

INTRODUCTION

Sterols are essential components of eukaryotic membranes. They are either synthesized *de novo* from acetate or taken up from the environment. Cholesterol is the main sterol found in mammals; campesterol, sitosterol and stigmasterol are the most prominent in the plant kingdom; ergosterol (Fig. 1) is the characteristic sterol in the higher fungi (Ascomycetes, Basidomycetes and Fungi imperfecti). These are just a few examples of the great number of sterols identified. For example plant cells synthesize a cocktail of sterols. Regardless of the great number of sterols found in other organisms, sterols containing a $\Delta^{5,7}$-diene system, the characteristic feature of ergosterol, are restricted to higher fungi[1] and some protozoa, e.g. *Leishmania*.[2]

Sterols are known to interact with phospholipids and to maintain optimal membrane fluidity and integrity. They also modulate the activity of membrane-bound enzymes. Next to their bulk functions in membranes sterols are required for cellular proliferation and differentiation (for reviews see Ref. 1, 3-8).

A number of questions can be formulated:

1. Why should fungal cells contain ergosterol and not cholesterol?
2. Is ergosterol the best-suited sterol to maintain optimal membrane fluidity and integrity and to regulate the activity of membrane bound enzymes?
3. Is the 24β-methyl group (Fig. 1), that distinguishes ergosterol from cholesterol, required to regulate cellular processes in fungi?
4. Are ergosterol and/or its derivatives (e.g. steroids) involved in morphogenesis?

Numerous comprehensive review articles on the biosynthesis and on the role of ergosterol and other 24-alkylsterols in the maintenance of membrane integrity and activity of membrane-bound enzymes have been published recently.[1,3-9] The immediate need for repetition is not apparent. Therefore, in this paper focus will be on the two last questions only.

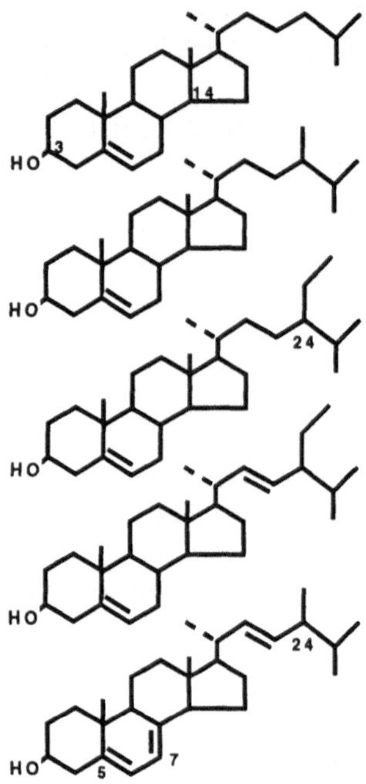

Fig. 1 Structures of mammalian, plant and fungal sterols

IS ERGOSTEROL INVOLVED IN THE YEAST-TO-MYCELIUM TRANSFORMATION?

Comparison of the sterols of a *Candida albicans* strain (MEN) with those of a mutant (MM 2002) unable to produce mycelia, both grown in a mycelium-promoting medium, showed no significant differences in sterol composition between the yeast and mycelial morphological forms.[10] However, determining the ergosterol content in a *C. albicans* isolate (ATCC 44859) grown either in a yeast-promoting or mycelium-promoting medium revealed that per mg protein, hyphae contain about 1.7 times more ergosterol than yeasts (Table 1). Sadamori[11] also found the ergosterol content to be higher in the mycelial form. This author also measured a higher lanosterol content and the squalene content was found to be about 10 times higher in the mycelial than in the yeast form.[11] It should be mentioned that the sterol composition is related to the developmental and nutritional stage of the analyzed strain and to the analytical procedures employed for sterol extraction, preparation and separation.[1,13] Therefore, the question whether or not hyphal and yeast forms differ in their sterol composition is not so important. It is more relevant to discuss the impact of ergosterol and other sterols on the cell cycle and morphogenesis.

Table 1. Morphology, protein and ergosterol content of *C. albicans* ATCC 44859*

Medium		NYP	CYG
Incubation	time (h)		
		24	24
	temperature (°C)	37	37
Morphology		mycelium	yeast
Protein content (µg/ml)		16.0 ± 5.5	632.9 ± 180.5
Ergosterol content (µg/ml)		0.6 ± 0.1	12.9 ± 1.0
Ergosterol (µg/mg protein)		36.3 ± 8.8	21.8 ± 5.8**
Ergosterol (% of sterols)		76.8 ± 5.8	80.7 ± 6.3

* NYP-medium: 1 mM N-acetylglucosamine, 3.35 g yeast nitrogen base, 1 mM L-proline, 0.1 M sodium phosphate pH 7.0 and 4.5 g NaCl per litre H_2O; CYG-medium: 5 g casein hydrolysate, 5 g yeast extract and 5 g glucose per litre H_2O. The protein content was determined by the Biorad® method with albumin as standard. The ergosterol content of heptane extracts of cell homogenates was determined by measuring the absorption difference between 292nm and 282nm. An ergosterol standard curve was used to calculate the ergosterol contents.[12] Results are mean values of 6 experiments ± SD.
** $p = 0.008$

Using polyene-resistant, hypha-deficient mutants of *C. albicans* Shimokawa *et al.*[14] showed that the mutants (KD 4700, KD 4900) contained only traces of ergosterol as compared with the wild strain (KD 14). Although the ergosterol content was low, KD 4700 still grew in its yeast form, albeit at a slower rate than the wild-type (mean mass-doubling times: 140 min versus 80 min). A revertant (KD 4734) showed wild-type characteristics with regard to polyene sensitivity, sterol composition, growth rate and hypha formation. These studies suggest that ergosterol depletion has more dramatic consequences for hypha than for yeast cell formation. Instead of ergosterol KD 4700 and KD 4900 accumulate 14-methylsterols. Therefore, it cannot be excluded that the block in hypha formation originates from the accumulating 14-methylsterols of which some have been shown to change the physicochemical properties of membranes.[5,15]

Two other studies confirm the results of Shimokawa *et al.* Bard *et al.*[16] and Lees *et al.*[17] used a nystatin-resistant, cytochrome P450-deficient mutant of *C. albicans* (strain D10). Since the 14α-demethylation of lanosterol or eburicol (24-methylene dihydrolanosterol) is a P450-dependent reaction, D10 does not form ergosterol and accumulates the following 14-methylsterols: 14-methylfecosterol (37 %), obtusifoliol (13.4 %), lanosterol (13.4 %), eburicol (25.7 %) and 24,25-dihydrolanosterol (19.8 %).[16] Omission of the nystatin selective pressure resulted in the revertant strain D10R which contained ergosterol (70.4 %), ergosta-7,22-dien-3ß-ol (9.7 %), ergosta-dien-3ß-ol (5.4 %) and lanosterol (14.4 %).[16] This revertant is sensitive to nystatin,[16] miconazole, clotrimazole and ketoconazole[17] and has a mean doubling time of 72 min as compared to the 114 min mean doubling time observed for strain D10. Strain D10 was shown to be defective in hypha formation: % hypha formation in *C. albicans* D10 is 4.8 as compared with 100 % in the revertant.[17] These results again suggest that hypha formation is dependent on the P450-dependent 14α-demethylase. The fact that D10 is synthesizing significant amounts of 14-methylfecosterol may be the origin of the residual growth observed under yeast-forming conditions. Indeed, studies of Orth and Sisler[18] indicate that *Ustilago maydis* sporidia treated with 14α-demethylase inhibitors accumulate 14α-methylfecosterol and show a residual slow growth rate, as does a 14α-demethylase-deficient *U. maydis* mutant (*erg-40*)[19] and the P450-deficient *C. albicans* strain D10.[16] In contrast, the terbinafine-resistant *U. maydis* mutant AR212 which, upon treatment with the 14α-demethylase inhibitor propiconazole, showed no accumulation of 14α-methylfecosterol, is much more sensitive to the azole antifungals etaconazole, ketoconazole and propiconazole than the wild-type, which accumulates 14α-methylfecosterol.[18]

Azole antifungals, such as miconazole, ketoconazole and itraconazole, are known to inhibit the P450-dependent 14α-demethylase (for reviews see Refs. 1, 4, 20, 21). If hypha formation is dependent on the 14α-demethylation of lanosterol or eburicol, *C. albicans* grown in mycelium-promoting medium supplemented with, e g., itraconazole should be defective in hypha formation. As shown in Fig. 2C, the addition of 30 nM itraconazole resulted in yeast growth only and at 100 nM itraconazole non-growing clusters of yeast cells were present (results not shown).

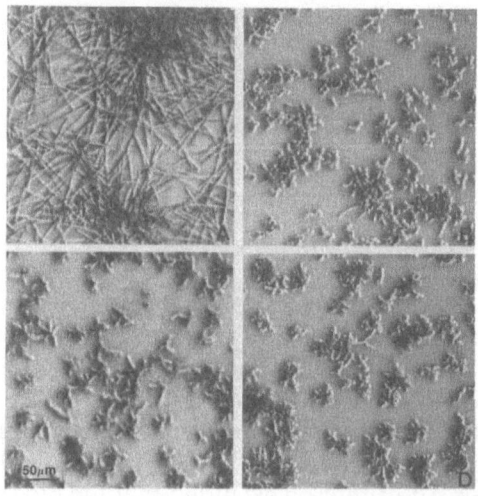

Fig. 2 Effects of amorolfine (B), itraconazole (C) and terbinafine (D) on hypha formation *C albicans* (ATCC44859) was grown for 24 h at 37 °C in the mycelium promoting NYP medium Prior to inoculation 0 1% DMSO (A), 3 µM amorolfine, 0 03 µM itraconazole or 10 µM terbinafine (D) were added to the cultures Photographs were taken after 24 h of incubation at 37 °C.

These results indicate that itraconazole-treated cells are not only blocked in their hypha forming capability but they also do not show the residual growth rate observed with the 14-methylfecosterol-accumulating *U. maydis*[18] and *C. albicans* strain D10.[17] Analysis of the sterols synthesized from acetate indicates that at 30 nM itraconazole 21.5 % of the extract still consists of ergosterol (Table 2).

Thus, even in the presence of this substantial ergosterol synthesis *C. albicans* is blocked in hypha formation (Fig. 2C). At 100 nM small amounts of ergosterol and

14-methylfecosterol are found (Table 2). The major accumulating sterol is 14α-methyl-ergosta-8,24-dien-3ß,6α-diol (3,6-diol; Table 2) which, at least in *Saccharomyces cerevisiae*, seems to be unable to support growth.[22,23] The other accumulating sterols are obtusifoliol, lanosterol and eburicol. These sterols are methylated both at position 4 and 14 and are considered as noxious for membrane function.[15,24,25] Thus, these sterols may also contribute to itraconazole's fungitoxic effects.

Table 2. Effects of itraconazole on ergosterol synthesis from $[^{14}C]$ acetate by *C. albicans* (ATCC 44859) grown in a mycelium-promoting medium.*

Sterol	% of Total		
	0	30 nM	100 nM
Ergosterol	76.8	21.5	2.8
14α-Methylfecosterol	1.4	3.7	5.4
Obtusifoliol	0	5.3	7.6
Lanosterol	0	2.7	3.2
Eburicol	0	5.3	5.8
14α-Methyl-ergosta-8,24(28)-diene-3β,6α-diol	5.9	35.7	48.6
14α-Methyl-ergosta-5,7,22,24(28)-tetraenol	0.4	1.6	2.6
Squalene	6.6	11.0	12.4
Unidentified sterols	8.7	12.1	10.2
4,14 dimethylzymosterol	0.4	3.0	1.7

* Cells were collected after 24 h of growth at 37 °C in NYP medium supplemented with $[^{14}C]$ acetate.[12] Separation of the sterols was performed by HPLC on a Zorbax C8 column. Sterols were eluted with a methanol:H2O mixture (95:5) at a flow rate of 1 ml/min for 25 min after which the column was eluted with pure methanol, in order to elute less polar metabolites. Sterols were identified according to their relative retention times and by gas chromatographic-mass spectrometric analysis.[12] Itraconazole (30 nM and 100 nM) and/or DMSO were added immediately before inoculation.

As far as the present authors are aware viable 14α-demethylase-deficient mutants have been described for yeasts only (see also Refs. 26-28). Thus, such a defect may be lethal for filamentous fungi and compounds that completely block ergosterol synthesis and induce the accumulation of 14-methylsterols that disturb membrane function should be cidal for filamentous fungi. Indeed, itraconazole at 1.4 µM proved to be cidal for *Trichophyton mentagrophytes*.[29] At this concentration small amounts of ergosterol (± 5 % of the control) are found only and eburicol and 3,6-diol represent more than 60 % of the sterols extracted (unpublished results).

The studies summarized so far suggest either that ergosterol is required for hypha formation or ergosterol depletion plus accumulation of membrane-disturbing 14-methylsterols block hypha formation.

How can We Discriminate Between Ergosterol and 14-Methylsterols?

One possibility to discriminate between ergosterol and 14-methylated sterols is to use ergosterol biosynthesis inhibitors that interfere with metabolic steps before

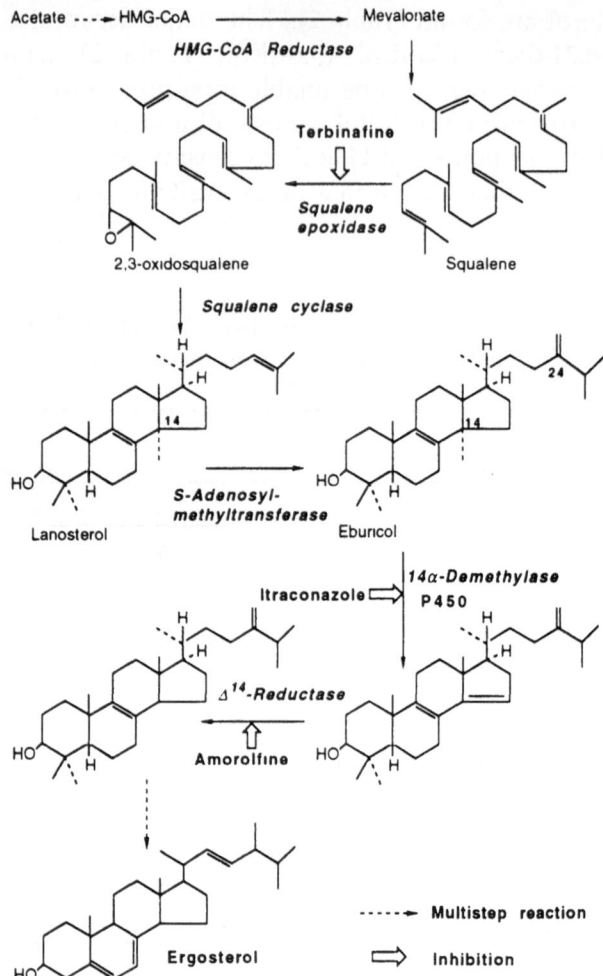

Acetate ···► HMG-CoA ─────────► Mevalonate

HMG-CoA Reductase

Terbinafine

Squalene epoxidase

2,3-oxidosqualene   Squalene

Squalene cyclase

Lanosterol   Eburicol

S-Adenosyl-methyltransferase

Itraconazole ⇨ 14α-Demethylase P450

Δ14-Reductase

Amorolfine

Ergosterol

·····► Multistep reaction

⇨ Inhibition

Fig. 3 Simplified ergosterol biosynthesis pathway with the target enzymes for terbinafine, itraconazole and amorolfine.

or after the 14α-demethylation of lanosterol or eburicol (Fig. 3). Terbinafine[30] is known to inhibit the squalene epoxidase and one of the target enzymes for amorolfine is the Δ14-reductase.[31] Inhibition of squalene epoxidase by terbinafine results in a decreased ergosterol synthesis and accumulation of squalene. So far there is no evidence for fungitoxic effects of squalene. A number of studies suggest the opposite:

1. Microgram amounts of squalene have been found in the vegetative cells of the plant pathogenic *Phytophthora cactorum*.[32] This fungus is unable to epoxidize and cyclize squalene to lanosterol[32,33] and sterols are not an absolute requirement for its vegative growth.[33,34]

2. Differential scanning calorimetry of multilamellar vesicles of dipalmitoylphosphatidylcholine containing 10 to 35 mole% squalene did not show any effect on the transition temperature or the enthalpy of melting.[35] Thus, squalene differs from lanosterol, which at 15 mole% decreased the enthalpy of melting by about 60 %.[36]

3. When *T. mentagrophytes* is grown for 48 h in PYG-medium [polypeptone (10 g/l), yeast extract (10 g/l) and glucose (40 g/l)] supplemented with increasing

concentrations of squalene (1-16 mg/100 ml), growth as measured by protein content increases with increasing concentrations of added squalene (Fig. 4A). The increased protein content coincides with an increased ergosterol content (Fig. 4B). However, most of the squalene accumulates within the cells (Fig. 4B). At a squalene concentration of 15 mg/100 ml medium the ratio of squalene:ergosterol is 6.8; this is similar to the ratio reached when *T. mentagrophytes* is grown for 48 h in the presence of terbinafine at its MIC value (i.e. 10^{-8} M).[30] These results suggest that the fungicidal activity for filamentous fungi is related more to the decreased ergosterol content than to the accumulation of squalene.

The efficacy of amorolfine may also originate from the lack of ergosterol rather than the accumulation of other sterols.[31] This hypothesis is supported by the elegant studies of Marcireau *et al.*[37] on the effects of the amorolfine analog fenpropimorph. Using the *S. cerevisiae* FKaux 30 mutant, which is permeable to exogenous sterols, although it is not auxotrophic, these investigators showed that the effects of fenpropimorph are not due to the accumulation of abnormal sterols (e.g. ignosterol) in treated cells, but are linked to the decreased ergosterol content. Inhibition of ergosterol synthesis leads to the arrest of cell proliferation in the unbudded G1 phase of the cell cycle.[37]

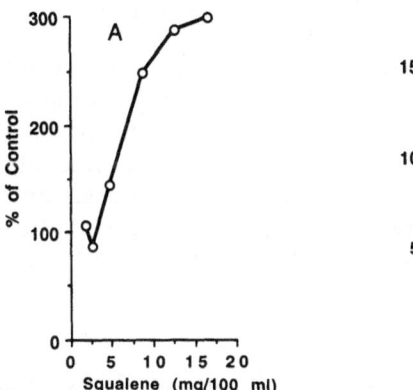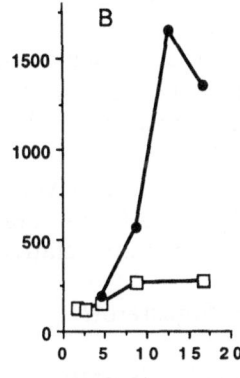

Fig. 4 Effects of exogenously added squalene on the protein content (A) and (B) intracellular ergosterol and squalene contents of *Trichophyton mentagrophytes*. Conidia of *T. mentagrophytes* B32663 were prepared according to Reinhardt *et al.*[64] A culture was grown for 3 weeks on potato-dextrose agar. A solution of 0.01% Tween-40® was pipetted on to the agar surface. The culture suspension was removed and poured into a sterile 500 ml Erlenmeyer flask containing sufficient sterile glass beads (6 mm diameter) to form a layer of at least 3 beads depth. The flask was shaken on a gyratory shaker at 250 rpm for 30 min to homogenize the culture and free the spores from the hyphae. After shaking, the homogenized suspension was filtered through sterile glass wool. The filtrate was centrifuged aseptically for 20 min at 5000 g and the pellet washed twice with sterile saline. 100 ml of polypeptone-yeast extract-glucose (10-10-40 g/l) medium was inoculated to a starting concentration of 10^6 conidia per ml. Cultures were grown for 48 h at 30 °C in a reciprocating shaker. Squalene was added before the medium was autoclaved. Cultures were collected on 8 μm filters and thouroughly washed with 50% ethanol. The mycelial mass was homogenized by means of glass beads. The protein content was determined by the Biorad® method with albumin as standard. Sterols were extracted with heptane after saponification. Separation of the sterols was performed by HPLC as described in Table 1. The ergosterol and squalene content of the heptane extracts of cell homogenates was determined after HPLC separation and according to a standard curve.

Both amorolfine (3 μM) and terbinafine (10 μM) suppress hypha formation by *C. albicans*, similar to the suppression seen with itraconazole (30 nM) (Fig. 2). Thus, these studies suggest that ergosterol is indeed required for hypha formation and that ergosterol depletion rather than the accumulation of precursors and/or abnormal sterols is the key to the action of antifungals on hypha forming capability and filamentous growth.

It should be noted that the impact of ergosterol on morphogenesis may be limited to fungi sensitive to ergosterol biosynthesis inhibitors. Indeed, some fungi exhibit a high degree of natural tolerance to sterol biosynthesis inhibitors.[1,38] The best known examples are species of *Phytophthora* and *Pythium*. These species do not produce sterols or do not require them for vegetative growth.[32,34] Natural tolerance of fungi to ergosterol biosynthesis inhibitors is not necessarily related to the *in vivo* response of the respective 14α-demethylases. Weete and Wyse[38] showed that the triazole antifungal propiconazole inhibited the 14α-demethylation of lanosterol in *Mucor rouxii*, a member of the Mucorales which shows a relatively high tolerance to 14-demethylase inhibitors.[39] This inhibition resulted in a dose-related decrease in ergosterol and an increase in 14-methylsterols such as eburicol.[38] By contrast with sensitive species[1,4,40] the decreased ergosterol synthesis did not result in an increase in linoleic acid. The results obtained by Weete and Wise[38] suggest that the quantitative and perhaps the qualitative nature of the requirement for sterols may be different in *M. rouxii*, and perhaps other tolerant Mucorales, than in the more sensitive fungi.

It is possible that in *Mucor* spp and other species tolerant to ergosterol biosynthesis inhibitors fatty acid synthetases, phospholipid synthesis and turnover are more essential than sterols for morphogenesis (for a review see Ref. 41). Indeed, cerulenin, a known inhibitor of fatty acid synthetases, added to *M. racemosus* cultures at sublethal concentrations did not affect the yeast growth rate, but the organism was not able to effect the yeast-to-hypha transformation. Cerulenin was also found to inhibit the increased rate of phospholipid synthesis that accompanies the *Mucor* yeast-to-hypha morphogenesis and cerulenin also significantly depressed the rates of turnover of phosphatidyl choline and phosphatidyl ethanolamine relative to the rates in untreated *Mucor*.

The Function of Ergosterol?

In most fungal cells ergosterol is not only the sterol best suited to maintain membrane integrity and activity.[1,3,5] There is evidence that this 24-alkylsterol is also involved in critical non-membrane-associated functions in the cell and this at extremely low concentrations. For example, a sterol auxothroph, *S. cerevisiae* RD5-R, does not grow on cholestanol till 1.2 nM ergosterol is added.[42] Kawasaki *et al.*[43] studied the yeast mutant GL7, a sterol auxotroph due to a defective squalene oxide to lanosterol cyclase and a lesion in the heme pathway. They showed that its relatively slow growth rate on medium containing cholesterol is markedly accelerated by supplementation with small amounts (250 nM) of ergosterol. Under these conditions cellular phospholipid synthesis was enhanced. One of the ergosterol-stimulated processes is the methylation of phosphatidylethanolamine to phosphatidylcholine (PC).[42] These investigators suggest that in yeast the metabolic effect of ergosterol on phospholipid synthesis is not generalized but is exerted specifically on certain key transformations, of which a transmethylation step producing PC may be one. Using the same sterol auxotroph of *S. cerevisiae* Dahl *et al.*[44] showed that sterol-depleted cells accumulate in an unbudded, G1 phase. Cell budding and proliferation are reinitiated upon addition of cholesterol and 250 nM ergosterol. Cholesterol alone is much less effective. Stimulation of a protein kinase antigenically related to pp60^{v-src} shows a positive correlation with exit from G1 phase following ergosterol addition.[44] Furthermore, a partially purified, membrane-associated protein kinase from GL7 is stimulated 2- to 3-fold *in vitro* by hormonal levels of ergosterol (1 nM). By contrast, the effective concentration for stimulation by cholesterol is 1 μM.[44] Ergosterol-stimulated cells also demonstrate an increase in phosphatidylinositol kinase activity.[44] The data obtained by Kawasaki *et al.*[43] and Dahl *et al.*[44] suggest that hormonal levels of ergosterol participate in a signalling process associated with a transmethylase, protein kinase and/or phosphatidylinositol kinase possibly involved in yeast cell cycle control. Thus ergosterol can serve as a signal for membrane-associated metabolic events and do so in amounts too small to influence the physical state of the membrane. As suggested by Bloch[45] ergosterol may therefore prove to be the most primitive of the steroid hormones.

Unfortunately, knowledge of the details of ergosterol's signal function in dimorphic and filamentous fungi is almost entirely unavailable. In an interesting study Nes *et al.*[46] show that regulation of sterol C-24 transalkylation may be a mechanism to mediate life cycle events of *Gibberella fujikuroi* (*Fusarium moniliforme* in the imperfect stage). Indeed, when cultures of *G. fujikuroi* are incubated with 24-epiiminolanosterol the introduction of a methyl group into sterol side chains at C-24 is blocked inducing a mycelial accumulation of lanosterol and other 24-desalkylsterols. This altered sterol composition leads to aberrant mycelial membranes resulting in growth inhibition. These investigators suggest that the changes in the sterol profile rather than the amount produced in 24-epiiminolanosterol-treated cultures was presumably the deleterious feature for growth inhibition. An analysis of the sterols present shows that the amount of 24-desalkylsterols synthesized is high enough to substitute for ergosterol's bulk function i.e. to determine the physical state of the membranes. However, these sterols seem to be unable to replace the sparking sterol i.e. ergosterol or another 24ß-methylsterol. Thus, the 24ß-alkyl substituent is obligatory for some of the sterol-controlled functions necessary to support mycelial growth. This is similar to the results obtained in yeast sterol auxotrophs. The latter will grow on bulk levels of lanosterol only when nanomolar concentrations of ergosterol are present.[47] However, there is still a controversy as to the specific structural requirements of the yeast "sparking sterol". Efforts to ascertain these specific structural requirements necessary to provide the "hormonal" or "sparking" regulating function have recently been investigated by means of gene disruption techniques.[48] These studies have shown that the C-24 sterol transmethylase (*ERG* 6), the Δ^8-Δ^7 isomerase (*ERG* 2), and the $\Delta^{5,6}$ desaturase (*ERG* 3) are not required for viability and the investigators conclude that the genes (*erg* 2, *erg* 3 and *erg* 6) are not required for the sparking function.[48] However, it should be noted that these studies were done with *S. cerevisiae* and there are many publications describing membrane structure alterations in these mutants furthermore, mutant *erg* 6 showed significant decreases in growth on glucose. When ethanol or glycerol was substituted for glucose, mutant *erg* 6 showed virtually no growth. With growth capability as a parameter, the lack of a C-24 transmethylase function has a more deleterious effect than the lack of the Δ^8-Δ^7 isomerase or the $\Delta^{5,6}$ desaturase.[48] These results indicate that indeed the 24ß-methyl group is necessary for normal growth as already shown in 1983 by Pinto and Nes.[49]

In summary, it must be said that most of the experimental support for the "steroid hormone" function is indirect and comes mostly from studies with *Saccharomyces* mutants. The potential relationship between morphogenesis and ergosterol is intriguing but still open to speculation.

IS THERE A ROLE FOR STEROID HORMONES IN FUNGAL GROWTH AND MORPHOGENESIS ?

Although proliferating cultures of *C. albicans* were found to contain constitutive 3α-, 3ß-, 20α- and 20ß-hydroxysteroid dehydrogenase activities they lack steroid hydroxylases and 4-ene-hydrogenases.[50] One can only speculate on the physiological role of these enzyme activities in *C. albicans*. It is possible that these enzymes are used to modify host steroid hormones or have a function in the metabolism of fungal sterols. If steroids or steroid-like structures play a role in morphogenesis they should be synthesized by the fungus. Indeed, the *N*-acetylglucosamine-yeast nitrogen base-proline (NYP)-medium we use to grow *C. albicans* in its mycelium form[51] does not contain steroids.

Parks *et al.*[42] tested aqueous extracts from a sterol wild-type, *S. cerevisiae* X2180-1A, and from commercial dried yeast, to determine if ergosterol was further metabolized to a compound that could fulfil the sparking requirement. In this study it was found that a sparking ergosterol replacement factor (SERF) is indeed present. They speculated that if SERF was derived directly from ergosterol, it would constitute a new end product of sterol metabolism that may have non-membrane-associated functions.

C. albicans contains a corticosteroid-binding protein (CBP) which has high affinity for corticosterone, progesterone and 11-deoxycorticosterone but low affinity for estrogens and ergosterol.[52,53] The results of Loose *et al.*[53] suggest that *C. albicans* possesses an endogenous, ethanol extractable ligand which appears to bind to both *Candida* CBP and mammalian glucocorticoid receptors. Thus far, the structure of this putative steroid hormone has not been elucidated. But these findings have led Loose *et al.*[53] to speculate that *C. albicans* has a hormone-receptor system that may modulate the physiology and/or pathogenicity of the fungus. Ketoconazole competitively displaces [³H]corticosterone from the *Candida* CBP.[54] This imidazole antifungal is by far the most potent competitor, inhibiting [³H]corticosterone binding by 50 % at 0.25 μM. Miconazole and econazole are considerably weaker competitors, whereas clotrimazole is not inhibiting binding by 50 % at concentrations up to 100 μM. However, there is no apparent correlation between the binding potency and antifungal potency[54] and inhibition of germ tube formation in *C. albicans*.[55]

Next to CBP the cytosol of *C. albicans* also contains an estrogen-binding protein (EBP).[56,57] Scatchard analysis of the [³H]estradiol equilibrium binding data of *C. albicans* yielded an apparent dissociation constant of 12.3 ± 2.1 nM and a maximal binding capacity of 753 ±145 fmol/mg protein. Binding competition experiments showed high specificity and stereoselectivity of EBP, demonstrating the following order of potency in displacing [³H]estradiol: 17ß-estradiol> estrone > estriol >17α-estradiol.[57] Negligible competitive potency was found for diethylstibestrol, tamoxifen or fungal hormones such as antheridiol,[57] which is a sterol hormonal substance in the water mold *Achlya*.[3]

Powell *et al.* also identified a 17ß-estradiol-binding protein in *C. glabrata*[56] and *Coccidioides immitis*[58,59] but not in *Candida parapsilosis*, *C. pseudotropicalis*, *C. krusei* and *C. stellatoidea*.[56] Burshell *et al.* characterized an EBP in the cytosol of *S. cerevisiae*[60] and Loose *et al.*[61] found a receptor-like EBP in the cytosol of

Paracoccidioides brasiliensis. The physiological function of EBP in *S. cerevisiae* is not defined. Investigation of steroid hormone actions in *P. brasiliensis* indicated that estradiol inhibited the fungal transformation from mycelial form to yeast form, the initial step of infection.[61,62] (see also G. San Blas, this book).

The studies of Kinsman *et al.*[63] indicate that 17ß-estradiol promotes germination in *C. albicans.* Increased hydroxylation of the estrogen (estriol) and pregnane (pregnanetriol) steroids increases the ability of steroids to promote germination.[63] Addition of luteinizing hormone (LH) to the culture medium also promotes germination.[63] LH increases P450-mediated sterol synthesis and steroidogenesis in mammalian cells. Thus, it is possible that LH stimulates ergosterol synthesis, the conversion of a precursor steroid or of ergosterol to a yeast steroid derivative involved in the yeast-to-mycelium transformation. However, so far our experiments do not show any effect of physiological concentrations of LH on ergosterol synthesis, ergosterol content and hypha formation of *C. albicans* in the mycelium promoting NYP-medium (unpublished results). More studies are needed to evaluate the effects of human gonadotropins or other glycoprotein hormones on morphogenesis.

CONCLUSIONS

A number of fungal species contain steroid-binding proteins which appear to bind vertebrate steroid hormones and endogenous fungal ligands. Vertebrate steroids and even luteinizing hormone seem to affect fungal morphisms. The question whether or not these mammalian hormones mimic endogenous steroid-like compounds involved in metabolic control is still open. Further research is certainly needed to characterize the endogenous fungal ligands and to elucidate the role of this "hormonal system" in yeast cell cycle control and morphogenesis.

Increasing evidence indicates that ergosterol is not only involved in membrane associated functions but also in critical non-membrane associated functions. There is evidence that hypha formation in *C. albicans* is dependent on the availability of ergosterol and this 24-alkylsterol seems to be obligatory for the growth of filamentous fungi.

The existing, fragmentary data suggest an abundance of research possibilities. A more complete understanding of the "regulatory effects" of ergosterol and other fungal sterols is needed and may result in the identification of novel targets for the control of fungal diseases.

REFERENCES

1. W. Köller, Antifungal agents with target sites in sterol functions and biosynthesis, in: "Target Sites of Fungicide Action", W. Köller, ed., CRC Press, Boca Raton (1992).
2. D.T. Hart, .W.J. Lauwers, G. Willemsens, H. Vanden Bossche and F.R. Opperdoes, Perturbation of sterol biosynthesis by itraconazole and ketoconazole in *Leishmania mexicana mexicana* infected macrophages, *Mol. Biochem. Parasitol.* 33: 123 (1989).
3. P.Z. Margalith, "Steroid Microbiology", Charles C. Thomas, Springfield (1986).
4. H. Vanden Bossche, Mode of action of pyridine, pyrimidine and azole antifungals, in: "Sterol Biosynthesis Inhibitors - Pharmaceutical and Agrochemical Aspects", D. Berg and M. Plempel, eds., Ellis Horwood Ltd., Chichester (1988).
5. H. Vanden Bossche, Importance and role of sterols in fungal membranes, in: "Biochemistry of Cell Walls and Membranes in Fungi", P.J. Kuhn, A.P.J. Trinci, M.J. Jung, M.W. Goosey and L.G. Copping, eds., Springer-Verlag, Berlin (1990).

6. M.A. Gealt, A. Abdollahi and J.L. Evans, Lipids and lipoidal mycotoxins of fungi, in: "Current Topics in Medical Mycology", Vol. 3, M.R. McGinnis and M. Borgers, eds., Springer Verlag, New York (1989).

7. P. Mishra and R. Prasad, An overview of lipids of *Candida albicans*, *Prog. Lipid Res.* 29: 65 (1990).

8. P. Mishra, J. Bolard and R. Prasad, Emerging role of lipids of *Candida albicans*, a pathogenic dimorphic yeast, *Biochim. Biophys. Acta* 1127: 1 (1992).

9. S.R. Parker and W.D. Nes, Regulation of sterol biosynthesis and its phylogenetic implications, in: "Regulation of Isopentenoid Metabolism", W.D. Nes, E.J. Parish and J.M. Trzaskos, eds., American Chemical Society, Washington DC (1992).

10. R.D. Cannon and D. Kerridge, Correlation between the sterol composition of membranes and morphology in *Candida albicans*, *J. Med. Vet. Mycol.* 26: 57 (1988).

11. S. Sadamori, Comparative study of lipid composition of *Candida albicans* in the yeast and mycelial forms, *Hiroshima J. Med. Sci.* 36: 53 (1987).

12. H. Vanden Bossche, P. Marichal, G. Willemsens, D. Bellens, J. Gorrens, I. Roels, M.-C. Coene, L. Le Jeune and P.A.J. Janssen, Saperconazole: a seleletive inhibitor of the cytochrome P-450-dependent ergosterol synthesis in *Candida albicans*, *Aspergillus fumigatus* and *Trichophyton mentagrophytes*, *Mycoses* 33: 335 (1990).

13. C. Arnezeder and W.A. Hampel, Influence of growth rate on the accumulation of ergosterol in yeast-cells in a phosphate limited continuous culture, *Biotechn. Lett.* 13: 97 (1991).

14. O. Shimokawa, Y. Kato and H. Nakayawa, Accumulation of 14-methyl sterols and defective hyphal growth in *Candida albicans*, *J. Med. Vet. Mycol.* 24: 327 (1986).

15. P. Marichal, H. Vanden Bossche, H. Moereels and R. Brasseur, Mode of insertion of azole antifungals and sterols in membranes, in: "Molecular Desscription of Biological Membranes by Computer Aided Conformational Analysis", Vol. 2, R. Brasseur, ed., CRC Press, Boca Raton (1990).

16. M. Bard, N.D. Lees, R.J. Barbuch and D. Sanglard, Characterization of a cytochrome P450 deficient mutant of *Candida albicans*, *Biochem. Biophys. Res. Commun.* 147: 794 (1987).

17. N.D. Lees, M.C. Broughton, D. Sanglard and M. Bard, Azole susceptibility and hyphal formation in a cytochrome P-450-deficient mutant of *C. albicans*, *Antimicrob. Ag. Chemother.* 34: 831 (1990).

18. A.B. Orth and H.D. Sisler, Mode of action of terbinafine in *Ustilago maydis* and characterization of resistant mutants, *Pestic. Biochem. Physiol.* 37: 53 (1990).

19. H.D. Sisler, R.C. Walsh and B.N. Ziogas, Ergosterol biosynthesis: a target of fungitoxic action, in: "IUPAC Pesticide Chemistry: Human Welfare and the Environment", Vol. 3, "Mode of Action, Metabolism and Toxicology", S. Matsunaka, D.H. Hutson and S.D. Murphy, eds., Pergamon Press, Oxford (1983).

20. Y. Yoshida and A. Aoyama, Cytochromes P-450 in the ergosterol biosynthesis, in: "Frontiers in Biotransformation", Vol. 4, "Microbial and Plant Cytochromes P-450: Biochemical Characteristics, Genetics Engineering and Practical Implications", K. Ruckpaul and H. Rein, eds., Akademic Verlag, Berlin (1991).

21. H. Vanden Bossche, P. Marichal, M-C. Coene, G. Willemsens, L.Le Jeune, W. Cools and H. Verhoeven, Cytochrome P450-dependent 14α-demethylase- Target for antifungal agents and herbicides, in: "Regulation of Isopentenoid Metabolism", W.D. Nes, E.J. Parish and J.M. Trzaskos, eds., American Chemical Society, Washington DC (1992).

22. P.F. Watson, M.E. Rose, S.W. Ellis, H. England and S.L. Kelly, Defective sterol C5-6 desaturation and azole resistance: a new hypothesis for the mode of action of azole antifungals, *Biochem. Biophys. Res. Commun.* 15: 1170 (1989).

23. S.L. Kelly, J. Rowe and P.F. Watson, Molecular genetic studies on the mode of action of azole antifungal agents, *Biochem. Soc. Trans* 19: 796 (1991).

24. W.R. Nes, B.C. Sekula, W.D. Nes and J.H. Adler, The functional importance of structural features of ergosterol in yeast, *J. Biol. Chem.* 253: 6218 (1978).

25. A.K. Lala, H.K. Lin and K. Bloch, The effect of some alkyl derivatives of cholesterol on the permeability properties and microviscosities of model membranes, *Bioorganic Chem.* 7: 437 (1978).

26. B.N. Ziogas, H.D. Sisler and W.R. Lusby, Sterol content and other characteristics of pimaricin-resistant mutants of *Aspergillus nidulans*, *Pestic. Biochem. Physiol.* 20: 320 (1983).

27. D. Barug and A. Kerkenaar, Resistance in mutagen-induced mutants of *Ustilago maydis* to fungicides which inhibit ergosterol biosynthesis, *Pestic Sci.* 15: 78 (1984).

28 J Guan, A Kerkenaar and M A De Waard, Studies on mechanism of resistance to imazalil in *Penicillium italicum, Tag Ber Acad Landwirtsch Wiss Berlin* 291 115 (1990)

29 J Van Cutsem, The *in vitro* antifungal spectrum of itraconazole, *Mycoses* 32 (Suppl 1) 7 (1989)

30 N S Ryder, Terbinafine mode of action and properties of the squalene epoxidase inhibition, *Brit J Dermatol* 126 (Suppl 39) 2 (1992)

31 A Polak, Amorolfine, RO 14-4767/002, Loceryl, in "Recent Progress in Antifungal Chemotherapy", H Yamaguchi, G S Kobayashi and H Takahashi, eds , Marcel Dekker Inc , New York (1991)

32 D G Gottlieb, R J Knaus and S G Wood, Differences in the sterol synthesizing pathways of sterol-producing and non-sterol-producing fungi, *Phytophatology* 68 1168 (1978)

33 C G Elliott, M E Hendrie, B A Knights and W Parker, A steroid growth factor requirement in a fungus, *Nature* 203 427 (1964)

34 W D Nes, G A Saunders and E Heftman, Role of steroids and triterpenoids in the growth and reproduction of *Phytophthora cacturum, Lipids* 17 178 (1982)

35 H Vanden Bossche and P Marichal, Mode of action of anti-*Candida* drugs focus on terconazole and other ergosterol biosynthesis inhibitors, *Am J Obstet Gynecol* 165 1193 (1991)

36 H Vanden Bossche, W Lauwers, G Willemsens, P Marichal, F Cornelissen and W Cools, Molecular basis for the antimycotic and antibacterial activity of *N*-substituted imidazoles and triazoles the inhibition of isoprenoid biosynthesis, *Pestic Sci* 15 188 (1984)

37 C Marcireau, M Guilloton and F Karst, *In vivo* effects of fenpropimorph on the yeast *Saccharomyces cerevisiae* and determination of the molecular basis of the antifungal property, *Antimicrob Ag Chemother* 34 989 (1990)

38 J D Weete and M L Wise, Effects of triazoles on fungi V Response by a naturally tolerant species, *Mucor rouxii, Exptl Mycol* 11 214 (1987)

39 M Sancholle, J D Weete and C Montant, Effect of triazoles on fungi I Growth and cellular permeability, *Pestic Biochem Physiol* 21 31 (1984)

40 H Vanden Bossche, Ergosterol biosynthesis inhibitors, in "*Candida albicans*, Cellular and Molecular Biology", R Prassad, ed , Springer Verlag, Berlin (1991)

41 M Orlowski, Mucor dimorphism, *Microb Rev* 55 234 (1991)

42 L W Parks, R J Rodriguez and C Low, An essential fungal growth factor derived from ergosterol a new end product of sterol biosynthesis in fungi? *Lipids* 21 89 (1986)

43 S Kawasaki, M Ramgopal, J Chin and K Bloch, Sterol control of the phosphatidylethanolamine-phosphatidylcholine conversion in the yeast mutant GL7, *Proc Natl Acad Sci USA* 82 5715 (1985)

44 C Dahl, H -P Biemann and J Dahl, A protein kinase antigenically related to pp60^{v-src} possibly involved in yeast cell cycle control positive *in vivo* regulation by sterol, *Proc Natl Acad Sci USA* 84 4012 (1987)

45 K Bloch, Lipid structure and function, in ' Lipids and Membranes Past, Present and Future", J A F Op den Kamp, B Roelofsen and K W A Wirtz, eds , Elsevier Science Publishers B V , Amsterdam (1986)

46 W D Nes, P K Hanners and E J Parish, Control of fungal sterol C-24 transalkylation importance to developmental regulation, *Biochem Biophys Res Commun* 139 410 (1986)

47 R J Rodriguez, C Low, C D K Bottema and L W Parks, Multiple functions for sterols in *Saccharomyces cerevisiae, Biochim Biophys Acta* 837 336 (1985)

48 N D Lees, B A Arlington and M Bard, Genetics and molecular biology of the genes functioning late in the sterol biosynthetic pathway of *Saccharomyces*, in "Regulation of Isopentenoid Metabolism", W D Nes, E J Parish and J M Trzaskos, eds , American Chemical Society, Washington DC (1992)

49 W J Pinto and W R Nes, Stereochemical specificity for sterols in *Saccharomyces cerevisiae, J Biol Chem* 258 4472 (1983)

50 R Ghraf, E R Lax, S Oza and H Schriefers, Transformation of C_{18}-, C_{19}- and C_{21}-steroids by cultures of *Candida albicans, J Steroid Biochem* 6 1531 (1975)

51 P Marichal, J Gorrens, J Van Cutsem and H Vanden Bossche, Culture media for the study of the effects of azole derivatives on germ tube formation and hyphal growth of *C albicans, Mycoses* 29 76 (1985)

52 D S Loose, D J Schurman and D Feldman, A corticosteroid binding protein and endogenous ligand in *C albicans* indicating a possible steroid-receptor system, *Nature* 293 477 (1981)

53. D.S. Loose and D. Feldman, Characterization of a unique corticosterone-binding protein in *Candida albicans, J. Biol. Chem.* 257: 4925 (1982).

54. E.P. Stover, D.S. Loose, D.A. Stevens and D. Feldman, Ketoconazole binds to the intracellular corticosteroid-binding protein in *Candida albicans, Biochem. Biophys. Res. Commun.* 117: 43 (1983).

55. M. Schaude, H. Ackerbauer and H. Mieth, Inhibitory effect of antifungal agents on germ tube formation in *Candida albicans, Mycoses* 30: 281 (1987).

56. P.L. Powell, C.L. Frey and D.J. Drutz, Identification of a 17ß-estradiol binding protein in *Candida albicans* and *Candida (Torulopsis) glabrata, Exptl. Mycol.* 8: 304 (1984).

57. R. Skowronski and D. Feldman, Characterization of an estrogen-binding protein in the yeast *Candida albicans, Endocrinology* 124: 1965 (1989).

58. B.L. Powell, D.J. Drutz, M. Huppert and S.H. Sun, Relationship of progesterone- and estradiol-binding proteins in *Coccidioides immitis* to coccidioidal dissemination in pregnancy, *Infect. Immun.* 40: 478 (1983).

59. B.L. Powell and D.J. Drutz, Identification of a high-affinity binder for estradiol and a low-affinity binder for testosterone in *Coccidioides immitis, Infect. Immun.* 45: 784 (1984).

60. A. Burshell, P.A. Stathis, Y. Do, S.C. Miller and D. Feldman, Characterization of an estrogen-binding protein in the yeast *Saccharomyces cerevisiae, J. Biol. Chem.* 259: 3450 (1984).

61. D.S. Loose, E.P. Stover, A. Restrepo, D.A. Stevens and D. Feldman, Estradiol binds to a receptor-like cytosol binding protein and initiates a biological response in *Paracoccidioides brasiliensis, Proc. Natl. Acad. Sci. USA* 80: 7659 (1983).

62. A. Restrepo, M.E. Salazar, L.E. Cano, E.P. Stover, D. Feldman and D.A. Stevens, Estrogens inhibit mycelium-to-yeast transformation in the fungus *Paracoccidioides brasiliensis*: implications for resistance of females to Paracoccidioidomycosis, *Infect. Immun.* 46: 346 (1984).

63. O.S. Kinsman, K. Pitblado and C.J. Coulson, Effect of mammalian steroid hormones and luteinizing hormone on the germination of *Candida albicans* and implications for vaginal candidosis, *Mycoses* 31: 617 (1988).

64. J.H. Reinhardt, A.M. Allen, D. Gunnison and W.A. Akers, Experimental human *Trichophyton mentagrophytes* infections, *J. Invest. Derm.* 62: 419 (1974).

FACTORS REGULATING MORPHOGENESIS IN *COCCIDIOIDES IMMITIS*

Garry T. Cole[1], David Kruse[1], Kalpathi R. Seshan[1],
Shuchong Pan[1], Paul J. Szaniszlo[2], Jon Richardson,[3]
and Buming Bian[3]

[1] Department of Botany
[2] Department of Microbiology
[3] Center for High Performance Computing
University of Texas
Austin, Texas 78713, USA

ABSTRACT

The parasitic cycle of *Coccidioides immitis* is unique among the human systemic fungal pathogens. However, at the level of cell wall biosynthesis and modification *C. immitis* demonstrates features which are shared by other fungal pathogens. Three distinct events in the morphogenesis of parasitic cells, or spherules, of *Coccidioides* are examined in this study. These include the diametric growth phase of round cells (young spherules), spherule segmentation, and endosporulation. Three enzymatic products of developing spherules have been suggested to participate in regulation of these successive morphogenetic stages of *C. immitis*. Preliminary evidence is presented that a β-1,3-endoglucanase contributes to plasticization of the round cell wall, and consequently plays a role in diametric expansion of young spherules. Results of earlier studies of a 34 kDa proteinase are reviewed which suggest that this wall-associated enzyme may function in development of the segmentation wall. Data from recent studies of a 100 kDa chitinase are presented which suggest that this enzyme participates in endosporulation of the parasitic cells. Morphogenetic studies of *Coccidioides* may lead to the identification of regulatory factors which are common to other pathogenic fungi, and therby, to the characterization of molecular targets for development of future antifungal reagents.

INTRODUCTION

Coccidioidomycosis is a human respiratory fungal disease which is endemic to the southwestern United States and certain arid regions of Mexico, Central and

Dimorphic Fungi in Biology and Medicine, Edited by
H. Vanden Bossche *et al.*, Plenum Press, New York, 1993

South America.[1] Drutz and Huppert[2] have estimated that 25,000 to 100,000 new cases of this mycosis are reported each year on the basis of skin test reactivity. The saprobic phase of *Coccidioides immitis* inhabits alkaline desert soils. The arthroconidia produced by mycelia are air-dispersed and if inhaled can establish a respiratory infection. The conidia are small enough to be carried in the airstream to the alveoli. BALB/c mice (males, 23-27 g) inoculated intranasally with different numbers of viable conidia have been shown to develop pulmonary lesions typical of coccidioidomycosis.[3] Death of the animals occurred in 7-12 days post-inoculation when the inoculum contained as few as 100 conidia.[4] The virulence of *C. immitis* is further underscored by the fact that this pathogen is one of the ten etiologic agents most frequently transmitted to laboratory workers when accidently exposed to the contents of Petri plate cultures of the saprobic phase.[2] Details of the morphogenetic conversion of arthroconidia to spherules and initiation of the parasitic cycle *in vitro* closely resemble developmental aspects reported from *in vivo* studies.[3] The parasitic cycle of *C. immitis* is unique among the human fungal pathogens.[5] Conidia undergo an extended growth phase, first becoming round cells and then large, multinucleate spherules. This increase in cell volume has been reported to be accompanied by rapid and apparently synchronous nuclear divisions.[6] This stage of the nuclear cycle is followed by segmentation of the cytoplasm, which occurs by centripetal growth of the innermost wall layer of the spherule envelope. Segmentation of the spherule continues until the protoplasm has been compartmentalized into a myriad of uninucleate endospores. The spherule envelope stretches as endospores undergo diametric growth, and finally ruptures to allow clusters of endospores to be released. The endospore clusters appear to be surrounded by an extracellular matrix of unknown composition which may contribute to passive resistance of the pathogen against host cellular immune defense.[2,3] Endospores may convert into secondary spherules *in vivo* while still part of the cluster. Individual endospores (2-5 μm diam.) which escape from the clusters and survive in the host function in hematogenous and lymphatic spread of the pathogen. The mycelial phase does not normally occur within the host, although a recent case of a patient whose tissue from a pleural peel showed extensive *C. immitis* hyphae has been described.[7]

In this paper we further characterize the morphogenetic events of the parasitic cycle of *C. immitis*. We have focused on three stages of spherule development: the diametric growth phase of multinucleate round cells which arise either from arthroconidia (1st generation) or endospores (2nd generation), the spherule segmentation phase, and the endosporulation phase. For each of these stages, we present preliminary evidence for production of a specific macromolecule which may participate in regulation of spherule morphogenesis.

MATERIALS AND METHODS

Developmental Analysis of the Parasitic Phase of *C. immitis*

Arthroconidia of strain C735 were used to inoculate Converse agar[8] contained in a thin glass culture chamber.[9] The chamber was purged with 20% CO_2/80% air and placed on a heated stage (39 °C) of a Zeiss photomicroscope II. Spherule development was recorded by time-lapse photomicrography.

Animal Inoculation and Histology

BALB/c mice (males, 23-27 g) were inoculated intranasally with arthroconidia of *C. immitis* grown on glucose-yeast extract (GYE) agar as previously described.[3] The animals were sacrificed at 7 days post-challenge, lungs were excised, and host tissue was fixed for light microscopy or electron microscopy as described below.

Electron Microscopy

Arthroconidia, spherules and endospores of *C. immitis* were prepared for scanning and transmission electron microscopy as previously described.[5,10]

Immunoelectron Microscopy and Immunofluorescence

The three antigenic products examined in this study (120-kilodalton[kDa] β-1,3-glucanase, 34-kDa chymotrypsin-like serine proteinase, and 100-kDa chitinase) were localized in parasitic cells of *C. immitis* by immunogold labeling methods.[11] Round cells of *C. immitis* were exposed to 1-deoxynojirimycin, an inhibitor of the β-1,3-glucanase.[11] Localization of the inhibitor in live and fixed cells was performed by immunofluorescence microscopy.[11]

Cytofluorometric Analysis of Genome Size of *C. immitis*

Microspectrophotometry was used to estimate the total nuclear DNA content of *C. immitis* by measurement of the fluorescence intensity of stained nuclei of saprobic and parasitic cells.[12,13] Dormant arthroconidia were obtained from 60 day GYE plate cultures. Parasitic cells grown in Converse liquid medium[8] were exposed to 0.2 M hydroxyurea to inhibit DNA synthesis and thereby induce cells to enter the G_1 phase of the cell cycle.[13] The saprobic and parasitic cells were incubated with 0.1% ribonuclease A (wt/v;Sigma Chemical Co., St. Louis, Mo.), stained with ethidium bromide, and examined by fluorescence microscopy as described.[13] *Saccharomyces cerevisiae* haploid strain (ATCC 24851) and diploid strain (ATCC 24860) were treated in an identical manner and used as controls in the cytofluorometric analysis. The intensity of fluorescence of the nuclear stain is considered to be approximately proportional to DNA content.[12] Incubation of cells with RNAase minimized background staining due to intercalation of ethidium bromide with dsRNA.[13] Samples were examined with a Zeiss Universal Photomicroscope equipped with a 100 W mercury lamp and 395 nm filter (Zeiss III RS) for epifluorescence illumination. Intensity of fluorescence was determined with a model MTI 65 Newicon camera (DAGE-MTI Inc., Albuquerque, N.M.) interfaced with an IBAS digital image processing system (Kontron Co., Munich, Germany).

Measurement of pH Flux in *C. immitis*

Converse liquid medium or agar plate cultures[8] were inoculated with arthroconidia, incubated at 39 °C, and purged every 48 h with 20% CO_2/80% air. The pH of the liquid culture was recorded daily for 8 days. Converse agar medium contained 25 µg phenol red (Sigma) per ml as a pH indicator. Plate cultures were examined for color change in the medium without their removal from the CO_2-purged transparent incubators. Intracellular pH of parasitic cells of *C. immitis* was examined using the polar, dual-emission pH indicator SNAFL calcein AM (seminaphthofluorescein).[14,15] A 10 mM stock solution of the dye was prepared in anhydrous dimethyl sulfoxide (Sigma). Young spherules were incubated (1 h, 37 °C) in culture medium (pH 7.0) without amino acids and containing 10 µM dye solution (v/v). The cells were examined with a fluorescence microscope equipped with FITC and Texas red filter units.

Screening the cDNA Library

A cDNA expression library for *C. immitis* mycelial phase was constructed in lambda Zap II phage vector (Stratagene, La Jolla, Calif.) as previously described.[16] The cDNA was packaged with Gigapack (Stratagene) and *Escherichia coli* PLK-F (Stratagene) was used as the host for the recombinant phage. Antibody screenings of the library for expression of the β-1,3-glucanase, 34-kDa serine proteinase, and chitinase were performed using specific antibody raised in guinea pigs or rabbits as described.[11,16] Substrate screening of the library for expression of the β-1,3-glucanase was performed using X-Glu (5-bromo-4-chloro-3-indolyl-β-*D*-glucopyranoside; Calbiochem, La Jolla, Calif.). For the latter screening method, recombinant phage-infected bacteria were plated on LB medium (8ml plus 1% agarose; Sigma) which contained 20 µl of 0.5 M isopropyl-β-*D*-thiogalactopyranoside (IPTG; Ambion, Austin, TX) and 250 µl of X-Glu (diluted from stock of 25 mg per ml of *N,N*-dimethylformamide; Sigma). The plate cultures were incubated at 37 °C for 12 h. Blue plaques were isolated and rescreened to increase the frequency of blue (enzymatically-active) clones. The Bluescript phagemid with insert were treated by the *in vivo* excision method as described[16] and the plasmid plus insert was used to transform *E. coli* XL1-Blue (Stratagene).

Fusion Protein Isolation and Characterization

Fusion proteins (FPs) produced by single clones isolated from the expression library by anti-β-1,3-glucanase antibody or X-Glu substrate screening methods were isolated with an affinity column using mouse anti-β-galactosidase antibody as described.[17] The immunoaffinity-isolated FPs were added to wells of microtiter plates (50 µg/well) for examination in enzyme-linked immunosorbent assays (ELISA) using a guinea pig antiserum (diluted 1:200-1:400 in phosphate-buffered saline [PBS], pH 7.6) which was raised against the deglycosylated β-glucanase of *C. immitis*.[11] The selected FPs derived from clones isolated by antibody and substrate screening methods showed comparable reactivity with the anti-β-glucanase in the ELISA. *In vitro* assays of β-glucanase activity of the FPs were performed as described[11] with p-nitrophenol-β-*D*-glucopyranoside (pNP-β-*D*-glc; Sigma) at a concentration of 10 mM in enzyme buffer (0.02 M Tris, 0.6 mM $CaCl_2$, pH 8.0).

Northern Hybridizations

Each pBluscript SK plasmid (Stratagene) insert was used to hybridize with total poly(A)-containing RNAs from *C. immitis* which had been electrophoresed in a 1.0% formaldehyde-agarose gel and processed as reported.[16] The plasmid without insert was used in separate hybridizations as a control. The probes were prepared and labeled by the random hexamer primer method as described.[18]

Transfer to Zeta-Probe blotting membrane (Bio-Rad Labs., Richmond, Calif.) and the hybridization procedure were performed as described.[19]

Northern dot blot hybridizations were performed with total *C. immitis* RNA using four different DNA probes. Isolation of the 1.2 kb cDNA which encodes the 34-kDa serine proteinase has been described.[16] The 1.8 kb cDNA which partially encodes the *C. immitis* β–1,3-glucanase, and the 1.2 kb cDNA which encodes the chitinase were separately excised from pBluescript by EcoRI and XhoI restriction enzyme digestion. The cDNAs were separated by agarose gel electrophoresis, eluted from the gels and isolated using GeneClean II kits (BIO 101, La Jolla, Calif.) according to the manufacturer's instructions. The chitin synthase probe was obtained by polymerase chain reaction (PCR). A 600 base pair (bp) *C. immitis* DNA fragment that is homologous to a conserved region of the chitin synthase gene was amplified from *Coccidioides* genomic DNA using conserved PCR primers that have been reported.[20] The 600 bp DNA fragment from the PCR product was separated by agarose gel electrophoresis and isolated as above. The constitutive control DNA probe used in this study was a conserved ribosomal DNA obtained from K. J. Kwon-Chung (National Institutes of Health, Bethesda, MD). The pBIR6 plasmid which contained a *Cryptococcus neoformans* rDNA insert[21] was digested with HindIII to yield a 8.6 kb fragment. All DNA probes were labeled by the random hexamer primer method as described above. RNA was isolated from spherule cultures of *C. immitis* strain C735 at 12 h intervals after inoculation with arthroconidia. The cells were pelleted and immediately resuspended in 5 M guanidinium isothiocyanate in 5 mM Tris-HCl, pH 7.5, plus 1 mM disodium salt of EDTA and 5% β-mercaptoethanol. The pelleted cells were frozen in liquid nitrogen, ground, and stored at -70 °C. For hybridization, the samples were brought to room temperature, homogenized, and extracted with phenol:chloroform (1:1). The RNA in the aqueous phase was precipitated by addition of LiCl to 2 M (final conc.). The isolated RNA was resuspended in diethylpyrocarbonate (DEPC)-treated water containing 32.5 units of RNAsin (Promega Corp., Madison, Wis.). The suspended RNA was further purified by addition of 2 units of deoxyribonuclease (RNAase free) (United States Biochem., Cleveland, Oh.) and incubation at 37 °C for 1 h. The RNA samples were quantified by differential absorption at 280 nm and 260 nm, diluted to equal concentration in DEPC-H_2O, and blotted on Zeta-Probe membrane (Bio-Rad) with a Bio-Dot microfiltration apparatus (Bio-Rad). The DNA probes were labeled by the random hexamer primer method as described above. The Northern hybridization method used was as described in the Bio-Dot apparatus instructions. The hybridized membranes were exposed to Kodak XAR-5 x-ray film to produce the autoradiograms.

Serial Sectioning and Computerized 3-D Reconstruction of Segmented Spherules

Spherules were grown in liquid cultures for various periods, chemically fixed, embedded in plastic resin, and thick-sectioned as described.[16] Serial sections (each approx. 1 μm thick) were stained with wheat germ agglutinin (WGA) conjugated with FITC (Sigma) and examined by fluorescence microscopy as described.[16] The photomicrographs were scanned with a digital scanner and data were stored as 8 bits binary files. The Application Visualization System (AVS) software package (Advanced Visual Systems, Waltham, Mass.) was used for generation of the three-dimensional (3-D) images from assimilation of data from the serial sections. Four steps were used to incorporate the two-dimensional images (digital scans of photomicrographs) into 3-D volume: image enhancement, image painter, image transformation, and transfer of image to volume. Details of these procedures will be presented in a forthcoming communication.

Western Blot (Immunoblot) Analysis

The anti-chitinase antiserum was raised in rabbits against a gel filtration (GF)-isolated fraction of the mycelial culture filtrate of *C. immitis* Silvera strain. The molecular size of the fraction under reducing conditions was estimated by sodium dodecyl sulfate polyacrylamide gel electrophoresis (SDS-PAGE) to be 48 kDa. Gel filtration, SDS-PAGE, and immunization procedures were performed as reported.[22] The nonreduced GF fraction showed reactivity with complement fixation (CF) antibody from coccidioidomycosis patients in the immunodiffusion-CF (ID-CF) assay.[23] The gel filtration isolated fraction had the same molecular size by SDS-PAGE (reducing gel) as the CF antigen previously reported by Zimmer and Pappagianis.[24] The reference antigens and human reference sera used in the ID-CF and immunodiffusion-tube precipitin (ID-TP) assays have been reported.[23] Western blot analysis was performed by reaction of the anti-CF/chitinase antiserum with the mycelial culture filtrate plus toluene lysate (F+L fraction). The latter was prepared as described.[23] The Western blot procedure has been reported.[17] *In situ* chitinase detection in SDS-PAGE (nonreducing) gels was performed as described[11] using pNP-β-D-N,N'-diacetylchitobiose (20 mM; Sigma) in sodium phosphate buffer (pH 7.0). The *in vitro* activity of the *C. immitis* β-1,3-glucanase and chitinase over a pH range of 2.0 to 10.0 was compared using 10 mM of respective substrates (pNP-β-D-glc or pNP-β-D-N,N'-diacetylchitobiose, respectively) and appropriate buffers.[11] The pH profile was obtained with acetate buffer (20 mM, pH 2-5), phosphate buffer(20 mM, pH 6-8), and CAPS buffer (3[cyclohexylamine]-1-propanesulfonic acid; 20 mM, pH 9-10).

RESULTS AND DISCUSSION

Spherule Morphogenesis

The time-lapse sequence in Fig. 1 A-D shows the initial burst of diametric growth of the cylindrical arthroconidia (A), their differentiation into round cells (B; approx. 36 h), and subsequent development of the segmentation wall apparatus surrounding the central vacuole (C,D). As endospores begin to form within the spherule at approximately 60 to 70 h after inoculation of the cultures with arthroconidia, the central vacuole disappears and the segmentation wall breaks down (Fig. E). As the endospores begin to grow within the maternal spherule, the outer envelope ruptures (Fig. 1F). The released endospores continue to grow and differentiate into a second generation of spherules which is initiated approximately 96 to 108 h post-inoculation of the cultures (Fig. 1G). Some variation in duration of these stages of spherule development has been recognized between strains of *C. immitis*.[16] One factor affecting initiation of the 1st generation of spherules is the lag phase in arthroconidial germination. Conidia of strain C735 (Fig. 1) obtained from 60 day GYE plate cultures have a relatively short lag phase. This strain was used to obtain RNA fractions which were probed for temporal expression of the selected genes examined here.

Formation of the Round Cell Initial. Conversion of arthroconidia to round cells (Fig. 2 A,B) results in the appearance of many tiny cytoplasmic vacuoles and the initiation of mitotic divisions.

The nuclei shown in the round cell in Fig 2B are approximately of equal diameter. A still unexplained feature of the nuclear cycle of this pathogen is that

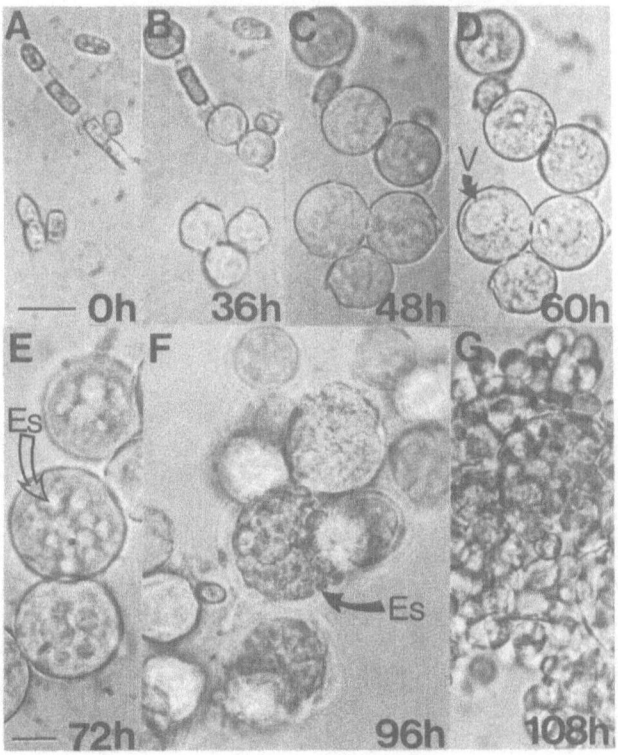

Fig. 1 A-G Time-lapse sequence (A-D) of early spherule development and stages of endosporulation (E,F) and second generation spherule formation (G) Times indicate duration after inoculation of cultures with arthroconidia of strain C735 ES, endospores, V, vacuole Bars represent 10 μm

the mature arthroconidia, which typically contain 2-5 nuclei, give rise to round cells after about 8 h incubation in Converse medium with a single, large nucleus. This nucleus is approximately twice the diameter of nuclei in the mature conidia and round cells at 24-36 h [4,5] Attempts are underway in our laboratory to synchronize production of the uninucleate round cells and examine the significance of this stage in the nuclear cycle of *C immitis*

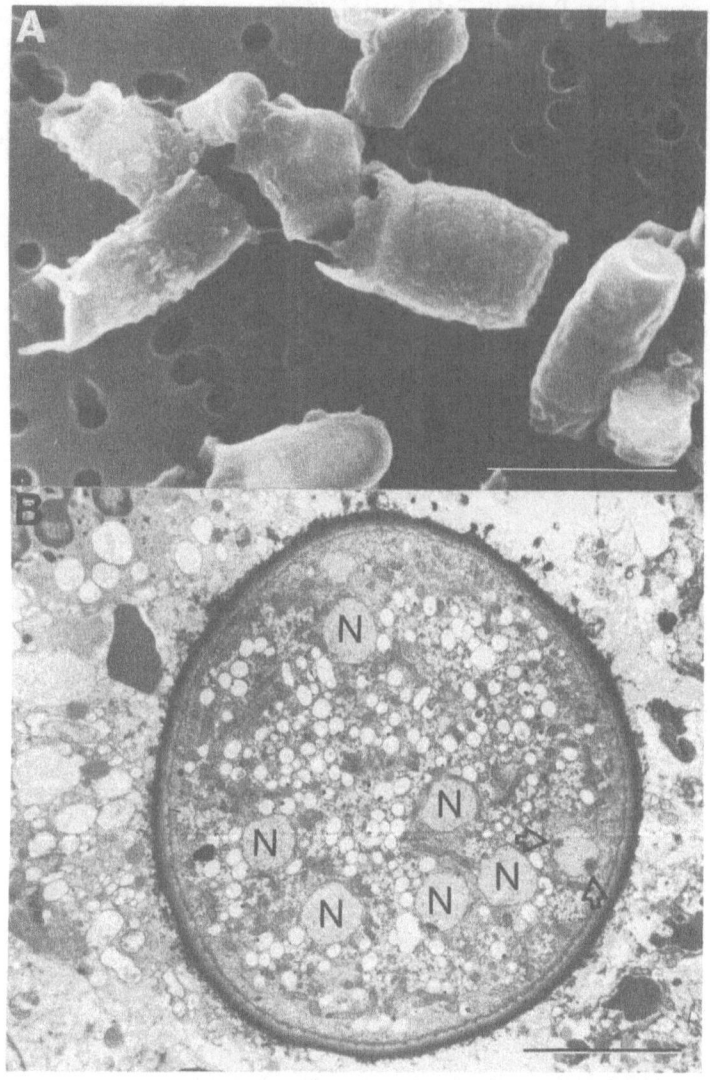

Fig 2 A,B Scanning electron micrograph of arthroconidia on Nucleopore membrane (A), and thin section of round cell initial (B) Arrows in (B) indicate mitotic figure N, nucleus Bars in (A) and (B) represent 10 μm and 5 μm, respectively

Ploidy of the Saprobic and Parasitic Cycles. We have recently examined the electrophoretic karyotypes of twelve clinical isolates of *C immitis* [19] We demonstrated that the chromosome number is four, and chromosomes range in molecular size from 3 2 to 11 5 megabases (Mb) Sun and Huppert[6] reported the presence of four homologous pairs of aceto-orcein-stained chromosomes in round cells produced from arthroconidia after incubation in Converse medium for 24 h This would suggest that *C immitis* is a diploid, dimorphic pathogen Confirmation of this ploidy awaits linkage studies and genetic analyses of selected mutants [25]

The genome size of *C. immitis* was estimated to be 29 Mb on the basis of total DNA content derived from addition of the average molecular size of chromosomes separated by contour-clamped homogeneous field (CHEF) gel electrophoresis.[13] This is approximately twice the haploid genome size of *Saccharomyces cerevisiae*.[26] This latter observation was supported by comparison of the electrophoretic estimate of genome size of *C. immitis* with that derived from cytofluorometric analyses of nuclear DNA content of *C. immitis* and *S. cerevisiae* (Fig. 3 A-E). Microspectrophotometry was used to estimate the total nuclear DNA content by ·measurement of fluorescence intensity of ethidium bromide-stained nuclei.[12] Arthoconidia were grown in Converse medium (39 °C, 20% CO_2/80% air) for 8 h and then exposed to 0.2 M hydroxyurea which inhibited DNA synthesis but was

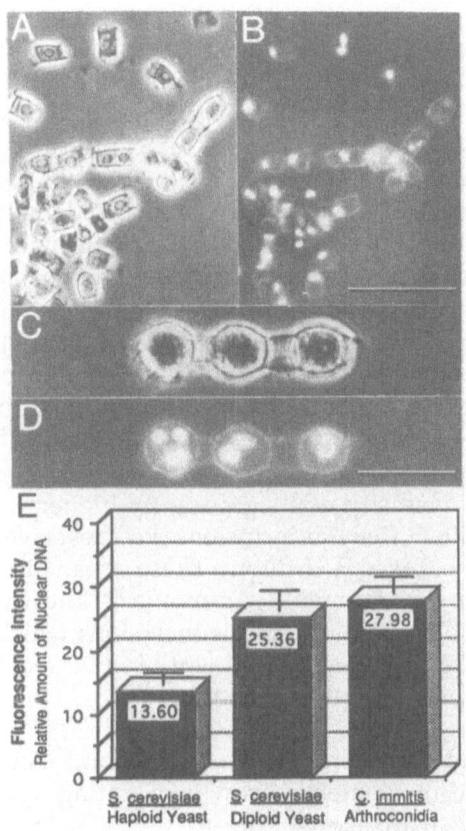

Fig. 3 (A-C). Arthroconidia isolated from 60-day-old GYE plate culture (A) and round cells isolated from Converse liquid medium plus hydroxyurea 14 h post-inoculation (C). The ethanol-fixed cells were stained with ethidium bromide after incubation with RNAase (B,D). Bars in (B) and (D) represent 10 µm.

(E). Relative amount of nuclear DNA (based on microspectrophotometric assays of fluorescence intensity of ethidium bromide-stained nuclei) of *Saccharomyces cerevisiae* haploid and diploid yeast cells, and round cells of *C. immitis*.

not toxic to the cells. The round cells continued to grow but were arrested in G_1 phase (Fig. 3 C,D). It is assumed that dormant arthroconidia obtained from 60 day GYE plate cultures (Fig. 3 A,B) were also in G_1 phase. Prior to staining, the ethanol-fixed cells were incubated in 0.1% ribonuclease (37 °C, 30 min) to minimize the amount of intercalation of ethidium bromide stain with dsRNA. *S. cerevisiae* haploid and diploid strains were treated in an identical manner. The validity of correlation of fluorescence intensity of the stained nuclei with total nuclear content was demonstrated by the 1:2 ratio of ethidium bromide intensity for these haploid and diploid strains, respectively. The fluorometric data and reported genome size of *S. cerevisiae* were used to estimate the genomic size of *C. immitis* by comparative microspectrophotometry. The calculated value (28.2 Mb) was comparable to the estimate derived from pulsed-field gel electrophoresis studies. These data also suggest that the ploidy of the saprobic and parasitic cycles is the same and the two phases are morphogenetic variations of the asexual cycle of this unusual pathogen.

Vacuolation and Ammonium Production during Round Cell Formation. The multitude of tiny vacuoles which were visible in the round cell initials (Fig. 2B) appear to coalesce as diametric growth proceeds. A large central vacuole surrounded by numerous smaller vacuoles (arrows in Fig. 4 A,B) is commonly visible in young spherules. Growth of parasitic cells of *C. immitis* on Converse agar which contained the pH indicator phenol red resulted in a visible color change (yellow to red) in the medium. This is indicative of a shift from pH 6.1 (fresh medium) to approximately pH 7.6-8.0. The pH profile of Converse liquid medium in the presence of the parasitic phase also demonstrated a shift from pH 6.3 (fresh medium) to 7.2 over a period of 6 days growth. This generation of an alkaline environment occurred in spite of the addition of CO_2 to the plate and flask cultures. To examine the intracellular pH of young spherules we used the SNAFL calcein AM indicator. This strain has been employed to determine intracellular pH in the physiological range by means of dual-emission or dual-excitation analyses. Under acidic conditions, SNAFL has been reported to fluoresce yellow-orange, while under basic conditions it fluoresces deep red.[14,15] Vacuolate spherules stained with this pH indicator fluoresced red using the Texas Red filter (emission 630 nm). Excitation intensity was significantly higher with the Texas Red filter than with the FITC filter. Cellular inclusions (vesicles?) were visible by fluorescent microscopy using the Texas Red filter. These organelles appeared to delineate intracellular alkaline environment. Ammonia is known to be produced by *C. immitis*.[27] It is possible that the spherule inclusions which react with the SNAFL stain contribute to the alkaline environment of *in vitro* cultures by transport and release of their contents at the plasma membrane.

Vacuolation and Diametric Growth of Round Cells. The central vacuole may also participate in diametric growth of young spherules. Spherule wall expansion, like hyphal tip elongation, may rely both on biosynthesis of wall polysaccharides and autolysis of preexisting components for growth.[28] The latter process provides for insertion of new polymers within the preexisting polysaccharide chains and maintenance of sufficient wall plasticity to permit turgor-driven expansion.[29] The internal turgor pressure during the period of rapid diametric growth of round cells may be provided by the central vacuole. The plasticity of the expanding wall of round cells may be maintained at least in part by a wall-associated, alkaline β-1,3-endoglucanase.[11] The β-glucanase has been localized in the cell wall and cytoplasmic (transport?) vesicles by immunoelectron microscopy with specific antiserum raised against the purified 120-kDa enzyme

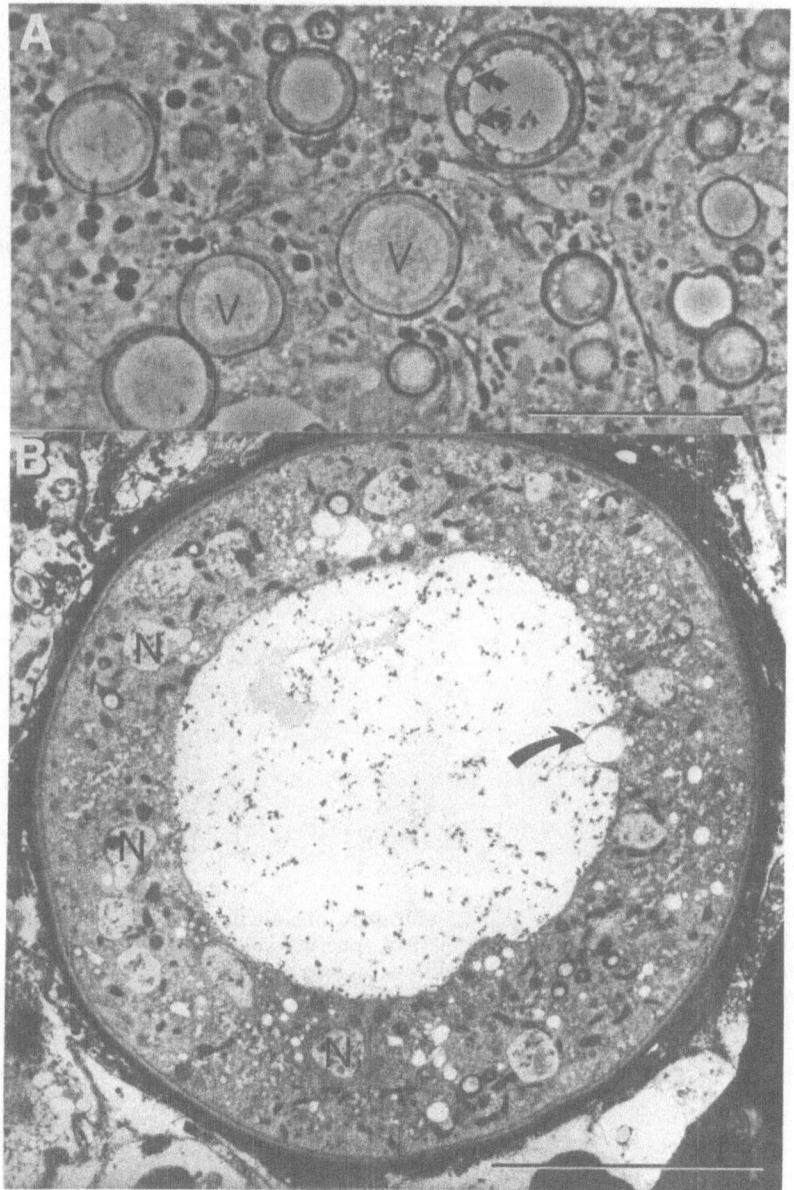

Fig. 4 A,B Thick section (A) and thin section (B) of vacuolate round cells in infected murine lung Arrows in (A) and (B) indicate coalescing vacuoles   N, nucleus, V, vacuole   Bars in (A) and (B) represent 20 µm and 10 µm, respectively

(Fig 5 A,B)   The boiled cell wall fraction isolated from young spherules was partially digested by the β-glucanase   β-glucans are reported to be structural components of the spherule wall [3]

Addition of 1-deoxynojirimycin to the parasitic phase culture medium at a concentration of 200 µM blocked or retarded diametric growth and conversion of arthroconidia to spherules [11] This glucose analogue is a potent inhibitor of the C.

immitis β-glucanase. Antibody was raised in guinea pigs against chromato-graphically-purified 1-deoxynojirimycin which was conjugated with bovine serum albumin. The inhibitor was localized by immunofluorescence in the wall of the 1-deoxynojirimycin-treated cells (Fig. 5 C-E).

Preliminary results are reported here of our isolation of the β-glucanase gene from a *C. immitis* expression library and examination of the temporal expression of the gene in the parasitic cycle. Four cDNAs have been isolated (Fig. 6A) on the basis of positive reactions of their respective clones with anti-β-glucanase. Two of these clones also expressed β-glucanase activity, as detected in the presence of the pNP-β-*D*-glc substrate. Each cDNA probe identified a 2.1-2.2 kb *C. immitis* RNA

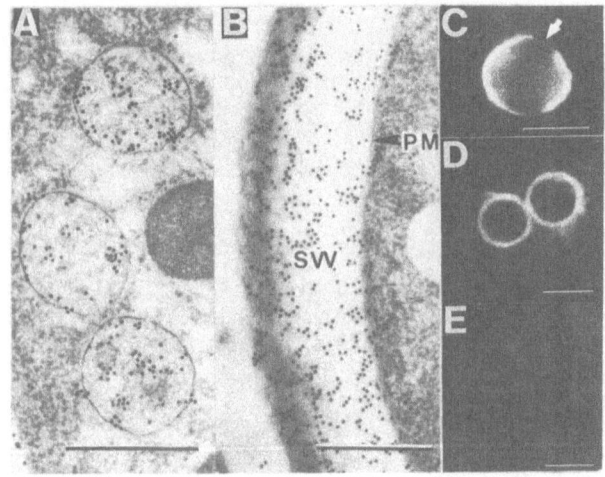

Fig. 5 (A,B). Immumolocalization of β-1,3-glucanase in transport vesicles and wall of young spherules (round cells). PM, plasma membrane; SW, spherule wall.
(C,E). Intact round cell (C) and thick sections of round cells (D,E) exposed to 200 μM 1-deoxynojirimycin in culture and then reacted with anti-deoxynojirimycin/FITC antibody (C,D) or control serum (E). Bars in (A) and (B) represent 0.5 μm. Bars in (C-E) represent 5 μm.

(Fig. 6B). Northern dot blots were performed using the 1.8 kb cDNA as a probe (Fig. 6C). Temporal expression of the gene was detected at 36 h and 108 h, which represent stages of diametric growth of 1st and 2nd generation round cells (Fig. 1). Early results of our sequencing analyses have revealed identity between the 1.8 kb, 0.4 kb, and 0.8 kb cDNA fragments (Fig. 7).

Spherule Segmentation. Compartmentalization of the coenocytic round cells is initiated by centripetal growth of the inner layer of the envelope (Fig. 8A).
A chymotrypsin-like, alkaline, serine proteinase of *C. immitis* with an estimated molecular size of 34 kDa has been shown by immunoelectron microscopy to be localized in the segmentation wall (SW; Fig. 8A) of the parasitic cells. The apparent high affinity of the 34 kDa proteinase for walls of the actively segmenting

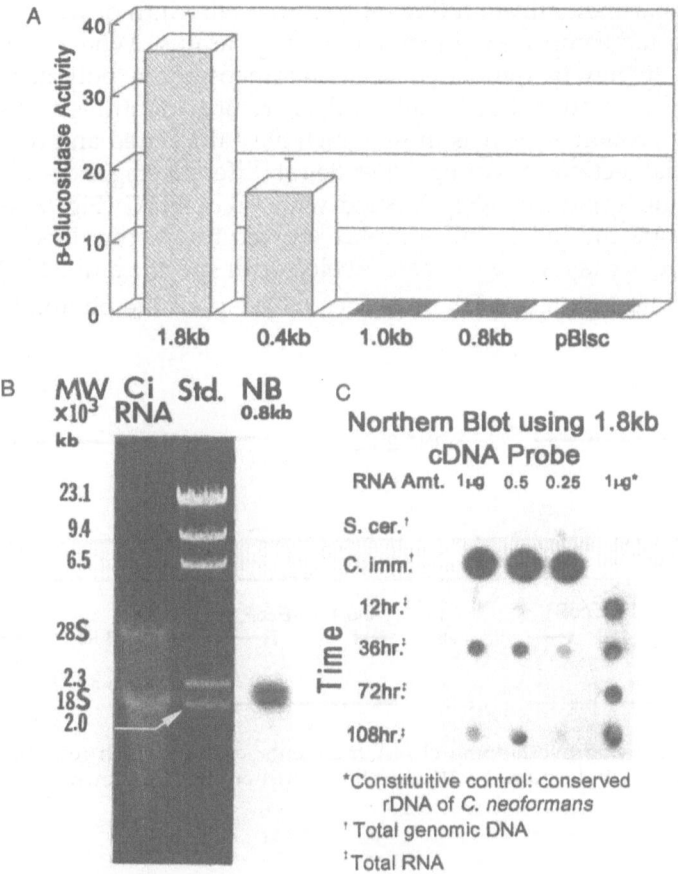

Fig. 6 (A). β-glucosidase activity of fusion proteins measured in the presence of pNP-β-D-glc and determined as micromoles of pNP released from the substrate per min. Different fusion proteins were identified by the size of the cDNA insert in kilobases (kb). The β-galactosidase expressed by the Bluescript plasmid alone (pBlsc) was used as a control in the assay of β-glucosidase activity. (B). Northern hybridization (NB) with labeled 0.8 kb cDNA probe of the electrophoretically separated total poly(A)-containing RNA of *C. immitis* (Ci). The molecular sizes of the standard mixture are indicated. (C). Northern dot blot with labeled 1.8 kb cDNA probe of total RNA of *C. immitis* spherules grown in Converse liquid medium for 12 h, 36 h, 72 h, and 108 h. Control hybridizations conducted between the same 1.8 kb probe and toal genomic DNA of *S. cerevisiae* and *C. immitis*, and between conserved rDNA probe and *C. immitis* spherule RNA samples.

spherules suggested that the enzyme may participate in the morphogenesis of these cells. We isolated a 1.2 kb cDNA from the expression library of *C. immitis* and presented evidence that a 927 kb open reading frame of the cDNA encoded the 34 kDa proteinase.[16] Expression of the proteinase gene was examined by Northern dot blot analysis of total RNA from different stages of parasitic cell cycle. Maximum levels of specific mRNA were detected during spherule segmentation and early endospore differentiation.[16] Thin sections of spherules which are in the process of compartmentalization (Fig. 8B), show a labyrinth of wall layers that are formed *de novo* and comprise the segmentation apparatus. Wheat germ agglutinin (WGA) binds to newly deposited *N*-acetylglucosamine polymers as well as crystallized chitin in fungal cell walls and, therefore, identifies walls that are in stages of active growth as well as those which have thickened and ceased to grow. Two sets of serial sections of young spherules at different stages of development of the segmentation apparatus were labeled with WGA-FITC (Fig. 9 A-F and J-N). Both the envelope and inner wall complex showed fluorescent label. The wall of these same developmental stages were labeled with specific anti-34 kDa-FITC and showed an identical pattern of fluorescence.[16] We used the photomicrographs of

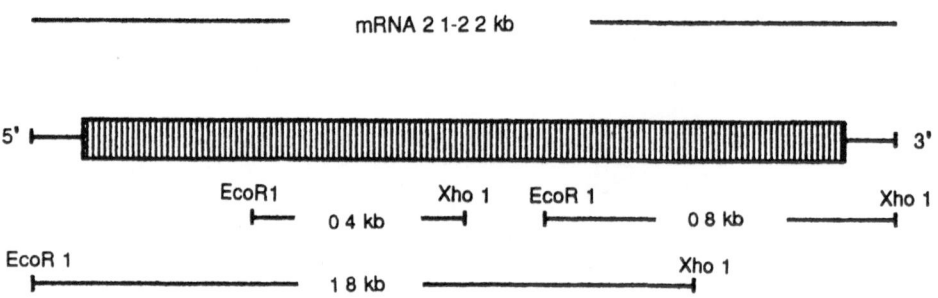

Fig. 7 Diagram showing overlapping cDNA fragments, each of which recognize a 2.1-2 2 kb mRNA by Northern hybridization. EcoRI and XhoI restriction sites are shown.

the WGA-FITC-labeled, serially-sectioned spherule in Fig. 9 A-F to produce a 3-dimensional computer reconstruction of half of the cell by means of the AVS software package (Fig. 9 G-I). The reconstructed images provide a better appreciation of the extensive wall synthesis involved in compartmentalization of the spherules prior to endospore differentiation. The central vacuole was still intact at this stage of development. The active, alkaline, 34 kDa serine proteinase has been isolated from viable, segmented spherules by mild detergent extraction.[31] The presence of the active enzyme in the segmentation wall which is undergoing intussusception of chitin microfibrils suggested an interrelationship between the proteinase and chitin synthase. Of possible significance is that the anti-34 kDa immunogold label in electron micrographs of segmented spherules is commonly located at the surface of the plasma membrane. A PCR-generated 600 bp DNA fragment, which we have indicated represents a conserved region of the chitin synthase gene of *C. immitis*,[20] was used as a probe to examine temporal expression of the synthase gene during spherule morphogenesis. As expected, an increase in level of specific message was detected in 1st generation spherule development as round cell expansion and subsequent segmentation wall synthesis occurred. This level of message was maintained through endospore differentiation and thus the

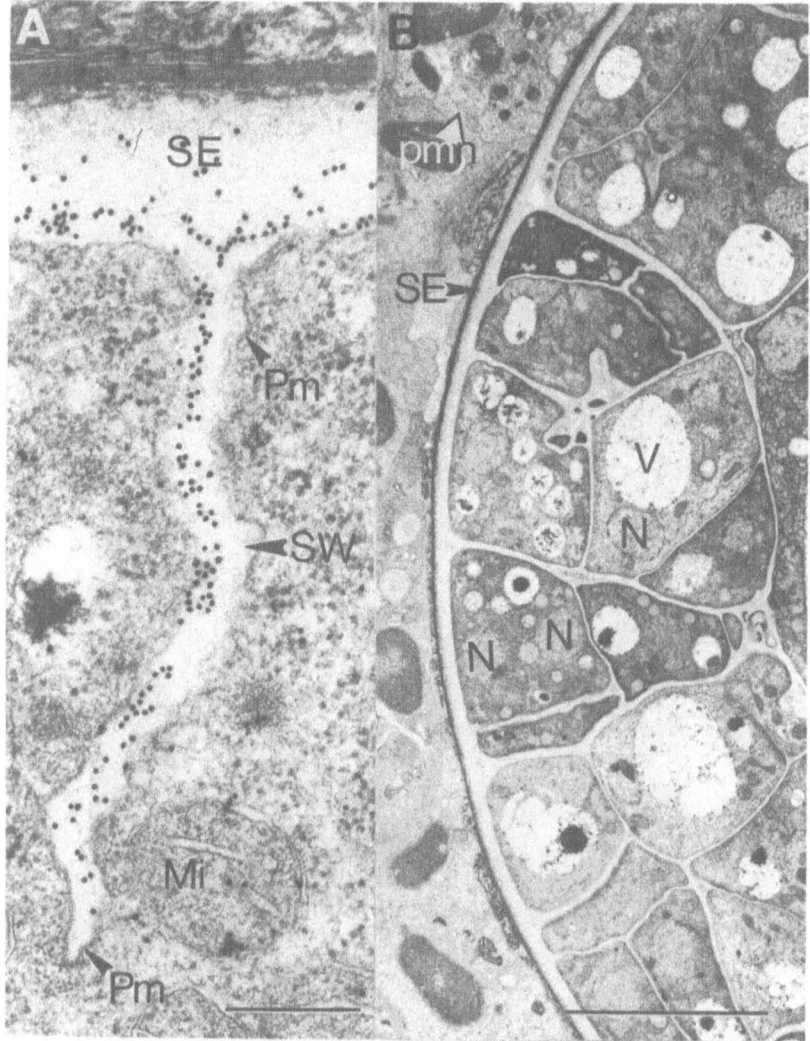

Fig. 8 Thin sections of early (A) and advanced (B) stages of segmentation of spherules. The immunogold label in (A) locates the 34 kDa proteinase in the initial stage of segmentation wall (SW) formation. Mi, mitochondrion; N, nucleus; Pm, plasma membrane; pmn, polymorphonuclear neutrophil of host; SE, spherule envelope; V, vacuole. Bars in (A) and (B) represent 0.5 μm and 3.0 μm, respectively.

chitin synthase gene appeared to behave as a constitutive gene (Fig. 12C). Additional studies are underway to determine whether a functional interrelationship exists between the 34 kDa proteinase and chitin synthase.

Endosporulation. As endospores differentiate within compartments of the segmented spherule, prominent vacuoles are visible in each of the newly-formed cells (Fig. 10A). At about this same stage, the central vacuole of the spherule ruptures. Based on preliminary examination of mature spherules with the fluorescent pH indicator, SNAFL, it appeared that the contents of the vacuole were neutral to slightly acidic. It is possible that release of the vacuolar central contents

205

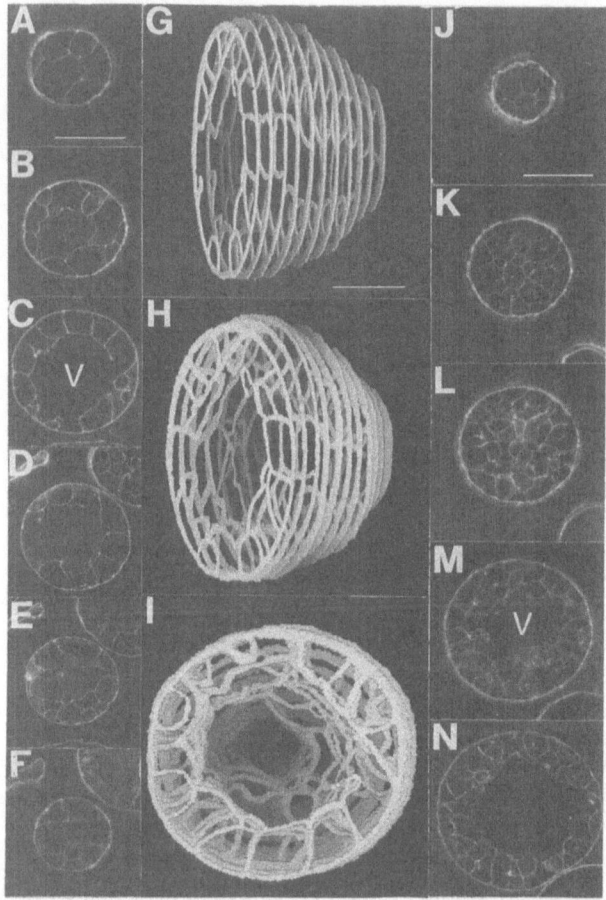

Fig. 9 (A-F) and (J-N). Two sets of serial sections of segmented spherules stained with WGA/FITC to reveal the spherule envelope and segmentation wall.
(G-I). Three dimensional reconstructions of the serial sections shown in (A-F) using a modified computer software package. Note that the image reconstruction made use of 11 sections of the spherule (not all WGA/FITC stained sections are shown). Bars in (A) and (J) represent 25 μm. Bar in (G) represents 12.5 μm.

into the region of the segmentation wall may lead to a slight but perhaps significant drop in pH The endospore initials which have probably already begun to produce ammonia, could at least buffer this effect. The endospores (2nd generation round cells) are eventually released from the maternal spherule upon rupture of the external envelope (Fig. 10B).   An important morphological difference between the early and late stages of endosporulation shown in Fig. 10 A and B, respectively, is the apparent disappearance of the segmentation wall

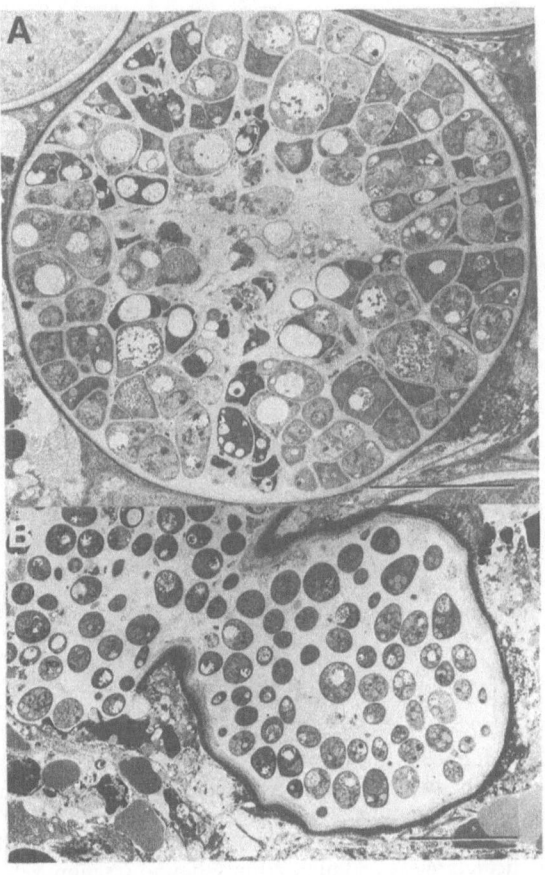

Fig. 10 A,B Thin sections of early (A) and advanced (B) stages of endosporulation of C immitis Note absence of central vacuole and cytoplasmic debris in mid-region of spherule in (A) Note absence of segmentation wall in (B) Bars in (A) and (B) represent 10 μm

apparatus On the basis of preliminary evidence that the segmentation wall contains chitin microfibrils,[30] and that *C. immitis* produces a chitinase,[32] we examined the possibility that the chitinase may be involved in digestion of the segmentation apparatus during the phase of endosporulation

Johnson and Pappagianis[32] have presented evidence that the CF antigen of *C. immitis* (48 kDa polypeptide in reducing SDS-PAGE gel) is a chitinase. Our rabbit antiserum which was raised against the nonreduced, gel filtration fraction of the mycelial culture filtrate (Silvera strain) recognized both a 100 kDa and 110 kDa band in nonreducing gels of the F+L material (Fig. 11A). The immunoaffinity-

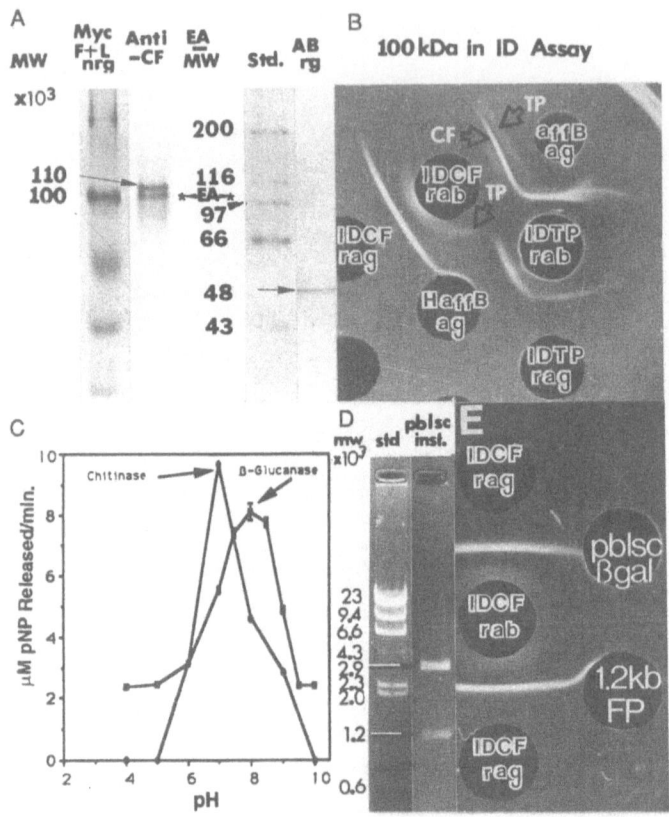

Fig. 11 (A) SDS-PAGE nonreducing gel (nrg) of mycleial culture filtrate + lysate (Myc/F+L) reacted with anti-CF rabbit antibody. EA, locates enzymatically-active band in presence of pNP-β-D-N,N'-diacetylchitobiose. AB/rg, immunoaffinity bound fraction in SDS-PAGE reducing gel. (B) Immunodiffusion assay of immunoaffinity-bound (affB) fraction using both ID-CF and ID-TP reference systems. HaffB, heated (60 °C, 30 min) affinity-bound fraction; rab, reference antibody; rag, reference antigen. (C) pH profile of 100 kDa chitinase using pNP-β-D-N,N'-diacetylchitobiose as substrate, and 120 kDa β-1,3-glucanase using pNP-β-D-glc as substrate. (D) Ethidium bromide-stained agarose gel separation of 1.2 kb cDNA which partially encodes fusion protein (FP) in Bluescript plasmid (pblsc) which was shown in the immunodiffusion assay to be weakly positive for ID-CF activity.

bound culture filtrate fraction, obtained using this same antiserum, was examined for *in situ* chitinase activity in a nonreducing SDS-PAGE with *p*NP-β-D-*N,N'*-diacetylchitobiose as substrate. A single band with molecular size of 100 kDa showed enzyme activity (Fig. 11A). When the immunoaffinity bound fraction was examined by SDS-PAGE under reducing conditions, a 48 kDa band and weak 110 kDa band were visible. The immunoaffinity-bound fraction showed high affinity for CF antibody from coccidioidomycosis patients in the ID-CF assay, as well as reactivity in the ID-TP assay (Fig. 11B). When the bound fraction was heated (60 °C, 30 min) ID-CF activity was lost, but ID-TP activity was retained. We have previously described a 110 kDa ID-TP antigen isolated from the mycelial culture filtrate of *C. immitis.*[23] We suggest that the 110 kDa component of the immunoaffinity-bound fraction in Fig. 11A is this same heat-stable, ID-TP antigen, while the 100 kDa component is the heat-labile, ID-CF antigen and chitinase.

Results of comparison of the pH profile of the β-1,3-glucanase described above and the 100 kDa chitinase are shown in Fig. 11C. As indicated in our earlier discussion, the optimal pH of the β-glucanase is approximately 8.0. The pH optimum for the chitinase, on the other hand, is 6.8-7.0. This difference may be significant in terms of localization and function of these two enzymes during spherule morphogenesis.

The anti-chitinase antibody was also used to screen the cDNA expression library of *C. immitis.* A 1.2 kb cDNA was isolated which encodes a fusion protein with ID-CF activity (Fig. 11 D,E). This same 1.2 kb cDNA recognized a single 1.8 kb mRNA band in a Northern blot (Fig. 12A). We have little evidence as yet that the 1.2 kb cDNA encodes the 100 kDa chitinase. However, we suggest if this is the chitinase gene, it should reveal maximum level of specific mRNA expression during early endospore differentiation, and the active enzyme should be present in the segmentation wall as endospores are released from the spherule compartments and the segmentation apparatus begins to autolyse. The anti-chitinase antibody, preadsorbed with the 110 kDa ID-TP antigen, was reacted with thin sections of 96 h endosporulating spherules which were prepared for immunoelectron microscopy (Fig. 12B). The immunogold label was localized in the endospore wall and remnants of the segmentation wall. We assume that the endospores secrete the chitinase which explains the presence of immunolabel in the endospore wall. Temporal expression of the 1.2 kb cDNA (chitinase gene?) was detected at maximum level at approximately 48 h of 1st generation spherule development, the level of message remained high through 84 h, and then sharply decreased at 96 h (Fig. 12C). One possible explanation of these data is that the inactive chitinase is synthesized and incorporated into the segmentation wall prior to endospore differentiation. Activation of the enzyme occurs only after endospore differentiation within the spherule compartments has occured. Perhaps the trigger for activation of the chitinase is a slight drop in pH when the central vacuole of the maternal spherule ruptures. This same trigger may result in the loss of specific message during the late stage of endosporulation. Internal pH-regulated genes have been described in bacteria.[33] It is not unreasonable that such genes may also exist in *Coccidioides.*

CONCLUSIONS

We have presented preliminary evidence for the association of a 120 kDa β-1,3-endoglucanase, 34 kDa chymotrypsin-like serine proteinase, and 100 kDa chitinase with separate morphogenetic events of the parasitic cycle of *C. immitis.* Confirmation that the isolated cDNAs used as probes in this study encode the respective enzymes (i.e., 1.8 kb cDNA encodes the β-glucanase, and 1.2 kb cDNA

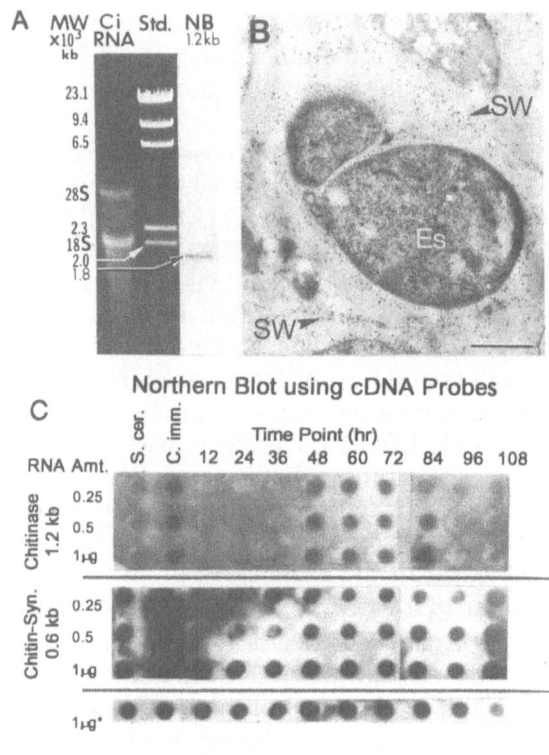

Northern Blot using cDNA Probes

* Constitutive control rDNA from *Cryptococcus neoformans*

Fig. 12 (A) Northern hybridization (NB) with labeled 1.2 kb cDNA probe of the electrophoretically separated total poly (A)-containing RNA of *C. immitis* (Ci). The molecular sizes of the standard mixture are indicated. (B) Thin section of endosporulating spherule reacted with anti-chitinase/CF antiserum preadsorbed with 110 kDa glycoprotein.[23] Note immunogold label localizes chitinase on segmentation wall (SW). Es, endospore. Bar represents 1 μm. (C) Northern dot blots labeled with the 1.2 kb chitinase probe or 0.6 kb chitin synthase probe of total RNA of *C. immitis* spherules grown in Converse liquid medium for 12 h to 108 h. Control hybridizations conducted between same probes and total genomic DNA of *S. cerevisiae* and *C. immitis*, and between conserved rDNA probe and *C. immitis* spherule RNA samples.

encodes the chitinase) is the focus of our current studies. Results of enzymological and immunolocalization investigations of these cellular products, and preliminary data from recombinant DNA studies suggest that these wall-associated enzymes may function as regulatory factors in the morphogenesis of *C. immitis*. Ultimately, we hope to develop a transformation system for *C. immitis* which, when combined with construction of β-1,3-glucanase, 34 kDa proteinase, or chitinase mutations derived from specific gene disruptions, will permit a rigorous assessment of the role of these wall-associated hydrolases in spherule morphogenesis.

ACKNOWLEDGMENTS

The authors are grateful to the Cell Research Institute for provision of the electron-microscopic facilities. This study was supported by Public Health Services Grant AI19149 from the National Institute of Allergy and Infectious Disease.

REFERENCES

1. D. Pappagianis, Epidemiology of coccidioidomycosis, *Curr. Top. Med. Mycol.* 2: 199 (1988).
2. D. J. Drutz and M. Huppert, Coccidioidomycosis: factors affecting the host-parasite interaction, *J. Infect. Dis.* 147: 372 (1983).
3. S. H. Sun, G. T. Cole, D. J. Drutz and J. L. Harrison, Electron-microscopic observations of the *Coccidioides immitis* parasitic cycle *in vivo*, *J. Med. Vet. Mycol.* 24: 183 (1986).
4. G. T. Cole and T. N. Kirkland, Conidia of *Coccidioides immitis*: their significance in disease initiation, in: "The Fungal Spore and Disease Initiation in Plants and Animals", G. T. Cole and H. C. Hoch, eds., Plenum Press, New York (1991).
5. G. T. Cole and S. H. Sun, Arthroconidium spherule-endospore transformation in *Coccidioides immitis*, in: "Fungal Dimorphism: With Emphasis on Fungi Pathogenic for Humans", P. J. Szaniszlo and J. L. Harris, eds., Plenum Press, New York (1985).
6. S. H. Sun and M. Huppert, A cytological study of morphogenesis in *Coccidioides immitis*, *Sabouraudia* 14: 185 (1976).
7. M. J. Dolan, C. P. Lattuada, G. P. Melcher, R. Zellmer, R. Allendoerfer and M. G. Rinaldi, *Coccidioides immitis* presenting as a mycelial pathogen with empyema and hydropneumothorax, *J. Med. Vet. Mycol.* 30: 249 (1992).
8. H. B. Levine, Purification of the spherule-endospore phase of *Coccidioides immitis*, *Sabouraudia* 1: 112 (1961).
9. G. T. Cole and W. B. Kendrick, A thin culture chamber for time-lapse photo-micrography of fungi at high magnification, *Mycologia* 60: 340 (1968).
10. G. T. Cole, Preparation of microfungi for scanning electron microscopy, in: "Ultrastructure Techniques for Microorganisms", H. C. Aldrich and W. J. Todd, eds., Plenum Press, New York (1986).
11. D. Kruse and G. T. Cole, A seroreactive 120 kilodalton β-1,3-glucanase of *Coccidioides immitis* which may participate in spherule morphogenesis, *Infect. Immun.* 60: 4350 (1992).
12. B. Wittmann-Meixner, E. Weber and A. Bresinsky, Different grades of correlation between relative nuclear DNA content, chromosome number, and ploidy levels in fungi, *Oper. Bot.* 100: 267 (1989).
13. G. T. Cole, *Graphiola phoenicis*: a taxonomic enigma, *Mycologia* 75: 93 (1983).
14. J. E. Whitaker, R. P. Haugland and F. G. Prendergast, Spectral and photophysical studies of benzo[c]xanthene dyes: dual emission pH sensors, *Anal. Biochem.* 194: 330 (1991).
15. R. P. Haugland, "Handbook of Fluorescent Probes and Research Chemicals", 5[th]ed., Molecular Probes, Inc., Eugene, Oregon (1992).
16 G. T. Cole, S. Zhu, L Hsu, D. Kruse, K. R. Seshan and F. Wang, Isolation and expression of a gene which encodes a wall-associated proteinase of *Coccidioides immitis*, *Infect. Immun.* 60: 416 (1992).
17. T. N. Kirkland, S. Zhu, D. Kruse, L. Hsu, K. R. Seshan and G. T. Cole, *Coccidioides immitis* fractions which are antigenic for immune T lymphocytes, *Infect. Immun.* 59: 3952 (1991).

18. A. Feinberg and B. Vogelstein, A technique for radiolabeling DNA restriction endonuclease fragments to high specific activity, *Anal. Biochem.* 132: 6 (1983).

19. S. Pan and G. T. Cole, Electrophoretic karyotypes of clinical isolates of *Coccidioides immitis*, *Infect. Immun.* 60: 4872 (1992).

20. A. R. Bowen, J. L. Chen-Wu, M. Momany, R. Young, P. J. Szaniszlo and P. W. Robbins, Classification of fungal chitin synthases, *Proc. Natl. Acad. Sci. U.S.A.* 89: 519 (1992).

21 B. I. Restrepo and A. G. Barbour, Cloning of 18S and 25S rDNAs for the pathogenic fungus *Cryptococcus neoformans*, *J. Bacteriol.* 171: 5596 (1989).

22. G. T. Cole, T. N. Kirkland and S. H. Sun, An immunoreactive, water-soluble conidial wall fraction of *Coccidioides immitis*, *Infect. Immun.* 55: 657 (1987).

23. G. T. Cole, D. Kruse and K. R. Seshan, Antigen complex of *Coccidioides immitis* which elicits a precipitin antibody response in patients, *Infect. Immun.* 59: 2434 (1991).

24. B. L. Zimmer and D. Pappagianis, Characterization of a soluble protein of *Coccidioides immitis* with activity as an immunodiffusion-complement fixation antigen, *J. Clin. Microbiol.* 26: 2250 (1988).

25. W. L. Whelan, The genetics of medically important fungi, *Crit. Rev. Microbiol.* 14: 99 (1987).

26. G. F. Carle, M. Frank and M. V. Olson, Electrophoretic separations of large DNA molecules by periodic inversion of the electric field, *Science* 232: 65 (1986).

27. W.S. Bump, Observations on growth of *Coccidioides immitis*, *J. Infect. Dis.* 36: 561 (1925).

28. S. Bartnicki-Garcia, Fundamental aspects of hyphal morphogenesis, in: "Microbial Differentiation. Twenty-third Symposium of the Society for General Microbiology", J. M. Ashworth and J. E. Smith, eds., Cambridge Univ. Press, London (1973).

29. N. A. R. Gow, Control of extension of the hyphal apex, *Curr. Top. Med. Mycol.* 3: 109 (1989).

30. R. F. Hector and D. Pappagianis, Enzymatic degradation of the wall of spherules of *Coccidioides immitis*, *Exptl. Mycol.* 6: 136 (1982).

31. L. Yuan, G. T. Cole and S. H. Sun, Possible role of a proteinase in endosporulation of *Coccidioides immitis*, *Infect. Immun.* 56: 1551 (1988).

32. S. M. Johnson and D. Pappagianis, The coccidioidal complement fixation and immunodiffusion-complement fixation antigen is a chitinase, *Infect. Immun.* 60: 2588 (1992).

33. J. L. Slonczewski, pH-regulated genes in enteric bacteria, *Am. Soc. Microbiol. News* 58: 140 (1992).

DIMORPHISM IN *HISTOPLASMA CAPSULATUM*: STUDY OF CELL DIFFERENTIATION AND ADAPTATION

George S. Kobayashi[a] and Bruno Maresca[a,b]

[a] Department of Internal Medicine
Washington University School of Medicine
St. Louis, Missouri 63110, USA
and
[b] Istituto Internazionale di Genetica e Biofysica
CNR
via Marconi 10,
80125 Napoli, Italia

ABSTRACT

Fungal morphogenesis, in particular dimorphism in pathogenic fungi, has intrigued clinicians and medical mycologists since its discovery at the turn of the century. Numerous reviews have been written on this subject. Using *Histoplasma capsulatum* as a model to study host-parasite interaction at the biochemical and molecular level, we have attempted to relate the clinical spectrum of histoplasmosis to natural variations in the characteristics of *H. capsulatum* and to the adaptations this organism must make as a saprobe and a parasite.

Fungi are ubiquitous in nature where they reside as free-living saprobes deriving no obvious benefits by parasitizing humans or animals. While we are constantly bombarded by infectious propagules, healthy immunologically competent individuals have a high degree of innate resistance against fungi. However, this obviously depends on the degree of exposure and dosage of organisms since most infections are asymptomatic or subclinical when they occur. The vast majority of fungi that have been implicated in the mycoses of humans require some form of trauma or an immunosuppressed state in order to infect and cause disease. Of the growing number of fungi implicated in the diseases of humans, only five, *Histoplasma capsulatum, Blastomyces dermatitidis, Paracoccidioides brasiliensis, Coccidioides immitis,* and *Cryptococcus neoformans* have the innate ability to infect and are considered primary pathogens of humans. The principal site of infection is the respiratory tract. Conidia, spores and other infectious propagules are inhaled and lodge on the mucous membranes of the respiratory tree or in the alveoli of the lung, where they encounter macrophages and are phagocytosed. To colonize the host successfully these organisms must be able to survive the elevated temperature of the body, elude phagocytosis, neutralize the hostile environment they encounter, or adapt in a manner that will allow them to multiply. Several factors contribute to infection and pathogenicity of these organisms. For example, four of the primary mycotic pathogens, *H. capsulatum*, *B. dermatitidis*, *P. brasiliensis*, and *C. immitis* are dimorphic, changing from a multicellular filamentous form to a unicellular morphology when they invade tissues.

In contrast, a morphologic transition does not play a role in the pathogenesis of *Cr. neoformans* infections since this organism is an encapsulated yeast at 25 °C, at 37 °C, and in host tissues. In the case of *Cr. neoformans*, several virulence factors have been identified. The capsular polysaccharide, for example, plays an important role by inhibiting phagocytosis. Encapsulated *Cr. neoformans* yeast are highly resistant to phagocytosis by human neutrophils, whereas acapsular mutants are effectively phagocytosed. However, since most of these acapsular strains have been generated by mutagenesis, it is difficult to rule out other genetic defects that contribute to the decreased pathogenicity. Another factor that appears to be important is phenoloxidase, an enzyme that is responsible for the formation of melanin from phenolic compounds. *Cr. neoformans* is the only species in the genus that possesses this enzyme, a characteristic that is used to identify the yeast rapidly in the clinical situation. Isolates of *Cr. neoformans* that lack phenoloxidase are avirulent. Although the role this enzyme plays in virulence is unknown, similar observations have been made with *Wangiella dermatitidis*. In addition to these putative virulence factors, the ability to survive and replicate at 37 °C is of paramount importance since isolates that cannot survive at this elevated temperature are avirulent.

The fungus *H. capsulatum* is widely distributed throughout the temperate, sutropical and tropical zones of the world. It is the etiologic agent of the most common respiratory fungal infection affecting humans. This fungus, like *B. dermatitidis* and *P. brasiliensis*, exists saprobically as a multicellular filamentous mould and parasitically as a unicellular budding yeast. This switch in morphology is reversible and can be reproduced under laboratory conditions by changes in oxidation-reduction potential, CO_2 tension, temperature of incubation, and nutritional factors. In the case of *H. capsulatum*, conversion to the yeast morphology is required for pathogenicity since isolates that are unable to undergo this transformation are avirulent. The uniqueness of this type of cellular differentiation in fungi is that it is distinct from embryogenesis in higher forms of life since the process is reversible and not an essential part of the life cycle of the organism.

The nutritional requirements for mycelial growth are simple; the organism can be grown in the filamentous phase by incubating cultures at 25 °C with glucose as sole source of carbon and ammonia as source of nitrogen. The yeast phase, on the other hand, is more fastidious, in addition to an incubation temperature of 37 °C, the organism requires sulfhydryl-containing compounds for initiation of yeast development and cysteine or cystine along with certain growth factors such as biotin, thiamine, or thioctic acid in order to maintain this morphology. The role of cysteine during morphogenesis has been analyzed by studying the effect of this amino acid on respiration of both phases and during the transition. We have shown that respiration, as in many other organisms, proceeds through a branched electron transport system consisting of two terminal oxidase pathways: one, the cytochrome system, is blocked by cyanide and antimycin; the other, an unidentified alternative oxidase, is insensitive to those inhibitors, and is specifically blocked by salicylhydroxamic acid (SHAM).

Three distinct stages have been distinguished following the temperature shift from 25 °C to 37 °C in the avirulent Downs strain of *H. capsulatum*. Stage I begins immediately after the temperature shift. It is characterized by a progressive decrease in cellular respiration rate of mycelia with a parallel two- to five-fold decrease in intracellular cysteine concentration, other amino acids, RNA synthesis and in the virulent temperature tolerant isolates there is a partial uncoupling of oxidative phosphorylation. This stage lasts between 24 and 40 h at which time cells enter stage II, which is characterized by a reduced respiration rate lasting 4 to 6 days at which time the cells enter stage III. In stage III there is a gradual recovery of cellular respiration to the level characteristic of yeast with a concominant increase in the concentration of cysteine and other amino acids. Similar changes occur in strains of *H. capsulatum* that are temperature tolerant and virulent for mice but they are less exaggerated than those observed in the Down strain.

Addition of cysteine or other sulfhydryl-containing compounds during stage II shortens the dormant stage and accelerates the mycelium-to-yeast transition. The requirement for cysteine or other sulfhydryl compounds to complete the mycelium-to-yeast transition is different from the yeast's requirement for cysteine. Cysteine appears to have at least two functions in the phase transition in *H. capsulatum*: (1) it is required in stage II for the mycelial cells to complete the transition to yeast; and (2) yeast phase cells but not mycelia contain a cysteine oxidase and yeasts have a nutritional need for cysteine that cannot be satisfied by other sulfhydryl-containing compounds. It has been suggested that this enzyme may regulate the level of intracellular free cysteine or provide a metabolic product of oxidation of cysteine that is important for mycelium-to-yeast transition. It should be emphasized that conversion of mycelium to the yeast morphology is required for pathogenicity. Table 1 summarizes some of the changes that occur during the phase transition.

The transition from one morphologic phase to the other can be reversibly triggered by shifting the temperature of incubation between 25 °C (mycelium) and 37 °C (yeast). This implies that each phase of growth is an adaptation to two remarkably different environments. The temperature-induced phase transition and the events in the establishment of infection are intimately interrelated and unlike morphogenesis in higher eukaryotes, the differentiation process in dimorphic fungi represents an adaptation to a new environment. As a pathogen the fungus must face challenges that may not be related strictly to the process of dimorphism such as higher temperatures, different oxidation-reduction potentials, and a hostile host environment.

An example of the capacity to change from a saprobic mycelial existence and adapt to a parasitic environment is the activation of heat shock (hs) genes during the induction of the yeast phase. These genes, among others, play an important role in adaptation to a new environment that dimorphic fungal pathogens encounter when they invade a human host. Since the transition from a saprobic to parasitic existence in *H. capsulatum* is induced by a sudden change in temperature from 25 °C to 37 °C it was suggested that the early events in the mycelium to yeast transition in this pathogen was a heat shock response which resulted in adaptation to the higher temperature. Thus in *H. capsulatum* a change in temperature serves not only to induce the *hs* phenomenon but it also triggers the morphologic transition. At between 34 °C and 40 °C after the shift up from 25 °C, *hs* proteins are induced in *H. capsulatum*. While the number of major heat shock proteins is similar, the pattern of synthesis differs depending on the strain tested and the temperature of induction. In the avirulent and temperature sensitive Downs strain *hs* proteins peaked at 34 °C and in the more virulent and temperature-tolerant strains G217B and G222B the response was highest at 37 °C. Along with these changes, partial uncoupling of respiration occured at 34 °C in the temperature-sensitive Downs strain and electron transport was decreased by less than 50% of the initial values as compared to the level measured at 37 °C. In the first 3 h after the temperature shift, maximal transcription of cloned *hsp70* and *hsp82* genes occured during *hs* at 34 °C for the thermosensitive Downs strain whereas for the temperature-tolerant G222B strains it occured at 37 °C. In these strains of *H. capsulatum*, we have shown that a close correlation exists between the level of *hs* gene expression, the level of thermotolerance and the degree of murine pathogenicity.

Table 1. Biochemical and physiological changes that occur during the three stages of the mycelium to yeast transition in *H. capsulatum*.

Stage I	a)	Partial uncoupling of oxidative phosphorylation
	b)	Decline in intracellular ATP
	c)	Induction of hsp70 and hsp83 genes (heat shock proteins appear)
	d)	Cysteine reductase appears
	e)	Intracellular amino acids decrease
	f)	Decrease in RNA
	g)	Decrease in net protein synthesis
	h)	Decrease in cAMP
	i)	Decrease in cytochromes b,c, and aa_3
Stage II	a)	Decrease in respiratory activity
	b)	Cysteine stimulates a "shunt pathway"
	c)	Cysteine oxidase appears late in Stage II
Stage III	a)	Cytochromes b,c and $aa3$ reappear
	b)	Respiratory activity increases
	c)	RNA and protein synthesis increases

Another feature of *H. capsulatum* is that it is an intracellular fungal pathogen whose pathogenicity relates to its ability to survive and replicate within macrophages. Studies with various pathogenic microbes indicate that the metabolic status of the mononuclear phagocyte plays an important role in host resistance to intracellular pathogens. Once internalized there is rapid fusion of the phagosome with lysosomes. Lysosomal enzymes and reactive oxygen metabolites elaborated during phagocytosis are involved in the microbiocidal activity of these cells. Organisms that survive within macrophages appear to have envolved mechanisms to escape these lethal effects. These include a failure to stimulate or inhibit production of reactive oxygen species, neutralization of their effects, or by preventing the interaction of parasite with the toxic substances.

Granulomatous inflammation is the hallmark of histoplasmosis where *H. capsulatum* is found as small budding yeasts almost exclusively within tissue macrophages. The importance of cell-mediated immunity in this disease has been well documented. Immune specificity is determined by T lymphocyte recognition and microbiocidal effector function is provided by macrophages. However, early during the course of infection in the non-immune host tissue macrophages play a permissive rather than defensive role. Our studies and those of others reveal a qualitative deficiency in the ability of host macrophages to generate an oxidative burst in response to *H. capsulatum*. Similar defects in oxidative burst have been reported with a variety of intracellular parasites such as *Toxoplasma gondii,* virulent *Salmonella typhi, Leishmania donovani, Yersinia pestis,* and *Mycobacterium leprae.*

Despite extensive phagocytosis, both live and dead *H. capsulatum* yeasts failed to trigger the generation of reactive oxidants. We ruled out the possibility that *H. capsulatum* exerted a lethal effect by showing that macrophages containing yeasts were still capable of phagocytosing latex and zymosan particles. Interestingly, macrophages that were first exposed to *H. capsulatum* and subsequently to zymosan or phorbol myristate acetate failed to elaborate oxidant in response to the later challenge. Similar observations have been made with macrophages that have ingested *Yersinia pestis.* It is also of interest that macrophages that have ingested *Toxoplasma gondii* or *Mycobacterium leprae* fail to trigger an oxidative burst but when they are subsequently exposed to zymosan or phorbol myristate acetate a vigorous oxidative response occurs. In addition to a failure to stimulate an oxidative burst, phagocytosed yeast cells of *H. capsulatum* either resist or inactivate the fungicidal activity of lysosomes.

Little is known about surface receptors of macrophages yeasts of *H. capsulatum.* Ward Bullock's group has shown that a family of receptors is implicated that are distinct from the fucose-mannose receptor responsible for soluble mannose.[1]

Unopsonized *H. capsulatum* cells are readily phagocytosed by macrophages but the receptor remains unknown: it may be closely linked to the variations in oxidative burst activity already described. Phagocytosis of immunoglobulin-opsonized particles occurs via macrophage IgG-Fc receptors. Opsonization of yeast cells of *H. capsulatum* with normal mouse serum or heat-inactivated normal mouse serum results in an insignificant oxidative burst whereas pretreatment of cells with specific antiserum results in an enhanced oxidative burst. It must be emphasized that C3b-mediated phagocytosis of particulate stimuli fails to stimulate an oxidatie burst. Opsonization with *H. capsulatum*-immune serum yields a significant oxidative burst, suggesting that the observed suppression can be overcome through binding and internalization via a different pathway (Fc receptor-mediated pathway).

An important consideration from an evolutionary point of view is that humans are exposed to only a small number of mycelial or microconidal propagules. These infectious propagules have the capacity to undergo the morphologic transition to the yeast phase. This poses major questions. How can developmentally regulated and phase specific genes, including those for virulence, be maintained in a fungal population without external selective pressures such as those provided by macrophages? Do they have different functions in the saprobic mycelial phase? In the mycelial phase do these genes code for different functions necessary for survival and are therefore uniformly maintained in the genome of this fungus?

ACKNOWLEDGEMENTS

Portions of the studies described in this report were supported by a contract from the Commission of the European Community, TSD2-0132-I, a contract from Ministero della Sanita-Instituto Superiore di Sanita, 4° Progetto AIDS 1991 6203-21, by National Institutes of Health PHS grant 29609, and NATO travel award 5-205/RG890437.

REFERENCES

1. W.E. Bullock and S.D. Wright, The role of adherence-promoting receptors, CR3, LFA-1 and P150,95, in binding of *Histoplasma capsulatum* by human macrophages. Abstract. *Fed. Proc.* 45: 247 (1986).

ADDITIONAL READINGS

L.G. Eisenberg and W.E. Goldman, *Histoplasma* variation and adaptive strategies for parasitism: New prospectives on histoplasmosis. *Clin. Microbio. Rev.* 4: 411 (1991).

B. Maresca and G.S. Kobayashi, Dimorphism in *Histoplasma capsulatum*: A model for the study of cell differentiation in pathogenic fungi. *Microbiol. Rev.* 53: 186 (1989).

J. Schwarz, *"Histoplasmosis"*, Praeger Science Press, New York (1981).

P.J. Szaniszlo, "Fungal dimorphism: with emphasis on fungi pathogenic for humans", Plenum Press, New York (1985).

BIOCHEMICAL AND PHYSIOLOGICAL ASPECTS IN THE DIMORPHISM OF *PARACOCCIDIOIDES BRASILIENSIS*

Gioconda San-Blas, and Felipe San-Blas

Instituto Venezolano de Investigaciones Científicas (IVIC)
Centro de Microbiología y Biología Celular
Apartado 21827
Caracas 1020A, Venezuela

ABSTRACT

Paracoccidioides brasiliensis is a dimorphic fungus pathogenic for humans. It shows a yeastlike phase at 37 °C and a mycelial phase at 23 °C. Thermal dimorphism has been related to sulfur metabolism, regulation of cAMP, hormone receptors, cell wall structure, and modulation of glucan synthetase activity through cytoplasmic proteinases. These aspects are reviewed herein.

INTRODUCTION

Paracoccidioides brasiliensis is a dimorphic fungus pathogenic for humans causing a systemic mycosis geographically confined to Latin America with endemic areas extending from Central America down to Argentina. *In vitro* and in tissue at 37 °C it grows in a yeast-like (Y) form whereas a mycelial (M) phase is produced at temperatures below 25 °C. Transformation to the Y phase is a key factor in the invasive process into the host, which is the reason why it is important to acquire a better knowledge of the metabolic events and regulatory mechanisms involved during the M to Y transition. It has been suggested that cAMP may play a regulatory role in the first hours of transition[1] and that sulfur metabolism is involved.[2-4] Hormones also play a role in the regulation of this process *in vivo*,[5] and cell wall modifications[6] and enzymatic regulation of wall glucan synthesis[7,8] are important in the expression of dimorphism. These aspects will be reviewed.

P. BRASILIENSIS DIMORPHISM AND CYCLIC AMP

The possible regulatory role of cAMP on the dimorphic process in *P. brasiliensis* has been studied by Paris and Durán.[1] Exogenous addition of cAMP

(10 mM) or its analogs to a medium in which Y cultures were put to transform to the M phase (or the reverse process) provoked a halt in transformation, an effect that was also achieved with phosphodiesterase inhibitors such as caffeine or theophilline. This effect did not correlate with changes in the intracellular levels of cAMP, which were not significantly different in both phases of the fungus. For this reason the authors were cautious in ascribing a definitive role of cAMP in the dimorphic process of *P. brasiliensis*.

P. BRASILIENSIS DIMORPHISM AND SULFUR METABOLISM

Comparative studies on the nutritional requirements of various strains of *P. brasiliensis* suggested that M cultures were prototrophic whereas Y cells required a sulfur-containing amino acid for growth.[2,3] Following studies on the role of cysteine on the M to Y transformation in *Histoplasma capsulatum*, Medoff et al.[4] tested the same effect on *P. brasiliensis*. According to their results, the temperature shift results in a process that can be divided into three stages. Stage 1 relates to the partial uncoupling of oxidative phosphorylation and a decrease in cellular ATP levels, respiration rates, and concentrations of electron transport components. The cells then enter a stage in which spontaneous respiration ceases (stage 2) and, finally, there is a shift into a recovery phase during which transformation to yeast morphology occurs (stage 3). Cysteine is required during stage 2 for the operation of shunt pathways which permit electron transport to bypass blocked portions of the cytochrome system. These results, however, may apply to the strain used by the authors and cannot necessarily be extrapolated to others, since under their experimental conditions for growing the fungus, it takes 21 days for M to Y transformation to complete, with yeasts appearing by day 12. Cysteine addition was also required to complete M to Y transformation. These results are different from those obtained by San-Blas and San-Blas[9] with strains Pb73, and Pb9, among others, where the initiation of the budding process 8 h after the change of temperature was reported, and a total conversion of the M culture to the Y phase by day 5, a process that needed no special nutritional requirement different from those necessary for growth.

P. BRASILIENSIS DIMORPHISM AND HORMONE RECEPTORS

The vertebrate hormone β-estradiol (E2) inhibits M to Y transition of *P. brasiliensis* in a dose-dependent manner.[10] In addition, *P. brasiliensis* possesses a cytosolic protein (EBP) that binds E2 with high affinity and stereospecificity.[11,12] This has led to the hypothesis that EBP acts as a receptor which modulates the behavior of the organism during M to Y transition. The possibility that E2 mediates inhibition as a result of altered protein synthesis was analyzed in polyacrylamide gels by Clemons et al..[13] They found 12 M-associated bands (range 30-140 kDa) and 18 Y-associated bands (range 22-127 kDa) and detected 5 novel bands (23-50 kDa) during M to Y transition. Treatment with E2 (2.6×10^{-7} M) altered protein profiles: 4 of 12 M-associated bands were maintained whereas the appearance of the 5 transition bands and 9 of 18 Y-associated bands were either blocked or delayed, suggesting that the functional responses of *P. brasiliensis* to E2 are related to regulation of protein expression, presumably mediated via a specific binding protein-ligand complex (Fig. 1). These results provide a biological explanation for the fact that paracoccidioidomycosis is more prevalent among the

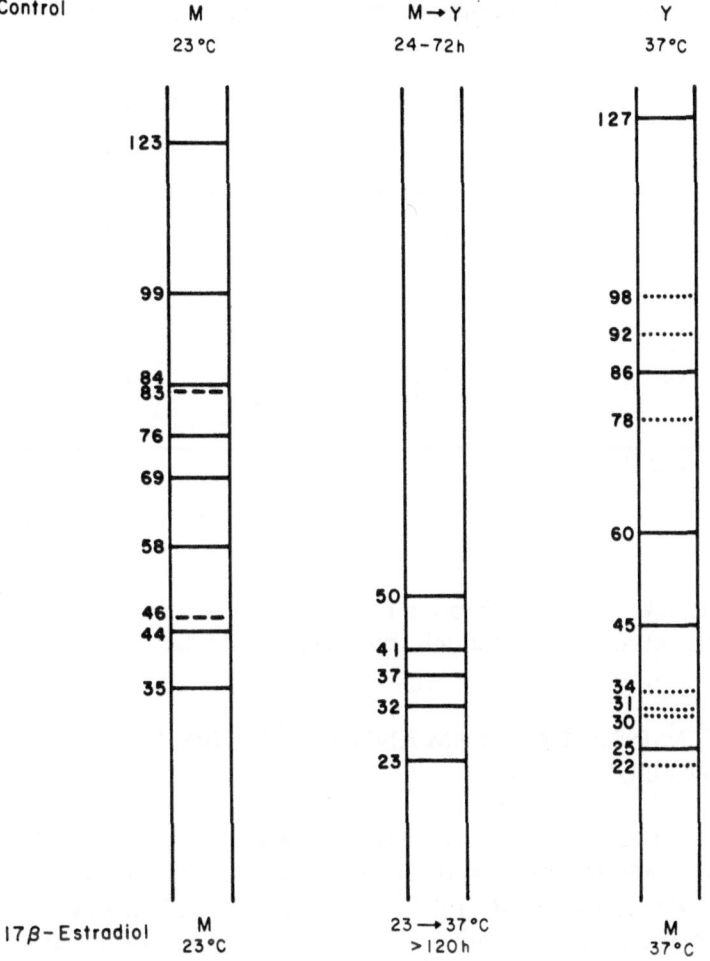

Fig. 1 Specific bands of cytosolic proteins from the Y and M phases of *P. brasiliensis* in the absence or in the presence of 17ß-estradiol. Bands shown as broken lines (- - -) represent those M proteins remaining for 120 h after the temperature shift from 23 ° to 37 °C in the presence of estradiol. Bands shown as spots (. . .) refer to those Y proteins either delayed or never shown in the presence of estradiol after the temperature shift. Transition bands appear at 24 to 72 h in controls but are delayed for more than 120 h in the presence of estradiol. Taken from Clemons *et al.*[13]

male population, inasmuch as circulating estradiol in females could block M to Y transformation of the infecting *P. brasiliensis*, the initial step in the development of the disease. However, it must be mentioned that although male prevalence is consistently reported in the natural human host, contradictory results come from experimental animal models. While female mice seem more resistant to paracoccidioidomycosis than male animals,[14,15] either the reverse[16] or no difference according to sex[17,18] were also reported. Since none of these studies considered the estrous cycle of female animals at the time of inoculation, Sano *et al.*[19] examined the susceptibilities of ddY female mice at the five stages of their estrous cycle and compared them with male animals. Their results (Table 1) suggested that female mice at proestrous, estrous and diestrous were less susceptible than male mice but equally susceptible at metestrous stages I and II.

Table 1. Estrous cycle of mice and susceptibility to infection by *P. brasiliensis*.

	Positive organs/total animals	
Sex or stage	Brain	Lung
Male	5/5	3/5
Female		
Proestrous	3/5	1/5
Estrous	3/5	2/5
Metestrous-I	6/6	4/6
Metestrous-II	5/5	3/5
Diestrous	3/5	2/5

Taken from Sano *et al.*[19]

These results, taken as a whole, put a note of caution on the possible extrapolations from experiments on animal models into the natural host.

P. BRASILIENSIS DIMORPHISM AND STRUCTURE AND BIOSYNTHESIS OF CELL WALL GLUCANS

Another aspect that concerns the regulation of dimorphism is the involvement of the cell wall and modulation of its biosynthesis during transformation. Earlier studies by Kanetsuna *et al.*[6] indicated that cell wall polysaccharides of *P. brasiliensis*, particularly α- and β-1,3-glucans, play a role in morphology and dimorphism since the former is only found in the Y form while the latter is found mainly in the M phase with only traces of it in the Y form (Table 2). Kanetsuna *et al.*[6] formulated a hypothesis to explain Y to M transformation, later modified by San-Blas and San-Blas,[9] in which the role of glucans and the regulation of protein disulfide reductase are decisive in the establishment of either morphology. According to them, when the Y cells are induced to form the mycelium, the synthesis of new RNA and proteins stops for about 8 h, and the synthesis of α-glucan decreases without initiation of synthesis of β-glucan. The latter begins only when synthesis of new proteins resumes, and mycelia start to build. When M to Y transformation occurs, the synthesis of β-glucan is drastically reduced through the inactivation of β-glucan synthetase. Together with these changes in glucan synthetase activity, a stimulation in protein disulfide reductase in the Y phase provokes the breakage of disulfide bonds in cell wall proteins, favoring the round shape. The reductase enzyme activity is substantially deactivated in the mycelial phase to allow for the establishment of long protein chains which can be easily accomodated in the mycelial cell wall. Further studies on the synthesis *in vitro* of glucan synthetase in *P. brasiliensis* suggest that cytosolic proteinases regulate the activity of β-glucan synthetase[7] since their addition to the glucan synthetase assay *in vitro* provoked a stimulation in glucan synthesis.To follow the activity of neutral and acid proteinases in the cytosol of the Y and M phases of the fungus, cytosolic fractions have been eluted through Sephacryl S-200 columns and fractions tested for proteinase activities (San-Blas *et al.*, unpublished data). Results indicated that neutral proteinase activities (two fractions of Mr 180 kDa and

Table 2. Main chemical features of *P. brasiliensis* cell wall.

	Y phase	M phase
Cell wall polysaccharides		
% α-glucan	44.4	0.0
% β-glucan	4.3	35.5
β-glucanase (nmol/mg/h)	78.0	39.4
Cell wall proteins		
% proteins	10.1	32.9
Cystine/cysteine (nmol/mg)	2.5	34.0
Activity of intracellular protein disulfide reductase (nmol/mg/h)	27.1	4.9

Taken from Kanetsuna *et al.*[9]

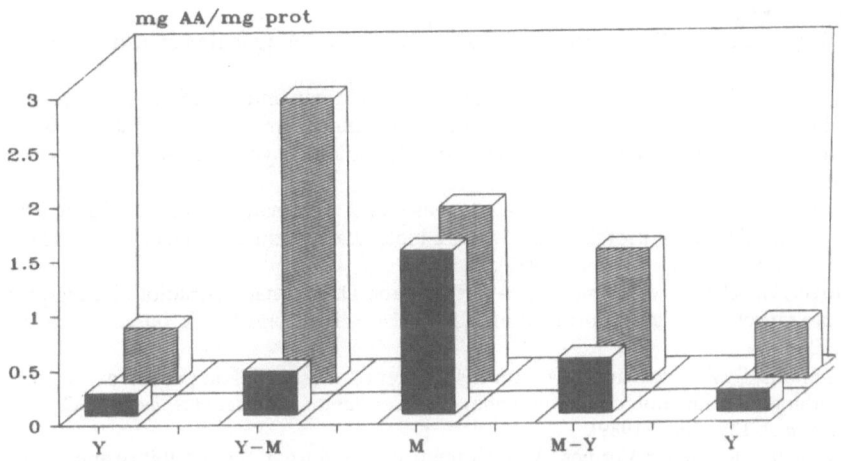

Fig. 2 Activities of cytoplasmic neutral proteinases in *P. brasiliensis* during the dimorphic process. Filled bars: proteinases of Mr 150-180 kDa; shadowed bars: proteinases of Mr 10-15 kDa.

15 kDa) were rather high during mycelial growth, substantially decreasing during transformation to the yeast form (Fig. 2).

These results have been incorporated into a working hypothesis[8,9] to explain the mechanisms of control of β-glucan synthetase activity in *P. brasiliensis*. When the Y phase is growing at 37 °C, b-glucan synthetase activity must be depressed to a minimum and the enzyme might be in a zymogenic state; in transforming to the mycelial phase, proteinases are strongly activated, causing the conversion of the inactive β-glucan synthetase into an active enzyme, and allowing for the permanent activation of β-glucan synthetase through the action of proteinases. In trying to transform hyphae to yeast cells, a new zymogenic enzyme is being synthesized, which cannot convert to an active form because of the poor activity of proteinases. In this way, a low rate of β-glucan synthesis is maintained while α-glucan begins to be synthesized to allow for the maintenance of the round shape of the yeast cells.

This model is currently used to provide a better insight into the molecular aspects of thermal dimorphism in *P. brasiliensis*.

REFERENCES

1. S. Paris and S. Durán, Cyclic adenosine 3',5' monophosphate (cAMP) and dimorphism in the pathogenic fungus *Paracoccidioides brasiliensis*, *Mycopathologia* 92: 115 (1985).
2. S. Paris, S. Durán and F. Mariat, Nutritional studies on *Paracoccidioides brasiliensis*: the role of organic sulfur in dimorphism, *J. Med. Vet. Mycol.* 23 : 85 (1985).
3. J.R. Ramírez-Martínez and J. Rodríguez, Nutritional studies on a methionine-requiring strain of *Paracoccidioides brasiliensis*, *Mycopathol. Mycol. Appl.* 46: 341 (1972).
4. G. Medoff, A. Painter and G.S. Kobayashi, Mycelial to yeast-phase transitions of the dimorphic fungi *Blastomyces dermatitidis* and *Paracoccidioides brasiliensis*, *J. Bacteriol.* 169 : 4055 (1987).
5. D. Stevens, The interface between endocrinology and mycology, *J. Med. Vet. Mycol.* 27 : 133 (1989).
6. F. Kanetsuna, L.M. Carbonell, I. Azuma and Y. Yamamura, Biochemical studies on the thermal dimorphism of *Paracoccidioides brasiliensis*, *J. Bacteriol.* 110 : 208 (1972).
7. G. San-Blas, F. San-Blas, L.E. Rodríguez and C.J. Castro, Un modelo de dimorfismo en hongos patógenos: *Paracoccidioides brasiliensis*, *Acta Cient. Venezol.* 38 : 202 (1987).
8. G. San-Blas and F. San-Blas, *Paracoccidioides brasiliensis*: current research in morphogenesis, *Jap. J. Med. Mycol.* 32 : 75 (1991).
9. F. San-Blas, and G. San-Blas, *Paracoccidioides brasiliensis*, in "Fungal dimorphism", P. Szaniszlo, ed., Plenum Press, New York (1985).
10. A. Restrepo, M.E. Salazar, L.E. Cano, E.P. Stover, D. Feldman and D.A. Stevens, Estrogens inhibit mycelium to yeast transformation in the fungus *Paracoccidioides brasiliensis* : implications for resistance of females to paracoccidioidomycosis, *Infect. Immun.* 46 : 346 (1984).
11. D.S. Loose, E.P Stover, A. Restrepo, D.A. Stevens and D. Feldman, Estradiol binds to a receptor-like cytosol binding protein and initiates a biological response in *Paracoccidioides brasiliensis*, *Proc. Natl. Acad. Sci.* 80 : 7659 (1983).
12. E.P. Stover, G. Schar, K.V. Clemons, D.A. Stevens and D. Feldman, Estradiol-binding proteins from mycelial and yeast-form cultures of *Paracoccidioides brasiliensis*, *Infect. Immun.* 51 : 199 (1986).
13. K.V. Clemons, D. Feldman and D.A. Stevens, Influence of estradiol on protein expression and methionine utilization during morphogenesis of *Paracoccidioides brasiliensis*, *J. Gen. Microbiol.* 135 : 1607 (1989).
14. V.L.G. Calich, L.M. Singer-Vermes, A.M. Siqueira, and E. Burger, Susceptibility and resistance of inbred mice to *Paracoccidioides brasiliensis* strains, *Br. J. Exp. Pathol.* 66 : 585 (1985).
15. V.L.G. Calich, E. Burger, S.S. Kashino, R.A. Fazioli and L.M. Singer-Vermes, Resistance to *Paracoccidioides brasiliensis* in mice is controlled by a single dominant autosomal gene, *Infect. Immun.* 55 : 1919 (1987).
16. J.G. McEwen, V. Bedoya, M.M. Patiño, M.E. Salazar and A. M. Restrepo, Experimental murine paracoccidioidomycosis induced by the inhalation of conidia, *J. Med. Vet. Mycol.* 25 : 165 (1987).
17. J. Defaveri, M.T. Rezkallah-Iwasso and M.F. Franco, Experimental pulmonary paracoccidioidomycosis in mice: morphology and correlation of lesions with humoral and cellular immune response, *Mycopathologia* 77 : 3 (1982).
18. M.A. Robledo, J.R. Graybill, J. Ahrens, A. Restrepo, D.J. Drutz and M. Robledo, Host defense against experimental paracoccidioidomycosis, *Am. Rev. Resp. Dis.* 125 : 563 (1982).
19. A. Sano, M. Miyaji and K. Nishimura, Studies on the relationship between paracoccidioidomycosis in ddY mice and their estrous cycle, *Mycopathologia* 115.: 73 (1991).

STUDIES ON PHASE TRANSITIONS IN *SPOROTHRIX SCHENCKII:* POSSIBLE INVOLVEMENT OF PROTEIN KINASE C

Wanda Colon-Colon, and Nuri Rodriguez-del Valle

Department of Microbiology and Medical Zoology
Medical Sciences Campus
University of Puerto Rico, PO Box 365067
San Juan, Puerto Rico 00936-5067

ABSTRACT

Protein kinase C (PKC) is a signal transducing enzyme, that has been related to the regulation of proliferative and morphogenetic processes in many eukaryotic systems. The yeast to mycelium transition in *Sporothrix schenckii* involves both morphogenesis and proliferation and has been reported by us to be stimulated by calcium ions but can occur in the absence of this cation. The work reported here suggests the involvement of PKC in the induction of the yeast to mycelium transition in *S. schenkii* in the absence of extracellular calcium ions. Phorbol-12-myristate-13-acetate (PMA), a tumor promoting agent and PKC activator was found to stimulate germ tube formation and germ tube growth by yeast cells induced to undergo transition to the mycelium form in a concentration dependent manner with an optimal stimulatory concentration of 2 μM. This same PMA concentration had a stimulatory effect on DNA and RNA synthesis in cells induced to undergo the yeast to mycelium transition and was found to inhibit cell duplication and bud formation in yeast cells induced to re-enter the budding cycle. Polymyxin B, an inhibitor of PKC, inhibited germ tube formation by yeast cells at concentrations of 50 and 100 μM, at all time intervals tested. This inhibition could be overcome if PMA 2 μM or calcium 1 mM were added to the medium, suggesting that the inhibition obtained in the presence of this antibiotic was due to an inhibition of PKC. These results support the involvement of PKC in the control of the dimorphic expression in *S. schenckii*. Other PKC inhibitors tested, H-7 and staurosporine gave unexpected results, probably because of their lack of specificity and the fact that they inhibit other protein kinases.

INTRODUCTION

Sporothrix schenckii is a dimorphic fungus which grows in the mycelium form with long branching filaments or in the yeast form as spherical or ovoid cells.

Dimorphic Fungi in Biology and Medicine, Edited by
H. Vanden Bossche *et al.*, Plenum Press, New York, 1993

Non-budding yeast cells can be selected by filtration and induced to re-enter a proliferative state which can also involve a change in morphology depending on the conditions of growth.[1,2] These cells will undergo morphogenetic transition to the mycelium form when they are inoculated into a buffered-salts medium with glucose and vitamins at pH 4.0 and 25 °C if the cell concentration is in the range of 2×10^5 cells ml^{-1}. They can also be induced to re-enter the budding cycle by cultivation in the same medium at pH 7.2 and 25 °C, with aeration, at an initial cell concentration of 2×10^7 cells ml^{-1}. The sequence of cellular and molecular processes that accompany these events have been described previously by us (in medium containing 4 μM Ca^{2+}). Both during the induction of the yeast to mycelium transition or the yeast cell cycle, protein, RNA and DNA synthesis precede the morphological events that characterize germ tube formation and bud formation. During the yeast to mycelium transition, germ tube formation is first detected 6 h after inoculation and is completed in almost all of the cells 12 h after inoculation, with the first nuclear division taking place 6 h after inoculation.[1] During the yeast budding cycle the first cell duplication occurs at 12 h of incubation[2] and occurs concomitantly with nuclear division.[2]

Calcium ions have been implicated in the control of proliferation and differentiation in many eukaryotic cells.[3-5] The yeast to mycelium transition in S. schenckii has also been observed to be stimulated by increasing the calcium concentration of the medium and it is inhibited by substances that affect calcium uptake.[6] Calcium uptake has been reported by us to accompany this transition and the kinetics of this uptake are comparable to those observed in other synchronized, proliferating cells. Uptake occurs very early following the induction of transformation, returning to basal levels and rising again at the time of DNA synthesis.[6]

Although calcium ions have been identified as stimulators of the yeast to mycelium transition in S. schenckii, this transition can take place even in the absence of extracellularly added calcium (in press). This observation suggests that these cells have enough intracellular calcium to allow germination to take place although with reduced kinetics through an alternate mechanism for the induction of germination. Protein kinase C (PKC) activation, which could take place in the absence of an increase in intracellular calcium is a possible alternate mechanism. PKC has been shown to be an important signal transducing enzyme in a number of systems.[7] Diacylglycerol generated by phospholipase activity and tumor-promoting phorbol esters, stimulate PKC by increasing its sensitivity to calcium ions.[7,8]

The following work was conducted to inquire into the possible involvement of PKC in the dimorphism of S. schenckii. We studied the effects of phorbol-12-myristate-13-acetate (PMA), a known activator of PKC,[9] on the kinetics of germ tube formation and germ tube growth in yeast cells induced to undergo the yeast to mycelium transition. In addition we studied the effect of this substance on the kinetics of cell duplication and bud formation in cells cultured under conditions that induce the re-entry into the budding cycle. The effects of PMA on the kinetics of RNA and DNA synthesis during the yeast to mycelium transition were also studied in order to determine if the observed stimulatory effects on the yeast to mycelium transition could be attributed to the stimulation of any of these two processes. Three inhibitors of PKC activity, the antibiotic polymyxin B,[10] 1-(5-

isoquinolinylsulfonyl)-2-methyl-piperazine (H-7)[11] and staurosporine,[12] and a less active PKC activator, phorbol-12,13-dibutyrate (PDBu)[10] were also tested for their effects on the kinetics of germ tube formation during the yeast to mycelium transition.

MATERIALS AND METHODS

Fungus and Inocula

Sporothrix schenckii originally isolated from a patient was used in all experiments (ATCC 58251). Conidial suspensions for obtaining the yeast form and yeast cell suspensions for the induction of the yeast to mycelium transitions were prepared as described previously.[1]

Medium

All reagents used were obtained from Sigma Chemical Co., except [^{3}H]-uracil which was obtained from the New England Nuclear Corp.

The medium used for all experiments was a buffered-salts medium with glucose and vitamins as previously described[1] except that calcium panthothenate was omitted. The pH of the medium was adjusted to 4.0 or 7.2 for the induction of the yeast to mycelium transition or the yeast cell cycle, respectively. Phorbol-12-myristate-13-acetate (1-8 µM), phorbol -12,13-dibutyrate (2 µM), H-7 (0.1-1 mM), staurosporine (0.1-1 µM), and polymyxin B (10, 50 and 100 µM) were added to the medium according to the experimental design. The medium was sterilized by filtration through Nalgene disposable units (0.20 µm). Fifty milliliters of medium were added to 125 ml flasks. All flasks was inoculated with the yeast cells suspension to give a final concentration of approximately 2×10^5 cells ml^{-1} or 2×10^7 cells ml^{-1} for the induction of the yeast to mycelium transition or the yeast cell cycle, respectively.

Determination of the Kinetics of Germ Tube Formation and Germ Tube Growth

Yeast cells were induced to form germ tubes in medium containing the substances to be tested for their effects on germ tube formation and at selected intervals they were examined microscopically for the presence of germ tubes as described previously.[1] Germ tube growth was determined by measuring the length of the germ tubes with a calibrated ocular micrometer as described previously,[1] in the presence and absence of 2 µM PMA.

Determination of RNA and DNA Synthesis

Incorporation of [5,6 ^{3}H]-uracil (0.4 µCi ml^{-1}, specific activity, 42.1 Ci mmol^{-1}) was used as an index of RNA and DNA synthesis in cells induced to form germ tubes as described previously.[1] The radioactivity incorporated into RNA and DNA was determined according to the method of Hatzfeld[13] as modified by Betancourt and collaborators.[1]

Determination of the Kinetics of Cell Duplication and Bud Formation

Yeast cells were allowed to re-enter the budding cycle as described previously[2] except that medium without added calcium was used. These cells were inoculated into medium with and without PMA (2 μM) and the kinetics of cell duplication and bud formation determined as described.[2]

Statistical Analysis

All the data are given as the average of at least three independent experiments ± one standard deviation. The significance of the data was determined by means of the Student t-test ($p<0.05$).

The percent stimulation of the incorporation of radioactivity into RNA and DNA was determined by dividing the counts per minute obtained in the presence of PMA (2 μM) by the controls (without PMA), multiplying by 100 and subtracting 100.

The percent stimulation or inhibition of germ tube formation was determined by dividing the percentage of cells with germ tubes in the experimental groups (in the presence of the compounds to be tested) by the percentage of cells with germ tubes in the controls, multiplying by 100 and subtracting 100.

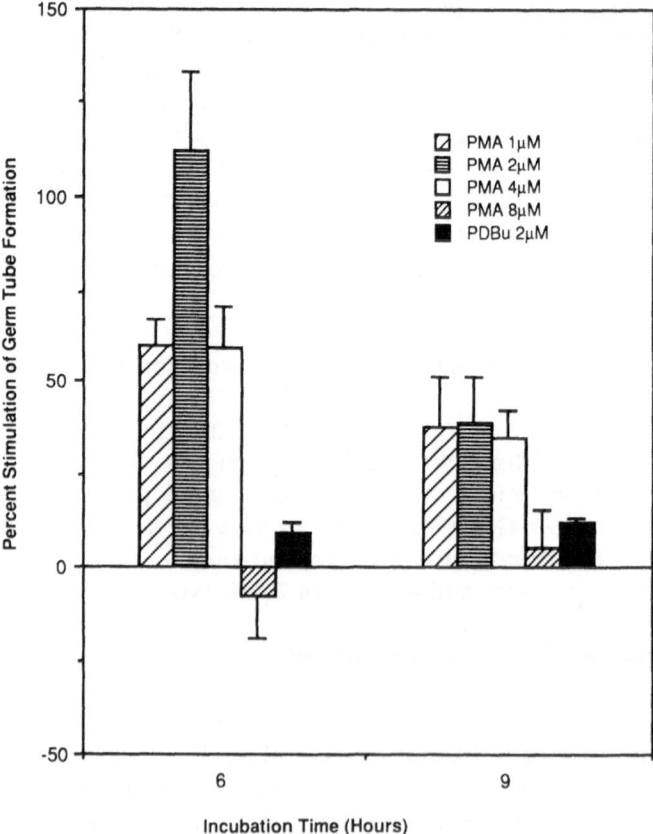

Fig. 1 Yeast cells were induced to form germ tubes in basal medium with glucose at pH 4.0 and 25 °C with varying concentrations of PMA (1-8 μM), PDBu (2 μM) and without these compounds (controls). At 6 and 9 h after inoculation they were examined microscopically for the presence of germ tubes. The average percent stimulation of germ tube formation of at least three independent experiments ± 1 SD is given.

RESULTS

Effects of PMA on Germ Tube Formation and Germ Tube Growth

Figure 1 shows the percent stimulation of germ tube formation of yeast cells induced to form germ tubes in the presence of PMA. This figure shows a concentration dependent curve with an optimal stimulation at 2 μM PMA at the 6 h interval after inoculation. However, 8 μM PMA had no significant effect at this same time interval. At 9 h after inoculation, a significant stimulation was observed which was approximately the same for all concentrations of PMA tested in the range of 1 to 4 μM. Furthermore, Fig. 2 shows that the average germ tube

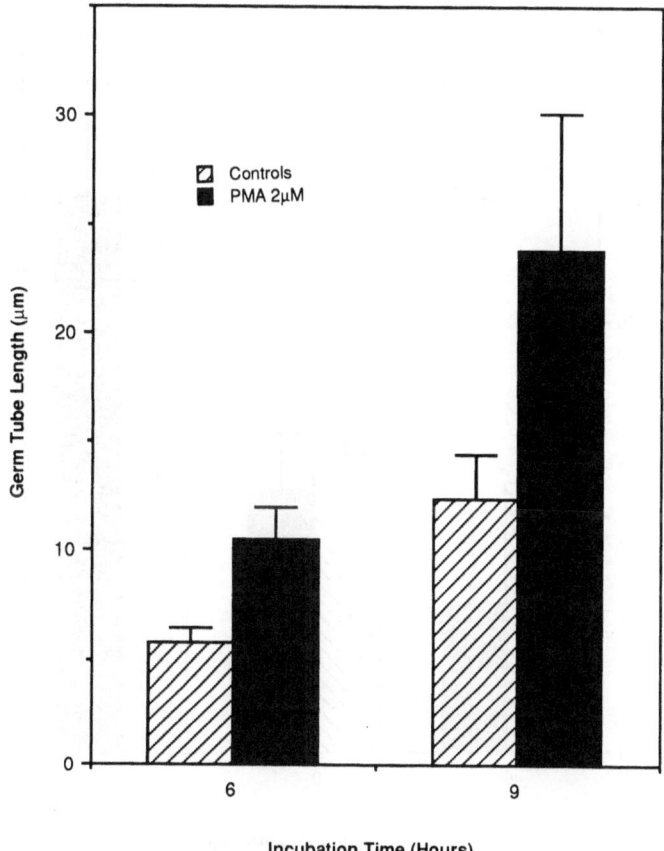

Incubation Time (Hours)

Fig: 2 Yeast cells were induced to form germ tubes in basal medium with glucose at pH 4.0 and 25 °C with and without PMA, 2 μM. At 6 and 9 h of incubation the cells were harvested and the germ tube length determined with a calibrated ocular micrometer. The average germ tube length of at least three independent experiments ± 1 SD is given.

length of cells undergoing the yeast to mycelium transition in the presence and absence of 2 μM PMA. was nearly twice that of cells in medium without PMA (controls) for the 6 and 9 h intervals after inoculation. These differences were significant for the 6 h interval.

However, 2 μM PDBu (Fig. 1) had only a slight but significant stimulatory effect on germ tube formation of 9±3% and 12±1% for the 6 and 9 h intervals after inoculation.

Effects of PMA on RNA and DNA Synthesis During the Yeast to Mycelium Transition

Figure 3 shows the effects of PMA (2 µM) on the incorporation of tritiated uracil into RNA and DNA. In the presence of this compound, there was a significant stimulation of the incorporation of radioactivity into RNA at the 6, 9 and 12 h intervals after inoculation. Figure 3 also shows the percent stimulation of the incorporation of radioactivity into DNA in the presence of PMA (2 µM). A significant stimulation was recorded for the 9 and 12 h intervals after inoculation.

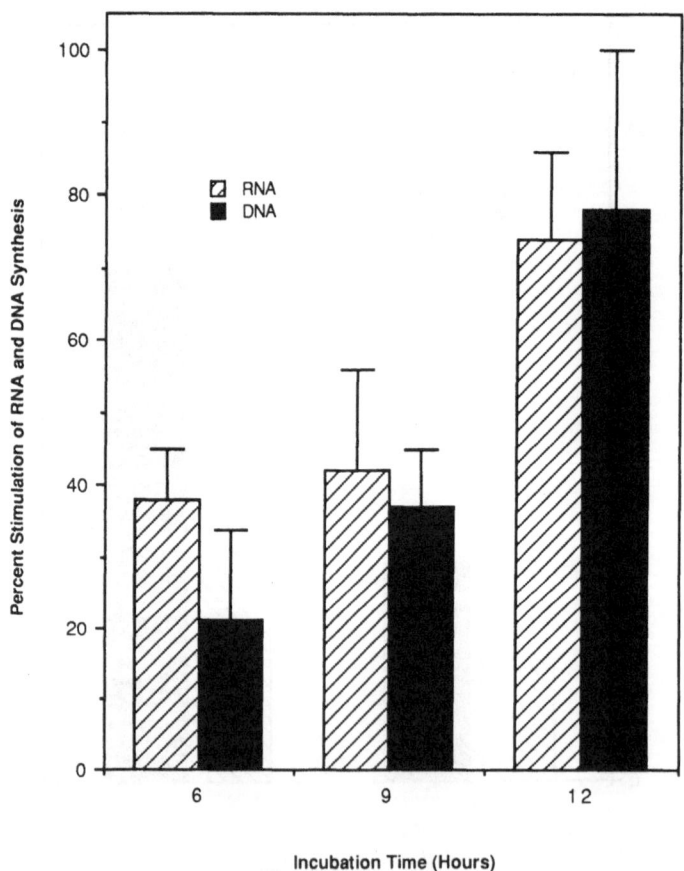

Fig. 3 Yeast cells were induced to form germ tubes in medium with and without 2 µM PMA in the presence of tritiated uracil. At 6, 9 and 12 h after inoculation the cells were harvested and the radioactivity incorporated into RNA and DNA determined as described.[1, 13] The results are given as the average percent stimulation ± 1 SD of at least three independent experiments.

Effects of Inhibitors of PKC Activity on the Yeast to Mycelium Transition

Figure 4 shows the percent inhibition by polymyxin B at 10, 50 and 100 µM concentrations of germ tube formation by yeast cells undergoing transition to the mycelium form. This figure shows that the 50 and 100 µM concentrations of polymyxin B inhibited germ tube formation at all intervals tested. Figure 5 shows that adding calcium chloride (1 mM) together with 50 μM polymyxin B, abolished the inhibition caused by polymyxin B giving a percentage of germ tube formation which is similar to that of the controls (without calcium and without polymyxin B) except for the value obtained at 6 h after inoculation which was significantly higher than the controls. This figure also shows that PMA 2 μM, significantly reduces the inhibition caused by polymyxin B at a 50 μM concentration, 6 and 9 h after inoculation.

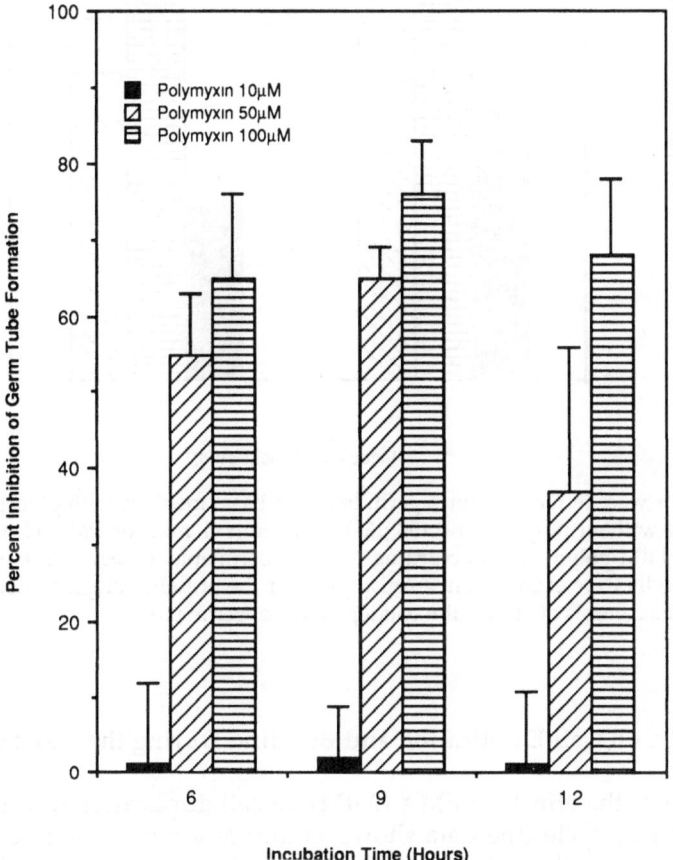

Fig. 4 Yeast cells were induced to form germ tubes in basal medium with glucose at pH 4.0 and 25 °C with varying concentrations of polymyxin B (10-100 μM) and without this compound (controls). At 6, 9 and 12 h after inoculation they were examined microscopically for the presence of germ tubes. The average percent inhibition of germ tube formation of at least three independent experiments ± 1 SD is given.

Concentrations of H-7 ranging from 0.1 mM to 1 mM were tested for their effects on germ tube formation by yeast cells for the 6, 9 and 12 h intervals after inoculation (Fig. 6). These data show that this compound had no significant effects on germ tube formation in these cells except for a slight, but significant stimulation at a concentration of 1 mM, 12 h after inoculation.

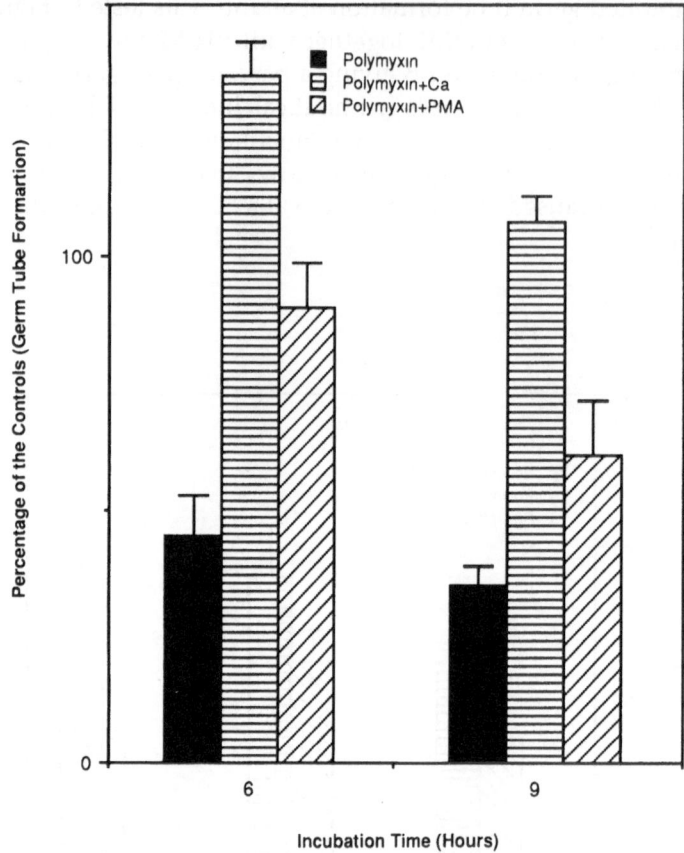

Fig. 5 Yeast cells were induced to form germ tubes in basal medium with glucose at pH 4.0 and 25 °C with and without polymyxin B (50 μM). Calcium (1 mM) or PMA (2 μM) were added together with 50 μM polymyxin B according to the experimental design. At 6, 9 and 12 h after inoculation the cells were examined microscopically for the presence of germ tubes. The average percentage of the controls of at least three independent experiments ± 1 SD is given.

Effects of PMA on Cell Duplication and Budding During the Yeast Cell Cycle

Fig. 7 shows the effects of PMA (2 μM) on cell duplication of cells induced to enter the budding cycle (the data shown is that of a representative experiment). This figure shows that the period before the first cell duplication was approximately 3 h longer for cells in medium with PMA than for the controls. The first cell duplication occurred 12 h after inoculation in the controls and 15 h after inoculation in the presence of PMA (2 μM). This figure also shows that PMA retarded cell division in these cells throughout the time course of the experiment. Fig. 8 shows the percentage of budding cells during the induction of the yeast cell cycle in the presence of PMA and the controls (without PMA). The percentage of budding cells in the presence of PMA was significantly less than that of the controls 12, 15 and 18 h after inoculation. In addition, in the presence of 2 μM PMA at 12 h after inoculation, 24±6% of the cells induced to re-enter the yeast cell cycle, instead of forming buds, formed germ tubes at conditions in which budding rather than germ tube formation was expected. The percentage of cells forming

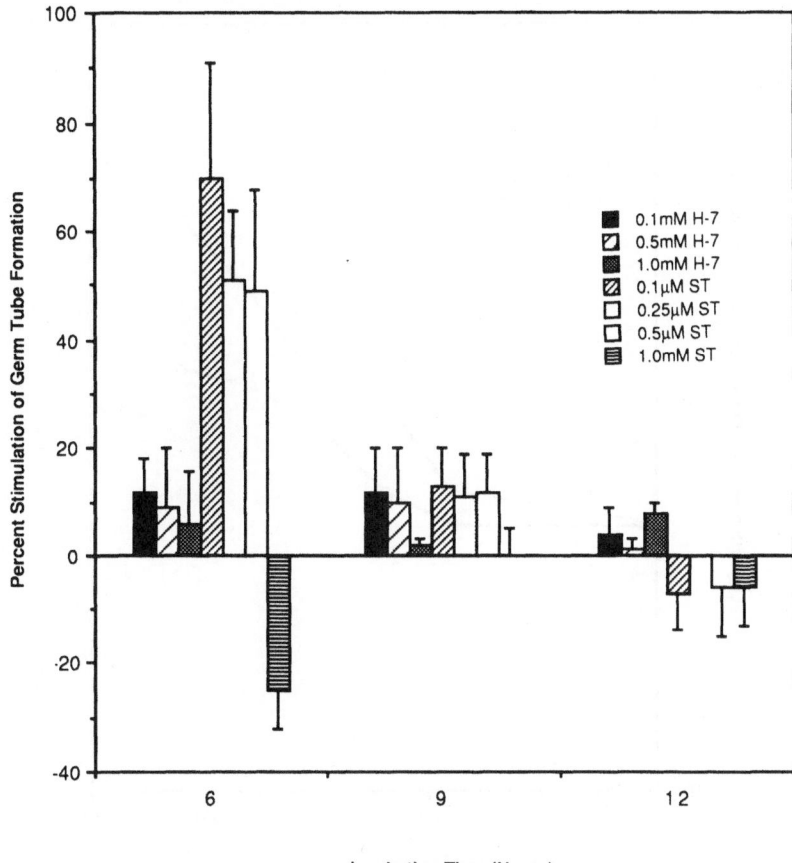

Fig. 6 Yeast cells were induced to form germ tubes in basal medium with glucose at pH 4.0 and 25 °C with varying concentrations of staurosporine (0.1 μM to 1 μM) or H-7 (0.1 mM to 1 mM) and without this compound (controls). At 6, 9 and 12 h after inoculation the cells were examined microscopically for the presence of germ tubes. The average percent stimulation of germ tube formation of at least three independent experiments ± 1 SD is given.

germ tubes at non-permissive conditions in the presence of 2 μM PMA increased slightly to 27±3%, 15 h after inoculation and remained stable thereafter throughout the time course of the experiment.

DISCUSSION

The activation of protein kinase C is believed to be of outmost importance in the induction of the proliferative and differentiation responses in many eukaryotic systems.[14-17] The present study was conducted in order to determine if PKC is involved in the control of the morphogenetic transition from the yeast form to the mycelium form in *S. schenckii* by studying the effects of known inhibitors and stimulators of this enzyme, primarily PMA.

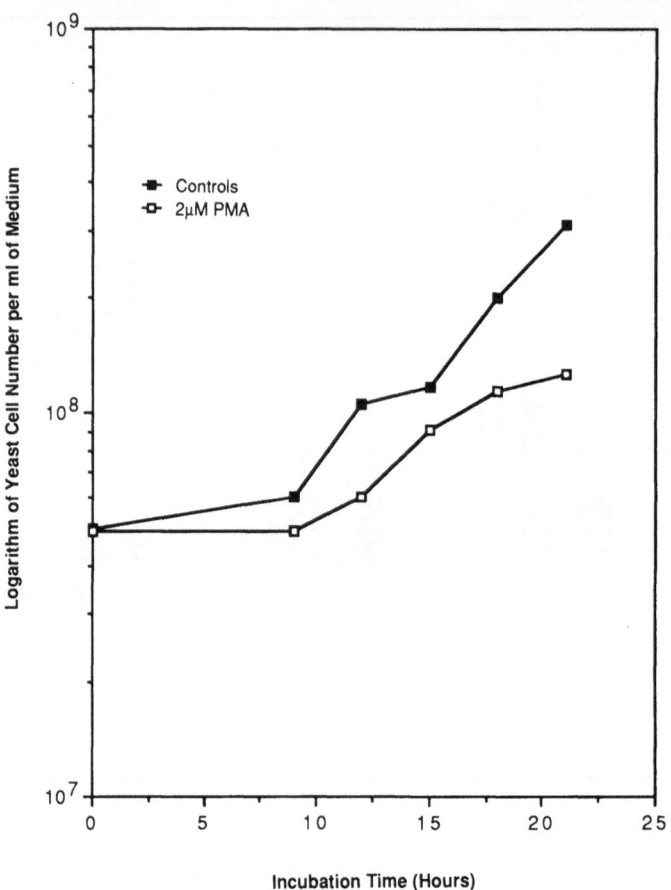

Fig. 7 Yeast cells were induced to form buds in a basal medium with glucose at pH 7.2 and 25 °C with and without 2 μM PMA. At intervals between 9 and 21 h after inoculation the cell concentration was determined with a Petroff-Hausser counting chamber. A bud was counted as an individual cell if it was approximately at least half the size of the mother cell. The results given are those of a representative experiment.

The effects of PMA on germ tube formation during the yeast to mycelium transition and the yeast cell cycle confirmed the possible involvement of PKC in the expression of the dimorphic potential of *S. schenckii*. Tumor-promoting phorbol esters are highly specific stimulators of protein kinase C and have been used extensively for probing the role of this enzyme in the regulation of cell growth and morphogenesis.[9,18] The addition of PMA to the medium stimulated germ tube formation and germ tube growth in cells induced to undergo the yeast to mycelium transition while delaying cell division and bud formation in cells induced to re-enter the budding cycle. Moreover, the addition of PMA to the medium induced the formation of germ tubes in approximately 25% of the cells under conditions where this transititon does not normally take place. This suggests that the effects of PMA are not circumscribed to a stimulation of proliferation which would only bring about an increase in the kinetics of germ tube formation and germ tube growth, but are directed towards favoring the establishment of a definite morphology, even at non-permissive conditions. The effects of PMA on the budding cycle includes both the induction of the alternate morphology and a lengthening of the cell cycle. Usually a lengthening of the cell

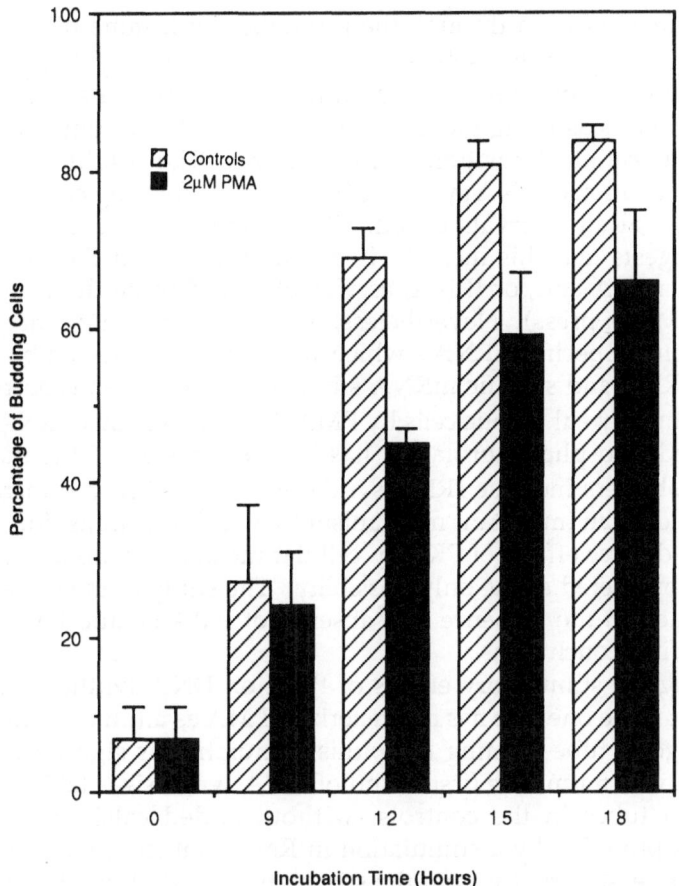

Fig. 8 Yeast cells were induced to form in basal medium with glucose at pH 7.2 and 25 °C with 2 μM PMA and without this compound (controls). At intervals between 9 and 18 h after inoculation they were examined microscopically for the presence of buds. The average percentage of budding cells of at least three independent experiments ± 1 SD is given.

cycle is associated with the induction of differentiation and in this case it could be the result of the induction of mycelial morphology. In many cell lines, phorbol-induced inhibition of proliferation is often accompanied by phenotypic modulation as in our case.[14-17] In other systems such as murine thymocytes, phorbol esters have been reported to induce a partial G_1-phase arrest[18] and in human colon carcinoma cells, PMA was reported to cause transient blocks at each phase of the cell cycle.[17] In our system a partial G_1-phase arrest could account for the lengthening of the lag period before the first round of cell duplication. No inquiry into the mechanism by which PMA brings about this changes were made, but further studies concerning the effects of PMA on macromolecular synthesis and protein phosphorylation will probably help us understand the mechanism involved.

The mechanism by which PMA brings about the stimulation of germ tube formation by yeast cells induced to undergo the yeast to mycelium transition could be very diverse. This phorbol ester has been observed to elicit a variety of cell responses through the stimulation of PKC activity such as the promotion of differentiation in a number of different human cell lines as mentioned previously, control of gene expression through the promotion of histone phosphorylation,[20] stimulation of RNA synthesis,[21-25] stimulation of DNA synthesis[26] and many others. The work reported here explored the possibility of a stimulation of RNA and DNA synthesis as being the mechanism by which the stimulatory effects on the yeast to mycelium transition were obtained. The results presented by us revealed a stimulation of RNA synthesis in germ tube forming cells in the presence of 2 µM PMA at all intervals tested. The magnitude of the increase in RNA synthesis suggests that this increase is probably in one of the major species of RNA, that is, rRNA and/or tRNA, as was observed to be the case with added calcium (1 mM) (in press). Nevertheless, this does not exclude the possibility of the stimulation of specific mRNA's which would not be detected by the methods used. Both rRNA and specific mRNA stimulation have been reported following stimulation of many eukaryotic cells by PMA.[20-23] Activation of RNA polymerase I by PKC mediated phosphorylation has been suggested as the mechanism by which phorbol esters increase rRNA synthesis.[25] The observed increase in RNA synthesis could be an important mechanism by which this transition is stimulated but when the diverse effects of PKC on cellular metabolism are taken into account it cannot be considered as the only possiblity. The study of the pattern of protein phosphorylation in the presence and absence of PMA should be studied before reaching any final conclusions.

Regarding the stimulatory effects of PMA on DNA synthesis we might say that this effect is the one that has characterized PMA as a tumor promoter in other systems. In germ tube-forming cells, this stimulation is significant after 9 h of incubation, at which time the first round of DNA synthesis is taking place in cells forming germ tubes in the controls (without added calcium). Because this stimulation is preceded by a stimulation in RNA synthesis, we could say that it might result at least partially from the stimulation in RNA synthesis, but it could also be an independent effect of PMA on DNA synthesis brought about through changes in cytosolic calcium or protein phosphorylation resulting from PKC activation.

The stimulation of germ tube formation in the presence of 2 µM PDBu offers further evidence as to the specificity of phorbol ester stimulation in this system. As expected, stimulation by PDBu was less than that of PMA: phorbol-12,13-dibutyrate is a less potent PKC stimulator.[9]

The study of the involvement of PKC in any system would not be complete without consideration of the effects of inhibitors of the enzyme. Nevertheless, these studies have been hindered by the lack of specificity of the inhibitors presently available, resulting in other effects which are independent of the effects on PKC.[27-30] One of the most specific inhibitors of PKC is the antibiotic polymyxin B. The specificity of this compound resides in the fact that it inhibits PKC by binding to the regulatory domain rather than to the catalytic domain.[10] Our results show an inhibition of germ tube formation by this compound at concentrations of 50 and 100 µM. The inhibition caused by polymyxin B seems to be specific for PKC because it can be overcome by PMA. The observation that 1 mM calcium can also overcome the inhibition by polymyxin B and even bring about a slight stimulation of germ tube formation not only supports the idea that polymyxin B is inhibiting germ tube formation by inhibiting PKC, but it also suggests the existence of two calcium-dependent pathways for mycelial conversion; one a PKC-dependent pathway that can function in the absence of extracellular calcium ions and another a calcium-calmodulin-dependent pathway, functioning in the presence of extracellular calcium.

The unexpected effects of other inhibitors of PKC on this system could be due to their lack of specificity as stated above or to differences in susceptibility of PKC isoenzyme to a particular inhibitor.[31] These seemingly contradictory effects are being reported more often as these compounds are used as a PKC inhibitor in a wider variety of biological systems. The compound H-7 has been reported to inhibit PKC in a competitive manner with respect to ATP.[11] In the presence of H-7 no inhibition of germ tube formation was observed, rather the percentage of germ tube-forming cells was slightly higher in the presence of this compound, although the differences were not significant except for a slight but significant stimulation 12 h after inoculation at the 1 mM concentration. The lack of inhibition by H-7 in systems where PKC stimulation has been documented could be due to many factors some of which include: concomitant inhibition of cyclic nucleotide dependent protein kinases[11] and a stimulation of actin polymerization.[32] In our system we have previously reported that dibutyryl derivatives of cyclic nucleotides and the cyclic nucleotide phosphodiesterase inhibitor, caffeine, inhibit germ tube formation by yeast cells. [33] This could explain the observed effects of H-7 on our system. If H-7 is inhibiting cyclic nucleotide dependent protein kinases, we would expect a stimulatory rather than an inhibitory effect on germ tube formation. The inhibition of both cyclic nucleotide and calcium dependent protein kinases could bring about the cancellation of the inhibitory effects of this compound.

Similarly contradictory results on germ tube formation were observed when staurosporine was added to the medium. This compound which has also been reported to inhibit PKC by interacting with the catalytic domain,[12] has been reported to induce both morphological changes and enzyme activities similar to those of other tumor promoters.[34] In our system, this compound was observed to have a concentration-dependent effect on germ tube formation with a stimulatory effect at nanomolar concentrations at 6 h of incubation and a slightly inhibitory effect at 1 μM. The results obtained in the presence of staurosporine are probably due, as in the case of H-7, to its lack of specificity as a PKC inhibitor. In addition to inhibiting PKC, it is also a potent inhibitor of a number of proteins such as cyclic nucleotide dependent protein kinases, myosin light chain kinase and tyrosine kinase (pp60 src).[35,36] As in the case of H-7 inhibition of cyclic nucleotide dependent protein kinases could account for the observed effects. The complex interaction of the different effects of protein kinase activities *in vivo* makes it very difficult to sort out the mechanism of action of staurosporine and H-7 in any system.

Our results suggest that PKC activation could be the mechanism by which yeast cells of *S. schenckii* can be induced to transform into the mycelium form of the fungus in the absence of extracellular calcium. An important subject for future studies will be to establish the pattern of phosphoproteins present during the induction of this transition and the yeast cell cycle in the presence of PMA and PKC inhibitors in order to shed some light on the intricate relationships between the protein phosphorylation pathways and dimorphism.

ACKNOWLEDGEMENT

The authors wish to thank Dr. Fernando Renaud for critically reviewing this manuscript. This work was supported by grant RR-8102 from the National Institutes of Health, U.S.A.

REFERENCES

1. S. Betancourt, L.J.Torres-Bauzá and N. Rodríguez-del Valle, Molecular and cellular events during the yeast to mycelium transition in *Sporothrix schenckii, Sabouraudia*, 23: 207 (1985).
2. S. Resto and N. Rodríguez-del Valle, Yeast cell cycle of *Sporothrix schenckii* , *J. Vet. Med. Mycol.* 26: 13 (1988).
3. C. Metcalfe, J.P.Moore, G.A.Smith and T.R. Hesketh, Calcium and cell proliferation, *Br. Med. Bull.* 42: 405 (1986).
4. M. Poenie and R.A. Steinhardt, The dynamics of $[Ca^{2+}]$ during mitosis, *Calcium and Cell Function* 1: 133 (1987).
5. R.B. Silver, Calcium and cellular clocks orchestrate cell division, *Ann. N.Y. Acad. Sci.* 582: 207 (1990).
6. S. Serrano and N. Rodríguez-del Valle, Calcium uptake and efflux during the yeast to mycelium transition in *Sporothrix schenckii, Mycopathol.* 112: 1 (1990).
7. Y. Nishizuka, Studies and perspectives of protein kinase C, *Science* 233: 305 (1986).
8. Y. Nizishuka, The role of protein kinase C on cell surface signal transduction and tumor promotion, *Nature* 308: 693 (1984).
9. U. Kikawa, Y. Takai, Y. Tanaka, R. Mikaya and Y. Nishizuka, Protein kinase C as a possible receptor protein of tumor-promoting phorbol esters, *J. Cell Biol.* 258: 11442 (1983).
10. G.J. Mazzei, N. Katoh and J.F. Kuo, Polymyxin B a more selective inhibitor for phospholipid sensitive Ca^{2+}-dependent protein kinase than for calmodulin-sensitive Ca^{2+}-dependent protein kinase, *Biochem. Biophys. Res. Commun.* 109: 1129 (1982).
11. H. Hidaka, M. Inagaki, S. Kawamoto and Y. Sasaki, Isoquinoline sulfonamides, novel and potent inhibitors of cyclic nucleotide dependent protein kinase and protein kinase C, *Biochem.* 23, 5036 (1984).
12. P.D. Davis, C.H. Hill, E. Keech, G. Lawton, J.S. Nixon, A.D. Sedgwick, J. Wadsworth, D. Westmacott and S.E. Wilkinson, Potent selective inhibitors of protein kinase C, *FEBS Lett.* 259: 61 (1989).
13. J. Hatzfeld, DNA labeling and its assay in yeast, *Biochim. Biophys. Acta* 229: 34 (1973)
14. E. Huberman, C. Heckman and R.Langenbach, Stimulation of differentiated functions in human melanoma cells by tumor-promoting agents and dimethylsulfoxide, *Cancer Res.* 39: 2618 (1979).
15. E. Huberman and M.F. Callahan, Induction of terminal differentiation in human promyelocytic leukemia cells by tumor-promoting agents, *Proc. Natl. Acad. Sci. USA* 76: 1293 (1979).
16. G. Rovera, T.G. O'Brien and L. Diamond, Induction of differentiation in human promyelocytic leukemia cells by tumor promoters, *Science* 204: 868 (1979).
17. P.L.Baron, M.J. Koretz, R.A. Carchman, J.M. Collins, A.S.Torkarz and G.A. Parker, Induction of the expression of differentiation-related antigens on human colon carcinoma cells by stimulating protein kinase C, *Arch. Surg.* 125: 344 (1990).
18. A.A. Farooqui, T. Farooqui, A.J. Yates and L.A. Horrocks, Regulation of protein kinase C activity by various lipids, *Neurochem. Res.* 13:499 (1988).
19. R.C. Howe and H.R. MacDonald, Distinct effects of phorbol esters and exogenous diacyl-glycerols in the induction of murine thymocyte proliferation, *J. Cell. Physiol.* 151: 8 (1992).
20. G.J. Pasktan and C. S. Baxter, Specific stimulation by phorbol esters of the phosphorylation of histones H2B and 1.14 in murine lymphocytes, *Cancer Res.* 45: 667 (1985).
21. J.L. Tilly, and A.L. Johnson, Effect of a phorbol ester, a calcium ionophore and 3',5'-adenosine monophosphate production on hen granulosa cell plasminogen activator activity, *Endocrinol.* 123: 1233 (1988).
22. E.G. Levin, K.R. Marotti and L. Santell, Protein kinase C and the stimulation of tissue plasminogen activator release from human endothelial cells, *J. Biol. Chem.* 264: 16030 (1989).
23. T. Haneda and P.J. Mc Dermott, Stimulation of ribosomal RNA synthesis during hyper-trophic growth of cultured heart cells by phorbol ester, *Mol. Cell. Biochem.* 104: 169 (1991).
24. S.N. Allo, P.J. Mc Dermott, L.L. Carl and H.E. Morgan, Phorbol ester stimulation of protein kinase C activity and ribosomal DNA transcription, *J. Biol. Chem.* 266: 22003 (1991).
25. D. Baranes and E. Razin, Protein kinase C regulates proliferation of mast cells and the expression of the mRNA of fos and jun proto-oncogenes during activation by IgE-Ag or calcium ionophore A23187, *Blood* 78: 2354 (1991).
26. M. Issandou and J. Darbon, Activation of protein kinase C by phorbol esters induces DNA synthesis and protein phosphorylation in glomerular mesangial cells, *FEBS Lett.* 281: 196 (1991).

27. J.M.Hebert, E. Seban and J.P. Maffraud, Characterization of specific binding sites for ^{3}H-staurosporine on various protein kinases, *Biochem. Biophys. Res. Commun.* 171: 189 (1988).

28. C.D. Smith, J.L. Glickman and K. Chang, The antiproliferative effects of staurosporine are not exclusively mediated by inhibition of protein kinase C, *Biochem. Biophys. Res. Commun.* 156: 1250 (1988).

29. A.G.H. Ederveen, S.E. Van Ernst-De Vries, J.J.H.H.M. De Pont and P.H.G.M. Willems, Dissimilar effects of the protein kinase C inhibitors, staurosporine and H-7 on cholecystokinin-induced enzyme secretion from rabbit pancreatic acini, *Eur. J. Biochem.* 193: 291 (1990).

30. C.A.Bedoy and P.L.Mobley, Astrocyte morphology altered by 1-(5-isoquinolinylsulfonyl)-2 methyl piperazine (H-7) and other protein kinase inhibitors, *Brain Res.* 490: 243 (1989).

31. C. Schachtele, R. Seifert and H. Osswald, Stimulus-dependent inhibition of platelet aggregation by the protein kinase C inhibitor polymyxin B, H-7 and staurosporine, *Biochem. Biophys. Res. Commun.* 151: 542 (1988).

32. H.U. Keller, V. Niggli and A. Zimmermann, The protein kinase inhibitor H-7 activates human neutrophils: effect on shape, actin polymerization, fluid pinocytosis and locomotion, *J. Cell Sci.* 96: 99 (1990).

33. M. Watanabe, T. Tamura, M. Ohashi, N. Hirasawa, T. Ozeki, S. Tsurufuji, H. Fujiki and K. Ohuchi, Dual effects of staurosporine on arachidonic acid metabolism in rat peritoneal macrophages, *Biochim. Biophys. Acta* 1047: 141 (1990).

34. N. Rodríguez-del Valle, N. Debs-Elías and A. Alsina, Effects of caffeine, cyclic 3',5' adenosine monphosphate and cyclic 3',5' guanosine monophosphate in the development of the mycelium form of *Sporothrix schenckii*, *Mycopathol.* 86, 29 (1984).

35. H. Kase, K. Iwahasi, S. Nakanishi, Y. Matsuda, K. Yamada, M. Takahashi, C. Murakata, A. Sato and M. Kaneko, K-252 compounds, novel and potent inhibitors of protein kinase C and cyclic nucleotide dependent protein kinases, *Biochem. Biophys. Res. Commun.* 142: 436 (1987).

36. T. Sako, A.I. Tauber, A.Y. Jeng, S.H. Yuspa and P.M. Blumberg, Contrasting actions of staurosporine a protein kinase C inhibitor on human neutrophils and primary mouse epidermal cells, *Cancer Res.* 48: 4646 (1988).

CLUES ABOUT CHROMOBLASTOMYCOTIC AND OTHER DEMATIACEOUS FUNGAL PATHOGENS BASED ON *WANGIELLA* AS A MODEL

Paul J. Szaniszlo,[1] Leonel Mendoza,[2] and Sankunny M. Karuppayil[1]

[1] Department of Microbiology
[2] Division of Biological Sciences
   University of Texas at Austin
   Austin, Texas, USA

ABSTRACT

Wangiella dermatitidis is an excellent model for discovering biologically and medically relevant information about other dematiaceous agents of mycosis. The major attribute of this phaeohyphomycotic fungus, which allows it to serve as a useful model, is its vegetative polymorphism. This polymorphism is expressed both *in vivo* and *in vitro* and programs a number of developmental choices leading to polarized growth by budding and hyphal apical extension or to a type of non-polarized, isotropic growth that produces spherically enlarged cells and multicellular forms. The latter phenotypes are virtually identical to the sclerotic bodies of chromoblastomycotic fungi. Study of the phenotypic switching between the polarized and isotropic growth forms has been possible because *Wangiella* can be induced to form the multicellular bodies from yeasts or hyphae *in vitro* by manipulation of culture medium pH, or by the incubation of certain cell-division-cycle mutants at elevated temperatures. It is evident that this particular type of dimorphism involves considerable rearrangement of cell-wall synthetic and deposition patterns, and results from the inhibition of polarized budding or hyphal apical extension, without the inhibition of isotropic growth, nuclear division or cytokinesis. It is also becoming evident that Ca^{2+} may have a major regulatory role in the switching, because new results showed that at acidic pH a low, but critical, concentration of Ca^{2+} was crucial for multicellular-body development, higher concentrations allowed maintenance of polarized growth, and that a Ca^{2+}/proton exchange mechanism might be involved. Support for this important role for Ca^{2+} has also been provided by other new results with EGTA. At high concentrations and at near neutral pH, this Ca^{2+} chelator caused a stage-specific, cell-cycle arrest

Dimorphic Fungi in Biology and Medicine, Edited by
H. Vanden Bossche *et al.*, Plenum Press, New York, 1993

in yeast cells, but at lower concentrations induced increasing numbers of cells to convert to multicellular forms. Parallel studies with the chromoblastomycotic fungi showed that Ca^{2+} also maintained the polarized growth of hyphae at very acidic pH, and that withholding Ca^{2+} by low concentrations of EGTA at more neutral pH induced isotropism and the formation of sclerotic bodies. These preliminary results with Ca^{2+} and the chromoblastomycotic fungi further validate the concept that *W. dermatitidis* continues to be an important model for dematiaceous pathogens of humans.

INTRODUCTION

Many fungal pathogens of humans adapt to different tissue environments in a manner that their dominant modes of vegetative growth are altered.[1] With most dimorphic pathogenic fungi, this phenotypic switching is between two polarized forms of growth leading to yeast budding and to hyphal apical extension. However, some pathogenic fungi exhibit a different type of dimorphism, which replaces polarized hyphal or yeast development with a nonpolarized, isotropic mode of growth. The tissue phases of the agents of chromoblastomycosis and coccidiodomycosis are most associated with this type of vegetative growth, although it is also characteristic of the dimorphic or polymorphic fungi responsible for mycoses as diverse as phaeohyphomycosis, adiaspiromycosis, and sometimes even sporotrichosis. These unique tissue phases are identified as sclerotic bodies for the chromoblastomycotic fungi, spherules for *Coccidioides immitis*, the agent of coccidioidomycosis, adiaspores for the agents of adiaspiromycosis and as planate cells and multicellular forms for certain phaeohyphomycotic agents, including the polymorphic fungus *Wangiella dermatitidis*.

Most of what is currently known about the possible mechanisms responsible for transitions in pathogenic fungi between polarized and nonpolarized growth forms, and also about the development of the multicellular phenotype comes from studies of *W. dermatitidis*, even though this type of development is most associated with the sclerotic bodies of the chromoblastomycotic fungi. These studies have established this melanized fungus as a simple model for dimorphic switching between these very different forms of vegetative growth among fungi in general, and as a model for the dematiaceous pathogenic fungi in particular. The use of *W. dermatitidis* as a model for elucidating common biological attributes among diverse dematiaceous pathogens is easily rationalized after consideration of the numerous findings already made using this approach. These include similarities in certain cytological features, such as septal ultrastructure,[2,3] similarities in cell-wall compositions,[4,5] the phylogenetic clustering of the few dematiaceous fungi investigated with regard to a conserved sequence associated with their chitin synthase genes,[6] the uniform nature of the melanin biosynthetic pathway first described as via the dihydroxynaphthalene (DHN) pathway in *W. dermatitidis* and later found to represent a virulence factor in that fungus and to be present in every vegetatively melanized dematiaceous pathogen investigated since,[7,8] and finally to a number of inferences or conclusions about the physiological and genetic control of the cell division cycle and its relationship in this fungus to dimorphic and polymorphic phase transitions. Each of these correlations was made by extending our efforts, albeit in a limited way, to a small repertoire of other dematiaceous pathogens, after some discovery was made with *Wangiella*. Our recent preliminary results suggesting that Ca^{2+} may have a role in similarly regulating phenotypic switching between polarized and nonpolarized modes of growth in both *Wangiella* and the chromoblastomycotic fungi further strengthens the concept that this fungus is a useful model for the pathogenic Dematiaceae.

Polymorphism and The Yeast Cell Cycle in *Wangiella*

Wangiella dermatitidis manifests its vegetative polymorphism in three distinct ways (Fig. 1). First, this fungus may grow in its predominant morphology as a

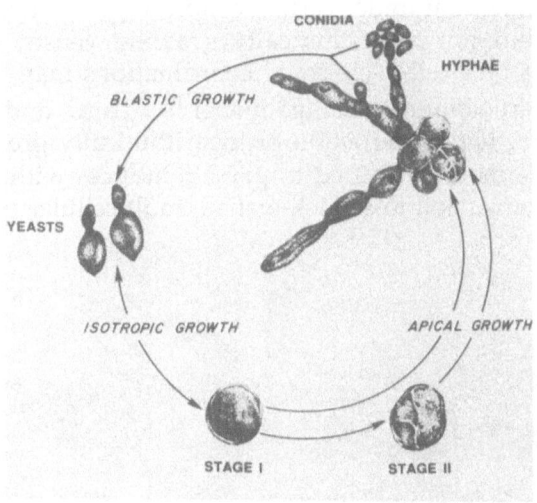

SCLEROTIC-LIKE BODIES

Fig. 1 Life cycle of *W. dermatitidis* (from Ref. 9, Courtesy of P. J. Szaniszlo).

yeast characterized by intermittent periods of polarized growth resulting in budding single cells and less frequently as pseudohyphae, which represent chains of polarly budding yeasts that complete the cell-cycle event of cytokinesis, but not cell separation.[9] The septum that brings about cytokinesis between a mother cell and a daughter bud in either of these yeast phenotypes is complete and nonperforate.[10] Second, it may alternatively exist as a hyphomycetous, conidiogenous, mycelial fungus typified either by polarized continuous or somewhat intermittent apical growth leading to true hyphae and moniliform hyphae, respectively.[10] The true hyphae are distinguished by their parallel side walls in optical microscopic section. In contrast, moniliform hyphae are virtually indistinguishable from pseudohyphae by light microscopy.[10] However, by transmission electron microscopy, both true hyphae and moniliform hyphae are easily identified as typical ascomycetous hyphal structures, because their septa are perforated with a central septal pore that has Woronin bodies nearby.[2] Third, *W. dermatitidis* may form isotropically enlarged, thick-walled cells that grow in volume by nonpolarized mechanisms. These forms often become multicellular by the production of both intersecting perforate and nonperforate septa in yeast or hyphal mother cells. It has been suggested that these multicellular bodies may be phenotypically arrested between yeasts and hyphae, because they retain the septum building capacity of both, but without the capacity to grow by either's polarized mechanisms.[11]

Three early critical discoveries have important relevance to the current understanding of vegetative polymorphism in *W. dermatitidis* and its relationship to its yeast cell cycle. First was the finding that logarithmically growing yeasts seem not to have the capacity for direct phenotypic switching.[2] Instead only aged, usually unbudded and thick-walled yeasts tend to undergo transitions to hyphal forms, and then only after acquisition of more G_0, spore-like, cytoplasmic characteristics. Yeast cells aged for shorter periods, which have not acquired this spore-like cytoplasm, usually resume budding when returned to fresh medium. Second was the finding that under certain conditions only more spore-like yeasts or some hyphae, but not logarithmically growing yeasts, form anastomosis bridges, suggesting that cell-cycle-stage coordinations may be required before selected cell types are able to exchange nuclei.[12] Third, and most importantly, was the finding that the introduction of logarithmically growing yeasts into a number of complete media acidified to pH 2.5, induces within a few days their near quantitative conversion to a thick-walled, multicellular phenotype (Fig. 2a).

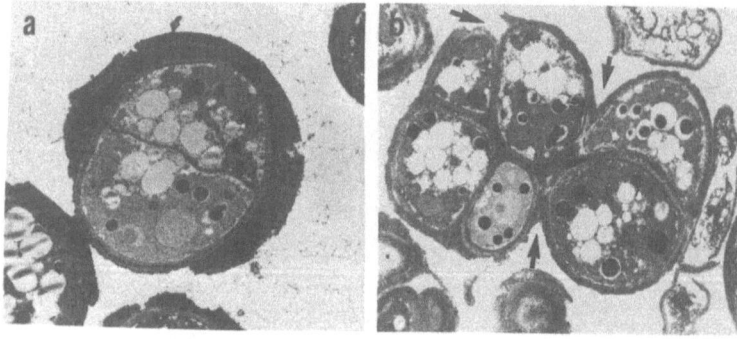

Fig. 2 Multicellular forms of *W dermatitidis* formed after prolonged culture in pH 2.5 nonsynthetic, semisolid medium (a) or liquid medium (b). Note the extremely thick wall that develops on the planate form from the solidified medium and the tendency for the wall to slough during isotropic growth in the liquid medium. The arrows in Fig. 2b indicate points where separation is occurring, which allows limited reproductive growth (Marlowe and Szaniszlo, unpublished micrographs).

Culture for longer periods under these conditions reveals that the multicellular forms are capable of a degree of reproductive growth as indicated by the tendency of some cells to separate from others (Fig. 2b). Return of the multicellular bodies, or their enlarged thick-walled precursors, to media having a more neutral pH allows hyphal outgrowth.[11] The resulting hyphae initially have the characteristics of moniliform hyphae that then may become true hyphae. Both hyphal types produce lateral hyphal buds that disarticulate and regenerate the yeast phase by budding, and less frequently produce phialo- or annello-type conidiophores that give rise to conidia with poorly characterized phenotypic potentials.[2,13,14]

Because conversion of yeasts to the multicellular forms appeared to result from the cessation of bud emergence without the inhibition of growth, nuclear division or cytokinesis, it was hypothesized that the acidic conditions were inhibiting an essential cell-cycle event leading to normal yeast growth.[11] An analogous switching from polarized budding to isotropic developmental processes had been reported previously for a certain temperature-sensitive (ts), cell-division-cycle (*cdc*) mutant of *Saccharomyces cerevisiae*; e.g. a strain possessing a mutation in the *CDC24* bud-emergence gene.[15] A search for similar *cdc* strains of *W. dermatitidis* yielded three ts, multicellular (Mc) strains that grow at pH 6.5 like the parental wild type (wt) as yeasts at the permissive temperature (25 °C), but unlike the wt under the same media conditions, do not continue to grow by budding at the semirestrictive temperature (37 °C).[16] Instead at this temperature these *cdc* strains stop growing polarly by budding and begin to enlarge isotropically. Eventually the isotropic forms become multicellular by the formation of transverse and vertical intersecting septa in the yeast mother cell.[16] In contrast, at restrictive temperatures (42 °-45 °C) few multicellular forms develop, even though most cells arrest as isotropically enlarged, unbudded cells that are still viable.[16] This suggested that the temperature-critical event that could not occur in these *cdc* mutants at semirestrictive and restrictive temperatures (37 °C and above, but not 35 °C) is bud formation.[16] This also suggested that the inhibition of bud formation in the mutants does not obligatorily commit cells to the muticellular phenotype.

Kinetic studies of the transition of yeasts to multicellular forms in the *cdc* mutants of *W. dermatitidis* incubated at 37 °C indicated that some Mc strains commit to conversion at different times in their yeast cell cycle. This prompted a detailed characterization of the cell-cycle events associated with normal yeast development at the permissive temperature, as a prerequisite to determination of the execution point (that time after which a shift to the restrictive temperature does not inhibit the progression of the current cell cycle) of each *cdc* mutation. Flow cytometry documented that yeast cells of *Wangiella*, like cells of other eukaryotes, have a cell cycle consisting of the four phases; G_1, S (the period of DNA synthesis), G_2 and M (mitosis) of durations somewhat longer than those of *S. cerevisiae*.[17,18] Subsequent microscopic analyses of synchronously and asynchronously dividing cells revealed that bud emergence in *W. dermatitidis* occurs in G_2 under conditions of slow growth.[19] However, under conditions of more rapid growth bud emergence is most often associated with late G_1 or S phase cells (unpublished results). Other studies documented that bud emergence is not dependent on DNA synthesis, and cells inhibited in DNA synthesis arrest as budded forms with undivided nuclei in the isthmus between a mother cell and daughter bud.[19] These observations collectively led to the proposal that the yeast cell cycle of *W. dermatitidis* is basically similar to that of *S. cerevisiae*, and consists of a DNA-nuclear division pathway and a bud formation-nuclear migration pathway, which diverge in G_1 and converge at cytokinesis.[17,19] Further studies of the effects of microtubule inhibitors suggested that the yeast cell cycle in *W. dermatitidis* consists of at least three major pathways (Fig. 3): the two already reported and a third responsible for nuclear migration and chromosome segregation.[20] A minor pathway was also suggested to be associated with the event of yeast mother-cell, daughter-bud separation, because under some conditions yeast cells continue budding growth and cytokinesis, but without cell separation, which results in the production of pseudohyphae. The final localization of the execution points for the mutations in two Mc strains to two slightly different positions before bud emergence in the bud emergence-bud growth pathway provided strong evidence that the mutations in the Mc strains were in different *CDC* genes.[20] Subsequent complementation studies using protoplast fusion methodology have tended to confirm this hypothesis (Cooper and Szaniszlo, unpublished data).

Conversion of yeasts to the multicellular phenotype in the *cdc* mutants of *W. dermatitidis* mutants follows the same two-stage process noted for multicellular-form development induced by acidity.[11,16] Stage I is arbitrarily marked by the formation of swollen, unbudded cells having one or more than one nucleus and thickened cell walls, whereas Stage II forms have at least one internal septum.[21] This morphogenesis is accompanied by marked changes in the cell-wall polymers chitin, β-glucan, and DHN melanin. Upon conversion of yeasts to multicellular forms, the relative amount of chitin and melanin in the cell wall dramatically increases, whereas β-glucan decreases.[4] Chitin deposition also becomes delocalized, shifting from the bud/birth scar regions to an inner wall layer during Stage I development, and to the inner wall layers and transverse septa during Stage II development.[9,21,22] Inhibition of the biosynthesis of chitin by polyoxin or of β-glucan by papulocandin during yeast-to-multicellular form transition causes

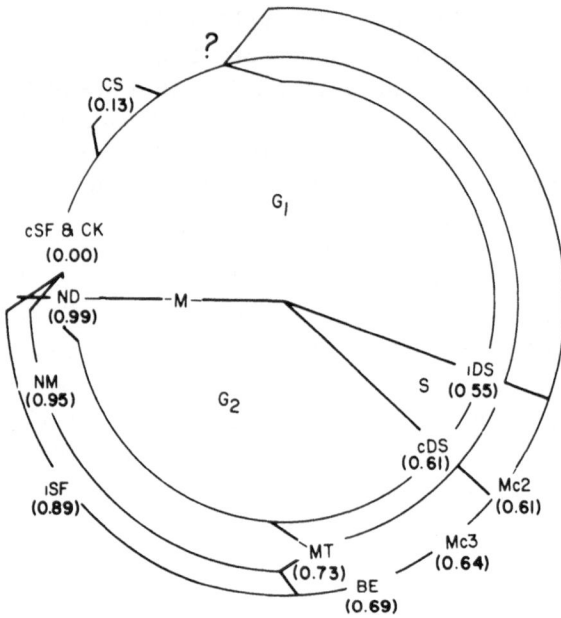

Fig. 3 Yeast cell cycle of *W. dermatitidis* showing independent pathways of dependent events. The cell-cycle events mapped and the approximate time (as a percentage of a cell cycle of 4.5 h duration in parentheses) each event occurs are as follows: cSF, completion of septum formation; CK, cytokinesis; CS, cell separation; iDS, initiation of DNA synthesis; cDS, completion of DNA synthesis; Mc2, execution point of *cdc2* mutation; Mc3, execution point of *cdc3* mutation; BE, bud emergence; MT, microtubule polymerization; iSF initiation of septum formation; NM, nuclear migration; ND, nuclear division (Modified from Refs. 4 & 20. Courtesy of P.J. Szaniszlo).

increased cellular death due to lysis.[21] By comparison, when wt yeasts are incubated in the presence of the chitin synthesis inhibitor polyoxin, chains of yeasts form which have aberrant septa similar to those sometimes observed in *chs* mutants of *S. cereviseae*, but little lysis occurs.[21,23] These results, together with the identification of multiple chitin synthases in *S. cerevisiae* and *Candida albicans*,[24-26] suggested that multiple chitin synthase genes, coding for multiple isoenzymes, might be involved in the chitin rearrangements occurring during the phenotypic switching in *W. dermatitidis*. Support for this hypothesis has been provided by recent experiments involving PCR technology that allowed the amplification, cloning and sequencing of three distinct 600 bp chitin synthase (*CHS*) homologous fragments from *W. dermatitidis*.[27] These three distinct *CHS* DNA fragments have been shown to fall into the same three distinct *CHS* classes detected during parallel studies with other fungi.[6]

Calcium and Yeast-to-Multicellular Form Conversion in *Wangiella*

Calcium is essential for progression of the cell division cycle in eukaryotes, and plays numerous roles in signal transduction leading to cell differentiation. Thus, preliminary studies have been initiated to investigate the role of Ca^{2+} in *W. dermatitidis* polymorphism (unpublished data of authors). These studies are particularly relevant because others have found that yeast actin assembly is sensitive to calcium concentrations, a tight association has been established between actin cytoskeletal structures and regions of cell-wall growth, and the *CDC24* gene product that is involved in bud emergence in *S. cerevisiae* is a Ca^{2+} binding protein.[28-31] This suggests that Ca^{2+} and actin activities may be partly responsible for the site-specific insertion of chitin and other polymers into cell walls and septa of *W. dermatitidis* during cell-cycle-related phenotypic switching. Our preliminary results have been surprising and have demonstrated that multicellular forms can be induced from logarithmically growing yeasts in a totally synthetic, pH 2.5 medium devoid of any organic supplements other than thiamine, which must be supplied as a vitamin requirement (Hsieh and Szaniszlo, unpublished data), and glucose. Previous attempts by us and others to induce significant yeast-to-multicellular-form conversion in this species in simple, synthetic media had resulted in general failure, although many cells often arrested as isotropically enlarged, unbudded forms (Hsieh and Szaniszlo, unpublished data; B. H. Cooper, personal communications). The single new factor added to the Czapek salts-ammonium chloride-glucose medium utilized in these experiments was 0.1 mM Ca^{2+}, which allowed a rapid rise in yeast populations of the percentage of cells that stopped budding, and instead began to isotropically enlarge and often become multicellular (Fig. 4a). Even more surprising was the finding that 1 to 5 mM additions of Ca^{2+} to the same synthetic medium were able to reverse the effects of the acidic medium conditions (Fig. 4b).

In a concentration dependent fashion, Ca^{2+} additions tended to maintain the growth of increasing percentages of the inoculum as budding yeasts and pseudohyphae. The extent of this reversal of the cell-cycle-inhibitory effects of acidic culture was demonstrated by dry weight measurements that showed biomass production in the presence of 5 mM Ca^{2+} at pH 2.5 was equivalent to that in the same medium adjusted to neutral pH (unpublished data of authors).

To rule out the possibility that the increase in yeast growth promoted by the higher concentrations of Ca^{2+} added to the pH 2.5 synthetic medium was not the result of medium neutralization, the final pH of the acidic cultures was measured (unpublished data of authors). A small but consistent, concentration-dependent, decrease in pH was noted in the acidic cultures, which was most pronounced in cultures with 5 mM Ca^{2+}. In addition, and to insure that these results with Ca^{2+} were specific, and not simply a generalized divalent cation effect, manganese, magnesium, zinc and cobalt additions were also investigated. Of these cations, only Mn^{2+} was able to act similarly to Ca^{2+} in its ability to reverse the effect of pH 2.5 culture (unpublished data). However, total biomass production at 5 mM Mn^{2+} concentration was only equivalent to that supported by 1 mM addition of Ca^{2+}, suggesting that although Mn^{2+} could substitute for Ca^{2+} to maintain yeast polarity and some yeast growth, it was not as efficient as Ca^{2+} for allowing yeast growth at rapid rates. These new results suggest that a proton-calcium antiport system in the may be involved. Alternatively, the effect may be related to a general reduction in cellular pH, which occurs during culture at pH 2.5. These and other hypotheses are under current investigation.

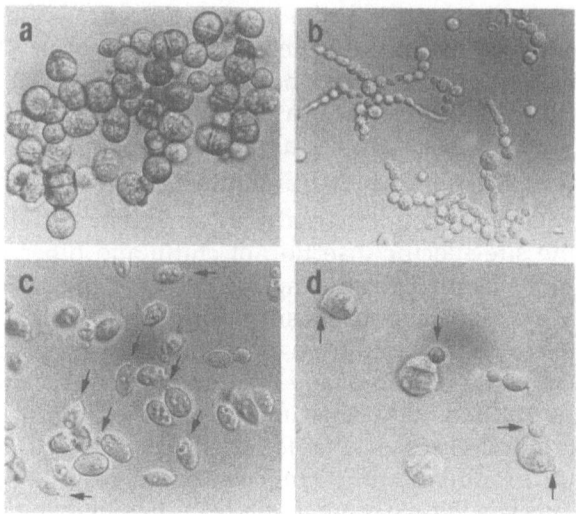

Fig. 4 Effects of Ca^{2+} additions to acidic (pH 2 5) synthetic medium (a,b) or EGTA additions to near neutral, buffered (pH 6 5) synthetic medium (c,d) on the morphology of *W dermatitidis*. Note tendency of 0.1 mM Ca^{2+} (a) to promote multicellular development at pH 2.5 whereas 5.0 mM Ca^{2+} (b) maintains polarized budding or hyphal apical extension. Also note that the 30 mM EGTA tends to arrest cells stage specifically with small bud initials (c), whereas 0 5 mM allows multicellular form development. The arrows in (c) and (d) identify bud initials (unpublished micrographs of authors).

The importance of Ca^{2+} to the cell-cycle relationship between polarized and nonpolarized growth mechanisms in *W. dermatitidis* has been confirmed by other recent preliminary experiments involving EGTA. This Ca^{2+} chelator was added to a synthetic medium similar to that used in the pH 2.5 experiments, but differed from that medium by being buffered against pH change at 6.5 (unpublished data of authors). In these experiments asynchronous populations of logarithmically growing yeasts exposed to high (30 mM) concentrations of EGTA, arrested within 24 hours in very high percentages (40-80%) as yeast mother cells with single bud initials (less than one quarter the size of the mother cell) that tended not to enlarge (Fig. 4c). In some cases more than one bud initial could be observed on a mother cell suggesting that possibly even at 30 mM EGTA, the bud initiation process in *W. dermatitidis* can be uncoupled from other bud growth processes. At lower concentrations of EGTA in the pH 6.5 buffered medium, yeast inocula had an increasing tendency to grow either as a budding yeast or to convert into isotropically enlarged, planate forms (mother cells with a single internal septum) or multicellular bodies (Fig. 4d). As with similar isotropic forms developing in pH 2.5 media, or after shifts of the *cdc* mutants to 37 °C, these forms tended to be multinucleate and exhibit delocalized chitin deposition as indicated by calcofluor staining (unpublished data of authors) These tendencies were also most pronounced in buffered medium having an EGTA concentration of ~0.5 mM, although even at 30 mM EGTA some chitin delocalization was apparent. Most interestingly however was the finding that at the ~0.5 mM EGTA concentration, about one half of the cell population initially arrested as normal size yeast mother cells with a bud initial identical to or only slightly larger than those on cells arrested by the 30 mM EGTA. The remainder of the yeasts in these cultures were either unbudded or had enlarged buds. By the end of the sixth day, the percentage of multicellular forms with one or more septa had increased to 20% of the final populations and together with the isotropically enlarged cells constituted about 40-50% of the phenotypes observed when the experiment was terminated (data not shown). These results with EGTA suggest that Ca^{2+} may be required for the same yeast cell-cycle events requiring this ion in *S. cerevisiae*.[31] It is known that mutants of this monomorphic yeast, which are defective in the synthesis of calmodulin, a calcium binding protein, cannot undergo mitosis. Also strains with mutations in *CDC24* and *CDC31*, both of which encode calcium-binding proteins, either cannot make a bud or cannot duplicate their spindle-pole bodies, respectively.[31,32] Perhaps processes of spindle-pole-body duplication, nuclear division, bud initiation and bud growth are regulated similarly in the polymorphic fungus *W. dermatitidis*. In environments where Ca^{2+} is extremely limited at neutral pH and not present in sufficient quantity to bind required amounts of calmodulin or other calcium-binding proteins, cell cycles might be totally arrested resulting in a terminal phenotype with a bud initial.[32] However, under more semirestrictive conditions represented by the presence in the environment of concentrations of Ca^{2+} too low to inhibit normal polarized growth, but not too low to allow nonpolarized growth, nuclear division or cytokinesis, *W. dermatitidis* by virtue of its polymorphism, produces isotropic forms, planate cells and occasionally multicellular forms, in addition to yeasts and hyphae.

Wangiella as a Model for the Chromoblastomycotic Fungi and Other Dematiaceae.

At first glance *Wangiella* might seem an inappropriate model for the dematiaceous pathogens because it is polymorphic. However, this potential liability of vegetative complexity is *Wangiella*'s main attribute. By virtue of its polymorphism, this single species expresses all of the morphologies of any dematiaceous pathogen observed *in vivo*, and can be induced to produce *in vitro* homogeneous populations of each for study. This property is particularly important for a model of the dematiaceous pathogens, because some such as *Wangiella* exist predominantly as yeasts in nature, but as mixtures of yeasts, isotropic forms and hyphae in tissue, others are hyphal both in nature and *in vivo*, and still others, like the chromoblastomycotic fungi, are hyphal in nature, but exist almost exclusively as sclerotic bodies in tissue.[9,33,34] In fact, although many aspects of their respective clinical and histopathologies differ, medical mycologists have traditionally tended to distinguish chromoblastomycosis and phaeohyphomycosis by the presence in tissue of sclerotic bodies in the former and their absence in the latter. A sclerotic body is defined as an isotropically enlarged, thick-walled, muriform (multicellular) unit, with muriform referring to the presence of intersecting vertical and horizontal septa. Yet agents of phaeohyphomycosis in tissue also frequently form thick-walled bodies that sometimes become septate in at least one plane, and resemble an early stage of sclerotic body development.[33-37] Thus, agents like *W. dermatitidis* have the genetic potential for multicellular body production, although this morphology is only infrequently expressed *in vivo*. It was the ability of *W. dermatitidis* to undergo transitions from yeasts or hyphae to multicellular forms that first prompted our attempts to induce these transitions *in vitro* as a simple model of cellular differentiation, and also as a model for sclerotic body development by chromoblastomycotic fungi.[11,16] Earlier attempts by others to cultivate the sclerotic bodies of the chromoblastomycotic fungi *in vitro*, as a prerequisite to an understanding of their potential as a virulence factor, had been only marginally successful.[38]

By taking particular advantage of the ability of *Wangiella* to grow as a yeast, which greatly aids exact quantitation of growth and inoculum levels, derivation of mutants, fusion of protoplasts for genetic analysis, and the evaluation of pathogenicity studies, a number of facts about *Wangiella* have been established at relatively sophisticated levels.[9] These have then guided additional studies of other dematiaceous fungi, albeit in a limited manner, which have been extremely rewarding. In particular, is our finding that like *Wangiella*, numerous other dematiaceous pathogens of humans, including all the chromoblastomycotic fungi, are melanized by polymerized 1,8-dihydroxynaphthalene. In our view, this finding is extremely significant because in *Wangiella* this cell-wall pigment is an important virulence factor.[7] Of possible equal potential importance is our more recent finding that a number of the dematiaceous pathogens, also like *Wangiella*, possess in their genomes multiple chitin synthase homologous DNA sequences.[6] These homologues fall into three classes based on derived amino acid sequences, and share an amazing level of conservation in each class. The phylogenetic clustering of *Wangiella* and the other asexual dematiaceous fungi, with a number of known or suspected ascomycetes by FITCH analyses of the conserved chitin synthase sequences, adds validity to the concept that many dematiaceous pathogens are relatively closely related. The clustering of the dermatiaceous pathogens in branches separated from the cleistothecial and perithecial fungi analyzed suggest they are closely enough related to obviate the need to study each dematiaceous form-species at the level of the model.[6]

Our current preliminary results, concerning the importance of Ca^{2+} to phenotypic switching from polarized to nonpolarized growth modes in *Wangiella*, suggest that through a circuitous route we may have found a factor with more direct relevance than extreme acidity to the morphogenesis of some pathogenic fungi in the human host. This factor is the Ca^{2+} concentration in the immediate vicinity of the fungus in the host, which may be only secondarily influenced by localized tissue acidic conditions, although both could affect intracellular Ca^{2+} homeostasis in the fungal pathogen. This hypothesis prompted us to extend our model studies with *Wangiella* to the chromoblastomycotic fungi, because of their almost exclusive production of sclerotic bodies *in vivo*. As expected, the results confirmed (unpublished data of authors) that representative strains of the chromoblastomycotic fungi *Cladosporium carrionii*, *Fonsecae pedrosoi* and *Phialophora verrucosa* could be induced to convert their hyphae to large numbers of sclerotic bodies by inoculation into pH 2.5 synthetic medium, containing sufficient, but low levels of Ca^{2+} (Fig. 5b,f,j). When no Ca^{2+} was included in this same medium, some isotropic forms developed from the hyphal inoculum, but these only rarely became multicellular, or if they did they mostly had only one septum (Fig. 5a,e,i). In contrast, the addition of higher levels of Ca^{2+} significantly reversed, in a concentration dependent manner, the tendency of these chromoblastomycotic fungi to form sclerotic bodies in the acidic medium (Fig. 5c,d,g,h,k,l). Also as

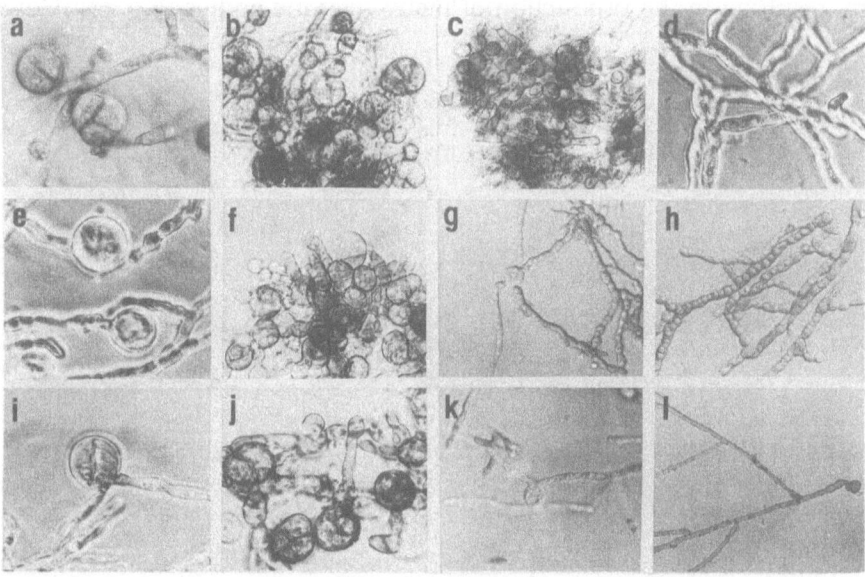

Fig. 5 Effects of 0 0 mM (a,e,ı), 0 1 mM (b,f,ȷ), 1 0 mM (c,g,k) and 5 0 mM (d,h,ı) Ca²⁺ additions to acidıc (pH 2 5) synthetic medıum on the morphology of *C carrıonıı* (a,b,c,d), *F pedrosoı* (e,f,g,h) and *P verrucosa* (ı,ȷ,k,l) Note tendency of the 5 0 mM Ca²⁺ concentratıons to maıntaın hyphal growth (unpublıshed mıcrographs of authors)

predicted, nuclear division continued during hyphal-to-sclerotic-body transitions, and chitin delocalizations occurred as indicated by staining with DAPI and calcofluor respectively. Confirmation that hyphal growth was commenserate at pH 2.5 and at the higher Ca^{2+} levels with that of controls grown at pH 6.5 was established by dry weight measurements. Finally, other measurements demonstrated a slight, but measurable, decrease in pH that was inversely correlated with the increasing Ca^{2+} concentrations, suggesting again that polarity as in *Wangiella* may be maintained by the action of a proton-calcium pump.

Other parallel studies involving EGTA and the same chromoblastomycotic fungi are yielding equally gratifying results (unpublished data of authors). At high concentrations of this Ca^{2+} chelator, hyphae of these fungi are inhibited in apical growth and in conversion to sclerotic bodies, probably because, as with *Wangiella*, nuclear division and other Ca^{2+} functions have been inhibited. At lower concentrations, EGTA is more semirestrictive than restrictive and allowed growth and nuclear division to continue, but in the absence of extensive polarized hyphal growth, resulting in the production of many sclerotic bodies (Fig. 6). However,

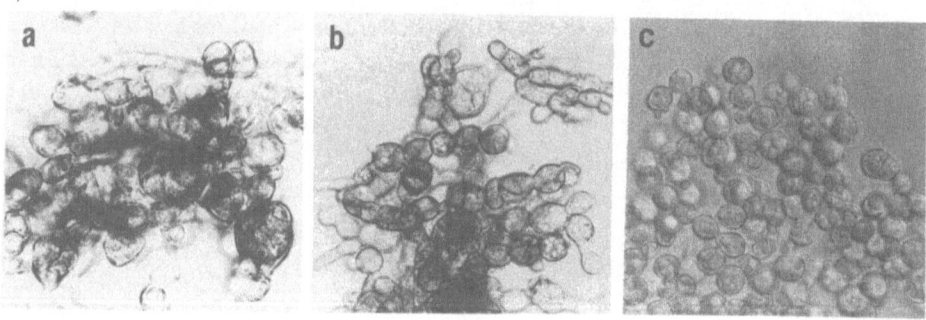

Fig. 6 Effects of 1 mM (a) 2 mM (b) and 8 mM EGTA additions to buffered (pH 6.5) synthetic medium on conversion of *C carrionii, P verrucosa* and *F pedrosoi,* respectively (unpublished micrographs of authors).

each form-species investigated appeared to be sensitive to a different concentration of EGTA. For example, *C. carrionii* was the most sensitive to EGTA. Concentrations as low as 0.5-1.0 mM induced sclerotic-body formation and inhibited hyphal growth in this fungus (Fig. 6a). In contrast, higher concentrations of EGTA (2.0-8.0 mM) were required with *P. verrucosa* and *F. pedrosoi* to achieve the same effect (Fig. 6b,c). This indicated that at least the strain of *C. carrionii* used in these studies, is more sensitive to Ca^{2+} deprivation than the two strains of *P. verrucosa* and *F. pedrosoi* being investigated. These findings might explain the ability of *P. verrucosa* and *F. pedrosoi* to cause cerebral phaeohyphomycosis, whereas *C. carrionii* is only associated with typical subcutaneous granulomatous lesions.[33,39] Perhaps some specialized tissues, like those of the central nervous system (CNS), withhold Ca^{2+} for their own functions less efficiently than the granulomatous tissue associated with subcutaneous chromoblastomycosis, thus allowing the more calcium-deprivation-tolerant fungi like *P. verrucosa* and *F. pedrosoi* to continue to grow in a phaeohyphomycotic manner in those tissues. This speculation is consistent with the reports of systemic (CNS and other tissues) involvement of *F. pedroisoi*[40-43] and *P. verrucosa*[44,45] but not *C. carrionii*.[39]

CONCLUSIONS

Our new results on the importance of Ca^{2+} to the phenotypic switching between polarized and nonpolarized growth in *Wangiella* and the chromoblastomycotic fungi seem to suggest a rational for the types of morphologies these fungi form in human tissues during infections. This rationale predicts that Ca^{2+} deprivation is regulating phenotypic expression of these fungi *in vivo*. Thus, possibly in the chronic, generally subcutaneous granulomatous lesions associated with chromoblastomycosis, Ca^{2+} withholding conditions might be more semirestrictive than those of the cystic lesions associated with phaeohyphomycosis caused by a fungus like *W. dermatitidis*, thereby inducing only sclerotic bodies in the former at the expense of the hyphal phenotype. In contrast, in the latter, calcium fluxes may be slight, allowing yeast growth predominantly, or variable, inducing both transient or permanent periods of semirestrictive cell-cycle arrests, resulting in the formation of mixtures of yeasts, hyphae, planate cells and infrequent multicellular forms. This scenario is highly plausible because many immunophysiological characteristics of granulomatous tissue reactions involve calcium-requiring processes. Thus, from the point of view of the chromoblastomycotic fungus, sustained Ca^{2+} withholding conditions may often be encountered. For example, Ca^{2+} is required for opsonization and during the classic complement cascade,[46] the regulation of phagocytosis in macrophages and other phagocytic cells,[47] the triggering and degranulation of mast cells and eosinophils in chronic lesions,[48] the polymerization of the enzymes involved in the formation of channels by cytotoxic killer cells,[46] and for many other inflammatory processes. It is possible that these mechanisms and that host inflammatory cells compete for Ca^{2+} in host tissue with the chromoblastomycotic fungi, and possibly even other pathogenic fungi that phenotypically switch between polarized to nonpolarized growth mechanisms. Our hypothesis is that after entry of a dematiaceous fungus into tissue, different critical levels of Ca^{2+} would be required by different strains and species of these fungi in order to proliferate or to switch to other phenotypes, particularly during early stages of infection. This, coupled with the fact that micro-environments of very low pH (2.5 to 5.9 pH) recently have been reported at sites of contact between macrophages and a foreign surface before and during phagocytosis,[49] lead us to believe that a process similar to the ones described earlier (low concentrations of calcium plus acidic conditions), acts as a transient or prolonged trigger in some dematiaceous fungi for the conversion of yeasts or hyphae to isotropic forms.

REFERENCES

1. P. J. Szaniszlo, An introduction to dimorphism among zoopathogenic fungi, in: "Fungal Dimorphism: With Emphasis on Fungi Pathogenic for Humans," P. J. Szaniszlo, ed., Plenum Press, New York (1985).
2. K. B. Oujezdsky, S. N. Grove and P. J. Szaniszlo, Morphological and structural changes during yeast-to-mold conversion of *Phialophora dermatitidis*, *J. Bacteriol.* 113: 468 (1973).
3. B. H. Cooper, S. Grove, C. Mims and P. J. Szaniszlo, Septal ultrastructure in *Phialophora pedrosoi*, *Phialophora verrucosa* and *Cladosporium carrionii*, *Sabouraudia* 11: 127 (1973).
4. P. J. Szaniszlo, P. A. Geis, C. W. Jacobs, C. R. Cooper and J. L. Harris, Cell wall changes associated with yeast-to-multicellular form conversion in *Wangiella dermatitidis*. In: "Microbiology-1983", D. Schessinger, ed., American Society for Microbiology, Washington, D C. (1983).

5. P. J. Szaniszlo, B. H. Cooper and H. S. Voges, Chemical compositions of the hyphal walls of three chromomycosis agents, *Sabouraudia* 10: 94 (1972).

6. A. R. Bowen, J. L. Chen-Wu, M. Momany, R. Young, P. J. Szaniszlo and P. W. Robbins, Classification of fungal chitin synthases, *Proc. Natl. Acad. Sci. USA* 89: 519 (1992).

7. B. E. Taylor, M. H. Wheeler and P. J. Szaniszlo, Evidence for pentaketide melanin biosynthesis in dematiaceous human pathogenic fungi, *Mycologia* 79: 330 (1987).

8. D. M. Dixon, A. Polak and P. J. Szaniszlo, Pathogenicity and virulence of wild-type and melanin deficient *Wangiella dermatitidis*, *J. Med. Vet. Mycol.* 25: 97 (1987).

9. P. A. Geis and C. W. Jacobs, Polymorphism in *Wangiella dermatitidis*, In: "Fungal Dimorphism: With emphasis on fungi pathogenic for humans", P. J. Szaniszlo, ed., Plenum Press, New York (1985).

10. S. N. Grove, K. B. Oujezdsky and P. J. Szaniszlo, Budding in the dimorphic fungus *Phialophora dermatitidis*, *J. Bacteriol.* 115: 323 (1973).

11. P. J. Szaniszlo, P. H. Hsieh and J. D. Marlowe, Induction and ultrastructure of the multicellular (sclerotic) morphology in *Phialophora dermatitidis*, *Mycologia* 68: 117 (1976).

12. K. B. Oujezdsky and P. J. Szaniszlo, Conjugation in the dimorphic chromomycosis fungus *Phialophora dermatitidis*, *J. Bacteriol.* 114: 1356 (1973).

13. M. R. McGinnis, *Wangiella*, a new genus to accommodate *Hormiscium dermatitidis*, *Mycotaxon* 5: 353 (1977).

14. G. S. DeHoog, Survey of the black yeasts and allied fungi. In: "Studies in Mycology", Vol. 15, "The Black Yeasts and Allied Hyphomycetes". G. S. DeHoog and E. J. Hermanides-Nijof, ed., Centraalbureau voor Schimmelcultuur, Baarn, The Netherlands (1977).

15. L. H. Hartwell, R. K. Mortimer, J. Culotti and M. Culotti, Genetic control of the cell division cycle in yeast, *Genetics* 74: 267 (1973).

16. R. L. Roberts and P. J. Szaniszlo, Temperature-sensitive multicellular mutants of *Wangiella dermatitidis*, *J. Bacteriol.* 135: 622 (1978).

17. R. L. Roberts and P. J. Szaniszlo, Yeast-phase cell cycle of the polymorphic fungus *Wangiella dermatitidis*, *J. Bacteriol.* 144: 721 (1980).

18. M. L. Slater, S. O. Sharrow and J. J. Gart, Cell cycle of *Saccharomyces cerevisiae* in populations growing at different rates, *Proc. Natl. Acad. Sci. USA* 74: 3850 (1977).

19. R. L. Roberts, R. J. Lo and P. J. Szaniszlo, Induction of synchronous growth in the yeast phase of *Wangiella dermatitidis*, *J. Bacteriol.* 141:981 (1980).

20. C. W. Jacobs and P. J. Szaniszlo, Microtubule function and its relation to cellular development and yeast cell cycle in *Wangiella dermatitidis*, *Arch. Microbiol.* 133: 155 (1982).

21. C. R. Cooper, Jr., J. L. Harris, C. W. Jacobs and P. J. Szaniszlo, Effects of polyoxin AL on cellular development in *Wangiella dermatitidis*, *Exp. Mycol.* 8: 349 (1984).

22. J. L. Harris and P. J. Szaniszlo, Localization of chitin in walls of *Wangiella dermatitidis* using colloidal gold labeled chitinase, *Mycologia* 78: 853 (1986).

23. J. A. Shaw, P. C. Mol, B. Bowers, S. J. Silverman, M. H. Valdivieso, A. Duran and E. Cabib, The function of chitin synthases 2 and 3 in the *Saccharomyces cerevisiae* cell cycle, *J. Cell. Biol.* 114: 111 (1991).

24. J. Au-Young and P. W. Robbins, Isolation of a chitin synthase gene from *Candida albicans* by expression in *Saccharomyces cerevisiae*, *Mol. Microbiol.* 4: 197 (1990).

25. C. E. Bulawa, M. Slater, E. Cabib, J. Au-Young, A. Sburlati, W. L. Adair and P. W. Robbins, The *S. cerevisiae* structural gene for chitin synthase is not required for chitin synthesis *in vivo*, *Cell* 40: 213 (1986).

26. J. Chen-Wu, J. Swicker, A. R. Bowen and P. W. Robbins, Expression of chitin synthase genes during yeast and hyphal growth phases of *Candida albicans*, *Mol. Microbiol.* 6: 497 (1992).

27. P. J. Szaniszlo and M. Momany, Chitin, chitin synthase and chitin synthase conserved region homologues in *Wangiella dermatitidis*, In: "NATO Workshop Proceedings on Molecular Biology and Its Application to Medical Mycology", B. Maresca, G. Kobayashi, H. Yamaguchi, ed., Springer-Verlag, New York, in press.

28. Adams, A.E.M., D. Botstein and D. G. Drubin, Requirement of yeast fimbrin for actin organization and morphogenesis *in vivo*, *Nature* 354: 404 (1991).

29. B. F. Sloat, A.E.M. Adams and J. R. Pringle, Roles of *CDC24* gene product in cellular morphogenesis during *Saccharomyces cerevisiae* cell cycle, *J. Cell Biol.* 89: 395 (1979).

30. C. Greer and R. Schenkman, Calcium control of *Saccharomyces cerevisiae* actin assembly, *Mol. Cell. Biol.* 10: 1279 (1982).

31. Y. Anraku, Y. Ohya and H. Iida, Cell cycle control by calcium and calmodulin in *Saccharomyces cerevisiae*, *Biochim. Biophys. Acta.* 1093: 169 (1991).

32. P. Baum, C. Furlong and B. Byers, Yeast gene required for spindle pole body duplication: homology of its product with Ca^{2+} binding proteins, *Proc. Natl. Acad. Sci. USA* 83: 5512 (1988).

33. K. J. Kwon-Chung and J. E. Bennett, "Medical Mycology," Lea and Febiger, Philadelphia (1992).

34. K. E. Greer, G. P. Gross, P. H. Cooper and S. A. Harding, Cystic chromomycosis due to *Wangiella dermatitidis*, *Arch. Dermatol.* 115: 1433 (1979).

35. T. Matsumoto, A. A. Padhye, L. Ajello and M. R. McGinnis. *Sarcinomyces phaeomuriformis*: a new dematiaceous hyphomycete, *J. Med. Vet. Mycol.* 24: 395 (1986).

36. A. A. Padhye, W. B. Helwig, N. G. Warren, L. Ajello, F. W. Chandler and M. R. McGinnis, Subcutaneous phaeohyphomycosis caused by *Xylohypha emmonsii*, *J. Clin. Microbiol*, 26: 709 (1988).

37. G. Koshi, V. Anandi, M. Urien, M. G. Kirubakaran, A. A. Padhye and L. Ajello, Nasal phaeohyphomycosis caused by *Biopolaris hawaiiensis*, *J. Med. Vet. Mycol.* 25: 397 (1987).

38. B. H. Cooper, *Phialophora verrucosa* and other chromoblastomycotic fungi, In: "Fungal Dimorphism: With Emphasis on Fungi Pathogenic for Humans", P. J. Szaniszlo, ed., Plenum Press, New York (1985).

39. J. W. Rippon, "Medical Mycology: The Pathogenic Fungi and The Pathogenic Actinomycetes," W. B. Saunders (1988).

40. H. O. Duque, Meningo-encephalitis and brain abscess caused by *Cladosporium* and *Fonsecaea*, *Am. J. Clin. Pathol.* 36: 505 (1961).

41. H. Ari and J. Endo, Deep mycoses (*Fonsecaea pedrosoii* and *Paecilomyces lilacinus*) after liver transplant, *Trop. J. Clin. Dermatol.* 31: 481 (1977).

42. R. D. Azulay and J. Serruya, Hematogenous dissimination of chromoblastomycosis. Report of a generalized case, *Arch. Dermatol.* 95: 57 (1967).

43. P. A. Wackym, C. F. Graz, R. F. Richie and C. R. Gregg, Cutaneous chromomycosis in renal transplants patients, *Arch. Intern. Med.* 145: 1036 (1985).

44. T. Iwatsu and M. Miyaji, Subcutaneous cyst caused by *Phialophora verrucosa*, *Mycopathologia* 64: 165 (1978).

45. Y. Kameda and T. Sasaki, A case of phaeomycotic cyst caused by *Phialophora verrucosa*, *Japan J. Med. Mycol.* 25: 379 (1979).

46. I. Roitt, J. Brostoff and D. Male, "Immunology," 2nd ed., Medical Publishing (1989).

47. M. W. Berhese and R. Snyderman, "Human monocytes and macrophages," Academic Press, London (1989).

48. M. Lichtenheld, A. Hameed and R. Podack, Structure of perforin and of esterases (granzymes) in human cytolytic granules, In: "Cellular Basis of Immune Modulation," J. K. Kaplan, D. R. Green and R. C. Beakley, eds., Alan R. Liss, Inc., New York, (1988).

49. I. A. Silver, R. J. Murrilo and D. F. Etherington, Microelectrode studies on the acid microenvironment beneath adherent macrophages and osteoclasts, *Exp. Cell. Res.* 175: 206 (1988).

DIMORPHISM IN *MUCOR* SPECIES

José Ruiz-Herrera

Departments of Genetics and Molecular Biology
Centro de Investigacion y de Estudios Avanzados del IPN
Unidad Irapuato
Km 9.6, Libramiento Norte Carretera Irapuato-Leon
Apartado Postal 629
Irapuato, GTO, Mexico

ABSTRACT

Dimorphism in Mucorales can be regarded as a differentiation process involving a series of events by which internal stimuli are perceived by cellular sensors and adequately transduced to produce as a response, a morphological change of the cell. As a generalized hypothesis it is proposed that the most important effector which triggers the mycelial-to-yeast transition is the establishment of a microaerobic environment. These conditions change the redox potential of the cell where a sensor distinct, but associated with members of the respiratory pathway, unchains the morphogenetic response. In *Mucor* species where microaerobiosis is not sufficient to induce yeast-like growth, it is suggested that the sensor has such a low potential that either additional catabolite repression or an unknown alteration brought about by a high CO_2 tension is necessary to initiate the transition. As members of the transduction chain cAMP and polyamines appear as likely candidates. The latter compounds probably regulate differential expression of specific genes, affecting the state of DNA methylation. The specific gene products synthesized under each condition lead in turn to the establishment of the growth pattern of the cell wall, probably affecting its composition and structure, but mainly affecting the guidance mechanisms responsible for the mobilization of vesicles and chitosomes to specific sites of the cell surface where wall synthesis takes place. As a result a spherical or a cylindrical wall is made up in response to a change in the environmental conditions.

INTRODUCTORY REMARKS

Several Mucorales have the capacity to grow as mycelial- or yeast-like forms depending on the environmental conditions, i.e. they are dimorphic. Dimorphic capacity of this group of fungi, probably reflects an adaptive response to their particular ecological niches. Taking into consideration the natural environmental factors known to favour one or the other morphology, it may be suggested that accumulation of nutritional sources or metabolic indicators of their presence, all stimulate colonial growth (the epitome of which is yeast growth), whereas scarcity of foodstuff favours mycelial development. In the laboratory the dimorphic response of Mucorales has attracted the attention of different research groups for a long time, since it represents an apparently simple morphogenetic system to be studied as a model of cell differentiation in eukaryotic organisms. In practice, however, the model has proved to be more complex than anticipated. Complexity can be explained on grounds of different reasons which may be summarized as follows: variability in the morphogenetic effectors and the impossibility of finding a common mode of action; occurrence of alterations in the metabolic pathways which accompany the morphogenetic phenomena; absence of clear patterns linking the effectors and the response of the cell; and differences in response among distinct species. All these problems are enhanced by the absence of a genetic system of analysis, and difficulties associated with the only transformation system reported (should I add, shortage of support because of the scarce practical importance of Mucorales...?).

For these reasons, in the present review I would like to analyze the problem from a simple perspective. I have preferred to explore the similarities existing among different species, rather than their differences, and to proppose some general ideas which may help to discern the morphogenetic thread. I hope that this approach may be useful to generate discussion in the field and to develop schemes capable of being subjected to experimental analysis.

The first idea that I would like to recall, is the concept that the normal habit of Mucorales is mycelial. Some species have the capacity to grow yeast-like only when subjected to an adequate stimulus. According to this concept, dimorphic Mucorales grow yeast-like only as long as morphogenetic effectors are present. When their stress is eliminated, the fungus goes back to its normal mode of growth.

The second concept refers to the nature of the morphogenetic effectors. Revision of the literature discloses that they are extremely varied, and with different modes of action. Nevertheless, I would like to focus on the one effector that appears to be the most important one: microaerobic conditions (I prefer to use the term microaerobic since *Mucor* species are unable to grow under strict anaerobiosis). According to their response to this effector, I have divided dimorphic Mucorales into two groups:[1] (1) species which respond to microaerobiosis by growing yeast-like; and (2) species where microaerobic environment induces yeast-like growth only when enriched with high CO_2 or glucose concentrations. It may be anticipated that if we can understand the mode of action of this single morphogenetic effector, microaerobiosis, we may construct a morphogenetic scheme where the rest of the effectors can be engaged. In my opinion, many of the described effectors act as general metabolic inhibitors which stress the cell, whereas others affect not the initial morphogenetic events, but cell response.

The final general concept that I wish to recall is that the morphogenetic transition can be regarded as a series of discrete events. For the sake of simplicity (but accepting in advance the existence of complex ancillary metabolic alterations), the following can be recognized: (1) perception of the external stimuli by specific cellular sensors; (2) transduction of the signal received into a biochemical message; (3) alteration in the genomic expression of the cell; and (4) reorganization of the elements responsible for polarization of the cell growth. Our knowledge of these steps is not uniform; we have more information on some of them than others. In the following pages I will proceed to analyze some with more detail than others, trying to offer a perspective of the whole process. Special emphasis is made on the event of stimuli perception, and a unifying hypothesis is proposed.

MICROAEROBIOSIS AS A MORPHOGENETIC EFFECTOR

The nature of the real manifestation of microaerobiosis as an effector and therefore the nature of the receptor, have remained obscure because the information gathered has been contradictory. Strong pieces of evidence suggest that mitochondria are involved in the sensing reaction. Among these I may cite that all yeast monomorphic mutants isolated from *Mucor bacilliformis* exhibit serious alteration in the respiratory chain.[2,3] A detailed analysis of a number of these monomorphic mutants revealed that they were pleiotropic and displayed variable cytochrome defects.[3] In some species it has been demonstrated that addition of respiratory inhibitors[4,5] or chloramphenicol[6] induces yeast-like growth. Accordingly, a link between respiratory activity and morphogenesis has been entertained. Nevertheless, contradictory evidence exists. Firstly, respiratory activity of monomorphic mutants of *M. bacilliformis* may be as high as in the wild type. Secondly, a respiratory-competent mutant from *Mucor racemosus* which grows yeast-like when incubated with glucose as carbon source has been isolated.[7] Thirdly as indicated above, there are species of *Mucor* that grow as mycelium under microaerobiosis.

Recently we have made a reappraisal of the problem and obtained further evidence that respiration is not directly linked to morphogenesis in *Mucor*. We observed that different inhibitors of respiration delayed germination of *M. rouxii* spores (Table 1); but once these overcame a long lag phase, they produced germ tubes (Fig. 1). These germinated spores had high respiratory activity, but their

Table 1. Effect of mitochondrial inhibitors on spore germination of *Mucor rouxii*.

Inhibitor	Time at which 50% germ tube formation occurred (h)
None 4	
Oligomycin (0.4 mg/ml)	7-9
Rotenone (4μM)	8-9
Antimycin (1μM)	7-8
KCN (1mM)	7-8
CO (stream)	7-8

* Respiration of inhibitor-treated cells, in contrast to control, was sensitive to SHAM

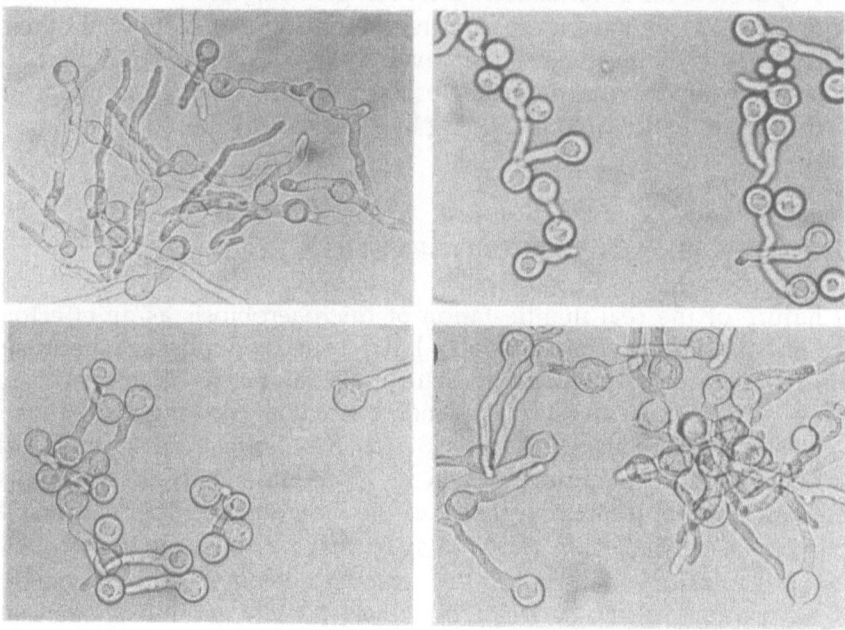

Fig. 1 Germination of *Mucor rouxii* spores in the presence of respiration inhibitors Upper left, control cells (4 5 h), upper right, cells incubated with 4 μM rotenone (9 h); lower left, cells incubated with 1 μM antimycin A (9 h), lower right, cells incubated with 1 mM KCN (9 h)

respiration in the presence of the inhibitors proceeded through a bypass that did not utilize cytochrome oxidase, but a SHAM-sensitive component. This bypass was not energy-linked, and the cells grew at the expense of a fermentative metabolism. This same bypass is utilized by the monomorphic mutants of *M. bacilliformis* described above [3] Further evidence against respiration state as the morphogenetic inducer was provided by the isolation of a mutant of *M. rouxii* with low levels of cytochrome oxidase, which nevertheless grows as normal mycelium (Table 2).

Table 2. General characteristics of a mycelial respiratory mutant of *Mucor rouxii*.

Parameter	Observation
Colonial morphology	Mycelial
Growth in non-fermentable substrates	Negative
Reduction of TTC* in plates	Negative
Growth rate in aerobiosis	Decreased
Growth rate in microaerobiosis	Normal
Respiratory properties	Sensitive to SHAM
Levels of b and c cytochromes	Normal
Level of aa_3 cytochromes	<30% of wt**
K_i for cyanide	Increased
K_m for cytochrome c	Decreased
V max for cytochrome c	Decreased

* triphenyl tetrazolium chloride, ** wt wild type

The above described results regarding the induction of the SHAM-sensitive bypass, rule out also the possibility that energy coupling is the role played by mitochondria in mycelial growth. Moreover, energy coupling as well as transmembrane potential may also be ruled out in the light of results obtained with oligomycin. This compound dissipated energy conservation in the mitochondria of *M. rouxii* spores and delayed germination, but not the final formation of germ tubes. According to these accumulated data, it may be proposed that the real morphogenetic effector of microaerobiosis is the change in the redox potential of the cell. This change would be sensed not by members of the respiratory chain, but probably by a mitochondrial component associated with them. According to this hypothesis, microaerobic conditions or a block in the respiratory chain - either by addition of an inhibitor or by a mutation - would be equivalent maintaining the sensor in a reduced state. What would be then the difference between members of dimorphic classes 1 and 2? The most likely possibility is the redox potential of the sensor which would be so low for class 2 members that even the above conditions would not keep it in a reduced state. In this sense it is interesting to recall that even the minute amounts of oxygen permeating through Teflon tubing in N_2-flushed cultures of *M. racemosus* are enough to permit their mycelial growth.[8] At the other extreme, shaken cultures of *M. bacilliformis* grow partially yeast-like. Only when flushed with a vigorous air stream is a homogeneous mycelial population obtained. Accordingly, class 2 members require an additional metabolic alteration brought about by CO_2 or high glucose concentrations. It is highly probable that the latter provokes catabolite repression. We have observed that high glucose concentrations inhibit several mitochondrial activities.[9] This hypothesis is strengthed by the behavior of the mutant of *M. racemosus* described above which grows yeast-like when supplemented with glucose.[7] Additionally, we have observed that the phenotype of LEV mutants of *M. bacilliformis* is affected in solid media by glucose concentrations: the higher the sugar content, the more homogeneous the yeast growth. To these results I may add the effect of cAMP on morphogenesis. It has been observed that exogenously added cAMP inhibits the yeast-to-mycelium transition in *M. racemosus* induced by a decrease in the flow of N_2 flushed through liquid cultures of the fungus; this reduction in gas flow permitted the permeation of small amounts of air into the culture through the Teflon tubing.[8] Counteraction of the morphological effect of these low oxygen tensions by cAMP, can be explained on the grounds that in the Mucorales (and probably in other fungal systems) high cAMP concentrations are associated with a state of catabolite repression.[10]

The mode of action of CO_2 has remained elusive, but probably it is different from that of glucose. Induction of a change in intracellular pH is not a remote possibility. We have observed that variations in external pH when *M. rouxii* is supplemented with specific amino acids, may alter dramatically its morphogenetic response.[11] Alternatively, a role at a different step in the morphogenetic pathway is a likely possibility. We have observed that CO_2 inhibits *M. rouxii* ornithine decarboxylase (ODC), a key enzyme in the biosynthesis of polyamines. As described below, polyamines play an important role in the differentiation of Mucorales.

The need to invoke a redox sensor involved in the morphogenetic pathway (and not merely the change in energy charge in the form of ATP/ADP ratio or transmembrane potential or the ratio between NAD and NADH) arises because all these parameters have a metabolic signification, i.e. they respond also to the switching between respiration and fermentation. Regarding the interaction of the hypothetical sensor with the respiratory chain, the behavior of LEV mutants of *M. bacilliformis* which display high respiratory activities through the SHAM-sensitive pathway and grow yeast-like[3] is consistent with the concept that the redox sensor is distinct, but linked to cytochromes.

SIGNAL TRANSDUCTION

Almost nothing is known about the initial reactions involved in the transduction of the morphogenetic signal. A role for the operation of transducing pathways known to be operative in signalling including *ras* proteins, inositides or protein modification (phosphorylation, etc.), has not been explored in detail. The

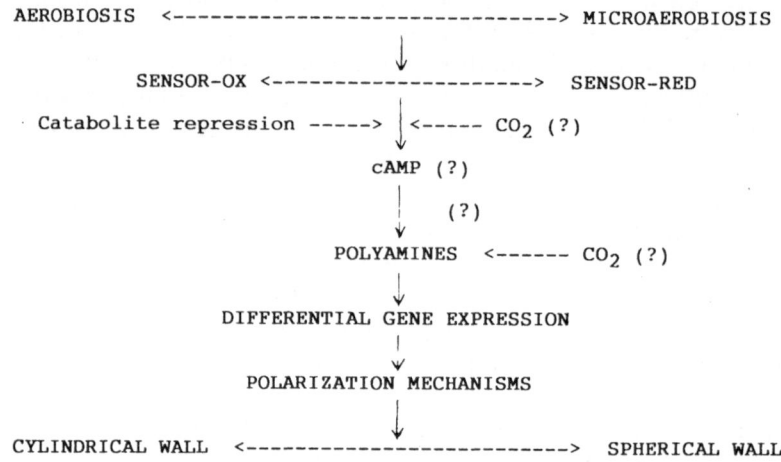

Fig. 2 Schematic representation of the morphogenetic scheme of *Mucor rouxii* in response to a change in O_2 tension. Although the scheme is represented as a unidirectional cascade, it should be anticipated that it involves additional complex metabolic alterations.

possibility that cAMP levels play a role in morphogenesis at this level is not remote. High cAMP levels have been found associated with microaerobic growth.[10] These levels drop before *M. racemosus* is transferred to aerobic conditions. Moreover, as indicated above, addition of cAMP prevents the corresponding yeast-to-hypha transition.[10] Incubation of spores of this fungus in the presence of high concentrations of the nucleotide prevented the outgrowth of buds. Instead the cells continued growing spherically and attained large volumes.[12,13]

The possibility that polyamines are involved at a further stage in signal transduction has received experimental support recently. Activity levels of the most highly regulated enzyme of the biosynthetic pathway: ODC, and concentrations of the two most abundant polyamines; putrescine and spermidine, all increase immediately after the transfer of microaerobic yeasts to aerobiosis, both in *M. racemosus*[14] and *M. rouxii*.[15] The evidence that this transitory increase is associated with morphogenesis and not with changes in the metabolic conditions of *M. rouxii* was provided by the observation that similar changes occurred when CO_2 was removed from the N_2 atmosphere,[16] a condition that induces the mycelial transition of this fungus. But probably the strongest piece of evidence comes from the observation that all monomorphic mutants of *M. bacilliformis* displayed extremely low levels of ODC.[3] Diaminobutanone (DAB), a competitive inhibitor of ODC, was able to block the mycelial transition of *M. rouxii* either under $N_2/CO_2 \rightarrow$ air, or $N_2/CO_2 \rightarrow N_2$ shifts. The drug also inhibited any increase in the level of polyamines pools during these shifts.[16] It is interesting to note that the inhibitor is also efficient at preventing defined steps in the cell cycle of Mucorales i.e., spore germination[17] and phorogenesis.[16] The results suggests a common mechanism in triggering all these processes, and reinforce the idea that hte dimorphic transition is a differentiative phenomenon useful as a general model.

The specific mode of action of polyamines in cell differentiation remains unknown. We have obtained evidence that suggests a relationship between polyamines and DNA methylation state, and therefore with differential gene expression.[18,19] Our observations may be summarized as follows: it is known that spores germination in *Mucor* occurs in three defined steps, the last of which (II) involves formation of germ tubes by swollen spores, i.e. a transition from an isodiametric to a polarized mode of growth. This last step requires DNA synthesis and is blocked specifically by DAB. Requirement for DNA synthesis was pinpointed to the necessity to synthesize hypomethylated chains. DAB did not inhibit DNA synthesis, but prevented its hypomethylation.[18] Using the *CUP* family gene from *M. rouxii* and *M. racemosus* as an appropriate indicator, it was demonstrated that *CUP* genes were all located within a hypermethylated region existing in the fungal genome of the spores, and they were not transcribed during the period of spherical growth of the spores. At the onset of the period of polarized growth the gene(s) became demethylated and transcribable. DAB inhibited both gene demethylation and transcription.[19] According it is not too risky to propose that polyamines play a similar role during the dimorphic transition of Mucorales, serving as the final step in the sensory transduction pathway. They would serve as the link between the morphogenetic signal and the initiation of the cell response. In this sense it is relevant that DAB inhibits the dimorphic transition of *Candida albicans*,[20] and that the DNA from the yeast form of the fungus contains higher levels of 5-methylcytosine than the DNA of the mycelial form.[21]

MORPHOGENETIC RESPONSE

It is almost certain that the genetic information expressed *de novo* by the cell as a response to the morphogenetic stimuli is involved in a change in its growth pattern. In a classical experiment Bartnicki-Garcia and Lippman[22] demonstrated beyond any reasonable doubt that whereas the yeast form of *Mucor rouxii* grows isodiametrically, the hyphal form grows at the apex, in the same way as with all mycelial fungi. Nevertheless, it is important to note that chemical and structural changes in the cell wall occur during the transition process, and that they may be important morphogenetic factors to shape the spherical or tubular cell walls. Use of the ODC inhibitor, DAB, described above has permitted demonstration that at least in two cases (one in *Mucor*[23] and the other in *Candida*[20]) the corresponding changes in chemical composition of the cell wall of the spherical and hyphal forms is related specifically to the differentiation processes, and not to changes in metabolic or environmental conditions.

A review of the mechanisms suggested to be responsible for cell shape is beyond the aims of this article. Nevertheless a few general ideas may be relevant at this point. Since morphology of fungal cells is due to the presence of a rigid wall, the great efforts made to understand the mechanisms underlying the formation of this structure are not surprising. Growth of the cell wall depends on the activity of the vesicular apparatus which carries biosynthetic enzymes, and the components that make it up. Coordinated synthesis and aggregation of the different components in the cytoplasm, at the membrane interphase and within the wall itself, lead to the organization of this structure. Cell shape therefore depends on the polarization mechanisms that displace cytoplasmic vesicles to specific areas of the cell surface. Among the vesicles two different populations have been recognized: macrovesicles (or plainly, vesicles) and microvesicles, among which the best characterized are chitosomes, responsible for the transport of chitin synthetase.[24] Spherical growth would involve a random mobilization of the vesicles to the whole cell surface, whereas cylindrical growth involves guidance mechanisms of the vesicles population to the hyphal apex.

Cytoskeletal components appear as the most logical candidates for guidance elements, although it remains unknown whether microtubules or microfilaments play the more important role in the phenomenon. At least in one different dimorphic system it has been shown that actin adopts a different organization in the yeast and hyphal forms: in *C. albicans* large buds, actin occurred in clusters distributed more or less homogeneously over the surface, whereas in mycelial cells actin clusters appeared more pronounced at the apex, with fibers distributed parallel to the growth axis.[25] Unfortunately, use of anti-actin or anti-tubulin drugs has produced contradictory information in different systems. The existence of electrical or ionic gradients across hyphal cells has been taken as suggestive that they may play some role in growth polarization, although the idea that they merely accompany hyphal growth has been favored more recently.[26] Without intention to enter into arguments, it must be indicated that nevertheless, evidence exists in different apically-growing systems that ion gradients play a role in growth polarization either by themselves or by stabilizing or affecting the cytoskeleton. In the related species *Phycomyces blakesleeanus*, we have demonstrated that dissipation of the calcium gradients by the calcium ionophore A23187 abruptly stopped hyphal growth and chitin biosynthesis. The effect of the ionophore was pinpointed to a drastic reduction in the number of apical vesicles.[27]

Independently of the nature of the real guidance system involved in cytoplasmic vesicles mobilization, this would be one of the main targets of the morphogenetic reaction of the cell. As a result, a spherical or a cylindrical cell wall would be synthesized in response to changes in the environmental conditions. As stated earlier, it may be anticipated that this adaptive mechanism plays an important ecological role for the best fitting of these organisms to their particular niches in nature.

ACKNOWLEDGEMENTS

Experimental work from the author's laboratory was partially supported by the Consejo Nacional de Ciencia y Tecnología, México. Thanks are given to Mrs. Carmen Medrano for careful typing of the manuscript.

REFERENCES

1. J. Ruiz-Herrera, Dimorphism in *Mucor* species with emphasis on *Mucor rouxii* and *Mucor bacilliformis*, in: "Fungal Dimorphism", P.J. Szaniszlo, ed., Plenum Press, New York (1985).
2. R. Storck and R.C. Morrill, Respiratory-deficient, yeastlike mutants of *Mucor, Biochem. Genet.* 5: 467 (1971).

3. J. Ruiz-Herrera, A. Ruiz and E. Lopez-Romero, Isolation and biochemical analysis of *Mucor bacilliformis* monomorphic mutants, *J. Bacteriol.* 156: 264 (1983).

4. B.E. Schulz, G. Kraepelin and W. Hinkelmann, Factors affecting dimorphism in *Mycotypha* (Mucorales): a correlation with the fermentation/respiration equilibrium, *J. Gen. Microbiol.* 82: 1 (1974).

5. M. Friedenthal, A. Epstein and S. Passeron, Effect of potassium cyanide, glucose and anaerobiosis on morphogenesis of *Mucor rouxii*, *J. Gen. Microbiol.* 85: 15 (1974).

6. J. Zorzopulos, A.J. Jobaggy and H.F. Terenzi, Effects of ethylenediamine-tetraacetate and chloramphenicol on mitochondrial activity and morphogenesis in *Mucor rouxii*, *J. Bacteriol.* 115: 1198 (1973).

7. P.T. Borgia, N.K. Gokul and G.J. Phillips, 1985, Respiratory-competent conditional developmental mutant of *Mucor racemosus*, *J. Bacteriol.* 149: 115 (1982).

8. G.J. Phillips and P.T. Borgia, Effect of oxygen on morphogenesis and polypepetide expression by *Mucor racemosus*, *J. Bacteriol.* 164: 1039 (1985).

9. C. Cano-Canchola, E. Escamilla and J. Ruiz-Herrera, Environmental control of the respiratory system in the dimorphic fugus *Mucor rouxii*, *J. Gen. Microbiol.* 134: 2993 (1988).

10. A.D. Larsen and P.S. Sypherd, Cyclic adenosine 3′, 5′-monophosphate and morphogenesis in *Mucor racemosus*, *J. Bacteriol.* 117: 432 (1974).

11. A. Leija, J. Ruiz-Herrera and J. Mora, Effect of *L*-amino acids on *Mucor rouxii* dimorphism, *J. Bacteriol.* 168: 843 (1986).

12. M. Orlowski, Changing pattern of cyclic AMP-binding proteins during hyphal germ tube emergence from sporangiospores of *Mucor*, *Biochem. J.* 182: 547 (1979).

13. K.F. Wertman and J.L. Paznokas, Effects of cyclic nucleotides upon the germination of *Mucor racemosus* sporangiospores, *Exp. Mycol.* 5: 314 (1981).

14. C.B. Inderlied, R.L. Cihlar and P.S. Sypherd, Regulation of ornithine decarboxylase during morphogenesis of *Mucor racemosus*, *J. Bacteriol.* 141: 699 (1980).

15. C. Calvo-Mendez, M. Martinez-Pacheco and J. Ruiz-Herrera, Regulation of ornithine decarboxylase activity in *Mucor bacilliformis* and *Mucor rouxii*, *Exp. Mycol.* 11: 128 (1987).

16. M. Martinez-Pacheco, G. Rodriguez, G. Reyna, C. Calvo-Mendez and J. Ruiz-Herrera, Inhibition of the yeast-mycelial transition and the phorogenesis of Mucorales by diamino butanone, *Arch. Microbiol.* 151: 10 (1989).

17. J. Ruiz-Herrera and C. Calvo-Mendez, Effect of ornithine decarboxylase inhibitors on the germination of sporangiospores of Mucorales, *Exp. Mycol.* 11: 287 (1987).

18. C. Cano, L. Herrera-Estrella and J. Ruiz-Herrera, DNA methylation and polyamines in regulation of development of the fungus *Mucor rouxii*, *J. Bacteriol.* 170: 5946 (1988).

19. C. Cano-Canchola, L. Sosa, W. Fonzi, P. Sypherd and J. Ruiz-Herrera, Developmental regulation of *CUP* gene expression through DNA methylation in *Mucor* spp., *J. Bacteriol.* 174: 362 (1992).

20. J.P. Martinez, J.L. Lopez-Ribot, M.L. Gil, R. Sentandreu and J. Ruiz-Herrera, Inhibition of the dimorphic transition of *Candida albicans* by the ornithine decarboxylase inhibitor 1,4-diamino butanone: alterations in the glycoprotein composition of the cell wall, *J. Gen. Microbiol.* 136: 1937 (1990).

21. P.J. Russell, J.A. Welsch, E.M. Raschlin and J.A. McCloskey, Different levels of DNA methylation in yeast and mycelial forms of *Candida albicans*, *J. Bacteriol.* 169: 439 (1987).

22. S. Bartnicki-Garcia and E. Lippman, Fungal morphogenesis: cell wall construction in *Mucor rouxii*, *Science* 165: 302 (1969).

23. A. Obregon, S. Monzalvo, C. Calvo-Mendez and J. Ruiz-Herrera, Ultrastructural and chemical alteration in germinating spores of *Mucor rouxii* (Zygomycetes), induced by two compounds which inhibit their developmental pattern, *Crypt. Bot.* 1: 323 (1990).

24. C.E. Bracker, J. Ruiz-Herrera and S. Bartnicki-Garcia, Structure and transformation of chitin synthetase particles (chitosomes) during microfibril synthesis *in vitro*, *Proc. Natl. Acad. Sci. USA* 73: 4570 (1976).

25. J.M. Anderson and D.R. Soll, Differences in actin localization during bud and hyphae formation in the yeast *Candida albicans*, *J. Gen. Microbiol.* 132: 2035 (1986).

26. F.M. Harold and J.H. Caldwell, Tips and currents: electrobiology of apical growth, in: "Tip growth in plant and fungal cells", I.B. Heath ed., Academic Press, New York (1990).

27. J. Ruiz-Herrera, C. Valenzuela, G. Martinez-Cadena and A. Obregon, Alterations in the vesicular pattern and wall growth of *Phycomyces* induced by the calcium ionophore A23187, *Protoplasma* 148: 15 (1989).

MORPHOLOGICAL VARIATION IN *MALASSEZIA* AND ITS SIGNIFICANCE IN PITYRIASIS VERSICOLOR

Gillian Midgley

Department of Medical Mycology
St John's Institute of Dermatology
St Thomas's Hospital
London, UK

ABSTRACT

Morphological variation among yeasts of the genus *Malassezia* has been recognized since their earliest descriptions. As the forms with oval and spherical cells are easily distinguished, it is proposed to designate them as two separate species, *M.f urfur* and *M. ovalis*, even though both forms can be associated with the condition, pityriasis versicolor, and both may exhibit filaments in skin and in culture. Isolates of these two yeasts have other distinguishing features such as their viability and the nature of their lipid requirements. *M. ovalis* can be divided further into three morphological variants, isolates of which have remained stable for numerous transfers. These, along with *M. furfur*, can all be recognized by their immunological reactions to hyperimmune rabbit antisera. It remains to be seen whether the divisions suggested here for the genus *Malassezia* become accepted, particularly as more information is gathered from molecular studies.

INTRODUCTION

The genus *Malassezia* includes organisms which exhibit polymorphism in that both yeasts and filaments are produced. This morphology has been documented since the first known description of these yeasts, which was in 1846 when Eichstedt[1] realised the fungal nature of the disease, pityriasis versicolor. He described the appearance of spherical yeasts and filaments present in the infected scales, representing the organism which was eventually named *M. furfur*.[2] Some years later in 1873, similar yeasts, without associated filaments, were observed by Rivolta[3] in scales from psoriasis, and contemporary reports revealed that they could in fact be present in various dermatological conditions, such as seborrheic dermatitis, and also in apparently healthy skin. A review of these early descriptions[4] reveals that not only spherical but oval yeasts were recognized and that some workers considered them to represent two separate yeasts while others incorporated the different forms into a single polymorphic species. Notable among the latter are Unna[5] who did not recognize these organisms as yeasts and referred to them as "Flaschen Bacilli" and also Sabouraud[6] who in 1904 actually created a new yeast genus, *Pityrosporum*, with his description of *P. malassezi*. Sabouraud's *Pityrosporum* later became known as *P. ovale*[7] but by the time of the publication in 1952 of the monograph "The Yeasts"[8] only oval forms were included in the description and a contemporary publication by Gordon[9] designated the spherical yeasts as a separate species, *P. orbiculare*. Investigations were now naturally including the study of laboratory cultures as well as comparing the appearances of the yeasts in skin.

A similarity between *Malassezia* and *Pityrosporum* was recognized by Sabouraud but the lack of filaments associated with his *Pityrosporum* prevented him from including them in the same genus. Although a common identity has been accepted now for several years, both names have persisted. Cultural isolates, and the yeasts seen *in vivo* without filaments have tended to be referred to as *Pityrosporum*, while *M. furfur* has been reserved for the microscopic appearance of the organism seen in pityriasis versicolor. Nevertheless, *Malassezia* and *Pityrosporum* are clearly synonymous and there appears to be no justification in maintaining both genera. As *Malassezia* has precedence, Baillon[2] having created the genus in 1889, taxonomists have accepted this as the correct name.[10] It is clear from the literature however, that this has not found universal acceptance, nevertheless *Malassezia* will be used throughout this report.

MORPHOLOGY OF *MALASSEZIA*

There are three species of Malassezia currently accepted by taxonomists; *M. furfur*, *M. sympodialis* and *M. pachydermatis*. *M. sympodialis* is described in a report of two isolates;[11] it is distinguished by showing sympodial budding but as it has not been reported elsewhere it has been excluded from the discussion here. I propose to add *M. ovalis*; this is to accommodate the lipid-dependent yeasts which have oval cells (i.e. those which were previously named *P. ovale*). This species is also recognized by Borelli,[12] but it was in fact created as long ago as 1927 by Acton and Panja.[13] *M. furfur* therefore represents only the yeasts with spherical cells and is a synonym of *P. orbiculare*. *M. pachydermatis* has always been considered as a distinct species due to its lack of dependence on lipids for growth. It is a commensal on the skin of animals and only an occasional transient inhabitant of human skin.

In all the forms of *Malassezia* illustrated here, cell multiplication takes place by monopolar budding, where successive blastoconidia are formed at the same point on the mother cell. This leads to the formation of a thickened area on the cell wall, a distinct bud scar which is a very characteristic feature of these yeasts.

Appearances in Skin

The illustrations below show yeasts that have been stained by treating the skin with Parker's stain (equal volumes of Parker's Blue/Black permanent Quink and 30% potassium hydroxide).

M. furfur. The familiar appearance in pityriasis versicolor scales shows spherical yeasts of up to 8 μm in diameter, frequently in clusters and associated with filaments (Fig. 1). The young blastoconidia are formed on a narrow base at the point of attachment to the mother cell. The hyphae consist of cells, 10-25 μm in length and 2-5 μm wide, often separated, or else aligned end to end, so forming septate angular elements of varying sizes. These may also show branching.

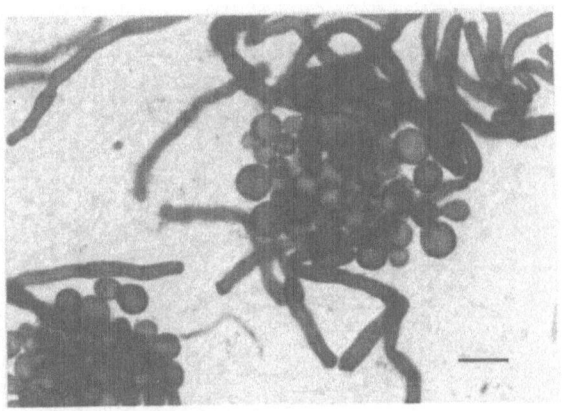

Fig. 1 *M. furfur* in material from pityriasis versicolor of the upper trunk. Bar = 10 μm.

M. ovalis. Ovoid yeasts associated with filaments (Fig. 2) may also be encountered in material from pityriasis versicolor.[12,14,15] It is possible that this form of *Malassezia* may occur more frequently in tropical areas. The yeasts measure from 4-6 μm long and 1.5-3 μm wide with the buds being formed on a broad base. The hyphae are slightly thinner than those seen in *M. furfur* and often form longer lengths with some filaments reaching 50 μm.

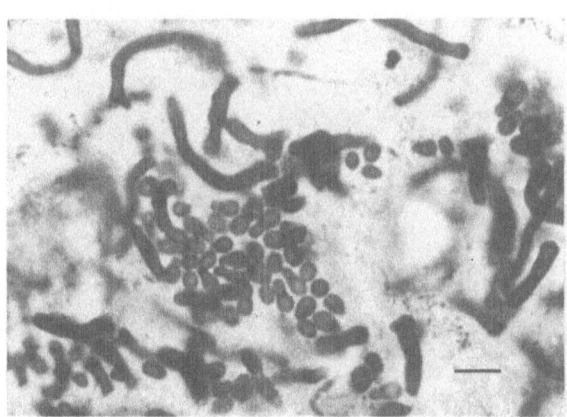

Fig. 2. *M. ovalis* in material from pityriasis versicolor of the thigh. Bar = 10 μm

In other conditions such as seborrheic dermatitis, or in normal skin, yeasts are present which vary considerably in their cell shape. These include spherical yeasts of similar dimensions to those typically seen in the classical form of pityriasis versicolor (Fig. 3). Easily distinguished from these are ovoid yeasts 1.5-6 μm long. Other forms are small cylindrical cells 1.5-3 μm in length and a further variant showing large cylindrical cells reaching 6 μm (Fig. 4). A sample of skin may reveal yeasts of a uniform shape, or several forms may be present at the same site.

Appearances in Culture

The different forms of *Malassezia* described above can also be recognized in cultural isolates. Not only can *M. furfur* be distinguished from *M. ovalis* but the latter can be divided further into three distinct variants. All of these have been documented previously and are shown here again as they appear when isolated on a modified Dixon agar incubated at 32 °C for 10 days.[15]

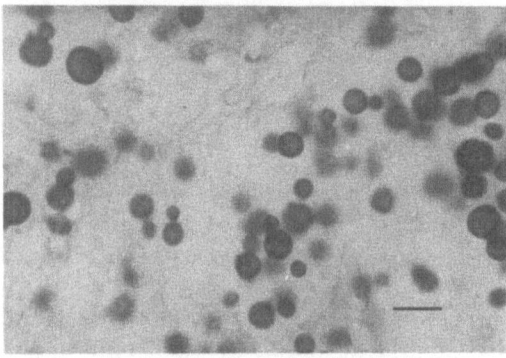

Fig. 3 Spherical yeasts of *M. furfur* in skin from the chest of a patient with seborrheic dermatitis. Bar = 10 μm.

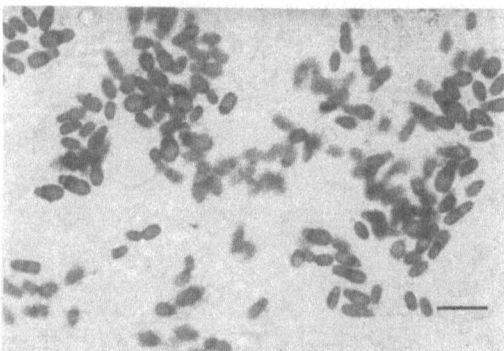

Fig. 4 Large cylindrical yeasts of *M. ovalis* in skin from a scaly lesion in the groin. Bar = 10 μm.

M. furfur. The colonies have a convex elevation with a rough surface and deep furrows (Fig. 5a). The texture is brittle and the cells are particularly difficult to emulsify when a suspension is prepared. The spherical yeasts can reach up to 8 μm in diameter with buds forming on a narrow base (Fig. 5b). Short filaments can be formed by the elongation of a bud although the mother cell may also elongate to form an ovoid cell (Fig. 5c). However, the base of the bud remains narrow and the cell morphology remains distinguishable from *M. ovalis*. The filaments exhibit branching and this often occurs near the point of origin from the mother cell. Careful examination of preparations for the light microscope will reveal crosshatching of the cell wall (Fig. 5c). This was observed as long ago as 1899 by

Matakieff[16] and it represents the distinct ridges on the inner surface of a multi-layered cell wall clearly demonstrated many years later by studies on the ultrastructure of both yeasts and filaments of the *Malassezia* genus.[17-19]

M. ovalis. Three variants have been placed within this species determined by the shape of the cell, which is ovoid or elongated, in contrast to the spherical cells of *M. furfur*.

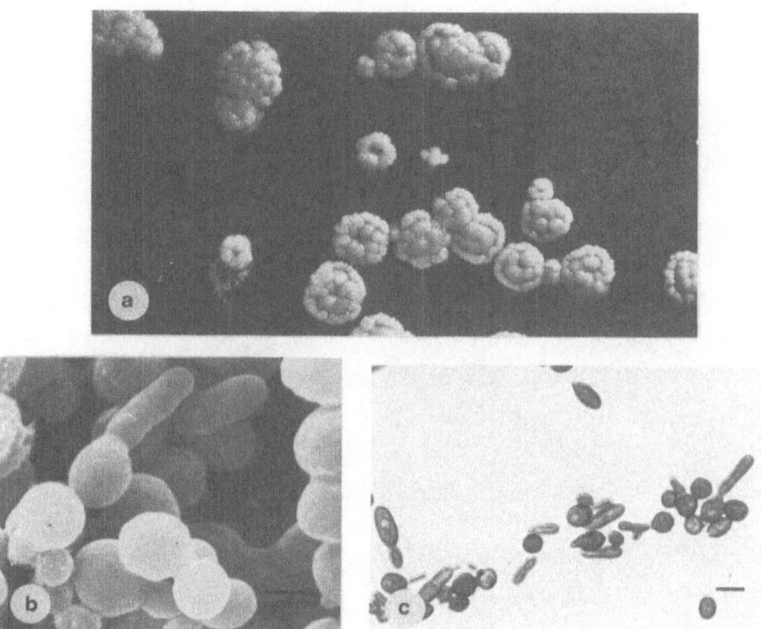

Fig. 5 The morphology of *M. furfur*: a) Rough, deeply folded colonies on Dixon agar. b) Scanning EM of spherical yeasts showing a narrow bud base. Bar = 5 μm. c) PAS-stained yeasts showing short filaments with initial branching and some elongated cells. Bar = 10 μm.

M. ovalis Form 1 has colonies with a slight convex elevation, rough surface and fine radial grooves (Fig. 6a). The texture is brittle and the cells again difficult to emulsify. The yeasts are small, 1.5-3.5 μm in length. They have an almost cylindrical shape with the buds formed on a broad base in relation to the cell width, so giving the characteristic "bottle" appearance (Fig. 6b). The yeasts are often seen in pairs and no filaments have been observed.

Fig. 6 The morphology of *M. ovalis* Form 1. a) Finely folded colonies on Dixon agar. b) Scanning EM of small cylindrical yeasts with buds forming on a broad base. Bar = 5 μm.

M. ovalis Form 2 has smooth dull colonies, often flat but may have a slight convex elevation (Fig. 7a). The texture is sticky rather than brittle. The yeasts are cylindrical in shape but much larger than Form 1. The mother cell reaches up to 6 μm and the bud 4 μm so that the whole cell can approach a length of 10 μm with an average size of 6 μm (Fig. 7b). Filaments may be formed either as an elongation of the bud or by the formation of a germination tube at another point on the cell surface (Fig. 7d). This latter feature has not been seen in cultures of *M. furfur*. The filaments can be fairly long, the maximum length observed being 65 μm and branching is also present (Fig. 7c).

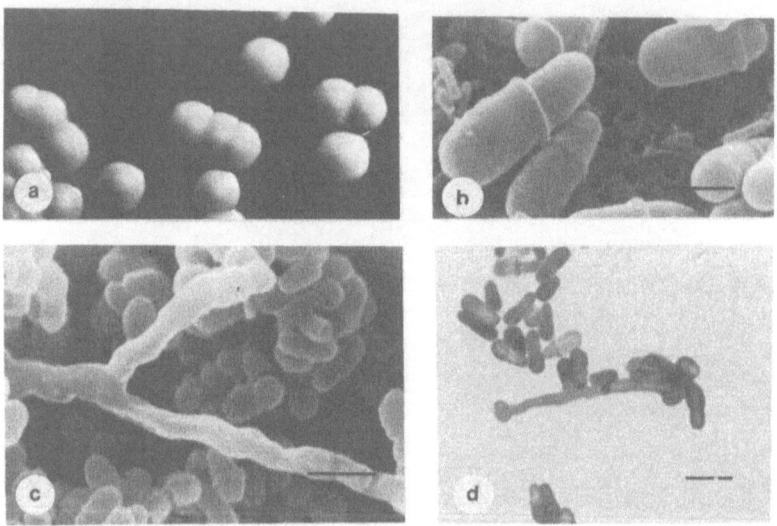

Fig. 7 The morphology of *M. ovalis* Form 2. a) Smooth colonies on Dixon agar. b) Scanning EM showing large cylindrical yeasts with buds forming on a broad base. Bar = 5 μm. c) Scanning EM showing a branching filament. Bar = 10 μm. d) Yeasts stained by PAS showing a filament arising from the lateral aspect of a cylindrical cell. Bar = 8 μm.

M. ovalis Form 3 has smooth, shiny colonies which are flat or with a slight elevation (Fig. 8a). They have a soft texture so that cells are easier to mount in a suspension. The cells are ovoid in shape, 2.5-6 μm long. Although the base of the bud is narrower than the maximum width of the mother cell, it is equal in width to the diameter of the bud and can still be considered as broad (Fig. 8b). Filaments are not a feature of this variant.

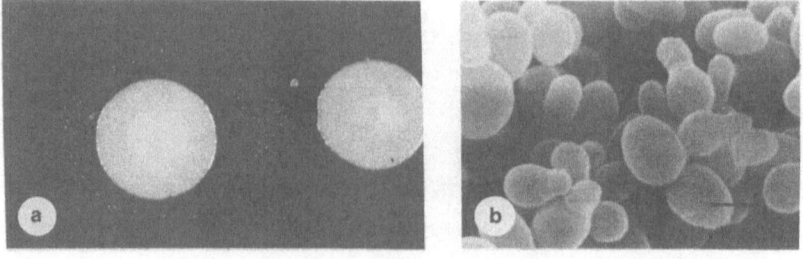

Fig. 8 The morphology of *M. ovalis* Form 3. a) Smooth flat colonies on Dixon agar. b) Scanning EM showing ovoid yeasts. The bud base, narrower than the maximum width of the mother cell but equal in width to the bud. Bar = 5 μm.

PHYSIOLOGY

Multiple purified isolates of *M. furfur* and the three variants of *M. ovalis* have been maintained for many transfers and have remained stable in their morphology. The reported differences in their physiological properties help to confirm these divisions.[15]

These include a slower rate of growth and a more rapid loss of viability for *M furfur* when compared with *M ovalis* isolates. To these may be added a more exacting requirement for lipids by *M. furfur*. The Dixon formula contains ox bile, oleic acid and Tween 40 to provide the necessary lipid for the development of *Malassezia* cultures and satisfactory growth develops with all the variants However, a simple medium, such as 2% glucose/1% peptone/1 2% agar with a supplement of 0.5% Tween 80, will support good growth of *M ovalis* but little or no growth of *M. furfur* (Fig 9)

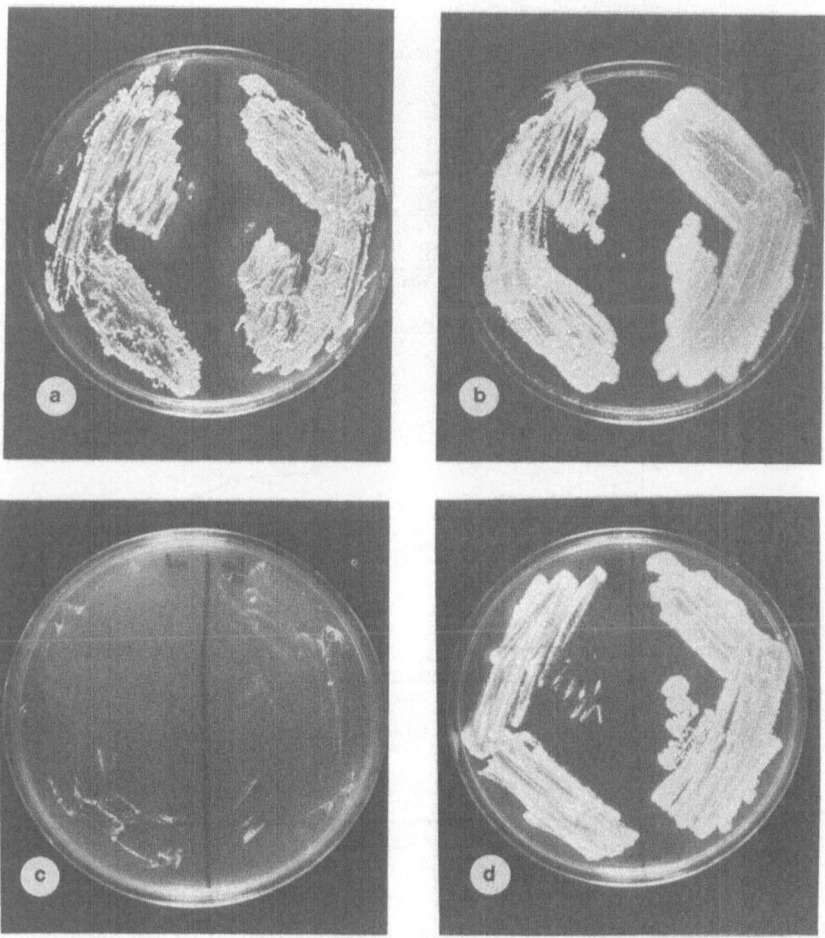

Fig. 9 Comparison of growth of *M furfur* and *M ovalis* on Dixon agar and on glucose/peptone/Tween agar (GPT) a) *M furfur* and b) *M ovalis* Form 1 on Dixon agar c) *M furfur* and d) *M ovalis* Form 1 on GPT agar Incubation for 7 days at 32 °C

Immunological studies have given conflicting reports. Using hyperimmune rabbit antisera and the sensitive technique of crossed immunoelectrophoresis, Bruneau and Guinet[20] showed that their isolates of P(M).orbiculare and P(M).ovale had the same complement of antigens. In our studies at the St.John's Institute of Dermatology, specific antibodies to two variants could be demonstrated by immunodiffusion after absorption experiments comparing antisera raised to M. furfur and M. ovalis Form 3. This has been confirmed for all of the variants by the more sensitive method of enzyme linked immunosorbent assay where whole cells were used as antigens. In one of a series of experiments, two isolates each of M. furfur and M. ovalis Form 1 were allowed to react with an antiserum to M. furfur. The serum was used before and after absorption with whole cells of three cultures, M. furfur, M. ovalis Form 1 and M. pachydermatis until there was no further reduction in titer (Fig. 10). In the system where the homologous cultures were used as antigens, only absorption with M. furfur removed all the antibodies. Some specific reactivity to M. furfur remained after absorption with the other cultures. With antigens of the heterologous M. ovalis, however, there was still a strong response to the unabsorbed serum, but absorption with any of the three cultures removed the antibody reaction. Similar findings have been reported by Cunningham et al[21] who were able to distinguish serovars of Malassezia by the more subjective technique of immunofluorescence.

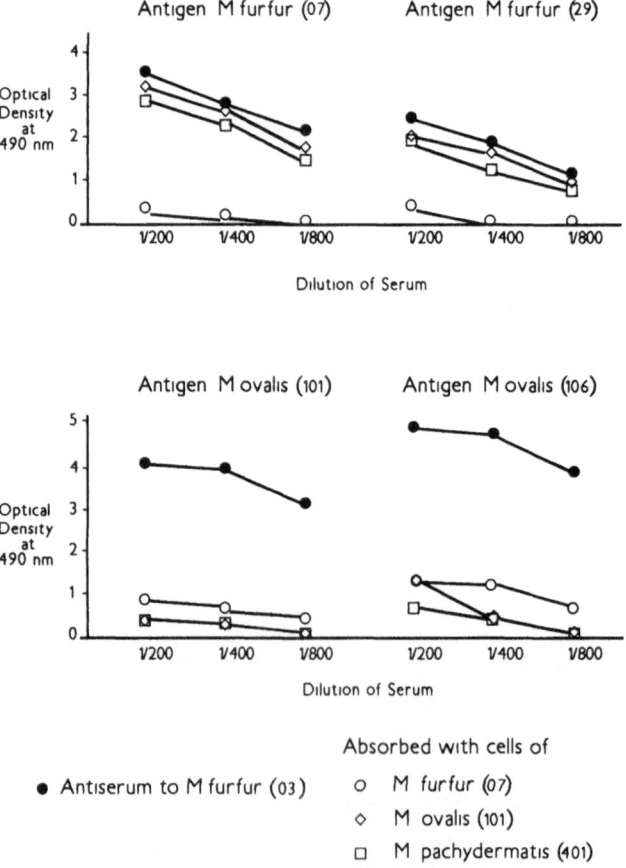

Fig. 10 ELISA antibody response in hyperimmune rabbit antiserum raised to a single isolate of M furfur using whole cells of two isolates each of M. furfur and M. ovalis .
Form 1 as antigens.

To return to the clinical relevance of the different *Malassezia* variants. All those described above can be both visualized in and isolated from normal skin. Their presence in material taken from cases of seborrheic dermatitis, and similar conditions cannot therefore be of diagnostic significance. Even the extent of colonization with these yeasts cannot consistently be related to clinical signs.[22] However, whether the organisms play a role in the pathogenesis of seborrheic dermatitis by some mechanism other than a straightforward infection is a question which is not addressed in this revue. Nevertheless, in the condition, pityriasis versicolor, the presence of filaments in clinical material is a diagnostic feature. *M. furfur* is almost always isolated from the classical form of the disease. The isolates may not be pure, and concurrent growth of *M. ovalis* variants, which are more vigorous yeasts under the cultural conditions described above, may mask *M. furfur*. So far, isolates of *M. furfur* from pityriasis versicolor do not appear to have a greater ability to form filaments in vitro than isolates from normal skin but the conditions for conversion of yeasts to filaments have not been fully determined although studies on this aspect have uncovered a number of factors.[23-25] No doubt any condition which did influence this change would arouse interest as a possible contribution to the mechanisms involved during the pathogenesis of pityriasis versicolor.

M. ovalis Form 2 also produces filaments in culture and it is this form which is usually isolated from pityriasis versicolor where ovoid yeasts are seen in the skin.

One interesting aspect when comparing the two populations of cases of pityriasis versicolor, is the distribution of lesions on the body. When *M. furfur* was present, the vast majority of cases were infected on the upper trunk, whereas with *M. ovalis* most of the lesions were distributed between the abdomen, groins and lower limbs.[15] Differences have also been noticed in the distribution of the two species in normal skin. It is generally found that on the face and scalp, the oval yeasts outnumber the spherical ones whereas it is the spherical yeasts which predominate on the trunk.[22,26] In a series of 60 subjects at our Institute, the observations reveal that *M. furfur* is apparently more affected by the local factors in that there were significantly more determinations of this species on the trunk when compared with the face or scalp (Table 1). In contrast, *M. ovalis* had a more even distribution between the sites.

DISCUSSION

The morphological variation seen in the yeasts of the genus *Malassezia* which may be regularly seen in, and isolated from human skin, has enabled them to be divided into four distinct variants. As the cultures have been shown to be stable in their morphology and also distinct in other properties, it seems appropriate to indicate this in their nomenclature. One has therefore been assigned to *M. furfur* and the remaining three can be considered as forms of *M. ovalis*. *M. furfur* and *M. ovalis* Form 2 are both able to develop filaments in culture. Apparent differences in the origin of the hypha need to be confirmed and will have to wait for more detailed studies on the induction of filamentous growth in both species. It is significant also that the yeasts of these two variants show similarities with those seen in pityriasis versicolor scales.

Table 1. A comparison of the prevalence of *Malassezia* yeasts at two sites.

		Numbers of specimens revealing *Malassezia* yeasts by microscopy or by culture		
60 Subjects	No. of Samples	*M. furfur*	*M.ovalis*	x^2 tests
Trunk	51	42 (82%)[a]	37 (73%)[b]	a: b P=.65
Face/Scalp	50	28 (56%)[c]	35 (70%)[d]	c: d P=.21
x^2 tests		a: c P=<.01	b: d P=.95	

Although any one form of *Malassezia* is not exclusive to any particular body site, surveys of the skin show that there is a different general distribution of the spherical and oval forms over the body. The question arises as to whether the two species, *M. furfur* and *M. ovalis* have a different ecology, or whether, if they represent different phases of a single species, are the varying conditions at each site influencing a change in morphology? Opinions which prefer to maintain a single species for these yeasts are influenced by morphological instability of cultural isolates[27,28] and by genome comparison.[29] It is difficult at the moment to resolve this question. Genetic studies at the St.John's Institute of Dermatology, involving karyotyping and RFLP analysis, are beginning to confirm the divisions suggested here, but more isolates need to be studied and cultures from diverse sources would add valuable information. Although the distribution of *M. pachydermatis* on animal skin is well documented,[28] *M.ovalis* can also be frequently isolated from a variety of warm blooded animals.[15] This fact is not fully appreciated and so a valuable source of isolates has hitherto been neglected. However, it is possible that as we acquire more data, additional variants will undoubtedly emerge and the dividing line between them may become less clear.

REFERENCES

1. E. Eichstedt, Pilzbildung in der Pityriasis versicolor, *Froriep Neue Notiz. a. d. Natur und Heilk.* 39: 270 (1846).
2. H. Baillon, "Traité de botanique médicale cryptogamique", Octave Doin, Paris (1889).
3. S. Rivolta, "Parassiti vegetali" Giulio Speirani, Torino (1873).
4. G. Midgley, "The Characterisation of Lipophilic Yeasts" Ph.D. Thesis, Univ. London (1990).
5. P.G. Unna, Die Farbung der Mikroorganismen im Horngewebe, *Monatsh. f. Prakt. Dermatol.* 13: 225 (1891).
6. R. Sabouraud, "Maladies du cuir chevelu. II.- Les maladies desquamatives", Masson et cie., Paris (1904).
7. A. Castellani and A.J. Chalmers, "Manual of Tropical Medicine", Baillière, Tindall & Cox, London (1913).
8. W.Ch. Slooff, Genus 6. *Pityrosporum* Sabouraud, in "TheYeasts. A Taxonomic Study", J. Lodder, ed., North-Holland Publishing Company, Amsterdam (1970).
9. M.A. Gordon, The lipophilic mycoflora of the skin. 1. *In vitro* culture of *Pityrosporum orbiculare* n.sp., *Mycologia* 43: 524 (1951).

10. D. Yarrow and D.G. Ahearn, Genus 7. *Malassezia* Baillon, in"The Yeasts. A Taxonomic Study", N.J.W. Kreger-van Rij, ed., Elsevier Science Publishers BV, Amsterdam (1984).

11. R.B. Simmons and E. Guého, A new species of *Malassezia, Mycol. Res.* 94: 1146 (1990).

12. D. Borelli, Pitiriasis versicolor por *Malassezia ovalis, Mycopathologia* 89: 147 (1985).

13. H.W. Acton and G. Panja, Seborrhoeic dermatitis or pityriasis capitis: a lesion caused by the *Malassezia ovale, Indian Med. Gaz.* 62: 6013 (1927).

14. T. Piamphongsant, Pityriasis pigmentosa: the clinical features of pathogenic *Pityrosporum ovale, J. Dermatol.* 10: 355 (1983).

15. G. Midgley, The diversity of *Pityrosporum (Malassezia)* yeasts *in vivo* and *in vitro, Mycopathologia* 106: 143 (1989).

16. E. Matakieff, "Le pityriasis versicolor et son parasite", Thèse Fac. Méd. Univ. Nancy (1899) cited by Slooff.[8]

17. M. Barfatani, M.A. Munn and O.A. Schjeide, An ultrastructure study of *Pityrosporum orbiculare, J. Invest. Dermatol.* 43: 231 (1964).

18. A.S. Breathnach, M. Gross and B. Martin, Freeze-fracture replication of cultured *Pityrosporum orbiculare, Sabouraudia* 14: 105 (1976).

19. E. Guého, Ré-evaluation du genre *Malassezia* à l'aide de lamicroscopie électronique et des comparaisons génomiques, *Bull. Soc. Fr. Mycol. Méd.* 17: 245 (1988).

20. S.M. Bruneau and R.M.F. Guinet, Quantitative immuno-electrophoretic study of genus *Pityrosporum* Sabouraud, *Mykosen* 27: 123 (1984).

21. A.C. Cunningham, J.P. Leeming, E. Ingham and G. Gowland, Differentiation of three serovars of *M. furfur, J. Appl. Bact.* 68: 439 (1990).

22. D.C. Clift, H.J. Dodd, J.D.T. Kirby, G. Midgley and W.C.Noble, Seborrhoeic dermatitis and malignancy. An investigation of the skin flora, *Acta Derm. Venereol. (Stokh.)* 68: 48 (1988).

23. M. Nazzaro Porro, S. Passi, F. Caprilli and R. Mercantini, Induction of hyphae in cultures of *Pityrosporum* by cholesterol and cholesterol esters, *J. Invest. Dermatol.* 69: 531 (1977).

24. M. Dorn and K. Roehnert, Dimorphism of *Pityrosporum orbiculare* in a defined culture medium, *J. Invest. Dermatol.* 69: 244 (1977).

25. J. Faergemann, R. Aly and H.I. Maibach, Growth and filament production of *Pityrosporum orbiculare* and *P. ovale* on human stratum corneum *in vitro, Acta Derm. Venereol.* 63: 388 (1983).

26. S.O.B. Roberts, *P. orbiculare* : incidence and distribution on clinically normal skin, *Br. J. Dermatol.* 81: 264 (1969).

27. M. Randjandiche, Polymorphisme de *Pityrosporum ovale* (Bizzozero) Castellani & Chalmers *in vivo* et *in vitro, Bull. Soc. Fr. Mycol. Méd.* 5; 79 (1976).

28. I.F. Salkin and M.A. Gordon, Polymorphism of *Malassezia furfur, Can. J. Microbiol.* 23: 471 (1977).

29. M. Randjandiche, "Le genre *Pityrosporum* Sabouraud 1904. "Dissertation Fac. Méd. Vét. Univ. Liège (1979).

Dimorphism and Pathogenesis

COMPLEMENT RECEPTORS IN *CANDIDA ALBICANS* AND OTHER YEASTS: STRUCTURE, FUNCTION AND ROLE IN PATHOGENESIS

Margaret K. Hostetter[1,2], Eric J. Michael[1], and Catherine M. Bendel[1]

Departments of Pediatrics[1] and Microbiology[2]
University of Minnesota
Minneapolis, Minnesota, USA

ABSTRACT

Surface proteins which bind C3 fragments C3d and iC3b are present in *Candida albicans* and related species. This paper discusses the C3d and iC3b receptors in *C. albicans* and other yeasts and evaluates their relationships to mammalian receptors for C3d (complement receptor type 2) and iC3b (complement receptors type 3/4). In the case of the mammalian receptors, relationships to extended protein families are indicated, in that CR2 is a member of the RCA family (regulators of complement activation family), while CR3/CR4 are classified as β_2 integrins. The existence of C3-binding proteins in yeasts bears important implications for an understanding of yeast pathogenesis and for the larger scope of evolutionary biology.

Although monoclonal antibodies recognizing mammalian CR2 bind to *C. albicans*, structural and functional parallels between the C3d binding protein in *C. albicans* and human CR2 are somewhat less compelling. Isolation of a gene encoding a C3d-binding protein in yeast will be helpful in resolving these issues.

In contrast, the iC3b receptor in *C. albicans* shares considerable antigenic, structural, and functional homologies with the β_2 integrins α_M and α_X. *C. albicans* exhibits surface fluorescence with several monoclonal antibodies recognizing epitopes of α_M and α_X; surface fluorescence is also evident with other *Candida* species but is completely absent in *Saccharomyces cerevisiae*. Structural parallels include a polypeptide structure of 165±15 kD isolated from the cytosol of both *C. albicans* and *S. cerevisiae*. Functional analogies in *C. albicans* include the ability to bind iC3b with identical affinity and the mediation of adhesion to host tissues. As with the mammalian integrins α_M and α_X, adhesive mechanisms in *C. albicans* involve attachment to arginine-glycine-aspartic acid (RGD) sites in extracellular matrix proteins; in human epithelial cell lines, such attachment can be inhibited by peptides and polypeptides which contain the RGD sequence. Preliminary studies of a gene encoding a putative integrin analog in *S. cerevisiae* provide intriguing evidence that the iC3b receptor in yeast may be a primitive eukaryotic precursor of the mammalian integrins.

INTRODUCTION

The observation that *Candida albicans* and other *Candida* species express binding sites for active fragments of the third component of complement, C3, has important implications for our understanding of yeast pathogenesis and evolutionary biology. In the first report of this phenomenon,[1] sheep erythrocytes bearing C3 fragments iC3b and C3d rosetted with hyphal elements from *C. albicans* and *C. stellatoidea*. Because such "immune adherence", (i.e., rosetting of complement-coated sheep erythrocytes) had long been interpreted as qualitative evidence for cellular expression of complement receptors, these results suggested that surface receptors for C3d and iC3b were present in *Candida* species.

The biologic import of complement receptors derives not only from their ability to bind complement fragments, but also from their position within extended protein families (Table 1). From this standpoint, evidence of a significant relationship between complement receptors in yeast and their counterparts in mammalian cells will provide new insights into the evolution of these complex and highly specialized proteins. In particular, yeast systems are particularly suited to answer questions regarding gene organization and expression, as well as to explore the consequences of gene deletion.

Table 1. Terminology for complement receptors.

Receptor for	Complement Nomenclature	Leukocyte Nomenclature	Superfamily
C3d	CR2	CD21	Regulators of Complement Activation (RCA)
iC3b	CR3 CR4	CD11b/CD18 CD11c/CD18	Integrins

This paper will discuss two types of C3-binding proteins in yeast: the C3d receptor and the iC3b receptor. In order to maintain some equilibrium amidst a confusing and expanding terminology, the reader is referred to Table 1. For the purposes of this report, the C3d receptor will be called the C3d binding protein, since a relationship to the RCA family (regulators of complement activation) is not yet well established; the iC3b receptor, in contrast, will be referred to as the integrin analog.

Complement Receptors in Mammalian Cells

Table 2 lists the current roster of mammalian receptors for the third component of complement, C3, and their respective ligands. During complement activation, native C3, a disulfide-linked dimer with an α-chain of 115 kD and a β-chain of 75 kD, is cleaved to successively smaller, biologically active fragments:

C3b, iC3b, and C3d. C3b, shortened by the removal of a 10 kD fragment from the amino terminus, consists of a 105 kD α-chain in disulfide linkage with an uncleaved β-chain. iC3b has two α-chain fragments of 67 kD (amino terminus) and 40 kD (carboxy terminus), themselves joined by an intra-chain disulfide bond; the β-chain remains uncleaved at 75 kD. C3d, the smallest ligand, encompasses 33 kD of the midportion of the α-chain polypeptide. The specific structure and conformation of each individual fragment permits it to serve as a highly specific ligand for a particular complement receptor.

Table 2. Roster of mammalian receptors for fragments of the third component of complement.

Receptor	Primary Ligand
CR1	C3b
CR2	C3d
CR3	iC3b
CR4	iC3b

Complement receptor type 1 (CR1), a monomer of 205-260 kD, recognizes the ligand C3b and is widely distributed in mammalian cells such as leukocytes, erythrocytes, glomerular podocytes, and follicular dendritic cells.[2] It promotes attachment of phagocytes to cells bearing covalently bound C3b and also participates in the degradation of immune complexes. There is no evidence for the presence of CR1 in *Candida* species.

Complement receptor type 2 (CR2), a single-chain polypeptide of 140 kD, is expressed predominantly on B lymphocytes[3] and some malignant epithelial cell lines.[4] CR2 binds fluid-phase C3d at reduced ionic strength (Ka=3.2×10^4 L/M), recognizes C3d when covalently bound to target cells, and attaches to infective Epstein-Barr virus *in vitro*, thereby facilitating cellular infection of B lymphocytes. CR2 is also intimately involved in B cell proliferation and the initiation of the antibody response.

Complement receptor type 3 (CR3), a dimer with an α-chain of 165 kD and a β-chain of 95 kD, is also known as Mac-1, Mo1, or CD11b/CD18.[5] The dimeric form on mammalian leukocytes constitutes an essential adhesive protein and directs a number of adhesion-dependent functions, such as adherence to glass or plastic, chemotaxis, or diapedesis.

Complement receptor type 4 (CR4), also known as p150,95 or CD11c/CD18, consists of an α-chain of 150 kD and a β-chain of 95 kD, the latter identical in amino acid sequence to the β-chain of CR3.[6] The ligand for CR3 and its highly homologous cousin, CR4, is iC3b, although this interaction is of relatively low affinity, with a Ka of 2.45×10^6 L/M.[7] Because the α-chains of CR3 and CR4 contain divalent cation binding sequences,[8,9] these two complement receptors have been classified as β_2 integrins. Among the integrin superfamily, CR3 is known as $a_M\beta_2$; CR4 is called $a_X\beta_2$.

In order to assess the relationship between C3d- and iC3b-binding proteins in *C. albicans* and complement receptors in mammalian cells, antigenic expression, structure, and function must be systematically compared.

The C3d Binding Protein

The binding of C3d by *C. albicans*, most intensively studied by Calderone and colleagues,[10-15] has not yet been shown to conform to typical receptor kinetics, including the demonstration of specificity, reversibility, or saturability; thus, this antigen is perhaps best called a C3d-binding protein. Electrophoresis of homogenized hyphal extracts identified doublet bands at 60-62 kD and 68-70 kD on SDS-PAGE and Western blotting with the monoclonal antibody CA-A, which recognizes several candidal surface glycoproteins. Some, but not all monoclonal antibodies to the human C3d receptor, CR2 (CD21), also reacted with the 60 kD band on Western blot.[15] However, the reported reaction with an antibody to C3d[15] presents a conundrum, since antibodies to the ligand should not recognize its receptor. The function of the C3d binding protein in *C. albicans* is not understood, although the expression of the protein *in vivo* [14] and the proliferation of lymphocytes from infected mice to the *C. albicans* C3d binding protein *in vitro* [15] suggest that this protein may be a stimulus for B lymphocyte proliferation. However, its cellular function in yeast remains obscure. Given these substantial differences in structure and function, definitive demonstration of a relationship between mammalian CR2 and the C3d binding protein in *C. albicans* will require isolation and sequencing of the candidal gene.

Interestingly, other investigators[16] have recently characterized a laminin receptor on *C. albicans* germ tubes, with a dissociation constant of 1.3×10^{-9}M. As with the C3d binding protein, binding of laminin was inhibited by prior heating or trypsinization of cells, and an analysis of a DTT/iodoacetamide extract from germ tubes revealed 2 bands at 60-62 kD and 68 kD after Western blotting with a rabbit antiserum to laminin. Similarity in molecular weight between the laminin receptor and the C3d binding protein suggests both an antigenic identity and a function as an adhesin, although neither has yet been proved.

The iC3b Receptor (Integrin Analog) in *Candida albicans*

Considerably more is known about the iC3b receptor in *C. albicans*, which shares antigenic, structural, and functional homologies with α-chain epitopes of the human complement receptors CR3 (Mac-1; CD11b/CD18) and CR4 (p150,95; CD11c/CD18), themselves members of the mammalian integrin family. The mammalian integrins, heterodimeric integral membrane proteins which mediate cellular communication and adhesion, are grouped in subsets on the basis of their β-chains.[17] CR3 and CR4, members of the β_2 integrin subset, share a 60-70% identity in the amino acid sequences of their α-chains, and their β-chains are completely identical.

Antigenic Similarity with CR3 (α_M) and CR4 (α_X). Initial studies of antigenic similarity in *Candida* species demonstrated binding of the anti-CR3 monoclonal antibody anti-Mo1 to *C. albicans* hyphae, while monoclonal antibodies to human CR1 (CD35) and CR2 (CD21) were shown by direct immunofluorescence microscopy not to bind.[18] In subsequent studies with yeast-form *C. albicans*, specific binding of several IgG and IgM monoclonal antibodies recognizing α-chain epitopes of CR3 and CR4 - including OKM1, OKM10, M1/70, Mab 17, and Mab 44 - has been quantitated by flow cytometry.[19-20] Thus, results from several laboratories have confirmed the presence of an antigenic homolog of CR3/CR4 in *C. albicans* in both yeast and hyphal phase.[18-20] This protein in *Candida albicans* is now called the integrin analog.

In contrast, monoclonal antibodies recognizing the common β-chain (CD18) shared by CR3 and CR4 in mammalian cells failed to react with *C. albicans*, whether by direct immunoprecipitation,[19] by immunofluorescence, or by Western blotting (unpublished observation). Only a polyclonal antibody against a synthetic peptide from the carboxy terminus of the chicken β_1 integrin subunit has been shown to recognize a band of 95 kD on Western blotting of *C. albicans* extracts.[21] Whether *C. albicans* possesses or simply fails to express a β-subunit for the dimeric integrin proteins thus remains unclear. However, one intriguing possibility is that expression of the β-chain may be developmentally regulated, perhaps by the transition from yeast to hyphal phase or in response to environmental conditions such as steroids or hyperglycemia.[22]

Several factors can modulate antigenic expression. The incorporation of 10 mM D-glucose, as opposed to equimolar L-glutamate, in the growth media led to a 4 to 6-fold increase in surface expression of the integrin analog in *C. albicans*, as quantitated by flow cytometry; surface expression increased linearly as the concentration of D-glucose exceeded 10 mM.[19,23] It should be noted that these glucose concentrations are on a par with plasma glucose levels ≥180 mg/dl, a level of hyperglycemia which seems to foster *C. albicans* infections *in vivo*.[22,23] Other 6-carbon and 5-carbon sugars such as galactose and fructose stimulated modest increases in surface expression, while the integrin analog was completely undetectable after growth of *C. albicans* in the presence of 2-deoxy-D-glucose.[23] In addition, both trypsin[19] and Pronase® (MKH, unpublished observation) substantially decreased surface expression of the integrin analog. Transformation from yeast to hyphal phase augmented both iC3b binding and surface fluorescence, the latter by some 10-fold.[19] As would be expected from studies with hyphal transformation, expression of the integrin analog was maximal when *C. albicans* was grown at temperatures from 30-50 °C but was significantly decreased at temperatures above 56 °C.[20]

Structural Similarity with CR3 (α_M) and CR4 (α_X). Structural similarity between the integrin analog in *C. albicans* and its counterparts among the mammalian integrins is also striking. Among the mammalian integrins the a-chain of CR3 (α_M) has an M_r of 165 kD; the α-chain of CR4 (α_X), an M_r of 150 kD. Both share a common β-chain of 95 kD.[8,9] Cytosolic pools serve as depots for rapid up-regulation of these receptors to the surface of the mammalian cell.[24] In *C. albicans*, direct immunoprecipitation of [^{125}I] surface-labeled proteins with the anti-CR3 monoclonal antibody OKM1 revealed a single band at 130 kD.[20] Studies with *C. albicans* extracts have localized the integrin analog to both membrane and cytosol and have detected a single band of M_r 165 ± 15 kD under non-reducing conditions after Western blotting with a variety of monoclonal antibodies recognizing CR3 and CR4.[23] Immunoaffinity purification with the monoclonal antibodies OKM1 (anti-CR3) or BU15 (anti-CR4) have both identified a single band of 165 ± 15 kD under non-reducing conditions, while preliminary experiments conducted under reducing conditions have detected three bands at 56, 62, and 64 kD (MKH, unpublished observations).

Functional Similarity with CR3 (α_M) and CR4 (α_X). Two functions of the integrin analog in *C. albicans* have been described. In mammals, the C3 fragment iC3b serves as the preferred ligand for the integrins CR3 and CR4; the same is true in *C. albicans*.[19] Binding of purified human iC3b to yeast-phase isolates of *C. albicans* is specific, saturable, and reversible, with a Ka of 2.45 x 10[6] liters/mole - an affinity identical to that determined for the binding of iC3b to CR3 on the human neutrophil.[7,19] Neither the binding of iC3b[19] nor the adherence of iC3b-coated erythrocytes[20] is dependent upon divalent cations; however, such non-covalent binding of an otherwise opsonic C3 fragment by *C. albicans* subverts its recognition by phagocytic receptors.[23]

Considerably more important than iC3b binding, however, is the function of the integrin analog in *C. albicans* in mediating adhesion to human endothelium and epithelium. Confirmation of the role of the integrin analog in *C. albicans* in adhesion of this yeast to human endothelium has been published;[25] in these studies, preincubation of *C. albicans* with iC3b or with anti-CR3 monoclonal antibodies inhibited adhesion of *C. albicans* to human umbilical vein endothelium by 50%, while preincubation with albumin or C3d had no effect. These results confirm that the integrin analog in *C. albicans* mediates endothelial adhesion for this yeast and suggest that the RGD (arg-gly-asp) site in iC3b, which is absent in C3d, may be of central import for the adhesive interaction.

Preliminary experiments indicate that adhesion of *C. albicans* to human epithelium is also mediated by the integrin analog.[26] With HeLa cells as substrate, the adhesion of *C. albicans* and less pathogenic *Candida* species correlated directly with surface expression of the integrin analog on the yeast. Preincubation of *C. albicans* with iC3b or with RGD-containing peptides inhibited epithelial adhesion by 82%, an effect which implies that the integrin analog is the dominant epithelial adhesin for *C. albicans*. Because epithelial cells secrete iC3b,[4] the presence of this ligand on the epithelial surface presents an attractive site for *Candida* cells expressing the integrin analog. Other preliminary studies have demonstrated that expression of the integrin analog correlates directly with the ability of a variety of yeasts to adhere to epithelium; for example, expression and adhesion are highest for *C. albicans* but significantly decreased for *C. tropicalis*, *C. parapsilosis*, *C. krusei*, *C. glabrata*, and *S. cerevisiae*. The prominence of *C. albicans* in this respect further correlates with its predominance as the yeast most frequently isolated from infected hosts.[27] Once again, 12- to 15-mer RGD peptides inhibited the adhesive interaction by >50%, provided that they contained the correct flanking sequence.

Preliminary Studies of Putative Integrin Genes in *Saccharomyces cerevisiae*

Elucidation of the structure of the nucleotide sequences of integrin analogs in both pathogenic and non-pathogenic yeasts is underway. In preliminary studies, a gene that encodes a putative integrin analog in *S. cerevisiae* has been identified and sequenced (MKH, unpublished observations). This gene spans an open reading frame of 5.1 kb and includes a TATA box approximately 60 nucleotides upstream and a 600 bp untranslated region at the 3' end. Studies with a cDNA probe from the α_M transmembrane region demonstrate that the human probe and the gene encoding a putative integrin anolog in *S. cerevisiae* appear to hybridize to the same restriction fragments in yeast genomic DNA. The integrin analog in *S. cerevisiae* remains confined to the cytosol and does not appear on the suface of this yeast, at least as assessed by flow cytometry (MKH and CMB, preliminary observations).

Given these preliminary findings, we may develop a hypothetical model of integrin evolution (Fig. 1), with *S. cerevisiae* as the most primitive precursor: a gene for its integrin analog has been cloned and a protein is expressed with the structure of an integrin α-chain but remains entirely confined within the cytoplasm (Fig. 1, left panel). Several factors may contribute to this cytosolic localization, including the absence of a signal peptide, the incomplete but nonetheless intriguing similarity at the transmembrane domain, the absence of a β-chain, or the lack of an unidentified exporter protein.

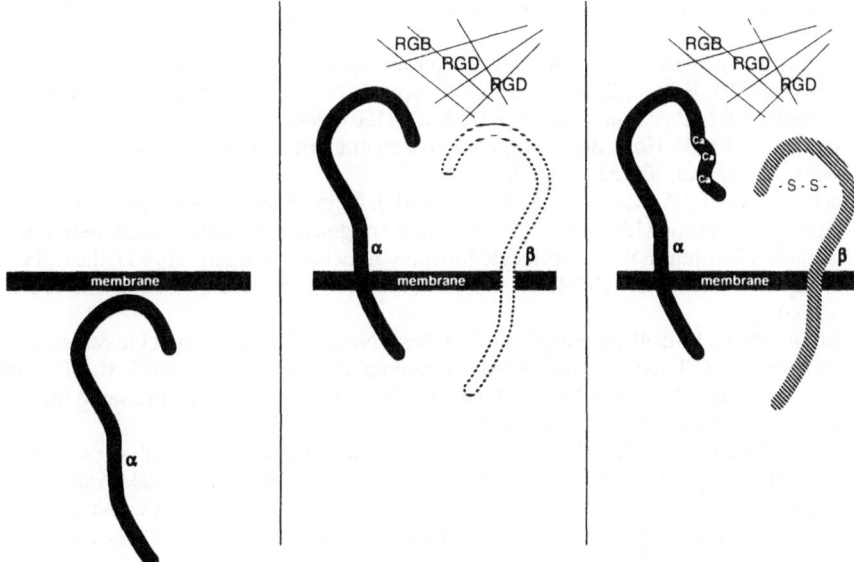

Fig. 1 A hypothetical model for integrin evolution: *S. cerevisiae*, left panel; *C. albicans*, center panel; ß2 integrins (mammals), right panel. RGD: arginine-glycine-aspartic acid sites in extracellular matrix proteins.

When integrin genes in *C. albicans* are successfully isolated, we would predict that the transmembrane domain shows an even greater similarity to its mammalian counterpart, thereby facilitating surface expression (Fig. 1, center panel). In addition, although we cannot detect a β-chain in yeast phase, its expression may be developmentally regulated, emerging perhaps only in hyphal phase, thereby accounting for the greater adhesiveness and invasiveness of this form of *C. albicans*. Nonetheless, expression of the α-chain itself allows recognition of RGD ligands such as iC3b and mediates adhesion to epithelium and endothelium as well.

Lastly, in the mammalian cell we have the full development of integrins as we know them with both α- and β-chains extending through the membrane to direct the wide variety of adhesive functions which have made the integrins such a versatile and fascinating biological family (Fig. 1, right panel).

Given this progression, it is tempting to speculate that *C. albicans* holds the key to our understanding of the structure and synthesis of integrin proteins in both primitive and sophisticated eukaryotic cells. Studies of the integrin analog in *C. albicans* at the genomic level should expand our understanding of the mechanisms underlying surface expression of this protein in yeasts, its role in pathogenesis, and its evolutionary relationship to the mammalian integrins.

ACKNOWLEDGMENTS

This work was supported by NIH grants AI24162 and 25827.

REFERENCES

1. F. Heidenreich and M.P. Dierich, *C. albicans* and *C. stellatoidea*, in contrast to other *Candida* species, bind iC3b and C3d but not C3b, *Infect. Immun.* 50: 598 (1985).
2. D.T. Fearon, Identification of the membrane glycoprotein that is the iC3b receptor of the human erythrocyte, polymorphonuclear leukocyte, B lymphocyte, and monocyte, *J. Exp. Med.* 152: 20 (1980).
3. J.D. Fingeroth, J.J. Weis, T.F. Tedder, J.D. Fingeroth, J.J. Weis, T.F. Tedder, J.L. Strominger, P.A. Biro and D.T. Fearon, Epstein-Barr virus receptor of human B lymphocytes is the C3d receptor CR2, *Proc. Natl. Acad. Sci. USA* 81: 4150 (1984).
4. E. J. Michael and M.K. Hostetter, Components of an autocrine loop for C3 synthesis by epithelial cells, *Pediat. Res.* 30: 153A (1992).
5. F. Sanchez-Madrid, J.A. Nagy, F. Sanchez-Madrid, J. Nagy, E. Robbins, P. Simon and T.A. Springer, A human leukocyte differentiation antigen family with distinct a-subunits and a common B-subunit: the lymphocyte function-associated antigen (LFA-1), the C3bi complement receptor (OKM1/Mac-1), and the p150,95 molecule, *J. Exp. Med.* 158: 2785 (1983).
6. B. L. Myones, J. G. Dalzell, N. Hogg and G.D. Ross, Neutrophil and monocyte cell surface p150,95 has iC3b receptor (CR4) activity resembling CR3, *J. Clin. Invest.* 81: 640 (1988).
7. D.L. Gordon, G.M. Johnson and M.K. Hostetter, Characteristics of iC3b binding to human polymorphonuclear leucocytes, *Immunology* 60: 553 (1987).
8. A.L. Corbi, T. Kishimoto, L.J. Miller, T.A. Springer, The human leukocyte glycoprotein Mac-1 (complement receptor type 3, CD11b) a Subunit, *J. Immunol*. 263: 12403 (1988).
9. A. L. Corbi, L.J. Miller, K. O'Connor, R.S. Larson and T.A. Springer, cDNA cloning and complete primary structure of the a subunit of a leukocyte glycoprotein, p150,95, *EMBO J.* 6: 4023 (1987).
10. R.A. Calderone, L. Linehan, E. Wadsworth and A.L. Sandberg, Identification of C3d receptors on *Candida albicans. Infect. Immun.* 56: 252 (1988).
11. L. Linehan, E. Wadsworth and R.A. Calderone, *Candida albicans* C3d receptor isolated by using a monoclonal antibody. *Infect. Immun.* 56: 198 (1988).
12. M.W. Ollert, E. Wadsworth and R.A. Calderone, Reduced expression of the funtionally active complement receptor for iC3b but not for C3d on an avirulent mutant of *Candida albicans*, *Infect. Immun.* 58: 909 (1990).
13. W.L. Whelan, J.M. Delga, E. Wadsworth, T.J. Walsh, K.Y. Kwon-Chung, R.A. Calderone and P.N. Lipke, Isolation and characterization of cell surface mutants of *Candida albicans*, *Infect. Immun.* 58: 1552 (1990).
14. T. Kanbe, R.-K. Li, E. Wadsworth, R.A. Calderone and J.E. Cutler, Evidence for expression of the C3d Receptor of *Candida albicans in vitro* and *in vivo* obtained by immunofluorescence and immunoelectron microscopy, *Infect. Immun.* 59: 1832 (1991).
15. M. Fukayama, E. Wadsworth and R. Calderone, Expression of the C3d-binding protein (CR2) from *Candida albicans* during experimental candidiasis as measured by lymphoblastogenesis, *Infect. Immun.* 60: 8 (1992).
16. J.-P. Bouchara, G. Tronchin, A. Annaix, R. Robert and J.-M. Senet, Laminin receptors on *Candida albicans* germ tubes, *Infect. Immun.* 58: 48 (1990).
17. R.O. Hynes, Contact and adhesive specificities in the associations, migrations, and targeting of cells and axons, *Cell*, 69: 11 (1992).
18. J.E. Edwards Jr., T.A. Gaither, J.J. O'Shea, D. Rotrosen, T.J. Lawley, S.A. Wright, M.M. Frank and I. J. Green, Expression of specific binding sites on *Candida* with functional and antigenic characteristics of human complement receptors, *Immunol.* 137: 3577 (1986).
19. B. J. Gilmore, E.M. Retsinas, J.S. Lorenz and M.K. Hostetter, An iC3b receptor on *Candida albicans*: structure, function, and correlates for pathogenicity, *J. Infect. Dis.* 157: 38 (1988).
20. A. Eigentler, T.F. Schulz, C. Larcher, E.-M. Breitwieser, B.L. Myones, A.L. Petzer and M.P. Dierich, C3bi-binding protein on *Candida albicans*: temperature-dependent expression and relationship to human complement receptor type 3, *Infect. Immun.* 57: 616 (1989).

21. E.E. Marcantonio and R.O. Hynes, Antibodies to the conserved cytoplasmic domain of the integrin b_1 subunit react with proteins in vertebrates, invertebrates, and fungi, *J. Cell Biol.* 106: 1765 (1988).

22. M.K. Hostetter, Handicaps to Host Defense. Effects of Hyperglycemia on C3 and *Candida albicans*, *Diabetes* 39: 271 (1990).

23. M. K. Hostetter, J.S. Lorenz, L. Preus and K.E. Kendrick, The iC3b receptor on *Candida albicans*: subcellular localization and modulation of receptor expression by glucose, *J. Infect. Dis.* 161: 761 (1990).

24. J.J. O'Shea, E.J. Brown, B.E. Seligmann, J.A. Metcalf, M.M. Frank and J.I. Gallin, Evidence For distinct intracellular pools of receptors for C3b and C3bi in human neutrophils, *J. Immunol.* 134: 2580 (1985).

25. K.F. Gustafson, G.M. Vercellotti, C. M. Bendel and M.K. Hostetter, Molecular mimicry in *Candida albicans*: Role of an integrin analog in adhesion of the yeast to human endothelium, *J. Clin. Invest.* 87: 1896 (1991).

26. C.M. Bendel and M.K. Hostetter, Inhibitory peptides and target proteins for epithelial adhesion in pathogenic *Candida* species, *Pediat. Res.* 30: 157A, (1992).

27. R. Horn, B. Wong, T.E. Kiehn and D. Armstrong, Fungemia in a cancer hospital: changing frequency, earlier onset, and results of therapy, *Rev. Infect. Dis.* 7: 646 (1985).

IMMUNOGENIC AND IMMUNOMODULATORY ROLES OF
DISTINCT MANNOPROTEIN CONSTITUENTS OF *CANDIDA ALBICANS*

Antonio Cassone

Department of Bacteriology and Medical Mycology
Istituto Superiore di Sanità
Rome, Italy.

ABSTRACT

Mannoproteins (MP) of *Candida albicans* are heterogeneous, molecularly complex, multifunctional constituents which exert a great impact on host response. This review particularly focuses upon MP recognition as antigens in cultures of human peripheral blood mononuclear cells (PBMC) and on their ability to activate an essential effector of host anti-*Candida* defense as the polymorphonuclear cell (PMN). PBMC of almost all normal human subjects indeed respond quite efficiently to a suitable MP preparation (MP-F2) *in vitro*, both in terms of lymphoproliferation and as transcription of genes for specific cytokines. In fact, IL-2 and IFN-γ, but not IL-4 or IL-5, are abundantly produced, suggesting that natural sensitization to *Candida* leads to a preferential expansion of a Th_1 lymphocyte subset. MP-F2-lymphocyte stimulation in human PBMC has a correlate in a delayed-type hypersensitivity (DTH) reaction in *Candida*-immunized mice. A distinctive constituent of 65 KDa appears as a predominant or even exclusive target of this cell-mediated immune response that is probably directed against the protein moiety of the molecule.

On the other hand, the mannan moiety is an effective stimulator of neutrophils *in vitro*. A remarkable consequence of MP-F2-stimulation is the production by neutrophils of a full set of cytokines such as IL-1β, TNF-α, IL-6 and IL-8 which are all able to activate neutrophils themselves and other cells. Thus, MP-F2 contains at least two distinct constituents which mediate two of those acquired or innate responses that play a central role in control of *Candida* growth *in vivo*.

INTRODUCTION

Mannoproteins (MP) are essential constituents in the architecture of the cell wall of *Candida albicans* as well as enzymatic and secretory constituents of the fungal cell.[1,2] They also play an important role in the host-parasite relationship in candidiasis.[3] Thus, MP are adhesins or virulence enzymes and immunogenic inducers of the host's response. There are molecular correlates to these multiple functions, as MP are structurally very complex, possessing distinct protein and saccharide moieties with both 0-linked and N-linked functional groups, as well as α- and β-linked oligomannosides.[4] MP are highly polydisperse and it has been so far practically impossible to separate them to homogeneity. Nonetheless, ion-exchange and gel permeation chromatographies give clear indications about the existence of a number of discrete MP "families", the separation and functional characterization of which would greatly improve our knowledge of the host-Candida interactions.[5,6] In this review, I will particularly focus on the immunogenic and immunomodulatory role played by cell wall mannoproteins in the host-Candida relationship, with emphasis on the response to Candida by normal human subjects, but in places, I will also mention some recent results obtained with experimental models.

A KEY QUESTION IN THE HOST-PARASITE RELATIONSHIP IN CANDIDIASIS

Man usually harbors C. albicans, mostly in the gastrointestinal (GI) tract, and Candida infections largely originate from an endogenous source.[1] So, what prevents Candida from leaving its ordinary state as commensal and invading the host superficially or systemically? The answer to this question turns out to be difficult because of the complex nature of adaptation of Candida to man adaptation as well as the mechanisms underlying the control of Candida growth in vivo. Recent research has, however, shed some light on the interplay of cellular and humoral factors in anti-Candida response. Cantorna and Balish[7,8] have shown that, in order to develop systemic candidiasis from GI colonization, mice must be immunodeficient both in the phagocytic/Natural Killer (NK) response (beige) and in T-cell mediated immunity (nude). These conclusions, emphasizing the dual aspects of the host-Candida relationship, i.e. natural and acquired immunity, are well in keeping with various observations concerning the ability of C. albicans to interact with all main effectors of innate immunity: PMN, monocyte/macrophage and NK lymphocyte.[9,10] Interaction with any of these effectors may activate all others through release of cytokines as IL-1β, TNF-α, IL-6, IL-8 and GM-CSF. Of particular relevance is that GM-CSF, a potent signal of PMN activation against Candida, can originate from Candida "recognition" by NK cells.[11] Although the experiments above were performed in vitro, it had previously been demonstrated that Candida is a powerful activator of NK cells in the peritoneal cavity of mice.[12] Finally, Bistoni and collaborators[13] developed a mouse model whereby chronic, non-lethal infection with low-virulence C. albicans confers to normal mice the capacity to develop resistance to systemic, lethal challenge with highly virulent C. albicans. The protection is accompanied (caused?) by up-regulation of Th-1 cytokine response (IL-2 and IFN-γ) and down-regulation of Th-2 cytokine response (Il-4, IL-6).[14] Importantly, if normal, uninfected mice were given a neutralizing anti-IL-4 antibody, the animals responded with increased secretion of IFN-γ/IL-2 and were amply protected against a systemic, lethal challenge with C. albicans.[15] Finally, antibody response to stress proteins have also been advocated as distinctive for invasive candidiasis in non-neutropenic hosts.[16] Thus, natural and aquired immune responses, mostly T-cell mediated, seem to act in concert to keep Candida under check. It seems appropriate to consider that regulated expression of cytokines produced by, and affecting, various cellular types, as well as extent and quality of antibody response play a central, critical role in the anticandidal defense.

Within this framework, the main purpose of our recent research has been the identification and characterization of *Candida* components that are recognized either as antigen against which a T cell-mediated immune response (CMI) is specifically elicited or as "immunomodulators" which are recognized "nonantigenically" by natural immunoeffectors and are able affect to non-specifically the degree of *Candida* multiplication *in vivo*. We have recently extracted, purified and characterized a mannoprotein fraction called MP-F2 which is able to mediate both antigenic and non-antigenic recognition of *C. albicans* by the immune system, in particular by T cells and neutrophils, two essential actors on the stage of anti-*Candida* response.[1,3,6,10] In HPLC, MP-F2 contained practically only mannose, with traces of glucose, and its protein content was about 5%. The [13C] NMR showed a typical profile of mannose carbons, with the presence of predominant α 1-6 and 1-3 bonds but also β 1-2 terminal bonds, recently demonstrated in *Candida* mannan.[4]

MANNOPROTEIN AS A TARGET OF CELL-MEDIATED IMMUNE RESPONSE IN MAN

When MP-F2 is given in sufficient doses (ranging from 20 to 50 µg/ml) to a culture of mononuclear cells from peripheral blood of normal human subjects a strong lymphocyte proliferative response usually occurs, as detected by DNA synthesis over a period of 7 days of culture.[17] This blastogenic response overlaps that achievable by adding to the culture whole inactivated cells of *Candida* whereas there is no significant proliferation with other mannoprotein fractions or with glucan (Table 1).

Table 1. Lymphocyte blastogenesis in response to *Candida* materials.

Material	Thymidine incorporation (cpm x 10^{-3} / 2 x 10^5 cells)	
	Subject 1	Subject 2
None	0.3 ± 0.07	0.45 ± 0.3
Whole *Candida*	24.7 ± 9.0	11.35 ± 1.3
DTT-Extracted *Candida*	1.2 ± 0.5	2.7 ± 0.8
DTT Extract	21.9 ± 0.2	14.9 ± 3.7
DTT Extract-Con A	2.4 ± 0.2	0.8 ± 0.4
Tetanus toxoid	40.7 ± 5.6	27.2 ± 5.1
Glucan	0.5 + 0.1	0.4 + 0.4

DTT: di-thiothreitol; DTT-extract-Con A: dithiothreitol extract passed through Concanavalin A.[6] All *Candida* materials were used at a final concentration corresponding to 50µg of polysaccharide per ml. Tetanus toxoid was as reported elsewhere.[19] Glucan is an alkali-acid insoluble preparation from the cell wall of *C. albicans* as reported elsewhere.[3]

A clear sign of T cell activation in the culture is transcription of genes for cytokines, the true regulators of the immune response. Both by nuclear S1 mapping and PCR of cDNA transcripts, Ausiello et al.[18] have recently demonstrated that MP-F2, is an effective inducer of the message for IL-1β, IL-2, IL-6, IFN-γ, TNF-α and GM-CSF, but not of IL-4, IL-5, and IL-10. Some messages, for instance, those of IL-6, TNF-α and IL-1, have particular kinetics as the genes were transcribed early and for a long time. Both early transcription and what is known about the source of cytokines themselves, suggest that monocyte-derived cytokines are abundant in the system, because of monocyte "activation" either as antigen-presenting cells or through mechanisms independent of this particular accessory function. As mentioned earlier[11] GM-CSF can also be produced by NK lymphocytes which are members of the mononucleate cell culture. Although cytokine-producing cells were not directly identified, it is reasonable to assume that the composite pattern of cytokines resulted from stimulation of T-lymphocytes, monocytes and NK-cells, all somewhat activated by the candidal material. Within T lymphocytes, the transcription of IL-2 and IFN-γ coupled to total absence of signals for IL-4, IL-5, and IL-10, strongly suggests that MP stimulates the Th_1 subset of lymphocytes but not the Th_2 subset.[18,19] Citokine messages were actually translated in the cultures and abundant amounts of IFN-γ, IL-6, TNF-α but not IL-4 were produced in MP-F2 or IL-2 stimulated, but not in unstimulated cultures. These experiments have been repeated with dozens of donors and, except for rare Candida-anergic subjects, there has been consistent, though individually variable, response to this mannoprotein in terms of lymphoproliferation and cytokine synthesis.

Past studies (summarized in Table 2) demonstrated that lymphoproliferative response to MP-F2 has all the characteristics of a classical response to a memory T-cell antigen like those elicited in the same system by other natural antigens such as the protein purified derivative of the tuberculous bacilli, (PPD) or artificial immunogens like the tetanus toxoid vaccine. Thus, from this consistency and also

Table 2. Evidence for the antigenic, non-mitogenic nature of T-cell activation by mannoprotein of Candida albicans.[6,17,18,19]

---

1. Kinetics of lymphoproliferation typical of response to microbial recall antigens (e.g. Tetanus toxoid).
2. Antigen-presenting cells and a functional CD3-T cell receptor required.
3. Activation MHC-restricted.
4. "Naive" lymphocytes from umbilical cord blood, fully responsive to PHA and IL-2, are not responsive to mannoprotein (or tetanus toxoid or PPD, an antigen from Mycobacterium tubercolosis.).
5. Early, functional loss of proliferation to mannoprotein by lymphocytes from HIV- infected subjects.

For details, see Refs 6, 17, 19.

from the strength of the quantitative response, MP-F2 seems to be or to contain a major antigen target of cell-mediated anti-Candida response in man. As already stated, the MP-F2 fraction is chemically fairly homogenous as for mannoprotein composition, but it is already known that these mannoprotein materials are polydisperse and heterogenous molecules.[4] By T cell immunoblotting and gel permeation chromatography, it has been shown that a 65 kDa mannoprotein is the most likely antigenic target of CMI in man.[20] This antigen accounted for less than 4% of the MP-F2 fraction, but contained about 30% of all protein moieties. Work on molecular cloning of MP-65 protein is in progress in our laboratory.

MANNOPROTEIN AS A TARGET OF DELAYED-TYPE-HYPERSENSITIVITY (DTH) IN MICE

We wondered whether the MP-F2 mannoprotein is also a main target of CMI in animal models. This aspect was recently investigated in collaboration with L. Romani and F. Bistoni in a particular model of chronic murine candidiasis.[13] It was shown that a strong delayed-type hypersensitivity (DTH) response is induced by the MP-F2 fraction, in *Candida*-sensitized mice. This response was comparable in potency to that of whole *Candida* cells, whereas an irrelevant mannoprotein did not elicit any DTH response. On the basis of these results we are now investigating whether normal mice could be immunized with MP-F2 against a lethal *Candida* challenge: preliminary results suggest that MP-F2 could indeed induce DTH-coupled protection against candidiasis.

MP-F2 AND NEUTROPHIL STIMULATION

In assessing the true relevance of MP-F2 induction of CMI in human PBMC and in mice, we must, however, consider another property
of MP-F2 that is also relevant to the problem of protection, that is the capacity to stimulate the anticandidal activity of neutrophils.

This has been amply shown in a series of experiments performed by Palma *et al*.[10, 21] demonstrating that preincubation of PMN cultures with the MP-F2 fraction greatly enhanced the capacity of PMN to kill the fungus, with an optimum at a dose of 10 µg/ml and with more manifest effects at the lowest neutrophils: *Candida* ratio. (Table 3). Note that an irrelevant mannoprotein fraction (MP-F1) obtained

Table 3. Effect of MP-F2 and MP-F1 on neutrophil ability to inhibit growth of *C. albicans in vitro*.

Treatment	*C. albicans* growth inhibition			I.U./10^7 PMN
	15:1	30:1	60:1	
None	29 ± 1	53 ± 4	67 ± 4	492
+ MP-F2	56 ± 4	83 ± 7	93 ± 8	1292
+ MP-F1	39 ± 2	59 ± 3	78 ± 5	671
None	18 ± 0	45 ± 1	67 ± 2	416
+ MP-F2	35 ± 1	73 ± 2	89 ± 8	813
+ MP-F1	23 ± 0	50 ± 2	69 ± 4	479
None	20 ± 2	36 ± 3	60 ± 7	385
+ MP-F2	45 ± 3	59 ± 5	75 ± 4	708
+MP-F1	21 ± 2	30 ± 7	63 ± 8	393

E:T Neutrophils: *Candida* ratio. Growth inhibition (% inhibition ± SE) and Inhibition Units (I.U.) calculations were as described in Ref. 10. MP-F1 is an irrelevant mannoprotein fraction collected from the same chromatographic column as that for MP-F2 (see Ref. 6). MP-F1 and MP-F2 were used at the same concentration (25 µg/ml).

from the same cell extract containing MP-F2 was totally inactive, indicating the specificity of MP-F2 activation. The intensity of neutrophil-activating effects by MP-F2 was comparable with those of GM-CSF and LPS, two potent neutrophil stimulators.[21,22] The mechanims for the enhanced PMN activity probably relied upon lactoferrin release, as anti-lactoferrin antibodies totally inhibited the potentiation of PMN anticandidalactivity induced by MP-F2. Interestingly, treatment of MP-F2 with a proteolytic enzyme to degrade the protein but leave much of the polysaccharide, resulted in the elimination of the capacity to induce T cell proliferation but maintained neutrophil stimulating activity. On the contrary, treatment, with α-mannosidase, to degrade much of the mannan moieties but apparently preserving the protein content of the fraction, resulted in the loss of neutrophil activation but maintenance of lymphocyte blastogenesis.[6,10] All this suggests that different moieties of MP-F2 are expressing the potential for CMI recognition and neutrophil activation and that mannan receptors are likely to be expressed on PMN surface.

NEUTROPHILS AND CYTOKINES

It has been demonstrated by several groups that PMN may produce, under suitable stimulation, much the same cytokines as are ordinarily produced by the activated macrophages, for instance IL-1β, IL-6 and TNF-α.[20,23] The stimulation with MP-F2 was a very good signal for the production of cytokines by the neutrophils, compared to the stimulation by LPS. Table 4 shows the quantity of the different cytokines detected in several experiments in the culture supernatant of PMN-stimulated with LPS or MP-F2.

Despite the obvious individual variations, MP-F2 consistently induced in remarkable quantities of IL-1β, TNF-α, IL-6 and IL-8.

Table 4. Cytokine production by human PMN

Cytokine	Expts	Production at 24th hour of culture (pg/ml ± SE)		
		Unstimulated	+ MP-F2	+ LPS
IL-1β	6	18 9 ± 7 0	151 5 ± 61.3*	187.2 ± 81.9*
IL-6	4	6 8 ± 3 4	430 9 ± 109.4*	379.8 ± 151.6*
IL-8	3	22 7 ± 5 8	720 5 ± 76.7*	-
TNF-α	3	15.9 ± 6 2	150.4 ± 27.3*	-

* P < 0 001, Student's t test, unstimulated vs MP-F2 or LPS-stimulated cultures
Non-significant difference between MP-F2 and LPS LPS is lipopolysaccharide from E coli (used at 100 ng/ml) MP-F2 was used at 25 µg/ml

Note that all these cytokines are capable of activating both neutrophils themselves and macrophages for their candidacidal mechanism. Although the relevance of each single cytokine in the mechanism by which neutrophil inhibit *Candida* is not known, we can speculate about the existence of autocrine and paracrine stimulation of PMN or PMN-Mf interactions. Experiments performed by Palma *et al.* have demonstrated that the supernatant of PMN activated with whole heat-killed *Candida* cells conferred to the fresh PMN a greater inhibitory potential against *Candida* and this resulted from a cumulative effects of cytokines as IL-1β, IL-6, IL-8, TNF-α and GM-CSF (manuscript in preparation).

CONCLUSIONS

The available evidence summarized above demonstrates that MP-F2, probably through a 65 kDa mannoprotein antigen, is a main target of CMI in human PBMC. The cytokine pattern associated with T cell response in anti MP-F2 CMI, characterized by IL-2 and IFN-γ but not IL-4 and IL-5 production, is of that kind believed to characterize a protective CMI response in animal models. MP-F2 also contains an as yet uncharacterized molecule(s) which is (are) a main target of strong DTH response in a murine model of *Candida* infection. DTH response has been clearly related to Th-1 cytokine response (IL-2, IFN-γ) and to anti-*Candida* protection in mice.[14,15] These initial results provide a basis for further experiments on the protective role of immunization with MP-F2 in animal models of *Candida* infections. Finally, MP-F2 contains one or more mannan moieties which strongly stimulate anti-*Candida* activity by PMN *in vitro*. The activity is apparently due to release of lactoferrin, but cytokines such as IL-1, IL-6, IFN-γ and IL-8 are likely to play an important amplifying role in neutrophil-neutrophil or even neutrophil-macrophage interactions. Progress in this field will require cloning of MP-F2 proteins as well as detailed characterization of their mannan moieties. Monoclonal antibodies directed against mannoside or peptide epitopes are being evaluated in an attempt to find specific reagents to employ both for MP purification and MP binding to, and processing by, mononucleate or polymorphonuclear cells. Finally, the capacity of distinct MP molecules or molecular families to stimulate distinctive cytokine patterns in man and experimental animals could provide an insight into the various, sometimes opposite, effects on the immune system exerted by these multifunctional constituents of *C. albicans*.

AKNOWLEDGEMENTS

The author's work reviewed here was partially granted by The National AIDS Project (Istituto Superiore di Sanità, Ministero della Sanità, Italy, Contract N° 720/X). The secretarial assistance of Mrs F. Baschieri is also gratefully acknowledged.

REFERENCES

1. F.C. Odds, *Candida* infections: an overview, *CRC Crit. Rev. Microbiol.* 15:1 (1987).
2. M.V. Elorza, A. Murgui, and R. Sentandreu, Dimorphim in *Candida albicans*: contribution of mannoproteins to the architecture of yeast and mycelial cell walls, *J. Gen. Microbiol.*131: 2209 (1985).
3. A. Cassone, Cell wall of *Candida albicans*, its functions and its impact on the host, *Curr. Top. Med. Mycol.* 3: 249 (1989).
4. R.D. Nelson N. Shibata, P. Podzorski, and M. J. Herron, *Candida* mannan: chemistry, suppression of cell-mediated immunity, and possible mechanisms of action, *Clin. Microbiol. Rev.* 4: 1 (1991).
5. J. Domer, K. Elkins, D. Ennist, and P. Baker, Modulation of immune responses by surface polysaccharides of *Candida albicans*, *Rev. Infect. Dis.* 10: 419 (1988).

6. A. Torosantucci, C. Palma, M. Boccanera, C.M. Ausiello, G.C. Spagnoli, and A. Cassone, Lymphoproliferative and cytotoxic responses of human peripheral blood mononuclear cells to mannoprotein constituents of *Candida albicans, J. Gen. Microbiol.* 136: 2155 (1990).

7. M.T. Cantorna, and E. Balish, Role of CD4+ lymphocytes in resistance to mucosal candidiasis, *Infect. Immun.* 59: 2447 (1991).

8. M.T. Cantorna, and E. Balish, Mucosal and systemic candidiasis in con-genitally immunodeficient mice, *Infect. Immun.* 58:1093 (1990).

9. J.Y. Djeu, D.K. Blanchard, A.L. Richards, and H. Friedman. Tumor necrosis factor induction by *Candida albicans* from human natural killer cells and monocytes, *J. Immunol.* 141: 4047 (1988).

10. C. Palma, D. Serbousek, A. Torosantucci, A. Cassone, and J.Y. Djeu, Identification of mannoprotein fractions from *Candida albicans* enhancing human PMN functions and involvement of lactoferrin in PMN inhibition of *Candida* growth, *J. Infect. Dis.* 167: in press (1992).

11. D.K. Blanchard, M.B. Michelini-Norris, J.Y. Djeu, Production of GM-CSF by large granular lymphocytes stimulated with *Candida albicans*: role in activation of human neutrophil function, *Blood* 33: 49 (1991)

12. A. Vecchiarelli, R. Mazzolla, S. Farinelli, A. Cassone, and F. Bistoni, Immunomodulation by *Candida albicans*: crucial role of organ colonization and chronic infection with an attenuated agerminative strain of *Candida albicans* for establishment of anti-infectious protection, *J. Gen. Microbiol,* 134: 2583 (1988).

13. F. Bistoni, A. Vecchiarelli, E. Cenci, P; Puccetti, P. Marconi, and A. Cassone, Evidence for macrophage-mediated protection against lethal *Candida albicans* infection, *Infect. Immun.* 51: 668 (1986).

14. E. Cenci, L. Romani, A. Vecchiarelli, P. Puccetti, and F. Bistoni, T cell subsets and IFN-γ production in resistance to systemic candidosis in immunized mice, *J. Immunol.* 144: 4333 (1990).

15. L. Romani, A. Mencacci, U. Grohmann, S. Mocci, P. Mosci, P. Puccetti, and F. Bistoni. Neutralizing antibody to interleukin 4 induces systemic protection and T helper type 1-associated immunity in murine candidiasis, *J. Exp. Med.* 176: 19 (1992).

16. R. Matthews, J. Burnie, D. Smith, I. Clark, J. Midgley, M. Conolly, B. Gazzard, *Candida* and AIDS: evidence for protective antibody, *Lancet* 263 (1988).

17. C.M. Ausiello, G.C. Spagnoli, M. Boccanera, I. Casalinuovo, F. Malavasi, C.U. Casciani, and A. Cassone, Proliferation of human peripheral blood mononuclear cells induced by *Candida albicans* and its cell wall fractions, *J. Med. Microbiol.* 22: 195 (1986).

18. C.M. Ausiello, F. Urbani, S. Gessani, G.C. Spagnoli, M.J. Gomez, and A; Cassone, Cytokine pattern in cultures of human peripheral blood mononuclear cells stimulated by a mannoprotein constituent from *Candida albicans*, submitted.

19. I. Quinti, C. Palma, E.C. Guerra, M.J. Gomez, I. Mezzaroma, F. Aiuti, and A. Cassone. Proliferative and cytotoxic responses to mannoproteins of *Candida albicans* by peripheral blood lymphocytes of HIV-infected subjects, *Clin. Exp. Immunol.* 85: 485 (1991).

20. A. Torosantucci, C. Bromuro, M.J. Gomez, C.M. Ausiello, F. Urbani, and A. Cassone, Identification of a 65-Kilodalton mannoprotein as a target of human cell-mediated immune response to *Candida albicans*, submitted.

21. C. Palma, A. Cassone, D. Serbousek, C.A. Pearson, and J.Y. Djeu, Lactoferrin release and IL-1, IL-6 and TNF production by human polymorphonuclear cells stimulated by various LPS: relationship to growth inhibition of *Candida albicans*, *Infect. Immun.* in press (1992).

22. D.B. Dubravec, D.R. Spriggs, J.A. Mannick, and M.L. Rodrick, Circulating human peripheral blood granulocytes synthesize and secrete tumor necrosis factor-α, *Proc. Natl. Acad. Sci. USA*, 87: 6758 (1990).

23. A. Lindemann, D. Riedel, W. Oster, S.C. Meuer, D. Blohm, R.H. Mer-tellsmann, and F. Herrmann, Granulocytes/macrophage colony-stimulating factor induces interleukin-1 production by human polymorphonuclear neutrophils, *J. Immunol.* 140: 837 (1988).

STRUCTURAL AND ANTIGENIC CORRELATES OF *HISTOPLASMA CAPSULATUM*

L. Jeavons, and A.J. Hamilton

Dermatology Unit, 18th Floor
Guys Tower, St. Johns Institute of Dermatology
Guys Hospital
London SE1 9RT, UK

ABSTRACT

We report on the isolation and partial characterisation of a novel yeast-phase specific glycoprotein from the dimorphic fungal pathogen *Histoplasma capsulatum*. This 79-82 kDa antigen was isolated with a monoclonal antibody raised using our previously described technique of cyclophosphamide ablation of shared B-cell responses. The glycoprotein is acidic, with a pI of 4.7 and is present only in the cytoplasm of the yeast phase. N-terminal sequencing demonstrated that this glycoprotein is homologous with a recently described 80 kDa antigen which appears to be immunologically significant.

INTRODUCTION

Histoplasma capsulatum is the causative agent of histoplasmosis, which represents one of the commonest respiratory mycotic infection of humans. The fungus is thermally dimorphic, existing as a multicellular mycelial form in the environment at 25 °C and as a unicellular yeast form in infected tissue at 37 °C. The infection is worldwide in distribution and has a particularly high prevalence in temperate and subtropical areas such as the Ohio Valley in North America and regions of South and Central America.[1,2] The primary infection is generally asymptomatic but the disease may present as an acute or chronic progressive pulmonary disease or as a disseminated infection. The latter form of the disease has recently received much attention due to the increasing prevalence of histoplasmosis in the acquired immune deficiency syndrome, particularly in areas where this fungal infection was previously endemic.[3]

The mechanisms underlying the yeast/mycelial transition have obviously attracted a great deal of interest[4] and it is self evident that the process would be underpinned by the appearance/disappearance of phase specific proteins involved in aspects of the cellular metabolic process and the morphology of the organism. Several of these proteins and their genes have now been characterized, including the yps-3 yeast specific gene[5] and various heat shock proteins such as hsp70 and hsp82.[6,7]

However it is difficult to isolate such genes or their products in the absence of, for example, homologous gene probes based on data obtained from heat shock proteins isolated from other organisms. Such problems are heightened by the fact that the majority of expressed proteins will be the same in both yeast and mycelia. In the last two years our group has developed a means of producing species specific monoclonal antibodies differentiating between antigenically related fungal pathogens by using the cyclophosphamide ablation of shared B-cell responses.[8,9] In this paper we report on the extension of this work which has allowed us to produce a yeast phase specific MAb against a 79-82 kDa glycoprotein. The methodology involved immunizing mice with *Histoplasma* mycelial antigen, treating them with the immunosuppressive cyclophosphamide to destroy the response to the latter antigen, and then inoculating with yeast antigen-B-cells responsive to shared epitopes having been destroyed, only those reacting to exclusively yeast antigens could now proliferate and produce antibody. The 79-82 kDa antigen has been purified, partially characterized and subject to N-terminal sequencing.

METHODS AND MATERIALS

Preparation of *H.capsulatum* Antigens

Two *H.capsulatum* isolates (NCPF 4119 and 4114) were obtained from Dr Colin Campbell, National Collection of Pathogenic Fungi, Mycological Reference Laboratory, Colindale, London, and grown up on either Sabouraud medium or conversion media[8,9] at 25 °C and 37 °C. After expansion on to the appropriate liquid media[8,9] the 37 °C cultures (yeast) were grown for 72 h prior to centrifugation (1500g for 20 min.) and washing with phosphate-buffered saline (PBS; 0.01 M, pH 7.4). The 25 °C cultures (mycelial) were grown for a total of 7 days and harvested by filtration. Cytoplasmic antigen production via homogenisation was as previously described.[8]

Immunization Protocol

This was a minor modification of previously published protocols dealing with the production of species specific MAbs.[8,9] Five female Balb/C mice were given intraperitoneal (i.p.) inoculations of mycelial antigen (50 µg/mouse) in Freund's complete antigen prior to 3 sequential i.p. doses of cyclophosphamide at 10 min, 24 h and 48 h intervals, at a dose of 50 mg/kg/mouse. The mice were then inoculated i.p. on day 15 and day 22 with yeast cytoplasmic antigen in Freund's Incomplete Adjuvant. After a test bleed on day 25 followed by an ELISA to determine the individual with the greatest differential response to the two antigens; this mouse received an intravenous boost with yeast antigen prior to fusion on day 29.

Fusion Protocol

This was as previously described using the myeloma cell line sp2/0,[8] as was the differential ELISA with yeast and mycelial antigen to determine those hybridoma colonies secreting specific antibody.[8,9] After two subclonings the relevant monoclonal cell line (69F) was used to produce ascitic fluid. The MAb was subclassed as previously described.[9]

ELISA, SDS-PAGE and Western Blot

These were all performed as previously described.[8,9] In the ELISA MAb 69F was used at a dilutions between 1: 250 and 1: 2000 and in the immunoblot it was used at 1: 1000. The goat anti-mouse IgG peroxidase-linked conjugate was used at 1: 2500 in ELISA and at 1: 1000 in immunoblot.

Periodate Oxidation/Alkaline Degradation of the 79-82 kDa Antigen

Three mg of yeast cytoplasmic antigen made up in 1.2 ml of PBS was treated with the same volume of 5mM sodium periodate for 30 min at 4 °C, then desalted on a Bio-gel P6 column (BioRad Laboratories, Hemel Hempstead, U.K.). The treated antigen was then subject to SDS-PAGE and immunoblot with MAb 69F.

The same quantity of yeast cytoplasmic antigen was treated overnight with 0.2 M KOH under a nitrogen blanket at 37 °C. The treated antigen was then analyzed as above after neutralization with 2 M acetic acid and dialysis against PBS.

Immunofluoresence Detection of the Antigen in Frozen Sections

This was performed as previously described on both mycelial and yeast cryostat sections.[8] MAb 69F was used at dilutions between 1: 50 and 1: 5000 and the fluorescein isothiocyanate conjugated goat anti-Mouse IgG probe at dilutions of 1: 10.

N-terminal Sequencing of 79-82 kDa Antigen

Twenty mg of yeast cytoplasmic preparation was subject to isoelectric focusing in the BioRad Rotofor system using ampholytes in the range 3-10. Fractions containing the 79-82 kDa antigen were detected by ELISA using 69F, and these were pooled, concentrated and subject to SDS-PAGE on a 10% gel using 2 mM thioglycolic acid in the electrode buffer. After western blot on to Immobilon-P (Millipore, Watford, U.K) and staining with Coomassie the relevant band was excised and subject to N-terminal sequencing reactions.

RESULTS

The differential ELISA performed on the mice used in the inoculation protocol revealed one individual that had a significantly higher serum reactivity to *Histoplasma* yeast antigen compared to mycelial antigen (Fig.1). The principle MAb resulting from the fusion, 69F, was shown to be of the IgG1 subclass and was highly specific by ELISA at a range of dilutions to the yeast phase antigen (Fig.2).

By Western blot the MAb was shown to recognize a major band at 79-82kDa, together with much less intense staining of bands at approximately 38,58,62 and 69

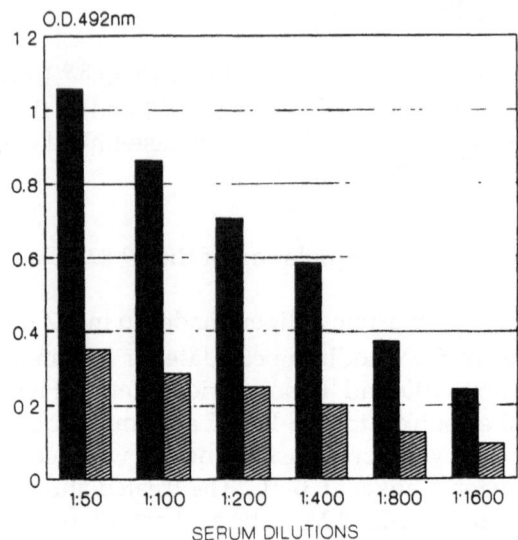

Fig. 1 Differential ELISA showing reactivity of polyclonal serum from the mouse used in the fusion. Solid bar: reaction against yeast phase antigen; hatched bar: reaction against mycelial antigen.

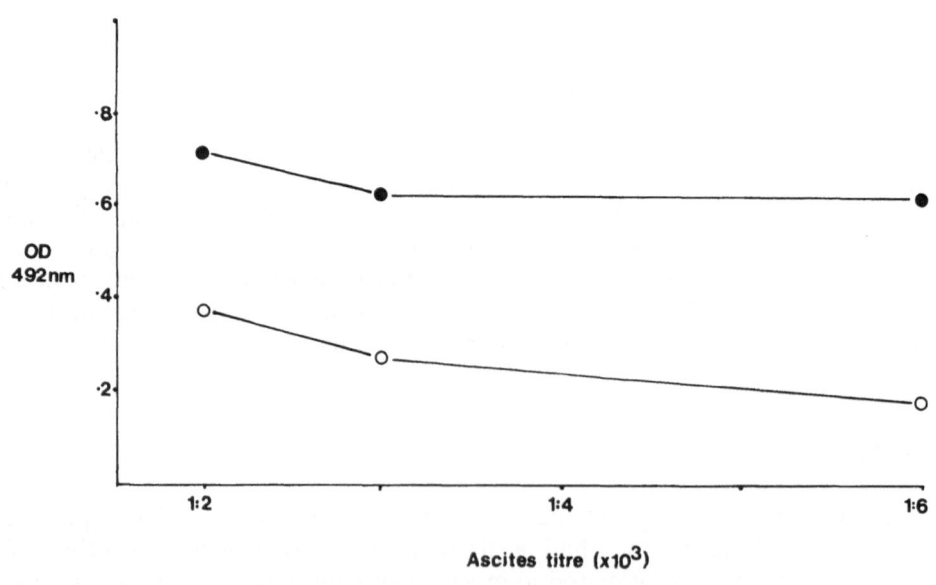

Fig. 2 ELISA illustrating the differential response of MAb 69F against yeast and mycelial antigens. Solid circles: reactivity to yeast antigen; open circles: reactivity to mycelial antigen

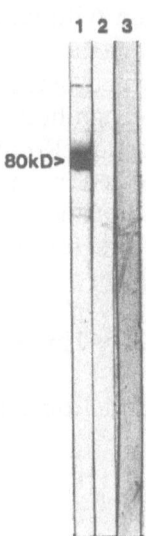

Fig. 3 Western blot analysis of MAb 69F. Track 1: reactivity to yeast cytoplasmic antigen. Track 2: reactivity to yeast cytoplasmic antigen after periodate treatment. Track 3: reactivity to yeast cytoplasmic antigen after treatment with KOH.

Fig. 4 a and b Comparative immunofluorescence reactivity of MAb 69F to yeast and mycelial cryostat sections respectively.

kDa (Fig.3). Treatment of yeast antigen with either alkali or periodate abolished recognition by 69F on immunoblots (same fig.).

Immunofluorescence studies revealed that 69F was reactive to the cytoplasm of the yeast phase with little obvious recognition of the mycelial phase (Fig.4).

Isoelectric focusing with the BioRad Rotofor system followed by ELISA revealed that the 79-82 kDa antigen located in fractions 5 and 6 of 20, equivalent to a pI of 4.7. After subsequent concentration and SDS-PAGE, followed by N-terminal sequencing of the 79-82kDa band, the following amino acid sequence was produced:

Ala-Pro-Ala-Val-Gly-Ile-Asp-Leu-Gly-Thr-?-Tyr.

This sequence proved to be almost identical to that of a recently described *H.capsulatum* antigen.[10]

DISCUSSION

We have successfully applied the technique of cyclophosphamide ablation of shared epitopes, previously used by our group to produce species-specific MAbs, to produce a yeast-phase-specific antibody. It would seem likely that antigenic differences between fungal species would be greater than those between the two morphological phases of the same fungi, and as such the production of MAb 69F represents a significant extension of the immunosuppressive method. There would seem to be no reason why a similar approach could not be used to produce MAbs differentiating between the two phases of the other tropical dimorphic fungi, and indeed the technique could also be used to look at the yeast/pseudohyphal transformation in *Candida*.

The effect of chemical cleavage on the ability of 69F to recognize the 79-82 kDa antigen clearly demonstrates that this molecule is a glycoprotein, and that the epitope to which 69F binds is itself glycosylated. The epitope clearly contains hexose rings, which are broken by periodate, and probably contains an O-glycosidic linkage between glycans and the β-hydroxyamino acids serine or threonine which are split by mild alkali treatment.

Immunofluorescence data showed clear labelling of the yeast cytoplasm with 69F; taken with the fact that the 79-82 kDa antigen is a glycoprotein this is suggestive of a membrane-bound molecule.The relatively low pI of the glycoprotein is not particularly unusual - we estimate from analysis of Rotofor fractions that up to 80% of cryptococcal proteins and glycoproteins have a pI of less than 7. It would seem likely that such low pI values are generated by the presence of acidic carbohydrate groups such as sialic acid as much as by the amino acid content.

Of the previously described phase-specific proteins from *H. capsulatum*, the glycoprotein we have described would appear to resemble hsp82 on the basis of similar molecular weight. However, the similarity would appear to be superficial since the N-terminal sequence of the 79-82 kDa glycoprotein that we have obtained does not show significant homology with any part of the deduced amino acid sequence of hsp82.[11] In contrast the N-terminal sequence we have described is almost fully homologous to a very recently described 80 kDa antigen, which has been shown to mediate protective immunity in mice and is homologous to hsp70.[10]

ACKNOWLEDGEMENTS

We would like to thank Lawrence Hunt, Protein Sequencing Unit, Department of Biochemistry, University of Southampton for performing N-terminal sequencing reactions. Thanks also to Dr Colin Campbell of the Mycological Reference Laboratory, PHLS, Colindale, London for supplying fungal isolates. This work was carried out with the aid of a grant from the Wellcome Trust.

REFERENCES

1. P.Q. Edwards and E.L. Billings, World-wide pattern of skin sensitivity to histoplasmin, *Am.J.Trop.Med.Hyg.* 20: 288 (1971).
2. W.F. Larrabee, L. Ajello and L. Kaufman, An epidemic of histoplasmosis on the isthmus of Panama, *Am.J.Trop.Med.Hyg.* 27: 281 (1978).
3. P.C. Johnson, G.A. Sarosi, E.J. Septimus and T.K. Satterwhite, Progressive disseminated histoplasmosis in patients with the Acquired Immune Deficiency Syndrome: a report of 12 cases and a literature review, *Semin.Respir. Infect.* 1: 1 (1986).
4. G.S. Kobayashi, G. Medoff, B. Maresca, M. Sacco and B.V. Kumar, Studies on phase transitions in the dimorphic pathogen *Histoplasma capsulatum*, in: P.J. Szaniszlo, ed., "Fungal Dimorphism with Emphasis on Fungi Pathogenic for Humans", Plenum Press, New York (1985).
5. E.J. Keath, A.A. Painter, G.S. Kobayashi and G. Medoff, Variable expression of a yeast-phase-specific gene in *Histoplasma capsulatum* strains differing in thermotolerance and virulence, *Infect. Immun.* 57: 1384-1390 (1989).
6. G. Minchiotti, S. Gargano and B. Maresca, The intron-containing hsp82 gene of the dimorphic pathogenic fungus *Histoplasma capsulatum* is properly spliced in severe heat shock conditions, *Mol. Cell Biol.* 11: 5624 (1991).
7. M. Caruso, M. Sacco, G. Medoff and B. Maresca, Heat shock 70 gene is differentially expressed in *Histoplasma capsulatum* strains with different levels of thermotolerance and pathogenicity, *Mol.Microbiol.* 1: 151 (1987).
8. A.J. Hamilton, L. Jeavons, P. Hobby and R.J. Hay, A 34-to 38-kilodalton *Cryptococcus neoformans* glycoprotein produced as an exoantigen bearing a glycosylated species-specific epitope, *Infect. Immun.* 60: 143 (1992).
9. A.J. Hamilton, M.A. Bartholomew, J.Figueroa, L.E.Fenelon and R.J. Hay, A murine monoclonal antibody exhibiting high species specificity for *Histopasma capsulatum*, *J.Gen.Microbiol.* 136: 331 (1990).
10. F.J. Gomez, A.M. Gomez and G.S. Deepe, An 80-kilodalton antigen from *Histoplasma capsulatum* that has homology to heat shock protein 70 induces cell-mediated immune responses and protection in mice, *Infect.Immun.* 60: 2565 (1992).
11. G. Minchiotti, S. Gargano and B. Maresca, Molecular cloning and expression of hsp82 gene of the dimorphic pathogenic fungus *Histoplasma capsulatum*, *Biochim. et Biophy. Acta* 1131: 103 (1992).

PATHOGENICITY OF *WANGIELLA DERMATITIDIS*

Annemarie Polak, [1] and Dennis M. Dixon[2]

[1] Pharma Division, Preclinical Research
F.Hoffmann-La Roche Ltd
CH-4002 Basel/Switzerland.
[2] National Institutes of Health
Institute of Allergy and Infectious Diseases
Bethesda, Maryland 20892, USA

INTRODUCTION

Wangiella dermatitidis is a polymorphic dematiaceous human pathogen causing phaeohyphomycosis.[2,5,12] The dark pigmentation associated with the majority of the dematiaceous fungi causing chromoblastomycosis and phaeohyphomycosis results from cell wall melanin which is synthesized via pentaketide metabolism and is termed DHN melanin.[6-8,16] *Wangiella dermatitidis* is strongly neurotropic in natural infection in man and in experimental infections in mice.[2,5] *Cryptococcus neoformans* is similarly neurotropic and also produces melanin. However the pigment is DOPA melanin and is derived by a different pathway from that of DHN melanin. DOPA melanin has been shown to be closely associated with virulence in *Cryptococcus neoformans*.[9-11] DHN melanin has been shown to play an important role in pathogenesis of selected phytopathogens.[4,6,16] Thus, it is logical to consider the role of DHN melanin in the pathogenesis of *W. dermatitidis* infections in mammals. This was investigated with the use of spontaneous and UV-derived melanin deficient mutants of *W. dermatitidis* as well as two wild type strains.

The following studies were performed:
- Induction of a fatal systemic infection is mice
- Quantitative culture study of various organs
- Histopathological studies of the brain
- Observation of CNS signs of infection
- Treatment of the wild type fungus with tricyclazole or feeding of the albino mutant with scytalone

All these studies unambigously showed that melanin is an important virulence factor for *W. dermatitidis* as it is for *Cr. neoformans* and for several plant pathogens.

Dimorphic Fungi in Biology and Medicine, Edited by
H. Vanden Bossche *et al.*, Plenum Press, New York, 1993

STUDIES OF VIRULENCE AND PATHOGENICITY[1-4,13,14]

The strains of *Wangiella dermatitidis* used were all derived from the same parent (DMD368, ATCC34100). Several albino mutants Mel3 = DMD 369 or 372, Mel2, UV10, 17, 22, 23 and a reddish brown mutant Mel1 were additionally used. A second wild-type strain CM26 was also tested.

The first experiments in an acute infection model showed a 100% mortality within 3 to 5 days following intravenous injection of the wild types CM26 or DMD368. These demonstrated a high degree of virulence since the mortality rate was comparable to those rates seen with *Histoplasma capsulatum* or *Cr. neoformans*. In sharp contrast to results with the wild type both strains of the Mel3 albino mutant tested were essentially non-lethal during the first 21 days after inoculation (Fig. 1). After a prolonged time, mice infected with the albino mutants began to exhibit bizarre behaviors showing some CNS signs of infection such as torticolis and ataxia. Also death began to occur in a linear fashion in these mice infected with the mutants. Thus, there was a dramatic decrease in the virulence of the naturally occuring Mel3 albino mutants relative to the black parental wild type, yet the albino mutant remained pathogenic (Fig. 1).

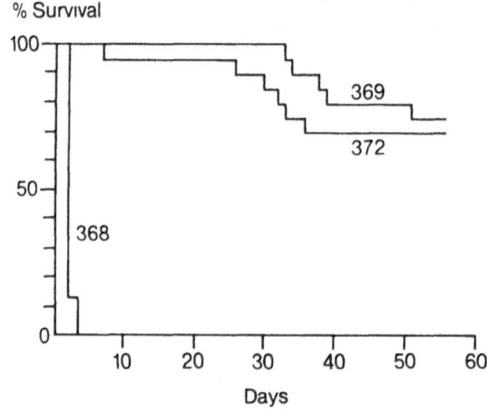

Fig. 1 Comparative survival of mice infected with 3 x 10^7 cells of *W. dermatitidis*. Wild type (DMD 368) vs albino mutants (DMD 369, DMD 372).

Additionally, melanin-deficient strains derived by UV-irradiation were tested. All exhibited significant reduction in mortality (Fig. 2). A reddish brown (dehydratase deficient) mutant[4] was also studied. This strain showed a clear reduced mortality in comparison to the wild type but it killed significantly more mice than any of the other mutants tested. The reddish brown pigments still produced by this mutant seem to replace partially the function of melanin as a virulence factor.

Kwon Chung described DOPA melanin as a virulence factor of *Cr. neoformans*. She observed a significant reduction in virulence with melanin-deficient strains of *Cr. neoformans*.[9-11,15] This reduction in virulence was defined on the basis of prolonged survival of mice infected with mutant strains relative to wild types and the inability of mutants of *Cr. neoformans* to multiply in the brain. The wild-type and albino mutants Mel3 grew exponentially in mouse brains, reached a plateau four days after inoculation, and declined thereafter. No significant quantitative differences were noted between the colony-forming units obtained from brains of mice infected with either wild type or Mel3 albino strains.[3] Thus, *W. dermatitidis* devoid of melanin is still able to persist and even multiply in the brain. These findings were unexpected and are in contrast to the findings of Kwon Chung with *Cr. neoformans*.[9-11]

Another observation was made. Melanized reverants were never recovered from the brains of mice infected with the albino mutants. Furthermore, histopathological sections of brains stained by Fontana Masson stain for melanin, indicated that Mel3 mutants remain melanin-deficient in vivo throughout the course of infection.

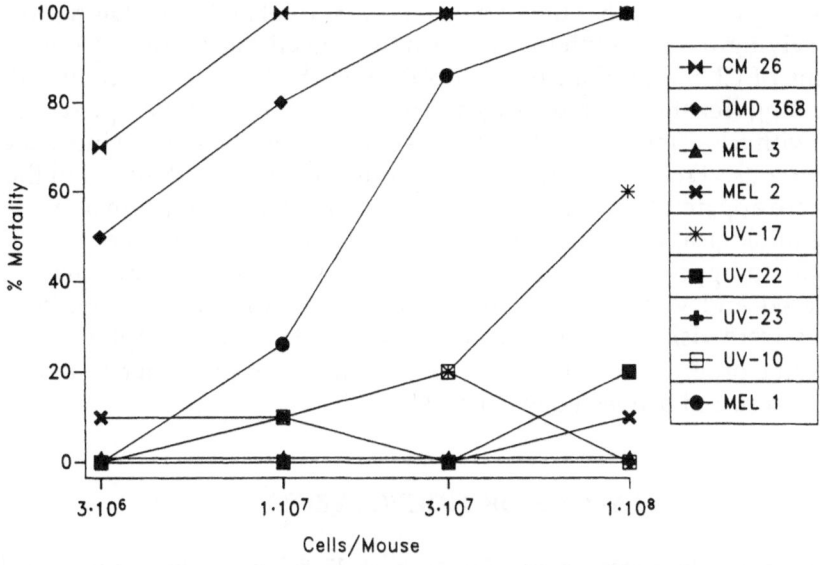

Fig. 2 Mortality response following intravenous injection of *W. dermatitidis*. Wild type (CM 26, DMD 368) and Mel⁻ mutants in Füllinsdorf mice. Data points represent mean values at day 21 after inoculation of groups of five mice per concentration.

Histopathological studies were conducted in conjunction with the quantitative culture studies. Mice were killed on day 1, 2, 7, 14 and 56 after the inoculation; brains were fixed and stained with hematoxylin and eosin (HE) or Grocott-Gomori methenamine silver (GMS). Individual yeast-like cells were seen in capillaries at the day 1 sample time. Extravascular clusters of yeast-like cells appeared in micro abscesses as discrete lesions 2 days after inoculation. On day 4 after inoculation the number of foci was 90 for the wild type, 50 and 120 respectively for the two albino mutants. The size of the fungal foci ranged from 50 to 500 µm with an average diameter of 175 µm and appeared also to be independent of melanin production. At day 14 after inoculation there was a dramatic decrease in the number of lesions, and this sequence correlated exactly with the trend observed by quantitative cultures of the organism from the brain. At this time there were only occasional foci seen at any of the sample times for any of the 3 fungal strains. In such areas there was indication of vasculitis and infrequent tissue regeneration.[1]

The histopathology response as determined from HE-staining showed an increasing number of polymorphonuclear leukocytes up through day 4, mononuclear cells were apparent by day 7. Fewer lesions (scarce) seen from day 14 onwards were of the granulomatous type and consisted of no more than 10 mononuclear cells each. Also in this respect there were no histological differences seen among the brains infected with the different fungal strains.[1,3]

Evaluation of the mycological appearance of the foci revealed that for both wild type and albino mutants a gradual transformation within the lesions from yeast-like cells to pseudohyphae and hyphal forms occured. This happened between days 1 and 4 after inoculation. Subsequently a decrease in the amount of fungus was seen and an increase in the amount of necrotic tissue and fungal debris within the lesion. The most exceptional histopathological finding was the appearance of hyphal growth extending beyond the periphery of the focal lesions to invade contiguous tissue. These invasive hyphae were found more often with the wild-type strain than with any albino strains. Thus it would be of interest to study the relationship between hyphal formation and virulence in this fungus.[1]

It seems, however, that the pathogenicity is unrelated to melanin production. Mice surviving the acute phase of infection showed later (after 21 days) typical symptoms of CNS disturbance. Torticolis and ataxia were noted, and the mice were running frenetically in the cages. These neurological signs appeared in mice infected with a low inoculum of wild type strain as well as in mice inoculated with albino mutants. From these findings one could conclude that although the albino mutants are apparently less virulent in respect of lethality they remain pathogenic. A histopathological basis for the CNS signs was unfortunately not found. Their manifestation appeared at the time when the lesions of the brain have already disappeared and viable fungal cells were greatly diminished, minimal tissue responses were visible but no correlation was seen to the symptoms. The CNS signs became progressively more frequent and more severe from day 21 onwards[1], and death occurred between days 21 and 56.

STUDIES WITH SCYTALONE OR TRICYCLAZOLE[13]

Wangiella cells were grown for 48 h at 37 °C in shaken Sabouraud broth cultures with or without the addition of either 50 µg scytalone/ml or 100 µg tricyclazole/ml. Cells were sedimented by centrifugation, washed in sterile physiological saline and resuspended to the desired inoculum concentration. Wild-type strains of *W. dermatitidis* were grown under the same conditions as controls.

Addition of 50 µg/ml scytalone during exponential growth in vitro resulted in melanized cells that reproducibly showed increased virulence relative to untreated melanin-deficient mutants (Fig. 3). Depending on the inoculum, this increase was 6 to 48%. These increases in virulence were always statistically significant. Treatment with scytalone in vivo did not increase the effect: scytalone most probably does not reach the brain tissue, where the fungus multiplies in sufficient amounts.

Tricyclazole treatment of wild type provided brown-pigmented, but not melanized fungus cells. These melanin-depleted cells showed decreased virulence relative to the untreated controls. This decrease in virulence was most obvious at the lowest inoculum dose (Fig. 3). There was again no additive effect by continuing treatment with tricyclazole in vivo, but this can be explained by insufficient pharmacokinetic properties. Treatment of *W. dermatitidis* wild type with other inhibitors of melanin synthesis also yielded a brown-pigmented *W. dermatitidis* but the reduction in virulence was not as significant as with tricyclazole therapy.

Tricyclazole therapy gives less impressive results than scytalone, which may be explained by the fact that scytalone treatment of albino mutants really gives black, wild-type cells with a stock of melanin in the cell wall. This melanin stock can be transferred to the daughter cell in the tissue and therefore a significant increase in virulence is observed. The tricyclazole therapy, however, does not produce white albino mutants. It produces cells with a brown pigmentation which can most probably replace some of the functions of the melanin as it was proven with the Mel1 mutant in earlier studies. These experiments with feeding scytalone and therapy with tricyclazole again support the involvement of melanin as a virulence factor in this acute infection of wangiellosis.

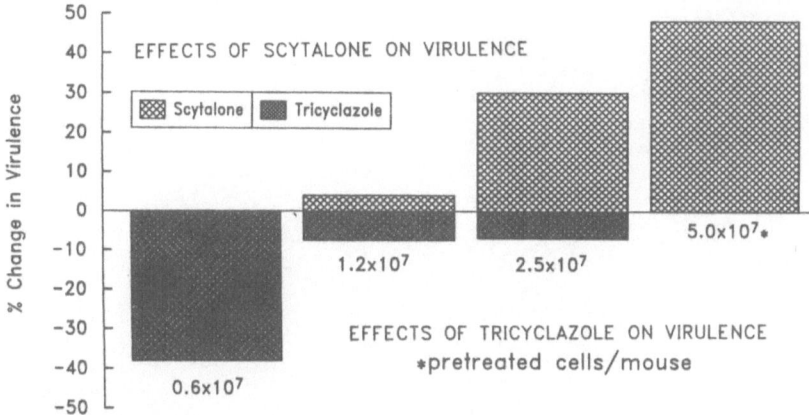

Fig. 3 Effects of scytalone or tricyclazole on the per cent mortality in mice injected intravenously with *W. dermatitidis* (change in per cent mortality to controls).

REFERENCES

1. D.M. Dixon, A. Polak and P.J. Szaniszlo, Pathogenicity and virulence of wild-type and melanin-deficient *Wangiella dermatitidis*, *J. Med. Vet. Mycol.* 25, 97 (1987).
2. D.M. Dixon, A. Polak and T.J. Walsh, Phaeohyphomycosis, in:"Proceedings of the Xth Congress of the International Society of Human and Animal Mycology", J.M. Torrez Rodriguez, eds., J.R. Prous, Barcelona (1988).
3. D.M. Dixon, A. Polak and G.W. Conner, Mel-mutants of *Wangiella dermatitidis* in mice, *J. Med. Vet. Mycol.* 27, 335 (1989).
4. D.M. Dixon, P.J. Szaniszlo and A. Polak, Dihydroxynaphthalene (DHN) melanin and its relationship with virulence in the early stages of phaeohyphomycosis, in:"The Fungal Spore and Disease Initiation in Plants and Animals", G.T. Cole and H.C. Hoch, eds., Plenum Press, New York (1991).
5. R.C. Fader and M.R. McGinnis, Infections caused by dematiaceous fungi: Chromoblastomycosis and phaeohyphomycosis, *Infect. Dis. Clin. N.A.* 2, 925 (1988).
6. P.A. Geis and C.W. Jacobs, Polymorphism in *Wangiella dermatitidis*, in:"Fungal Dimorphism", P.J. Szaniszlo, ed., Plenum Press, New York (1985).
7. P.A. Geis and P.J. Szaniszlo, Carotenoid pigments of the dematiaceous fungus *Wangiella dermatitidis*, *Mycologia* 76, 268 (1984).
8. P.A. Geis, M.H. Wheeler and P.J. Szaniszlo, Pentaketide metabolites of melanin synthesis in the dematiaceous fungus *Wangiella dermatitidis*, *Arch. Microbiol.* 137, 324 (1984).
9. K.J. Kwon-Chung and J.C. Rhodes, Encapsulation and melanin formation as indicators of virulence in *Cryptococcus neoformans*, *Infect. Immun.* 51, 218 (1986).
10. K.J. Kwon-Chung, W.R. Tom and H. Costa, Utilizaition of indole compounds by *Cryptococcus neoformans* to produce a melanin-like pigment, *J. Clin. Microbiol.* 18, 1419 (1983).
11. K.J. Kwon-Chung, I. Polacheck and T.J. Popkin, Melanin-lacking mutants of *Cryptococcus neoformans* and their virulence for mice, *J. Bacteriol.* 1501 1414 (1982).
12. T. Matsumoto, A.A. Padhye, L. Ajello and P.G. Standard, Critical review of human isolates of *Wangiella dermatitidis*, *Mycologia.* 76, 232 (1984).
13. A. Polak , Melanin as virulence factor in pathogenic fungi, *Mycoses.* 33, 215 (1989).

14. A. Polak and D.M. Dixon, Melanin as virulence factor in *Wangiella dermatitidis*: Addition and subtraction of melanin, in:"Program & Abstracts, Xth Congress of the International Society of Human and Animal Mycology, ISHAM", Montreal (1991).

15. J.C. Rhodes, I. Polacheck and K.J. Kwon-Chung, Phenoloxidase activity and virulence in isogenic strains of *Cryptococcus neoformans, Infect. Immun.* 36, 1175 (1982).

16. M.H. Wheeler and A.A. Bell, Melanins and their importance in pathogenic fungi, in:"Current Topics in Medical Mycology", M.R. McGinnis, ed., Springer-Verlag, Berlin (1987).

DO DIMORPHIC FUNGI MORE EASILY ESCAPE HOST DEFENSES AND TREATMENT?

COCCIDIOIDES IMMITIS AS MODEL

David A. Stevens

Division of Infectious Diseases, Department of Medicine, Santa
Clara Valley Medical Center, San Jose, California 95128-2699, USA
California Institute for Medical Research, San Jose, California
Division of Infectious Diseases and Geographic Medicine,
Department of Medicine, Stanford University School of Medicine

To answer the title question I will use as my model the endemic dimorphic pathogen *Coccidioides immitis*. Selected aspects of the interaction of *C. immitis*, its morphogenetic forms and its products with the host will be reviewed. The host interacts with the propagule and with the parasitic form, a total of three entities: arthoconidia of the mycelial form, the spherules and endospores of the parasitic form.

In our earliest studies, with H.B. Levine and others, we showed that spherules contain potent stimulants of cell mediated immunity. Spherulin, which is a lysate of spherules, was a potent detector of delayed-type hypersensitivity in epidemiologic studies in an endemic region.[1] This was further demonstrated, in subsequent studies,[2] in patients with the disease. In studies *in vitro*,[3] spherulin was better able to separate individuals with a prior coccidioidal experience from those unexposed than was the classic reagent, mycelial filtrate. However, in similar *in vitro* studies,[4] killed whole spherules were less potent than either of these two soluble reagents and both of the other morphological forms. This finding is relevant to later observations discussed here.

When we examined direct effects on leukocytes, in chemotaxis assays,[5] we found that both mycelial filtrate and spherulin are chemotaxinogens. Neither are chemotaxins. However, at high concentrations spherulin, unlike mycelial filtrate, inhibited chemotaxis. When we examined their interaction with complement,[6] both mycelial filtrate and spherulin were shown to activate independently both the classic and alternate pathways. At high concentrations spherulin remained active, thus its complement activating properties did not explain the inhibition of chemotaxis seen at high concentrations of spherulin. I will attempt to explain this.

We also studied the antigens produced when *C. immitis* was induced, by manipulation of the culture conditions, to multiply in its parasitic form *in vitro*.[7] We termed these endosporulation antigens. In lymphocyte blastogenesis assays, the endosporulation antigens distinguised skin test-positive and -negative individuals as did spherulin, but they were 500 times as potent as spherulin. Heating or dialysis of the materials increased their potency as a stimulant, whereas concentration of the materials decreased the potency. The endosporulation antigens of an avirulent *C. immitis* isolate were as potent a stimulant as these antgens from a virulent isolate, but were significantly less inhibitory as a concentrate. Thus the parasitic form of *C. immitis* contains both immune stimulatory and suppressive substances.

In the course of studies of patients with disseminated and progressive coccidioidomycosis we found that autologous serum of some patients depressed their lymphocyte blastogenic responses to coccidioidal antigens.[8] We later found, with Rebecca Cox, the presence of immune complexes in the sera of these patients.[9] The presence and the amount correlated with more severe disease. It was later shown these complexes contained parasitic phase substances that had immunosuppressive potential.

Several investigators studied the interaction of phagocytes with the infectious particles themselves.[10-19] It has been demonstrated that polymorphonuclear neutrophils attach poorly to arthroconidia. The neutrophils kill them poorly. Macrophages are unable to kill arthroconidia unless immune lymphocytes are present. With respect to the mechanisms involved, in infected alveolar macrophages (the defensive cells that initially encounter *C. immitis* infection) phagolysosome fusion fails to occur. Cationic peptides, however, can kill arthroconidia. Neutrophils can inhibit the necessary synthetic step of N-acetyl glucosamine incorporation, and this inhibition can be reproduced with hydrogen peroxide.

Arthroconidia have an outer wall layer. When this layer is stripped from the particle, phagocytosis is improved. Soluble components of the inner cell wall have been identified which suppress T cell proliferation and suppress superoxide production by macrophages.

With respect to spherules, as the spherule size increases, phagocytosis and killing by neutrophils decreases. There is a fibrillar material on the surface of spherules that inhibits leukocyte contact. Spherules blunt neutrophil chemiluminescence, and iodination from the neutrophils does not occur. Neutrophil degranulation does not occur in conjunction with phagosomes containing spherules. Neutrophils are inhibited in their ability to prevent spherule N-acetyl glucosamine incorporation. This is not merely a phenomenon of spherule size, because neutrophils can inhibit this function in arthroconidia of similar size. Moreover, as spherules increase in size they become more resistant to inhibition of incorporation by hydrogen peroxide, so the effect of size cannot be an antiphagocytic effect alone.

With respect to endospores, when spherules mature, rupture, and release their endospores the latter are released in a "packet". The matrix for this packet is a membranous viscous material derived from the inner spherule wall. This matrix inhibits neutrophil contact. Moreover, after ingestion, degranulation is poor. Endospores stimulate neutrophil chemiluminescence poorly. It is therefore probably not surprising that neutrophil killing of endospores is poor. Macrophage phagolysomal fusion fails to occur in response to endospores.

We have shown that neutrophils can be activated for fungal killing by the products of an immune reaction. This activation appears to be due mostly, if not entirely to gamma interferon.[20]   Activation in this manner can transform neutrophils into effective killers of endospores, but the cells are still ineffective against arthroconidia.[21] The increased killing of endospores after treatment is associated with an enhanced oxidative burst. However, the failure to kill arthroconidia is not due to a failure to achieve an oxidative burst. Some other property of arthroconidia prevents neutrophil killing of this fungal particle, even by activated neutrophils.

Thus my theme has been that *C. immitis* appears to have several weapons, unique to each morphological form, which thwart host defenses in several ways.

These morphological forms also appear to have unique susceptibilities to antifungals. The parasitic phase was shown to be more susceptible than arthroconidia to the polyene amphotericin B.[22,23] We showed endospores to be more susceptible *in vitro* to the imidazole, miconazole, than arthroconidia.[24] In contrast, resistance to the triazole, itraconazole, appeared to be greater in the parasitic phase. (However, both forms were susceptible to both miconazole and itraconazole). Moreover, synergy of itraconazole with rifampin in inhibition and in killing was greater for arthroconidia.[25] Cilofungin is a β-1,3-glucan synthetase inhibitor. In studies with John Galgiani, we showed inhibition of coccidioidal cell wall synthesis *in vitro* by inhibition of N-acetyl glucosamine uptake. However, *in vivo*, the drug had no effect on coccidioidal infection. This may relate to an almost nil drug effect on developing spherules, the parasitic form of *C. immitis*, as compared to a fairly substantial effect on arthroconidia.[26]

Therefore, targeting of antifungal drugs must take account of the differences between forms. With uncertainties about the meaning of differences in *in vitro* susceptibility between morphological forms, the necessity of using animal models in testing anticoccidioidal therapy is underscored.

REFERENCES

1. D.A. Stevens, H.B. Levine and D.R. Ten Eyck, Dermal sensitivity to two different doses of spherulin and coccidioidin, *Chest* 65: 530 (1974).
2. D.A. Stevens, H.B. Levine, S.C. Deresinski and L.J. Blaine, Spherulin in clinical coccidioidomycosis, *Chest* 68: 697 (1975).
3. S.C. Deresinski, H.B. Levine and D.A. Stevens, Soluble antigens of mycelia and spherules in the *in vitro* detection of immunity to *Coccidioides immitis*, *Infect. Immun.* 10: 700 (1974).
4. S.C. Deresinski, R. Applegate, H.B. Levine and D.A. Stevens, Cellular immunity to *Coccidioides immitis*: in vitro lymphocyte response to spherules, arthrospores and endospores, *Cell. Immunol.* 32: 110 (1977).
5. J.N. Galgiani, R.A. Isenberg and D.A. Stevens, Chemotaxigenic activity of extracts from the mycelial and spherule phases of *Coccidioides immitis* for human polymorphonuclear leukocytes, *Infect. Immun.* 21: 862 (1978).
6. J.N. Galgiani, P. Yam, L.D. Petz, P.L. Williams and D.A. Stevens, Complement activation by *Coccidioides immitis*: in vitro and clinical studies, *Infect. Immun.* 28: 944 (1980).
7. C. Brass, H.B. Levine and D.A. Stevens, Stimulation and suppression of cell-mediated immunity by endosporulation antigens of *Coccidioides immitis*, *Infect. Immun.* 35: 431 (1982).
8. R.P. Harvey and D.A. Stevens, In vitro assays of cellular immunity in progressive coccidioidomycosis: evaluation of suppression with parasitic phase antigen, *Am. Rev. Resp. Dis.* 123: 665 (1981).
9. R.A. Cox, R.M. Pope and D.A. Stevens, Immune complexes in coccidioidomycosis. Correlation with disease involvement, *Am. Rev. Resp. Dis.* 126: 439 (1982).
10. L. Beaman and C.A. Holmberg, Interaction of nonhuman primate peripheral blood leukocytes and *Coccidioides immitis* in vitro, *Infect. Immun.* 29: 1200 (1980).
11. L. Beaman and C.A. Holmberg, In vitro response of alveolar macrophages to infection with *Coccidioides immitis*, *Infect. Immun.* 28: 594 (1980).

12. L. Beaman, E. Benjamini and D. Pappagianis, Role of lymphocytes in macrophage-induced killing of *Coccidioides immitis* in vitro, *Infect. Immun.* 34: 347 (1981).

13. L. Beaman, E. Benjamini and D. Pappagianis, Activation of macrophages by lymphokines: Enhancement of phagosome-lysosome fusion and killing of *Coccidioides immitis*, *Infect. Immun.* 39: 1201 (1983).

14. M. Huppert, G.T. Cole, S.H. Sun, D.J. Drutz, P. Starr, C.L. Frey and J.L. Harrison, The propagule as an infectious agent in coccidioidomycosis, in: "Microbiology 1983", D. Schlesinger, ed., American Society for Microbiology, Washington, D.C. (1983).

15. J.N. Galgiani, C.M. Payne and J.F. Jones, Human polymorphonuclear leukocyte inhibition of incorporation of chitin precursors into mycelia of *Coccidioides immitis*, *J. Infect. Dis.* 149: 404 (1984).

16. K.V. Clemons, C.R. Leathers and K.W. Lee, Role of activated macrophages in resistance to experimental *Coccidioides immitis* infections, in: "Coccidioidomycosis: Proceedings of the 4th International Conference on Coccidioidomycosis", H. Einstein and A. Catanzaro, eds., National Foundation for Infectious Diseases, Washington, D.C. (1985).

17. G.P. Segal, R.I. Lehrer and M.E. Selsted, In vitro effect of phagocyte cationic peptides on *Coccidioides immitis*, *J. Infect. Dis.* 151: 890 (1985).

18. C.L. Frey and D.J. Drutz, Influence of fungal surface components on the interaction of *Coccidioides immitis* with polymorponuclear neutrophils, *J. Infect. Dis.* 153: 933 (1986).

19. G.T. Cole and T.N. Kirkland, Conidia of *Coccidioides immitis*, in: "The Fungal Spore and Disease Initiation in Plants and Animals", G.T. Cole and H.C. Hoch, eds., Plenum Press, New York (1991).

20. D.A. Stevens, Interferon-gamma and fungal infections, in: "The Anti-Infective Applications of Interferon-Gamma", H.S. Jaffe, L.R. Bucalo and S.A. Sherwin, eds., Marcel Dekker, New York (1992).

21. E. Brummer, L. Beaman and D.A. Stevens, Killing of endospores, but not arthroconidia, of *Coccidioides immitis* by immunologically activated polymorphonuclear neutrophils, in: "Coccidioidomycosis: Proceedings of the 4th International Conference on Coccidioidomycosis", H. Einstein and A. Catanzaro, eds., National Foundation for Infectious Diseases, Washington, D.C. (1985).

22. M.S. Collins and D. Pappagianis, Lysozyme-enhanced killing of *Candida albicans* and *Coccidioides immitis* by amphotericin B, *Sabouraudia* 12: 329 (1974).

23. M.S. Collins and D. Pappagianis, Uniform susceptibility of various strains of *Coccidioides immitis* to amphotericin B, *Antimicrob. Agents Chemother.* 11: 1049 (1977).

24. H.B. Levine, D.A. Stevens, J.M. Cobb and A.E. Gebhardt, Miconazole in coccidioidomycosis. I. Assays of activity in mice and in vitro, *J. Infect. Dis.* 132: 414 (1975).

25. R.M. Tucker, D.W. Denning, L.H. Hanson, M.G. Rinaldi, J.R. Graybill, P.K. Sharkey, D.Pappagianis and D.A. Stevens, The interaction of azoles with rifampin, phenytoin and carbamazepine: in vitro and clinical observations, *Clin. Infect. Dis.* 14: 165 (1992).

26. J.N. Galgiani, S.H. Sun, K.V. Clemons and D.A. Stevens, Activity of cilofungin against *Coccidioides immitis*. Differential in vitro effects on mycelia and spherules correlated with in vivo studies, *J. Infect. Dis.* 162: 944 (1990).

Diagnosis and Treatment of Mycoses

SIGNIFICANCE OF MORPHOLOGIC FORM IN PRODUCING IMMUNODIAGNOSTIC ANTIGENS FOR DIMORPHIC FUNGAL PATHOGENS

Leo Kaufman

Mycotic Diseases Branch, MS G-11
Division of Bacterial and Mycotic Diseases
National Center for Infectious Diseases
Center for Disease Control
Atlanta, Georgia, 30333, USA

ABSTRACT

Selection of the appropriate anamorph of a dimorphic fungal pathogen for the isolation and production of useful diagnostic antigens is essential to the development of immunologic tests for diagnosing systemic mycotic infections and identifying their etiologic agents. The routine immunodiagnosis of blastomycosis, coccidioidomycosis, and histoplasmosis continues to be achieved with crude or purified antigens derived from the growth of mycelial, yeast, and/or tissue forms of their respective etiologic agents.

Blastomyces dermatitidis yeast forms and A antigen, *Histoplasma capsulatum* var. *capsulatum* yeast forms and mycelial H and M antigens, and *Coccidioides immitis* tube precipitin and complement fixation test antigens derived from mycelial coccidioidin serve well as aids for diagnosis and prognosis. To improve further the sensitivity and specificity of the immunodiagnostic assays, current studies are directed to the identification, isolation, characterization, and purification of conventional, as well as new sensitive, specific and clinically significant antigens, and to relating antigen development and production to specific phases in the morphogenesis of the mycelial and tissue forms.

INTRODUCTION

The fungal pathogens *Blastomyces dermatitidis, Coccidioides immitis,* and *Histoplasma capsulatum* var. *capsulatum* (henceforth referred to as *H. capsulatum*) exhibit a hyphal-conidium to yeast or spherule dimorphism. The host animal is usually infected by the saprobic or mycelial form, which *in vivo* converts to the tissue form. Although immunodiagnosis is usually best accomplished with

Dimorphic Fungi in Biology and Medicine, Edited by
H. Vanden Bossche *et al.,* Plenum Press, New York, 1993

reagents homologous to the parasitic form, this is not always the case. Immunodiagnostic tests for many of the systemic mycotic infections caused by the dimorphic pathogens use antigens derived from either or both forms of the fungus. The antigens may be fungal elements, i.e., whole or broken yeast cells, or homogenized mycelia, or they may be soluble antigens present in culture filtrates, cytoplasmic components, or cell extracts. The selection of the morphologic form and type of antigen most useful for producing immunodiagnostic reagents is determined by various factors: their antigenicity, and specificity, the concentration and quantity of their epitopes, their sensitivity and specificity in a particular type of test, the physical nature of the antigen(s) required, or the nature and structure of the target antigen or antibody. The form's ability to grow readily in culture to an abundant mass is also important as is the need for and availability of appropriate biosafety level facilities for their growth. The significance of morphologic form for isolation and production of diagnostic antigens for the aforementioned dimorphic pathogens is discussed in this paper.

Blastomyces dermatitidis

As early as 1949[1] the yeast and mycelial (blastomycin) culture filtrate antigens of *B. dermatitidis* were noted to exhibit cross-reactivity in complement fixation (CF) tests, particularly with *H. capsulatum* antisera. The yeast-form antigens of the fungus, however, were considered more specific than those derived from mycelium. Subsequent CF and immuno-diffusion (ID) studies[2,3] supported this premise (Table 1). Test specificity, however, was far from desirable, as indicated by extensive independent studies. These studies showed that *B. dermatitidis* yeast cells shared antigens with the yeast, tissue and mycelial form elements of a variety of fungal pathogens including *Candida spp., Coccidioides immitis, H. capsulatum* varieties *capsulatum, duboisii,* and *farciminosum, Paracoccidioides brasiliensis,* and *Penicillium marneffei.*[2,4,5,6]

Comparative CF studies with yeast- and mycelial-form filtrate and homogenate antigens, derived from identical strains of *B. dermatitidis*, revealed that the mycelial antigens were relatively insensitive (Table 1).[3] Yeast-form culture filtrate and homogenized yeast cell antigens have been used for many years in CF tests, with antibodies usually detected in less than 50% of the sera from patients with proven blastomycosis. Analytical studies indicate that the major active CF antigens appear to be cytoplasmic in nature.[7]

Busey and Hinton[2] using micro-ID tests with phenolized yeast and mycelial-form filtrate antigens, detected 80% of 80 proven cases of blastomycosis with yeast antigens and 48% with the mycelial antigens (Table 1). Cross-reactions were evident, particularly with histoplasmosis case sera. Subsequently micro-ID tests in my laboratory[3] at CDC with 49 blastomycosis case sera and yeast culture filtrate antigens demonstrated a 59% sensitivity and 83% specificity. Cross-reactions occurred with coccidioidomycosis, histoplasmosis, and paracoccidioidomycosis case sera. Examination of the immunoprecipitates revealed that a single band, designated "A", occurred only with proven blastomycosis case sera.[3] The 49 blastomycosis case sera as well as a variety of heterologous case sera were retested in parallel with a reference serum containing the A precipitin. Fifty-seven percent of the blastomycosis case sera produced a band of identity with the reference A precipitate, compared with none of 104 heterologous case sera. In a more extensive evaluation of the A precipitinogen with 113 sera from proven cases of blastomycosis, 79% of the cases were specifically detected (Table 1). The use of the

Table 1. Value of yeast and mycelial-form antigens in complement fixation (CF) and immunodiffusion (ID) tests for blastomycosis.

Test	Yeast antigens			Mycelial antigens			References
	Type	Sens.[a]%	Spec.[b]%	Type	Sens.%	Spec.%	
CF	F[c]	51	ND[g]				
ID	F	80	X[h]	F	48	X	2
CF	H[d]	41	65	H,F	Poor	ND	3
ID	F 59	83					
ID	A[e]	57	100				
ID	A	79	100				
CF	PA[f]	40	100				9
ID	PA	65	100				

[a] Sensitivity. [b] Specificity. [c] Culture filtrate. [d] Homogenized. [e] *B. dermatitidis A.* [f] Purified A. [g] Not determined. [h] Cross reactive.

A antigen, contained in concentrated yeast culture filtrate, and the incorporation of the reference A precipitate in the micro-ID test[3] contributed to the test's increased sensitivity and specificity. In 1980 Green and coworkers[8] purified the A antigen by DEAE anion exchange column fractionation. This antigen missed 60% and 35% of the cases, respectively,[9] in CF and ID tests, although it was specific in both (Table 1).

The use of yeast-derived purified A antigen in enzyme immunoassays (EIA) resulted in the improved detection of blastomycosis cases (Table 2). Specificity varied from 98% to 100%, depending upon the test format.[9,10] Sodium dodecyl sulfate-polyacrylamide gel electrophoresis (SDS-PAGE) analysis of the purified yeast culture filtrate by Hurst and Kaufman.[11] showed it to consist of a minimum

Table 2. Value of yeast-form antigens in enzyme immunoassay, radioimmunoassay, and immunoblotting assays for blastomycosis.

Antigen Type	Test	Sensitivity %	Specificity %	References
PA[a]	I-EIA[b]	80	98	9
PA	S-EIA[c]	88	100	10
98 kDa	WB[d]	86	92	11
120 kDa	RIA[e]	85	100	13

[a] Purified A. [b] Indirect enzyme immunoassay. [c] Sandwich enzyme immunoassay. [d] Western blot. [e] Radioimmunoassay.

of 30 proteins. Five of these were reactive with blastomycosis case sera by Western blot. One of the bands, a 98-kDa predominantly protein antigen, demonstrated a sensitivity of 86% with blastomycosis case sera. Specificity, although high, was not absolute, and cross-reactions with histoplasmosis case sera continued to be a problem. The 98-kDa antigen was electroeluted and shown in ID tests to be identical to the A antigen. The exact cellular location of the A antigen is uncertain, although it has been found in the cytoplasm, cell wall, and culture filtrates of all *B. dermatitidis* strains tested.[12]

More recently, Klein and Jones[13] described a 120-kDa protein derived from the cell wall of the yeast form of *B. dermatitidis* (Table 2). Radioimmunoassays (RIA) with this antigen demonstrated a sensitivity of 85% and total specificity. The antigen is a heat stable protein with many properties similar or identical to the 98-kDa antigen described by Hurst and Kaufman.[11]

Kaplan and Kaufman[4] (Table 3) investigated the direct fluorescent antibody (FA) staining characteristics of fluorescein-labeled antiglobulins prepared against yeast-form cells of *B. dermatitidis*. They found that in addition to staining the yeast and mycelial forms of *B. dermatitidis*, the reagent also cross-stained the yeast and mycelial forms of *H. capsulatum*, *P. brasiliensis*, *P. marneffei*, and the cells of several other morphologically similar heterologous fungi. A conjugate specific only for the yeast form of *B. dermatitidis* was produced by adsorptions with the yeast cells of the *capsulatum* variety of *H. capsulatum* and *Geotrichum candidium*. Interestingly, this conjugate did not stain elements of the mycelial form of *B. dermatitidis*. In contrast, the yeast-derived A antigen has proved in exoantigen tests (Table 3) to be a sensitive and accurate marker for identifying both the mycelial and yeast anamorphs of *B. dermatitidis*.[6]

Table 3. Value of yeast form-derived reagents for immunoidentification of *Blastomyces dermatitidis*.

Antigen or Reagent Type	Test	Sens.[a] %	Spec.[b] %	Form identified	References
Antiglobulin to whole antibody yeast cells	Fluorescent	100	100	Yeast	4
A	Exoantigen	100	100	Mycelial & yeast	6

[a] Sensitivity

[b] Specificity

Coccidioides immitis

The dimorphic pathogen *C. immitis* exists in nature and in normal culture in a hyphal form with arthroconidia. *In vivo* and under special laboratory conditions, the arthroconidia form spherules which endosporulate. In the 1930s Smith and coworkers[14] speculated that spherule-endospore elements would be necessary for the effective serologic diagnosis of coccidioidomycosis. In the absence of methods to produce these elements, they subsequently proceeded to develop the valuable classic CF and tube precipitin (TP) tests for detecting coccidioidal antibody using the mycelial culture filtrate antigen coccidioidin (Table 4). Later Huppert and Bailey[15,16] devised the IDCF and IDTP tests that demonstrated in an agar gel milieu reactions which corresponded to those of the CF and TP tests, respectively. These ID tests have been shown to be more sensitive than their conventional counterparts. In 1968[17] a latex agglutination (LA) test, designed to detect antibody similar to the TP, was introduced. Interestingly, for all the aforementioned tests, coccidioidin and the more concentrated hyphal toluene autolysate have proved to be excellent sources of antigen.

Table 4 lists the tests most widely used for diagnosing coccidioidomycosis, the antigens used therein, the type of antibody generally detected, test sensitivity and specificity, and the morphotype producing the diagnostic antigens. The detection of IgG antibody is useful for diagnosing the acute and late stages of coccidioidomycosis. This IgG detection may be accomplished with the CF or qualitative IDCF tests, which can detect 85% or greater than 90% of the primary nondisseminating case sera, respectively. Nonheated coccidioidin is used in both tests since the CF antigen is heat labile. The CF antigen, a 110-kDa native protein, is present in the lysates of hyphae and mature spherules and is probably cytoplasmic in origin.[14] Upon reduction and SDS-PAGE immunoblotting, the CF antigen is evident as a 48-kDa band. Recent studies have shown the antigen to be a chitinase.[18] The CF test is subject to low titer nonspecific reactions, particularly with histoplasmosis case sera,[19] whereas the ID is totally specific when lines of identity are noted with mandatory reference IDCF precipitates.

In 1960 Levine et al.[20] perfected a medium for facilitating the in-vitro growth of the spherule-endospore form of C. immitis. This achievement allowed the comparative evaluation of spherulin, the aqueous lysate of spherules, with coccidioidin in CF tests with human sera, in which its sensitivity was noted to correlate with that of coccidioidin. The spherule-derived antigen, however, was found to be significantly less specific than the mycelial-derived antigen[21] and offered no diagnostic advantage over coccidioidin. Thus, cross-reacting CF antigens are produced by both the mycelial and spherule forms but more extensively by the latter.

The TP, IDTP, and LA tests, when used with coccidioidin, effectively detect early C. immitis IgM antibody and allow the serodiagnosis of over 90% of the early acute cases of coccidioidomycosis (Table 4). With the exception of the LA test, which may show a false-positive rate ranging between 6% and 15%, these tests are highly specific.[17] The antigen responsible for binding IgM is resistant to heat and is readily obtained from mycelial culture filtrates and lysates as well as from arthroconidia, hyphae, spherules, and endospores.

Recent studies suggest that TP reactivity resides in at least two antigens. Zimmer and Pappagianis[22] noted TP activity with coccidioidal glycoproteins estimated to be 140-kDa and 225- kDa, whereas Kruse and Cole[23] characterized the TP antigens as 120 and 110-kDa products, with the 120-kDa antigen yielding a 240-kDa dimer. Immunoelectron microscopic studies have localized the 120-kDa TP antigen on the walls of in vitro-grown arthroconidia and spherules.[24]

The IDTP antigen appears to be released early in the in-vitro endospore to spherule development and to persist during the formation and maturation of the spherules. In contrast, the IDCF antigen is produced at the time cleavage planes occur in immature spherules and thereafter.[14] Galgiani et al.[25] recently described a 33-kDa antigen found in abundance in mature spherules and endospores, and in lesser quantities in the inner wall of arthroconidia. This antigen appears to be very promising, since in preliminary EIAs for *C. immitis* IgG and IgM antibodies, it demonstrated a sensitivity of 95% and a specificity of 98%.

The mycelial form of *C. immitis* has been shown to share antigens with a variety of pathogenic and nonpathogenic fungi, including *Auxarthron* spp., a *Geotrichum* sp., *H. capsulatum*, and *Uncinocarpus reesii*.[26] Exoantigen studies indicate, however, that typical, atypical, and nonsporulating mycelial *C. immitis* cultures produce three antigens that are specific. These are the HS antigen, so designated because it is heat stable at 60 °C for 30 min, and two heat labile antigens designated HL and F. The IDHS antigen is distinct and different from the IDTP antigen. The latter, although specific for serodiagnosing coccidioido mycosis, cannot be used for immunoidentification of *C. immitis* because it is shared by such arthroconidium-forming fungi as *Arachniotus reticulatus*, *Auxarthron zuffianun*, and *Malbranchea filamentosa*. The IDHL is a distinct heat labile antigen, whereas the F antigen is identical to the serodiagnostic IDCF antigen.[27] The IDHS antigen (Table 5) is consistently the first of the exoantigens to be produced. Thus it is the main

Table 4. Value of mycelial form coccidioidin (CN) in immunologic tests for coccidioidomycosis.

Antigen	Detects	Tests	Sensitivity %	Specificity %	Antigen sources
CN	IgG	CF/IDCF	85/>90	>95/100	Mature spherules, endospores, hyphae, mycelial culture filtrates, and lysates
CN/heated CN	IgM	TP/IDTP	90/>90	99/>90	Arthroconidia, hyphae, spherules, endospores, mycelial culture filtrates, and lysates
Heated CN	IgM	LA	>90	<94	Mycelial toluene lysates or culture filtrates

antigen used to identify *C. immitis* mycelial form isolates. The antigen is present in both mycelial and spherule-endospore culture filtrates and lysates.[14] Exoantigen tests with the HS antigen are totally specific and sensitive and eliminate the need for time-consuming and risky demonstration of the dimorphic nature of a culture suspected of being *C. immitis*.[26]

FA studies with conjugates produced against arthroconidia and hyphae indicated that these structures share antigens not only with the tissue form of *C. immitis*, but also with the yeast and mycelial forms of *B. dermatitidis*, *H. capsulatum*, and *P. brasiliensis*, as well as with elements of other heterologous fungal species. The spherules share antigens with both forms of *H. capsulatum* and with the mycelial form of *B. dermatitidis*. Spherule conjugates, although less cross-reactive, were low titered and impractical. Fluorescein-labeled rabbit antiglobulins produced against arthroconidia, either adsorbed with *H. capsulatum* yeast cells or diluted, were most effective for specifically identifying *C. immitis in vivo* (Table

Table 5. Value of *Coccidioides immitis* antigens for immunoidentification.

Reagent	Detects %	Tests %	Sens.[a]	Spec.[b]	Antigen sources
HS antiserum[c]	HS antigen in *C. immitis* mycelial cultures	Exoantigen	100	100	Mycelial culture filtrates and toluene lysates; spherule/endospore filtrates, and lysates.
Fluorescein-labeled anti-globulin to viable or non-viable arthroconidia[d]	Endospore and spherule antigens in vivo	FA	88	100	Spherules, hyphae, mycelial culture filtrates, and lysates

[a] Sensitivity

[b] Specificity

[c] HS precipitin arcs used to produce antiserum

[d] Conjugates consist of antiglobulin to arthroconidia

5).[28] The adsorbed reagents also weakly stained elements of the mycelial form of *C. immitis*.

It is obvious from the various described studies that antigens from both anamorphs of *C. immitis* are reactive with precipitin-positive and CF-positive human coccidioidomycosis case sera. The mycelial form, however, continues to remain the standard source for reliable and potent serodiagnostic antigens. Smith's original hypothesis that tissue-form elements would provide the best antigens for serodiagnosing coccidioidomycosis, although not fulfilled, may become a reality as easily accessible, new, sensitive, specific and clinically relevant antigens are isolated from spherule-endospore elements.[25]

Histoplasma capsulatum

Immunologic tests often provide the earliest evidence for diagnosing histoplasmosis. Such evidence may be obtained through use of CF and ID tests for antibody, exoantigen and FA tests for immunoidentification of the *H. capsulatum* varieties, and the RIA for *Histoplasma* polysaccharide antigen (HPA) detection. Knowledge of these tests' specificity and use of appropriate controls or reagents are mandatory since many of the antigens used in or detected by these tests contain cross-reactive epitopes and are thus not entirely specific. Although both the yeast and mycelial anamorphs of *H. capsulatum* var. *capsulatum* share antigens with one another as well as with the anamorphs of *B. dermatitidis*,[29] they are important sources of diagnostic antigens or reagents and are used singly or in some combination to achieve maximal diagnosis.

The yeast-form antigens of *H. capsulatum* are important for serodiagnosing histoplasmosis. The CF test with whole yeast cells of *H. capsulatum* remains the single most sensitive test for *H. capsulatum* antibody. These antigens detect over 90% of the case sera, in contrast to mycelial histoplasmin, which is positive with about 83% of case sera (Tables 6 and 7).[30] False- positive reactions may occur in the low titer range since the yeast form shares epitopes with *B. dermatitidis, C. immitis, P. brasiliensis,* and other mycotic agents as well as with *Leishmania* spp. The CF antigens appear to reside in the yeast cell wall;[7] however, attempts to isolate them and to increase their specificity have been essentially unsuccessful.[31]

Subcellular antigens prepared from yeast-form *H. capsulatum* var. *capsulatum* ribosomes demonstrate excellent sensitivity but poor specificity. Anti-ribosomal antibodies have been detected in 97% of histoplasmosis patients, but the antigens were unsatisfactory since extensive cross-reactions occurred with sera from patients with aspergillosis, blastomycosis, candidiasis, coccidioidomycosis, cryptococcosis, paracoccidioidomycosis, and tuberculosis.[32]

A specific FA for the yeast form of *H. capsulatum* can be prepared by adsorption of labeled antiglobulins with yeast cells of *B. dermatitidis* (Table 6). This var. *capsulatum* conjugate lacks sensitivity, however, because it fails to react with one of the five recognized serotypes of *H. capsulatum*, the 1:4 serotype, as well as with the *duboisii* variety which share epitopes with the yeast form of *B. dermatitidis.*[33,34] The latter conjugate is specific for its yeast anamorph and will not react with mycelial-form elements of *H.* capsulatum. A polyvalent reagent, capable of staining the five serotypes of the *capsulatum* and *duboisii* varieties of *H. capsulatum* as well as *B. dermatitidis,* was prepared by adsorption with *C. albicans.* Differentiation of *B. dermatitidis* from the *Histoplasma* varieties was accomplished by application of a *B. dermatitidis*-specific FA.

The isolation of a yeast form *H. capsulatum*-specific antibody has proved elusive. Kamel *et al.,*[35] produced four monoclonal antibodies (MAbs) with a disrupted yeast cell homogenate of *H. capsulatum* var. *capsulatum.* Two of the MAbs recognized an epitope of *H. capsulatum* that is shared with the yeast form of *B. dermatitidis.* The third cross-reacted with the yeast antigens of *B. dermatitidis, P. brasiliensis, Sporothrix schenckii, and C. albicans,* while the fourth cross-reacted with the yeast form of *B. dermatitidis,* the mycelial antigens of *C. immitis* and *Aspergillus fumigatus,* and with *Mycobacterium tuberculosis.* Recently Hamilton *et al.,*[36] using cyclophosphamide to suppress the immune response to immunodominant cross-reactive epitopes, produced a MAb exhibiting a high specificity for a var. *capsulatum* yeast form cytoplasmic epitope by both ELISA and Western blot. The ability of this MAb to distinguish all the serotypes of var. *capsulatum* in their intact yeast form has yet to be determined.

A heat-stable HPA has proved valuable for diagnosing disseminated histoplasmosis in AIDS patients and in other immunosuppressed and nonimmunocompromised patients (Table 6).[37] Capture antibody for HPA is produced, albeit with difficulty[38], against whole yeast-form cells of *H. capsulatum.* The value of yeast or mycelial subcellular fractions for the consistent elicitation of high-titered capture antibody deserves investigation. HPA has been found in the urine of 90% and in the blood of 50% of patients with disseminated histoplasmosis when tested by RIA. False-positive results have been encountered with urine or serum specimens from patients with blastomycosis and with a CSF specimen from a patient with coccidioidal meningitis. The exact nature of the HPA is unknown; however, as suggested by Reiss and Bragg,[39] its properties are sufficiently similar to the heat stable polysaccharide common C antigen[40] shared by *B. dermatitidis, C. immitis,* and *H. capsulatum* to warrant their comparative analysis. Other glycoprotein antigens including the H and M antigens may also participate in the HPA reaction.[41]

Table 6. Value of *Histoplasma capsulatum* yeast-form antigens or reagents in immunologic tests for histoplasmosis.

Antigen	Test or reagent	Sensitivity %	Specificity
Whole yeast	CF	90	NS[a]
Whole yeast	Specific FA	50	S[b]
	Polyvalent FA	100	NS
Histoplasma polysaccharide	RIA-antiglobulin to yeast	90[c]	NS

[a] Nonspecific
[b] Specific
[c] Antigenuria tests

Standardized CF, exoantigen, and ID tests for histoplasmosis using mycelial histoplasmin antigens have proved to be of practical and of diagnostic value (Table 7).[42] These tests can be performed with crude histoplasmin, with purified or delineated reference antigens. In general, three mycelial antigens, namely C, H, and M, account for the variety of histoplasmin reactions recorded. Histoplasmin designated for use in CF, exoantigen, or ID tests must contain adequate concentrations of both the H and M proteins.[42]

Table 7. Value of *Histoplasma capsulatum* mycelial-form antigens in immunologic tests for histoplasmosis.

Antigen	Test	Sensitivity %	Specificity %
Histoplasmin	CF	83	NS[a]
	ID	83	S[b]
H	Exoantigen	99	100
	ID	12	100
M	Exoantigen	86	100
	ID	81	100
C	ID	ND[c]	NS
Non-M	ID	ND	100

[a] Nonspecific
[b] Specific
[c] Not determined

The CF and ID tests with histoplasmin will detect about 83% of serum specimens from immunocompetent patients with histoplasmosis. CF tests with histoplasmin, like those with the yeast-form antigen, can yield false-positive reactions with case sera from patients infected with *B. dermatitidis*, *C. immitis*, and *P. brasiliensis*, whereas histoplasmin ID tests with control H and M precipitates are totally specific.

Hamilton *et al.*[43] have suggested that the M antigen is a catalase with an epitope shared with catalases of a variety of dimorphic fungi. They expressed concern that this could lead to false-positive results in ID tests with sera from patients infected with *B. dermatitidis*, *P. brasiliensis*, and *S. schenckii*. Neither their data nor our cumulative experience over two decades has substantiated such cross-reactions in the histoplasmin ID test. The usefulness of histoplasmin H and M antigens in the CF test and other quantitative assays could possibly be improved by eliminating the closely bound carbohydrate C antigen. The M antigen detects most of the case sera reactive with histoplasmin (81%), whereas the H antigen detects only about 12% of the cases (Table 7). African histoplasmosis cases are not readily diagnosed by CF and ID tests; however, precipitins may on occasion be diagnostic.[44]

Another mycelial antigen designated non-M has been noted to produce a precipitin band close to the M but between it and the H band.[45] Non-M appears to be specific to *H. capsulatum*, and precipitins homologous to it, in my experience, occur at about the same time as those to the M precipitin are noted. The C antigen is a histoplasmin constituent which induces a cross-reacting antibody which precipitates close to the serum well. Because this epitope is shared by many of the dimorphic pathogens, it has little diagnostic value. Studies in my laboratory indicate that the mycelial *H. capsulatum* C, H, M, and non-M antigens are not form-specific since they are shared with the yeast form.[46] The yeast forms of both var. *capsulatum* and var. *duboisii* produce H and M but not as abundantly as the mycelial forms. FA studies with labeled H, M, and non-M antisera indicate that the respective homologous antigens are not located on the surface of yeast-form cells of *H. capsulatum* or *B. dermatitidis*.[45]

The monotypic genus *Histoplasma* consists of var. *capsulatum*, var. *duboisii* and var. *farciminosum*. The mycelial forms of these three varieties are indistinguishable from each other and are typically characterized by production of tuberculate macroconidia. A number of saprophytic fungi in the genera *Arthroderma*, *Chrysosporium*, *Corynascus*, *Renispora*, and *Sepedonium* have mycelial forms that grossly and microscopically resemble those of the *Histoplasma* varieties. FA and exoantigen studies indicate that the mycelial form elements of *H. capsulatum* share antigens with *Chrysosporium and Sepedonium* spp.[47,48] The soluble H and M antigens, however, are produced only by the three *Histoplasma* varieties. Both antigens are unique to *H. capsulatum* and are reliable for identification of this genus. The antigens however, do not allow differentiation of the three varieties. The H protein is the first and most abundant antigen released from the mycelial form in culture. In exoantigen tests it demonstrates a 99% sensitivity for detection of *H. capsulatum* (Table 7). The M antigen is released during subsequent growth and autolysis of the mycelial form and detects only about 86% of mycelial form *H. capsulatum* cultures.[48] Interestingly, yeast- form cultures, perhaps because they resist autolysis, are not good sources for the M antigen. Yet, *in vivo*, the M precipitin appears earlier and is more prevalent than the H precipitin.[30] Consequently the M antigen is more sensitive than the H antigen for detecting early histoplasmosis.

FA studies with fluorescein-labeled globulins against mycelial elements of *H. capsulatum* demonstrated that they strongly stained the hyphal elements, microconidia, and macroconidia of the homologous strains, as well those of *Chrysosporium* and *Sepedonium* spp. Adsorptions and antigenic analyses indicated that a stronger relationship existed between *H. capsulatum* and the *Sepedonium* spp. than with the *Chrysosporium* spp. A single-species-specific FA reagent for distinguishing the mycelial form of *H. capsulatum* from the saprobes could not be produced. From all indications, however, a specific epitope for the mycelial form elements of *H. capsulatum* exists, but in a quantity so low that it is of little or no practical diagnostic value.[47]

REFERENCES

1. S.B. Salvin, The serologic relationship of fungus antigens, *J. Lab. Clin.Med.* 34: 1096 (1949).
2. J.F. Busey and P.F. Hinton, Precipitins in blastomycosis, *Am. Rev. Respir. Dis.* 95: 95 (1967).
3. L. Kaufman, D.W. McLaughlin, M.J Clark and S. Blumer, Specific immunodiffusion test for blastomycosis, *Appl. Microbiol.* 26: 244 (1973).
4. W. Kaplan and L. Kaufman, Specific fluorescent antiglobulin for the detection and identification of *Blastomyces dermatitidis* yeast-phase cells, *Mycopath. Mycol. Applic.* 19 : 173 (1963).
5. M.S. Sudman and W. Kaplan, Antigenic relationship between American and African isolates of *Blastomyces dermatitidis* as determined by immunofluorescence, *Appl. Microbiol.* 27: 496 (1974).
6. L. Kaufman and P.G. Standard, Specific and rapid identification of medically important fungi by exoantigen detection, *Ann. Rev. Microbiol.* 41: 209 (1987).
7. F.C. Odds, L. Kaufman, D. McLaughlin, C. Callaway and S. Blumer, Effect of chitinase complex on the antigenicity and chemistry of yeast-form cell walls and other fractions of *Histoplasma capsulatum* and *Blastomyces dermatitidis*, *Sabouraudia*, 12: 138 (1971).
8. J.H. Green, W.K. Harrell, J.E. Johnson and R. Benson, Isolation of an antigen from *Blastomyces dermatitidis* that is specific for the diagnosis of blastomycosis, *Curr. Microbiol.* 4: 293 (1980).
9. S. Turner and L. Kaufman, Immunodiagnosis of blastomycosis, *Semin. Respir. Infect.* 1: 22 (1986).
10. C.Y. Lo and R.H. Notenboom, A new enzyme immunoassay specific for blastomycosis, *Am. Rev. Respir. Dis.* 141: 84 (1990).
11. S. Hurst and L. Kaufman, Western immuno blot analysis and serologic characterization of *Blastomyces dermatitidis* yeast form extracellular antigens, *J. Clin. Microbiol.* 30: 3043 (1992).
12. J.H. Green, W.K. Harrell, J.E. Johnson and R. Benson, Preparation of reference antisera for laboratory diagnosis of blastomycosis, *J Clin. Microbiol.* 10: 1 (1979).
13. B.S. Klein and J.M. Jones, Isolation, purification and radiolabeling of a novel 120-kD surface protein on *Blastomyces dermatitidis* yeasts to detect antibody in infected patients, *J. Clin. Invest.* 85: 152 (1990).
14. D. Pappagianis and B.L. Zimmer, Serology of coccidioidomycosis, *Clin. Microbiol. Rev.* 3: 247 (1990).
15. M. Huppert and J.W. Bailey, The use of immunodiffusion tests in coccidioidomycosis. I. The accuracy and reproducibility of the immunodiffusion test which correlates with complement fixation, *Am. J. Clin. Pathol.* 44: 364 (1965).
16. M. Huppert and J.W. Bailey, The use of immunodiffusion tests in coccidioidomycosis. II. An immunodiffusion test as a substitute for the tube precipitin test, *Am. J. Clin. Pathol.* 44: 369 (1965).
17. M. Huppert, E.T. Peterson, S.H. Sun, P.A. Chitjian and W. J. Derrevere, Evaluation of a latex particle agglutination test for coccidioidomycosis. *Am. J. Clin. Pathol.* 49: 96 (1968).
18. S.M. Johnson and D. Pappagianis, The coccidioidal complement fixation and immunodiffusion complement fixation antigen is a chitinase, *Infect. Immun.* 60: 2588 (1992).
19. L. Kaufman and M.J. Clark, Value of the concomitant use of complement fixation and immunodiffusion tests in the diagnosis of coccidioidomycosis, *Appl. Microbiol.* 28: 641 (1974).
20. H.B. Levine, J.M. Cobb and C.E. Smith, Immunity to coccidioidomycosis induced in mice by purified spherule, arthrospore, and mycelial vaccines. *Trans. N.Y. Acad. Sci.* 22: 436 (1960).

21. M.H. Huppert, I. Krasnow, K.R. Vukovich, S.H. Sun, E.H. Rice and L.J. Kutner, Comparison of coccidioidin and spherulin in complement fixation tests for coccidioidomycosis, *J. Clin. Microbiol.* 6: 33 (1977).

22. B.L. Zimmer and D. Pappagianis, Immunoaffinity isolation and partial characterization of the *Coccidioides immitis* antigen detected by the tube precipitin and immunodiffusion-tube precipitin tests, *J. Clin. Microbiol.* 27: 1759 (1989).

23. D. Kruse and G.T. Cole, Isolation of tube precipitin antibody-reactive fractions of *Coccidioides immitis. Infect. Immun.* 58: 169 (1990).

24. G.T. Cole, D. Kruse, S. Zhu, K.R. Seshon and R.W. Wheat. Composition, serologic reactivity, and immunolocalization of a 120-kilodalton tube precipitin antigen of *Coccidioides immitis. Infect. Immun.* 58: 179 (1990).

25. J.N Galgiani, S.H. Sun, K.O. Dugger, N.M. Ampel, G.G. Grace, J. Harrison and M.A. Wieden, an arthroconidial-spherule antigen of *Coccidioides immitis*: differential expression during in vitro fungal development and evidence for humoral response in humans after infection or vaccination, *Infect. Immun.* 60: 2627 (1992).

26. L. Kaufman and P. Standard, Improved version of the exoantigen test for identification of *Coccidioides immitis* and *Histoplasma capsulatum* cultures, *J. Clin. Microbiol.* 8: 42 (1978).

27. L. Kaufman, P.G. Standard, M. Huppert and D. Pappagianis, Comparison and diagnostic value of the coccidioidin heat-stable (HS and tube precipitin) antigens in immunodiffusion, *J. Clin. Microbiol.* 22: 515 (1985).

28. W. Kaplan and M.K. Clifford, Production of fluorescent antibody reagents specific for the tissue form of *Coccidioides immitis, Am. Rev. Respir. Dis.* 89:651 (1964).

29. L. Kaufman and W. Kaplan, Serological characterization of pathogenic fungi by means of fluorescent antibodies. I. Antigenic relationships between yeast and mycelial forms of *Histoplasma capsulatum* and *Blastomyces dermatitidis, J. Bacteriol.* 85: 986 (1963).

30. L. Kaufman and E. Reiss, Serodiagnosis of fungal diseases, in: "Manual of Clinical Laboratory Immunology", N.R. Rose, E.C. de Macario, J.L. Fahey, H. Friedman and G.M. Penn, eds., Am. Soc. Microbiol. Washington, D.C. (1992).

31. L. Pine, C.J. Boone and D. McLaughlin, Antigenic properties of the cell wall and other fractions of the yeast form of *Histoplasma capsulatum, J. Bacteriol.* 91: 2158 (1966).

32. C. Raman, N. Khardori, L.A von Behren, L.J. Wheat and R.P. Tewari, Evaluation of an ELISA for the detection of anti-*Histoplasma* ribosomal and antihistoplasmin antibodies in histoplasmosis, *J. Clin. Lab. Anal.* 4: 199 (1990).

33. L. Kaufman and W. Kaplan, Preparation of a fluorescent antibody specific for the yeast phase of *Histoplasma capsulatum, J. Bacteriol.* 82: 729 (1961).

34. L. Kaufman and S. Blumer, Development and use of a polyvalent conjugate to differentiate *Histoplasma capsulatum* and *Histoplasma duboisii* from other pathogens, *J. Bacteriol.* 95: 1243 (1968).

35. S.M. Kamel, L.J. Wheat, M.L. Garten, M.S. Bartlett, M.R. Tansey and R.P. Tewari, Production and characterization of murine monoclonal antibodies to *Histoplasma capsulatum* yeast cell antigens, *Infect. Immun.* 57: 896 (1989).

36. A.J. Hamilton, M.A. Bartholomew, L.E. Fenelon, J. Figueroa and R.J. Hay, A murine monoclonal antibody exhibiting high species specificity for *Histoplasma capsulatum* var. *capsulatum, J. Gen. Microbiol.* 136: 331 (1990).

37. L.J. Wheat, R.B. Kohler, R. Tewari, M. Garten and M.L. French, Significance of *Histoplasma* antigen in the cerebrospinal fluid of patients with meningitis, *Arch. Intern. Med.* 149: 302 (1989).

38. L.J. Wheat, Personal communication (1991).

39. E. Reiss and S.L. Bragg, Immunochemical analysis of histoplasmin proteins and polysaccharides, in: "Fungal Antigens: isolation, purification and detection", E. Drouhet, G.T. Cole, L. de Repentigny, J.P. Latge and B. Dupont, eds., Plenum Press, New York (1988).

40. D.C. Heiner, Diagnosis of histoplasmosis using precipitin reactions in agar gel, *Pediatrics*, 22: 616 (1958).

41. R.M. Zancope-Oliveira, S.L. Bragg, S.F. Hurst, J.M. Peralta and E. Reiss, Evaluation of cation exchange chromatography for the isolation of M glycoprotein from histoplasmin, *J. Med. Vet. Mycol.* (in press).

42. L. Kaufman, Experience in the standardization of histoplasmosis immunological tests and reagents, in: "Proceedings of the X Congress of the International Society for Human and Animal Mycology-ISHAM", J.M. Torres-Rodriguez, ed., J.R. Prous Science, Barcelona (1988).

43. A.J. Hamilton, M.A. Bartholomew, J. Figueroa, L.E. Fenelon and R.J. Hay, Evidence that the M antigen of *Histoplasma capsulatum* var. *capsulatum* is a catalase which exhibits cross-reactivity with other dimorphic fungi, *J. Med. Vet. Mycol.* 28: 479 (1990).

44. P. Peeters, G. Depre, F. Rickaert, J. Coremans-Pelseneer and E. Serruys, Disseminated African histoplasmosis in a white heterosexual male patient with the acquired immune deficiency syndrome, *Mykosen*, 30: 449 (1987).

45. J.H. Green, W.K. Harrell, S.B. Gray, J.E. Johnson, R.C. Bolin, H. Gross and G.B. Malcolm, H and M antigens of *Histoplasma capsulatum*: Preparation of antisera and location of these antigens in yeast-phase cells, *Infect. Immun.* 14: 826 (1976).

46. J. Brough, Serological techniques for the specific detection of antibodies to the yeast-form of *Histoplasma capsulatum*, Dissertation, University of North Carolina, School of Public Health, Chapel Hill, (1972).

47. L. Kaufman and B. Brandt, Fluorescent-antibody studies of the mycelial form of *Histoplasma capsulatum* and morphologically similar fungi, *J. Bacteriol.* 87: 120 (1964).

48. P.G. Standard and L. Kaufman, Specific immunological test for the rapid identification of the genus *Histoplasma*, *J. Clin. Microbiol.* 3: 191 (1976).

DIAGNOSIS OF HISTOPLASMOSIS - REVIEW OF CURRENT METHODS

Joseph Wheat

Indiana University School of Medicine and
Richard Roudebush Department of Veterans' Affairs Hospital
Indianapolis, IN 46202-2879, USA

INTRODUCTION

Histoplasmosis is the most common endemic mycosis in the United States. Over 500,000 cases are diagnosed annually. Additionally, histoplasmosis is common in Latin America, the Caribbean Islands, parts of Asia, and Africa. Uncommonly cases are diagnosed in the United Kingdom and Europe. Histoplasmosis is an important opportunistic infection in patients with impaired cellular immunity, particularly in those with AIDS.

Following inhalation of *Histoplasma capsulatum*, most individuals develop asymptomatic or self-limited pulmonary infection.[1] Illness occurs more commonly following heavy exposure and in immunocompromised hosts. Histoplasmosis causes chronic pulmonary infection in individuals with underlying lung disease, and widespread disseminated infection in those who are immunocompromised.

Currently, a variety of tests is available for diagnosis of histoplasmosis. In the last 6 years tests have been developed based on detection of a glycoprotein antigen in body fluids of infected patients. An understanding of the uses and limitations of these diagnostic tests is required to permit prompt and accurate diagnosis of histoplasmosis. Accurate diagnosis is important in self-limited cases so that work-up for other infections can be discontinued and in patients with chronic pulmonary or disseminated infection so that appropriate treatment can be initiated.

REVIEW OF ACCURACY OF STANDARD DIAGNOSTIC TESTS

Fungal Cultures

Isolation of *H. capsulatum* from sputum, blood or other tissues offers the strongest basis for diagnosis. Based on experience from Indianapolis,[2] cultures are positive in 10-15% of patients with self-limited disease, 66-85% with chronic pulmonary disease and 84-90% with disseminated disease (Table 1). In chronic pulmonary histoplasmosis, multiple sputum specimens must be obtained to achieve a sensitivity of 85%. While cultures are positive in most patients with disseminated disease, the 2-4 weeks required to identify growth of *H. capsulatum* often results in unacceptable delays in initiation of treatment.

Fungal Stain

Identification of organisms by fungal stain of blood smears, sputum or other respiratory secretions, or biopsies of tissue offers a means of rapid diagnosis, however the sensitivity for this method is relatively low[2-5]
(Table 1). Also, accurate identification of *H. capsulatum* by fungal stain is difficult, requiring technicians and pathologists experienced with fungal identification. *Pneumocysistis carinii* has been misidentified as *H. capsulatum* in many laboratories. *Candida* species and other yeasts may be mistaken for *H. capsulatum*. Additionally, staining artifacts may be misidentified as *H. capsulatum*.

Serologic Tests for Antibodies

Serologic tests for antibodies are very useful for diagnosis of histoplasmosis. By use of the immunodiffusion and complement fixation tests, high levels of antibodies can be demonstrated in nearly 90% of cases[2-5] (Table 1). The complement fixation test is more sensitive (79%) than the immunodiffusion test (73%), but both tests must be performed to achieve the maximum sensitivity. The sensitivity is lower, approximately 70%, in those who are immunosuppressed. Complement fixing antibodies typically appear before precipitating antibodies.

Table 1. Sensitivity of diagnostic tests for histoplasmosis.

| Test | Type of Histoplasmosis | | |
	Self-Limited	Chronic Pulmonary	Disseminated
	Sensitivity %		
Antigen	25-39	10-21	80-95
Culture	10-15	67-85	84-90
Fungal stain	9	17	43
Serology	80-90	95-100	70-80
Skin test	90	90	50-90

Drawbacks of serologic tests for antibodies include false-positive tests in patients with other diseases and delayed appearance of antibodies following acute infection. Although cross-reacting antibodies may develop in patients with blastomycosis, coccidioidomycosis and paracoccidioidomycosis, these tests are relatively specific. High levels of antibodies may persist for several years following recovery from histoplasmosis.[2] In such individuals, a misdiagnosis of histoplasmosis may be based on detection of antibodies in patients with other diseases. In endemic areas background positivity rates range from 0.5% by immunodiffusion to 5% by complement fixation.[2]

Antibodies cannot be detected during the first 4 weeks of infection. Also, antibodies may not be detected in up to 30% of immunosuppressed individuals.[4,5] Thus, clinical judgment must be used in interpreting serologic tests.

Antigen Detection

Tests for antigen offer important advances in diagnosis of histoplasmosis and complement other diagnostic tests.[6] Antigen may appear early in the course of infection, before development of antibodies. Tests for antigen can be performed within one work day, providing early diagnosis and allowing prompt treatment. Antigen levels correlate with extent of infection. Consequently, tests for antigen are most sensitive for diagnosing disseminated disease and extensive pulmonary infection with diffuse infiltration.

The sensitivity of antigen detection is greater using urine than serum specimens, 90% versus 70%, respectively, in disseminated disease.[6] Occasional patients, however, may have positive results in serum but negative results in urine. Also, antigen may be detected in the spinal fluid of over 40% of patients with *Histoplasma* meningitis[3,7] and in alveolar lavage fluid of patients with extensive pulmonary histoplasmosis.[8]

Table 2. Interpretation of *Histoplasma* antigen results.

Antigen Units	Interpretation	Implication	Recommendation
<1.0	Negative	Disseminated histoplasmosis unlikely	None
1.0 to 2.0	Weak positive	Probable histoplasmosis	Repeat
>2.0	Positive	histoplasmosis	Repeat if inconsistent with clinical findings

Tests for antigen are highly specific.[6,8] Cross-reacting antigens have been detected in specimens from patients with paracoccidioidomycosis and blastomycosis. These fungi contain a cross-reacting galactomannan antigen on their cell surface.

Antigen results are expressed quantitatively. The radioimmunoassay uses rabbit anti-*H. capsulatum* IgG antibodies as the capture and the [^{125}I] radiolabeled detector antibodies. Results of 1.0, as defined by comparison to normal controls, are positive; and results of 2.0 units and greater are highly suggestive of histoplasmosis (Table 2). Occasionally, patients with results of 1.0 to 2.0 cannot be proven to have histoplasmosis; tests should be repeated in such cases. To support the significance of positive antigen test results in patients without positive cultures or fungal stains, serologic tests should be ordered. Patients with consistent clinical findings and antigen levels between 1.0 and 2.0 units who also demonstrate positive serologic tests for antibodies should be regarded as having histoplasmosis. In those without positive serologic tests for antibodies the diagnosis of histoplasmosis remains less certain. In such cases, antigen tests and serologic tests should be repeated in 1 to 2 months.

Skin Tests

Histoplasmin has been used as a skin test antigen, but such testing has not been useful diagnostically.[1] Although skin tests are positive in most non-immunocompromised individuals with histoplasmosis, background positivity rates of 50-80% in endemic areas prevent use of this information for diagnosis of histoplasmosis. Skin test positivity persists for life in most reactive individuals who reside in endemic areas because of repeated exposure to *Histoplasma*. Thus, skin test reactivity fails to distinguish recent from remote infection. Also, skin tests may boost humoral reactivity, complicating interpretation of results of serological tests in histoplasmosis. Skin tests may be used in epidemiological studies to determine endemic rates and patterns of histoplasmosis, but should not be used diagnostically.

SPECIFICITY OF DIAGNOSTIC TESTS FOR HISTOPLASMOSIS

Specificity of the standard tests[2] and the antigen assay[6,8] has been established in prior studies. Growth of *H. capsulatum* in culture is assumed to be specific for histoplasmosis. Contamination can occur, but has not been reported. Although the specificity of fungal stains appears to be good, mistakes can occur in misidentification of *Pneumocystis carinii* or other fungi as *H. capsulatum*. Also, less skilled technologists may misidentify artifacts as fungal organisms.

The specificity of serologic tests for antibodies is good. Cross-reacting antibodies occur in patients with certain other fungal diseases.[9] Additionally, low titers of complement fixing antibodies may occur in up to 5% of individuals residing in endemic areas for histoplasmosis.

The specificity of antigen detection is greater than 98%.[6,8] Cross reacting antigens have been detected in patients with coccidioidomycosis, paracoccidioidomycosis and blastomycosis (unpublished observations).

Histoplasmin skin tests also may be positive in patients with coccidioidomycosis, paracoccidioidomycosis and blastomycosis.

RECOMMENDED APPROACH TO DIAGNOSIS OF HISTOPLASMOSIS

Based on the relative sensitivity of the different diagnostic tests in the specific clinical types of histoplasmosis, the following approach is recommended. (Table 3)

Self-Limited Disease

In patients with clinical findings of self-limited histoplasmosis, including acute pulmonary illnesses, pericarditis and rheumatologic syndromes, serologic tests for antibodies should be performed. Tests for antigen and cultures are uncommonly indicated in these patients. In patients with severe acute pulmonary

Table 3. Summary of suggested work-up for histoplasmosis.

	Self Limited	Severe Acute Pulmonary	Chronic Pulmonary	Disseminated
Serology	+++	+	+++	++
Antigen	+	+++	+	+++
Histology		+++	+	++
Culture		+	+++	+++

histoplasmosis, tests for antigen and fungal stain of alveolar lavage fluid or lung biopsy specimens are indicated; lavage fluid should also be tested for antigen, positive in up to 70% of cases.

Disseminated Disease

In suspected disseminated disease, tests for antigen allow for a rapid diagnosis. Antigen may be detected in the spinal fluid in over 40% of patients with meningitis caused by *H. capsulatum*. Additionally, cultures of blood are indicated with suspected disseminated disease. If tests for antigen are negative, examination of bone biopsy specimens by fungal stain and culture should be considered. Other sites, including cutaneous or mucosal lesions and enlarged lymph nodes, should be biopsied for fungal stain and culture. Serologic tests for antibodies are less useful in disseminated disease but should be considered if tests for antigen and fungal stains are negative. Positive serologic tests for antibodies may provide a basis for empiric treatment or for performance of additional biopsies in selected cases.

Chronic Pulmonary Disease

In chronic pulmonary histoplasmosis, serologic tests for antibodies and cultures of sputum should be performed in all cases. Tests for antigen should be considered in patients with clinical findings of concurrent disseminated disease.

Antigen Detection for Monitoring Therapy

Antigen concentrations fall in response to therapy.[10] Antigen levels are expressed quantitatively, and by testing prior and current specimens together, thus avoiding interassay variation, changes in concentrations can be monitored. In patients with AIDS who were treated for histoplasmosis with amphotericin B, antigen levels in serum fell at a rate of 0.7 units per week during the first 12 weeks of therapy, and stabilized at low levels during chronic maintenance treatment. Antigen levels in urine fell at a somewhat slower rate in urine, 0.5 units per week, during the initial intensive therapy and then stabilized during chronic maintenance therapy.

The slow and incomplete clearance of antigen in patients with AIDS is explained by persistent infection in this group. Although the clinical findings may be controlled by chronic antifungal treatment, viable organisms persist in reticuloendothelial tissues and relapse occurs if treatment is discontinued.

More rapid and complete clearance occurs in patients without AIDS. Antigen levels usually reach the negative range within 3 months of initiation of treatment, and remain undetectable during follow-up. Uncommonly, the infection may not be cured in patients without AIDS, behaving similar to that in patients with AIDS. In such patients, antigen may persist after completion of treatment, and relapse may follow. Such patients should be followed closely, and antigen levels should be monitored periodically.

USE OF ANTIGEN DETECTION TO IDENTIFY RELAPSE

In patients with AIDS, relapse occurred in 5-20% of patients despite chronic maintenance treatment to suppress infection. [3] Such relapses have occurred as a result of noncompliance, use of interacting medications which interfere with absorption or accelerate metabolism of antifungal agents, or resistance to the antifungal agent. Antigen levels increase by at least 2 units in such cases. [11] Of 20 patients who experienced a relapse during chronic suppressive therapy, antigen levels increased in the urine in 17 of 18 (94%) and in the blood of 12 of 14 (86%). Similar increases occurred in only 1 of 56 (2%) controls.

Antigen levels may increase before the clinical diagnosis is suspected and proven by culture, providing an early basis for intensification of therapy. However, increases in antigen may follow the appearance of clinical findings of relapse in some patients. Two patients have been identified with stable antigen levels at the time of relapse. In both, fungal blood cultures were reported to be positive 2 weeks later; and in each, antigen levels had increased by the time positive blood cultures were identified.

GUIDELINES FOR INTERPRETATION OF INCREASES IN ANTIGEN DURING THERAPY

The following guideline has been proposed for interpretation of increases in antigen during treatment (Table 4). Increases of less than 2 units are regarded as insignificant and do not require culturing or change in therapy. Increases of 2 to 4 units are classified as mild increases and suggest treatment failure. In such cases clinical and laboratory evaluation for treatment failure are indicated. The antigen tests should be repeated and details of treatment (dosage, interacting medications, compliance, absorption) should be evaluated. For patients with increases of more

Table 4. Guidelines for interpretation of antigen (HPA) increases during treatment of histoplasmosis.

HPA Increase	Interpretation	Implication	Recommendation
0 to 1.9	Stable	None	None
2.0 to 4.0	Mild increase	Probable failure	Culture Repeat antigen Review therapy
≥4.1	Mod-mark increase	Failure	Culture Intensify therapy

than 4 units, treatment failure is very likely. Cultures should be obtained and treatment should be intensified.

RECOMMENDATIONS FOR MONITORING THERAPY

Periodic measurement of antigen levels in blood and urine are useful for monitoring therapy and for identifying relapse. Although the assay is highly reproducible, results for individual specimens often change by more than 2 units when they are tested on different days. To be used to monitor therapy or identify relapse, **the prior and current specimen must be tested together to eliminate this interassay variability**.

Specimens must be tested periodically during and following therapy if they are to be used to identify treatment failure or relapse. If specimens are not tested periodically during therapy, the prior specimen may not reflect the recent past level of antigenemia or antigenuria. Ongoing studies monitoring antigen levels at monthly intervals during the initial 12 weeks of intensive induction therapy and during chronic maintenance therapy may define the optimal frequency of testing. Until results of those studies are available, a minimal frequency of testing would appear to be at initiation of treatment, completion of intensive treatment, and then every 2-3 months during chronic maintenance treatment.

After discontinuation of therapy in patients without AIDS, antigen levels should be followed until they become negative and then every 3-6 months during the first 2 years of follow-up after stopping therapy. Antigen levels should also be repeated at time of suspected relapse.

AVAILABILITY OF HISTOPLASMOSIS ANTIGEN TESTING

Histoplasma antigen testing is available by submitting specimens to my laboratory.[*] The test is performed 5 days weekly, and results are available within 1 work day. Positive results are reported by phone, and other reports are mailed. By use of overnight airmail, results are available for all specimens within 48-72 hours.

Commercial development of *Histoplasma* antigen testing will increase the availability of diagnostic approach. The accuracy, precision and reproducibility of the test are acceptable for commercial development. Conversion to enzyme immunoassay methodology [12] will improve its attractiveness for commercial development. Studies in progress suggest that enzyme immunoassay methods will be as sensitive as radioimmunoassay. Development of methods to produce large amounts of antibodies recognizing the antigen found in patients will increase the commercial potential of this assay. Monoclonal antibody technology should permit large scale production.

---

[*] L. Joseph Wheat, M.D., Wishard Memorial Hospital; Rm. OPW430; 1001 West Tenth Street, Indianapolis, IN 46202-2879.

ACKNOWLEDGEMENTS

Research grants from the Department of Veterans' Affairs and from the AIDS Program of the National Institutes of Allergy and Infectious Disease. I would like to thank Nancy Richey for typing this manuscript.

REFERENCES

1. L.J. Wheat, Histoplasmosis, *Infect. Dis. Clin.* NA.; 2 : 841 (1988).
2. L.J. Wheat, M.L.V. French, R.B. Kohler, S.E. Zimmerman, W.R. Smith, J.A. Norton, H.E. Eitzen, C.D. Smith and T.G. Slama, The diagnostic laboratory tests for histoplasmosis, *Ann. Intern. Med.* ; 97: 680 (1982).
3. L.J. Wheat, P. Connolly-Stringfield, R.L. Baker, M.F. Curfman, M.E. Eads, K.S. Israel, S.A. Norris, D.H. Webb and M.L. Zeckel, Disseminated histoplasmosis in the Acquired Immune Deficiency Syndrome: Clinical Findings, Diagnosis and Treatment, *Medicine* ; 69: 361 (1990).
4. B. Sathapatayavongs, B.E. Batteiger, L.J. Wheat, T.G. Slama and J.L. Wass, Clinical and laboratory features of disseminated histoplasmosis during two large urban outbreaks, *Medicine* ; 62: 263 (1983).
5. L.J. Wheat, J. Wass, J. Norton, R.B. Kohler and M.L.V French, Cavitary histoplasmosis occurring during two large urban outbreaks--analysis of clinical, epidemiologic, roentgenographic, and laboratory features, *Medicine* ; 63: 201 (1984).
6. L.J. Wheat, R.B. Kohler and R.P. Tewari, Diagnosis of disseminated histoplasmosis by detection of *Histoplasma capsulatum* antigen in serum and urine specimens. *New Engl. J. Med.* ; 314: 83 (1986).
7. L.J. Wheat, R.B. Kohler, R.P. Tewari, M.L. Garten and M.L.V. French, Significance of *Histoplasma* Antigen in the Cerebrospinal Fluid of Patients with Meningitis, *Arch. Intern. Med.* ; 149: 302 (1989).
8. L.J. Wheat, P. Connolly-Stringfield, B. Williams, K. Connolly, R. Blair, M. Bartlett and M. Durkin, Diagnosis of histoplasmosis in patients with the acquired immunodeficiency syndrome by detection of *Histoplasma capsulatum* polysaccharide antigen in bronco alveolar lavage fluid, *Am. Rev. Respir. Dis.* ; 145: 1421 (1992).
9. L.J. Wheat, M.L.V. French, S. Kamel and R.P. Tewari, Evaluation of cross reactions in *Histoplasma* serologic test, *J. Clin. Microbiol.* ; 23: 493 (1986).
10. L.J. Wheat, P. Connolly-Stringfield, R. Blair, K. Connolly, T. Garringer, B.P. Katz and M. Gupta, Effect of successful treatment with amphotericin B on *Histoplasma capsulatum* variety *capsulatum* polysaccharide antigen levels in patients with AIDS and histoplasmosis, *Am. J. Med.* ; 92: 153 (1992).
11. L.J. Wheat, P. Connolly-Stringfield, R. Blair, K. Connolly, T. Garringer and B.P. Katz, Histoplasmosis relapse in patients with AIDS: Detection Using *Histoplasma capsulatum* variety *capsulatum* antigen levels, *Ann. Intern. Med.* ; 115: 936 (1991).
12. S.E. Zimmerman, P. Connolly-Stringfield, L.J. Wheat, M.L.V. French and R.B. Kohler, Comparison of Sandwich Solid Phase Radioimmunoassay and Two Enzyme-Linked Immunosorbent Assays for Detection of *Histoplasma capsulatum* Polysaccharide Antigen, J. Infect. Dis. ; 160: 678 (1989).

ANIMAL MODELS OF MYCOSES CAUSED BY FUNGI THAT PRODUCE YEAST- AND HYPHAL FORMS

Jan Van Cutsem, Frans Van Gerven, and Jan Fransen

Janssen Research Foundation
B-2340 Beerse, Belgium

ABSTRACT

Various predisposing factors and immunodepressing agents are used in animal models to enhance fungal pathogenicity and to lead to more homogenous invasion of fungi. In this study the influence of cyclophosphamide, hydrocortisone acetate, prednisolone acetate, mechlorethamine hydrochloride, antibiotics, deferoxamine and others is described. The most drastic effects occurred with mechlorethamine hydrochloride pretreatment, which resulted in permanent neutropenia in the guinea-pig, but not in the mouse. Irreversible diabetes was obtained in mice and rats with streptozotocin. The injection of the iron chelator deferoxamine led rapidly to fatality in zygomycosis. Deferoxamine is also a potent antagonist of amphotericin B. Various experimental mycoses were induced by fungi presenting differences in morphology, including candidosis: systemic and in pregnancy in various animal species, intestinal, corneal and cutaneous in guinea-pigs and vaginal in rats. In *Fusarium* and *Aspergillus* keratitis, *Malassezia* infection, sporotrichosis, penicilliosis, paracoccidioidomycosis, cryptococcosis, zygomycosis and dermatophytosis in guinea-pigs at least two different morphological forms were present. This was also the case for chromoblastomycosis in mice and for histoplasmosis in mice and guinea-pigs. The broad-spectrum activity of itraconazole was clearly demonstrated in animal models where the morphology of the fungi in human infections was mimicked.

INTRODUCTION

Dimorphism and polymorphism have been described for a large number of fungi. Different forms are seen both *in vitro* and *in vivo*, but large morphological differences between fungi growing *in vitro* and *in vivo* exist too.[1]

The fungi generally considered to be dimorphic are: *Sporothrix*, *Blastomyces*, *Histoplasma*, *Paracoccidioides* and *Coccidioides*. A second series of fungi contains organisms that are certainly also dimorphic: *Candida, Malassezia,* some phaeohyphomycetes and hyalohyphomycetes, *Wangiella* and *Penicillium marneffei.* Finally, a large number of fungi present not only important differences between *in vitro* and *in vivo* aspects, but *in vivo* their development is often not uniform, and it is not exceptional to observe filamentous forms in cryptococcosis, or yeast-like forms in dermatophytosis, fusariosis, aspergillosis, zygomycosis or other mycoses.

To obtain homogenous infections in animals various factors are important. The aim of this study is to review some animal models of experimental infection with dimorphic or polymorphic fungi.

PREDISPOSING AND IMMUNOMODULATING AGENTS

A large number of agents has been used in animal models to mimic the predisposed or the immunodepressed status in man.[2-19]

Details regarding the influence on the leukocyte parameters of the intraperitoneal (IP) injection of cyclophosphamide at 10 and 20 mg/kg for 4 days, of seven intramusclar (IM) injections of hydrocortisone acetate and prednisolone acetate, as well as the IP administration of two dosages (0.25 and 0.5 mg/kg) of mechlorethamine hydrochloride to guinea-pigs are given in Table 1. Of these agents only mechlorethamine hydrochloride was able to induce neutropenia, persisting till the end of an experiment (in general 28 to 42 days).[20-22] In mice, however, neutropenia was only temporary.[23] A dosage of 0.5 mg/kg was in general too high, all guinea-pigs losing between 25 and 40 % of body weight (BW) and about 30 % dying because of the drug, but with 0.25 mg/kg body weight loss was between 5 and 15 % and an animal succumbed only rarely. The immunodepression caused by cyclophosphamide and corticoids was always reversible.

It is possible to induce chemical diabetes in guinea-pigs with alloxan. In mice and rats the results obtained with alloxan were poor. On the other hand the IP administration of streptozotocin at a total dose of 120 mg/kg over three days had no influence on the differential leucocyte count, nor on the serum glucose level of albino guinea-pigs (unpublished data). Swiss mice and Wistar rats[24] were highly sensitive to streptozotocin. The onset of leucocytosis and diabetes was slower in mice than in rats, both parameters being irreversible (Tables 2 and 3). The same parameters in the control groups, receiving only the solvent, remained unchanged during the 13 weeks of the experimentation, but 36 and 38 % of diabetic mice and rats respectively had succumbed. The survivors at that time were emaciated and cachectic.

Dialysis patients with ketoacidosis are highly susceptible to mucormycosis. Fatalities have been frequently reported in association with deferoxamine (DFO) usage.[25] Fungizone® Squibb (amphotericin B) is currently the only therapy that may have beneficial effects in severe mucormycosis, yet only exceptionally was a patient infected with a zygomycete and receiving DFO able to escape in spite of treatment with amphotericin B.[25]

In the guinea-pig the invasion by *Rhizopus oryzae* and *Rh. microsporus* var. *rhizopodiformis* was enhanced after a single dose of 50 mg of DFO by the IM route in an animal of 500 g. Moreover, the activity of amphotericin B was completely antagonized (Table 4).[26,27] The severity of zygomycosis and rapidity of lethal outcome increased in relation to the total dosage of DFO. DFO and iron are considered to be growth factors for zygomycetes.[28] The same phenomenon has been confirmed in mice by Japanese investigators.[29]

Table 1. Influence of predisposing factors on the differential leucocyte counts of the guinea-pig.

Parameter	Control data before treatment (n = 208)[1]	Cyclophosphamide (IP)		Hydrocortisone acetate (IM)	Prednisolone acetate (IM)	Mechlorethamine HCl (IP)	
		10 mg/kg 4 x (n=18)	20 mg/kg 4 x (n=17)	10 mg/kg 7 x (n=11)	10 mg/kg 7 x (n=12)	0.25 mg/kg 2 x (n=161)	0.5 mg/kg 2 x (n=47)
		% of leucocytes versus controls					
Leucocytes	7.8 ($\times 10^3$ per mm^3)	70***[2]	60***	98	132*	80***	63***
Neutrophils	42	172***	176***	197***	252***	62***	74***
Eosinophils	0.89	106	152	100	52	76*	26***
Basophils	0.08	25	71	71	33	175	213
Lymphocytes	56	71***	68***	57***	37***	123***	113*
Monocytes	1.2	75	90	140	80	342***	433***

[1] n: number of animals
[2] Mann-Whitney U test (two-tailed) probability: * $p \leq .05$; ** $p \leq .01$; *** $p \leq .001$
(From ref. 12)

Dermatophytosis and candidosis were not influenced by DFO. The severity of cryptococcal disease, however, was slightly more pronounced. *Aspergillus fumigatus* infection was enhanced to a small degree by increasing the number of DFO treatments and in relation to the inoculum size. When DFO was administered to guinea-pigs with systemic invasive aspergillosis the therapeutic dosage of itraconazole or saperconazole has to be increased in order to cure the disease (in preparation).

Table 2. White blood cell count in diabetic[1] Swiss mice and Wistar rats.

Week	Swiss mice WBC x 10^3 per mm^3	Wistar rats WBC x 10^3 per mm^3	
		o	o
Before	9.5	7.7	7.5
1	8.5	9.5	10.4
2	12.2	12.7	12.0
3	19.3	14.2	12.3
4	15.6	12.7	12.7
5	16.1	14.7	11.6

[1] Diabetes was induced by IP injection of 40 mg/kg of streptozotocin on three consecutive days (solution of 10 mg/ml in 0.01 M acetic acid).

CANDIDOSIS

Candida spp. are widely distributed in nature and are considered to be commensals in the gut of man and animals, but they are also responsible for various types of infection. Of the members of this genus, *C. albicans* is the most pathogenic, but nowadays non-*albicans* species are isolated from infections more frequently than previously.

Models of experimental candidosis have been reviewed by various authors.[3,4,7-9,11] *C. albicans* remains the preferred pathogenic species for experimentation in animal species.[11] More than 200 strains have been infected by various routes, leading to superficial, systemic or disseminated disease.[9-11]

Systemic Candidosis

To obtain systemic *Candida* infection the intravenous (IV) route is generally used. Susceptibility to such infection is largely dependent on various factors. One of those factors is the animal species used. When various animal species were infected by the IV route with blastospores of *C. albicans* ATCC44858, the severity of systemic disease in non-immunodepressed animals was almost comparable with doses of 2,000 colony forming units (CFU) per gram BW in the rabbit, 8,000 in the dog, guinea-pig, hamster and monkey, 40,000 in the mouse, 200,000 in the rat and 400,000 in the chicken (Table 5). In the rabbit and especially in the guinea-pig *Candida* in yeast- and/or mycelial form is widely distributed in various organs and more regularly present than in other species. Skin folliculitis followed by erythematous plaques developed in the guinea-pig and their number and severity was related to the injected inoculum (Table 6). *Candida* skin folliculitis was reduced or absent in other animal species. *Candida* eruptions occurred in the chicken on the comb, the wattles and the legs. They transformed rapidly into hyperkeratotic lesions covered with powdery squames (comb and wattles), and intertrigo of the toes. All lesions contained large amounts of yeasts and hyphae.

Table 3. Serum glucose in diabetic[1] Swiss mice and Wistar rats.

Week	Swiss mice				Wistar rats			
	Controls		Streptozotocin		Controls		Streptozotocin	
	n	glucose in mg/dl	n	glucose in mg/dl	n	glucose in mg/dl	n	glucose in mg/dl
Before	12	159	14	175	8	136	16	144
1	12	145	14	187	8	154	16	480
2	12	132	14	235	8	136	16	474
3	12	141	14	411	8	134	16	470
5	12	150	13	509	8	145	15	527
10	12	149	11	673	8	139	12	542
13	12	150	9	714	8	136	10	534

[1] Diabetes was induced by IP injection of 40 mg/kg of streptozotocin on three consecutive days (solution of 10 mg/ml in 0.01 M acetic acid).

Table 4. Effects of the use of deferoxamine in fungal infections in guinea-pigs

Mycosis	On severity of infections		On therapeutic outcome
	Number of dosages		
	1	2-4	
Dermatophytosis	none	none	none
Superficial candidosis[1]	none	none	none
Systemic candidosis	none	none	none
Cryptococcosis	none	slight	not known
Aspergillosis	none	inoculum dependent	slightly inoculum- and therapeutical dosage dependent
Zygomycosis[2]	very high	extremely high	complete antagonisation of activity of amphotericin B

1 Superficial candidosis also in vaginal candidosis in rats
2 Zygomycosis *Rhizopus oryzae* and *Rh microsporus* var *rhizopodiformis*

Table 5. Systemic candidosis in various animal species after IV infection with *C. albicans* ATCC 44858.

| Animal | | Inoculum | Number of animals | | | Positive cultures | |
Species	Breed or strain	x10^3 CFU per g BW	Total	Skin folliculitis[1] Moderate	Pronounced	Skin	Kidneys
Rabbit	New Zealand	2	16	3	7	11	16
Dog	Canis vulgaris	8	6	2	0	2	5
Guinea-pig	Pirbright	8	1760	2	1758	1759	1756
Golden hamster	Mesocricetus auratus	8	8	0	0	1	8
Monkey	Cercopithecus aethiops	8	6	2	1	3	5
Mouse	Swiss	40	80	2	0	5	80
Rat	Wistar	200	24	1	0	4	24
Chicken	Arbor acre	400	24	8	16	24	3

[1] *Candida* lesions on the comb, the wattles and intertrigo for chickens, folliculitis for others.
(Adapted from ref. 9)

Table 6. Systemic candidosis in guinea-pigs after IV infection with C. *albicans* ATCC 44858.

Inoculum N. of CFU per g BW	N. of Animals	Mortality %	Number of animals				
			Skin folliculitis		Positive cultures		
			Moderate	Pronounced	Skin	Kidneys	
32,000	18	89	0	18	18	18	
16,000	30	67	0	30	30	30	
8,000	1760	31	2	1758	1759	1756	
6,000	156	19	0	156	156	154	
4,000	40	13	1	39	39	37	
2,000	18	0	5	13	6	14	
1,000	12	0	7	3	4	9	
500	6	0	4	1	2	2	
250	12	0	6	0	2	3	
125	6	0	4	0	3	1	
63	12	0	2	0	0	0	
20	6	0	2	0	0	0	

(Adapted from ref. 9)

By preadministration of immunodepressing agents such as mechlorethamine hydrochloride to the guinea-pig with systemic candidosis, *Candida* invasion was aggravated, but susceptibility of these neutropenic animals to oral treatment with the lipophilic itraconazole was not affected.

Brown heroin is also considered to be a predisposing agent for *Candida* folliculitis. Three groups of 6 guinea-pigs were inoculated with 20 and three groups with 2,000 CFU of *C. albicans* per g. BW. In each series one group was a control receiving subcutaneous (SC) distilled water from the day of infection for 12 consecutive days. The second regimen was SC injection of pure morphine: 10 mg/kg for 4 days, followed by 20 mg/kg for 8 days. The third regimen consisted of SC injection of brown heroin 10 and 20 mg/kg for 4 and 8 days, respectively. (Seizure by the office of Counsel of prosecution of Antwerp No. 24543/87, followed by judgement and authorization of the Court of Justice of Antwerp No. 28, K No. 2960 on January 1, 1988). For both inoculum sizes skin folliculitis was more pronounced in guinea-pigs treated with pure or with brown heroin than in the controls. Clinically, eye lesions were not different and cultures for *C. albicans* from skin, left eye, right eye and kidney were comparable between both heroin groups and the controls.[10]

Several aspects of the anatomopathology of systemic candidosis in the guinea-pig have been described covering the invasion of various organs. The keratin layer of the epidermis became invaded by yeasts transmigrating through the epidermis and/or by direct transmission via the growing hair shaft. The yeasts grew out to fungal filaments in epidermal keratin (Fig. 1a). In the liver small necrotic foci with fungi were observed (Fig. 1b). Important histological changes with fungal invasion were observed in the kidney. Cortical tubular and interstitial lesions with presence of mycotic granulomas finally resulted in a diffuse, descending pyelitis

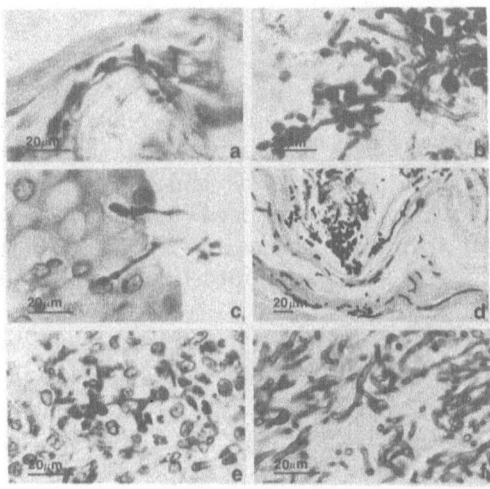

Fig. 1 *C. albicans* infection

a. Guinea-pig: skin: yeasts and hyphae in epidermal keratin. Periodic Acid-Schiff (PAS).
b. Guinea-pig: liver: invasion in necrotic foci. PAS.
c. Guinea-pig: urinary bladder: fungal development in epithelium. PAS.
d. Guinea-pig: esophagus with fungal elements in keratin. Gomori methenamine silver nitrate (GMSN).
e. Guinea-pig: mucosa of the stomach with *Candida* invasion. PAS.
f. Rat: vaginal candidosis with yeasts and hyphae in vaginal keratin layer. PAS.

and a pronounced fungal development with hyphae and blastospores. Due to the descending infection the urinary bladder became invaded by the fungus (Fig. 1c). Histological changes after *Candida* invasion have been described in detail.[30]

Gastrointestinal Candidosis

Gastrointestinal candidosis without administration of compromising agents resolved spontaneously in mice, rats, rabbits and guinea-pigs. However, oral administration of antibiotics stimulated the colonization in the gut of mice, but isolation of *C. albicans* was more irregular than in guinea-pigs and may be inhibited by excessive growth of *Torulopsis pintolopesii*. Guinea-pigs were immunocompromised by IM injection of prednisolone acetate at 10 mg/kg, five days weekly beginning 8 days before infection in combination with oral chloramphenicol at 50 mg/kg and streptomycin at 40 mg/kg on alternate days.[6] Six doses of antibiotics per week were given, starting 2 days before infection. An inoculum of 2×10^7 CFU of *C. albicans* blastospores was administered by gavage. Five days after oral inoculation the infection was well established and treated orally. Fecal monitoring was performed microscopically and with cultures. In the feces only yeasts were observed but tongue and esophageal samples contained blastospores and hyphae. The hyphae were also seen at necropsy in the keratinous epithelial layer of the tongue, in the esophagus, in the stomach and in the intestine. Hyphae of *C. albicans* especially in the lumen of the stomach were clearly distinguishable from the small yeasts of *Torulopsis pintolopesii*. *Candida* invasion of the keratin layer of the esophagus is shown in figure 1d. Mucosal glandular degeneration was observed in the stomach and yeasts and hyphae were present (Fig. 1e). Gallinaceous birds but especially young turkeys were highly susceptible to candidosis of the crop- and the glandular stomach after infection by gavage with *C. albicans*.

Skin Candidosis

Chemical diabetes in the adult albino guinea-pig was induced by IM injection of 200 mg alloxan per kg BW 24 h before infection. The intact clipped skin was inoculated with 8×10^6 blastospores of *C. albicans*.[5]

After 3 to 4 days small papules appeared. In one week vesicles were present containing clear exudate with blastospores and hyphae. There was erythema and oedema, followed by hyper- and parakeratosis with spontaneous cure in 5 to 6 weeks.

C. albicans was the most pathogenic species on the skin of guinea-pigs with induced chemical diabetes, but other non-*albicans* species also produced minute eruptions.[11]

Vaginal Candidosis

Vaginal candidosis was studied in mice, rats, guinea-pigs and rabbits. Normally cycling laboratory animals will recover completely from a vaginal inoculation with *C. albicans*. By administering estrogens to cycling animals the infection could be prolonged for a few days, but in general the hyphae disappear within three to seven days and only blastospores persist for a few more days. Apparently the mouse and the rat are slightly more sensitive than the guinea-pig and the rabbit. After ovarectomy and hysterectomy and SC injection of 0.1 mg estradiol undecylate in 1 ml sesame oil permanent pseudo-oestrus was produced in Wistar rats. Infection with 8×10^5 CFU of blastospores of *C. albicans* in 0.2 ml aqueous dilution resulted in rapid transformation of yeasts to hyphae in the vaginal keratin layer. The weekly administration of estrogens kept the rat in pseudo-estrus and allowed the infection to persist. In an experiment when vaginal candidosis was allowed to persist for 11 months, 60 castrated female rats received a weekly dosage of estrogens. The animals were monitored weekly for 11 months and 71 % of the rats were positive for *C. albicans* at each sampling during and at the end of the experiment (Table 7). After one month the presence of blastospores became more apparent, but at the end hyphae and yeasts were still present.

Table 7. Vaginal infection with *C. albicans* in ovarectomized and hysterectomized rats in permanent pseudo-estrus (group of 60 rats)[1]. Observation for 11 months.

Control at stated month	% of positive rats
0.25	100
0.5	100
1	98
3	90
6	75
9	73
11	71

[1] Weekly injection of estrogens.

The administration of streptozotocin (40 mg/kg for three days) to rats in permanent pseudo-estrus was responsible for irreversible diabetes. The serum glucose levels remained stable in the control group but rose from 144 mg per dl before streptozotocin to 480 mg per dl one week after the induction of diabetes. These high levels persisted until death. After 13 weeks the surviving diabetic rats were in very poor condition.[27] In diabetic rats vaginal candidosis was more pronounced than in non-diabetic animals and the balance of hyphae/blastospores was in favour of the hyphae. The cumulation of predisposing factors also makes vaginal infection with non-*albicans* species possible, especially *C. (Torulopsis) glabrata*. It was possible to cure vaginal candidosis in the diabetic rats with orally administered itraconazole.

In figure 1f, the histological appearance of *C. albicans* invasion in the keratin layer of the vagina is shown.

Other *Candida* Infections

After administration by gavage of 2×10^8 to 2×10^9 blastospores of *C. albicans* to pregnant guinea-pigs, rats and mice, the majority of the animals aborted.[9] Folliculitis with abundant development of hyphae was observed on the skin of the fetuses. After IV infection of the same pregnant animal species with 10^6 to 10^7 blastospores of *C. albicans*, the infection of the fetuses was more irregular, but some *Candida* lesions with hyphae and blastospores were still present on the fetal envelopes, placenta and epidermis.[9]

FUNGAL KERATITIS

Fungal keratitis on the superficially abraded cornea of guinea-pigs was induced with various strains of *C. albicans*, *Fusarium oxysporum*, *F. moniliforme*, *Aspergillus flavus* and *A. fumigatus* .[31,32]

Candida infections with 5×10^5 blastospores were highly inflammatory with high yields of hyphae and yeasts in the cornea, anterior chamber, lens and corpus vitreum (Fig. 2a).

F. oxysporum with 2×10^5 to 10^6 spores was moderately pathogenic, but *F. moniliforme* with 10^6 spores was more invasive: both produced hyphae and some swollen yeast-like elements. In figure 2b these elements are shown in the necrotic magma of the corpus vitreum after infection with *F. oxysporum*. Infection with *A. flavus* and *A. fumigatus* produced severe kerato-retinitis.

Topical treatment with miconazole (1%), ketoconazole (1%), itraconazole (0.5%) and saperconazole (0.5%) b.i.d. for two weeks reduced the lesions and the *Candida* flora. However, oral therapy of *Candida* keratitis with ketoconazole (20 mg/kg per day) for 2 weeks and with itraconazole or saperconazole at 10 mg/kg per day responded better than topical treatment.

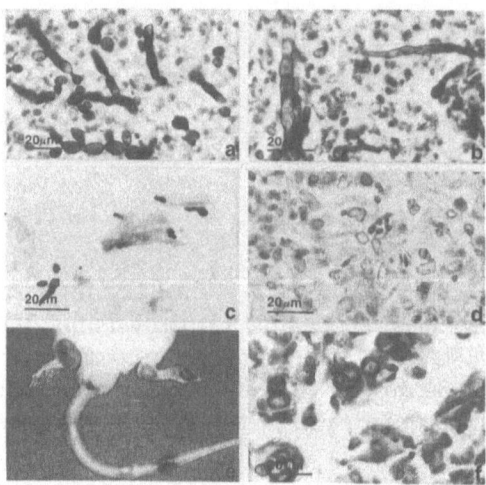

Fig. 2 a. *C. albicans*: eye infection in guinea-pig. Lens and corpus vitreum: inflammation and fungi. GMSN.
  b. *F. oxysporum*: eye infection in guinea-pig: necrotic magma in corpus vitreum, septated hyphae and swollen yeast-like element. GMSN.
  c. *M. ovalis*: skin infection in guinea-pig: yeasts and germ tubes in keratin layers. GMSN.
  d. *Sp. schenckii* infection in guinea-pig. Testicular granuloma with yeasts and cigar bodies. PAS.
  e. *F. pedrosoi* infection in mouse. Black pigmented granulomas on legs and tail.
  f. *F. pedrosoi* infection in mouse. Phalangial granuloma with hyphae and sclerotic cells. PAS.

Topical miconazole, enilconazole and ketoconazole at 1% and itraconazole and saperconazole at 0.5% were active in *F. moniliforme* infection. Oral treatment was highly efficacious with itraconazole and saperconazole at 10 mg/kg but only moderately effective with ketoconazole at 20 mg/kg. Topical treatment of kerato-retinitis caused by *A. flavus* resulted in moderate responses miconazole and ketoconazole, but the infection responded well to enilconazole, itraconazole and saperconazole. Ketoconazole was inactive by the oral route, but itraconazole and saperconazole administered at 10 mg/kg o.d. for two weeks considerably reduced the infection and the fungal flora.

MALASSEZIA INFECTION

Malassezia (Pityrosporum) *ovalis* or *M. orbicularis* (*M. furfur*) are present as episaprophyte yeasts on the human skin and are able to cause a number of pathologies such as pityriasis versicolor, dandruff, seborrheic dermatitis, folliculitis and others. The organism can be present in a yeast form or as short hyphae. The more animal-specific *M. pachydermatis* is present as a saprophyte or as a pathogen on the skin or in the cerumen of various animal species. *M. pachydermatis* has been often isolated from dermatitis and otitis externa where it was always present in the yeast form.[33,34] Inoculation on the intact electrically clipped skin of the back of guinea-pigs with 1.2×10^6 to 5.1×10^6 CFU of *M. ovalis*, *orbicularis* or *pachydermatis* under an occlusive dressing for five days produced folliculits, inflammation, hyperkeratosis and parakeratosis.[35] In direct microscopic preparations, yeasts and short hyphae of *M. ovalis* or *orbicularis* were seen and the organism was reisolated by cultures from skin samples. *M. pachydermatis* was always present in the stratum corneum as a yeast. The brandname of the occlusion dressing was Oclufol (Lohman KG Fahr, Rhein, Germany). It was maintained with Dermiclear tape (Johnson & Johnson). After daily inoculation with *M. ovalis* for 7 days , the infection became more inflammatory with severe erythema in some animals. After the climax the disease evolved to a seborrheic dermatitis-like syndrome. Ketoconazole and itraconazole were highly active by topical and oral treatment. Ketoconazole shampoo has been also very successfully evaluated. In histological sections of the epidermis, budding yeasts and germ tubes were present in the keratin layer (Fig. 2c).

SPOROTRICHOSIS

Guinea-pigs were highly susceptible to *Sporothrix schenckii*, when infected by the intratesticular or the IV route.[3,20,21,22]

After intratesticular infection numerous disseminations occurred in untreated and excipient-treated animals especially in the spleen, skin, liver and lungs.

Mycotic granulomas with degeneration and inflammation of tissues were found. Within the granulomas yeast cells and cigar-shaped elements were present (Fig. 2d). In an early-starting treatment schedule with 10 mg/kg of itraconazole per day for 28 days, 65% of the guinea-pigs were cured and free of fungi and 35% were improved. At higher dosages all animals became free of *Sp. schenckii*.

Guinea-pigs infected by the IV route with *Sp. schenckii* had rapid dissemination into various organs.

In a comparative therapeutic oral trial between itraconazole and fluconazole, the drugs were administered for 28 days o.d. starting 3 days after infection at dosages from 0.63 to 5 mg/kg. In the itraconazole group treated at 1.25 mg/kg no skin eruptions occurred, and 60% of testicles and 70% of prepuces were free of fungi. At 2.5 mg/kg of itraconazole all animals were mycologically cured. Fluconazole was only marginally active at 5 mg/kg and none of the animals treated with fluconazole was cured.

CHROMOBLASTOMYCOSIS

Experimental IV infection in guinea-pigs with *Fonsecaea pedrosoi* is poorly invasive. Small eruptions are produced, but spontaneous cure always occurs. Pathogenic effects in the rat are also unsatisfactory (unpublished data). The Swiss mouse, however, is more sensitive[16] and most of the strains tested were highly invasive leading to multiple disseminated black granulomas in nearly all organs (Fig. 2e). Two inoculum concentrations of 8 strains of *F. pedrosoi* at 25 and 37 °C were used for IV infection: 5×10^5 and 2×10^6 CFU per mouse. In all animals chromoblastomycosis was produced, but the highest and most uniform disease was obtained with the inoculum grown at 37 °C. Very large cutaneous and subcutaneous masses, rich in hyphae, sclerotic cells and yeast forms were seen (Fig. 2f). Degenerated and perforated bone structures and bone fractures were observed. Bone and cartilage resorption by numerous osteoclasts was present. Definite thrombosis with necrosis accompanied by fungal invasion was seen in the lungs, liver, kidneys and adrenals.

PENICILLIUM MARNEFFEI INFECTION[36,37]

Normal and neutropenic (mechlorethamine hydrochloride) albino guinea-pigs were infected IV with 12.5×10^3, 25×10^3 or 10^5 CFU of *Penicillium marneffei* per g BW. The strain was isolated from an AIDS patient, who also was a drug-addict. All animals were highly positive. The severity of the infection was obviously related to the size of the inoculum and to the immunological state of the animals. Skin eruptions were general about 14 days after infection. They were covered with scales. In microscopic preparations of the squames of the skin, long filaments were seen. At autopsy, necrotic foci were present in the lymph nodes and spleen of all immunocompetent animals, and in various organs (especially liver, testicle and epididymis) of some animals, but cultures were positive for *P. marneffei* in almost all organs. In the immunocompromised animals the lesions were more pronounced. Cultures grown from these organs were highly positive (skin, lungs, lymph nodes, liver, spleen, kidneys, epididymis and testicles). Microscopically, short, septate hyphae were observed in the dermis. Bicellular, sometimes tricellular yeast-like elements were seen in the internal organs. Histological examination of the guinea-pigs showed pneumonitic lesions and bronchiolitis without detectable fungal invasion. Within the liver perilobular infiltration, bile duct proliferation, hepatocytic vacuolation and necrotic foci were seen, but fungal elements could not be detected. The most prominent pathological changes and fungal invasion were seen in the spleen and inguinal lymph nodes and their supporting adipose tissue, the testis and epididymis and the skin. The underlying muscles of the skin showed granulomatous lesions with intact yeasts and inflammatory cells. Muscle fibres were degenerated and necrotized. Similar granulomatous changes were found in the hypodermis and dermis. The epidermis was acanthotic, hyperkeratotic and showed vesicle and crust formation. Migration of intact yeast-like cells from the dermis to the epidermis could be observed. Short fungal elements were detected within the crusts and the superficial hyperkeratotic layers (Fig. 3a).

Large granulomas with epithelioid- and giant-cells, neutrophils and many spores and a moderate number of short filaments were observed in the white and red pulps of the spleen and in the cortex and medullary sinuses of the inguinal lymph nodes. Their supporting adipose tissue showed granulomatous cellular reaction with spores and short filaments (Fig. 3b). Comparable granulomatous changes and fungal elements were seen in the testis and epididymis.

After oral treatment of the infected guinea-pigs with itraconazole, the disease was cured and the fungus eliminated.[38]

A patient treated with itraconazole was also completely free of penicilliosis (control by autopsy and cultures, after he died from AIDS), documenting the activity of itraconazole and good correlation between the activity of itraconazole in the patient and in guinea-pig, as reported in the literature by several investigators.[39-52]

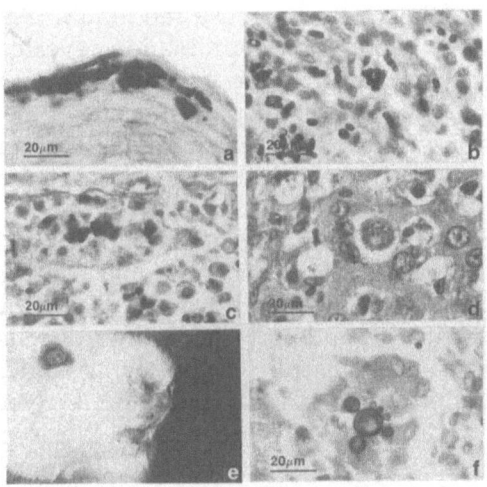

Fig. 3 Fungal IV infections.
   a. *P. marneffei*: guinea-pig: skin: short filaments in hyperkeratotic layer. GMSN.
   b. *P. marneffei*: guinea-pig: short and yeast-like fungal elements in inguinal lymph node. GMSN.
   c. *H. capsulatum*: mouse liver: yeasts within histiocytes in a vein. GMSN.
   d. *H. capsulatum*: mouse liver: yeasts within Kupffer cells in sinusoid. PAS.
   e. *P. brasiliensis*: guinea-pig: abscesses and erosions of nose and mouth mucosae.
   f. *P. brasiliensis*: guinea-pig: yeast with multipolar budding within giant cell in nose tissue. GMSN.

HISTOPLASMOSIS

Histoplasma farciminosum inoculated by the IV route did not produce infection in normal guinea-pigs or mice. The IV infection with *H. capsulatum* was more invasive in the mouse than in the guinea-pig. In the mouse, slight to moderate infection of the lungs, liver, spleen and kidneys occurred. *H. capsulatum* yeasts were mostly localized within histiocytes. Some yeasts showed germ tubes. The lungs revealed alveolitis and pneumonia. In the venous blood vessels endophlebitis and thrombus formation with histiocytes containing small yeasts were often observed. These vascular changes were also present in the liver. Yeasts were detected in larger hepatic veins, perilobular and centrilobular veins (Fig. 3c). Sinusoids were dilated and sometimes filled with histiocytes. Yeasts were found within the proliferating Kupffer cells of the sinusoidal linings (Fig. 3d). Hepatocytes were moderately vacuolized with some focal necrosis. The white and red pulps of the spleen presented slight degeneration, neutrophilic infiltration and often severe histiocytosis, with low numbers of fungi. The kidneys were scarcely infected. The cortical veins and venules, glomeruli and interstitial tissue contained yeasts within the histiocytes. *H. duboisii* was particularly pathogenic in the guinea-pig after intratesticular infection with multiple disseminations to various organs. Large budding yeast cells, often in chains, were present in pus. This infection responded dramatically to oral itraconazole.

PARACOCCIDIOIDOMYCOSIS

Intravenous infection of guinea-pigs with *Paracoccidioides brasiliensis* resulted in macroscopic lesions of the mucosae of nose and mouth. Abscesses and erosions were observed (Fig. 3e). Microscopic examination revealed individual and/or confluent fibrotic encapsulated mycotic granulomas within muscular tissue adjacent to the bones of legs and ribs, the posterior side of the ear and the tissue of nose and mouth. Lesions in the spleen, the inguinal lymph nodes, the testis and the skin occurred at a lower rate. Other internal organs were free of fungal invasion.

In the mucosae of the mouth and nose, granulomas with a caseous necrotic centre surrounded by many epithelioid and giant cells were observed. These giant cells often phagocytized living yeasts with multipolar buddings (Fig. 3f). The granulomas provoked muscle- and connective-fiber degeneration, fibrosis and inflammation. Scar tissue was present. The overlying epidermis of the nose and mouth granulomas had erosive and ulcerative lesions. Neutrophils and yeasts were seen within necrotic cell debris and crusts, outside the epidermis.

Dissemination of yeasts and ulcerative lesions at other locations of the skin were less pronounced. Only a few granulomas with epidermal ulcers and crusts containing some yeasts were seen. Some bone lesions were present.

CRYPTOCOCCOSIS

Cryptococcus neoformans is a saprobic yeast isolated with an increased frequency in predisposed and immunodepressed patients.[53,54] Intravenous infection with *C. neoformans* in guinea-pigs produced multiple cryptococcomas in various organs, including the skin, and also caused meningitis, encephalitis, conjunctivitis, choroiditis, pneumonia, interstitial inflammation of the kidneys and inflammation of the spleen and lymph nodes. These organs were seriously invaded by fungal elements. Inflammatory changes and fungal invasion were also observed in skin, testis and epididymis, heart and peritoneum.

In the dermis of the skin cryptococcomas with fungi were seen (Fig. 4a). Within the necrotic epidermal layers and crusts inflammatory cells and cryptococci were present.[15]

The gelatinous form of cryptococcosis was observed in tissues such as brain and meninges, in the cortex of the kidney, adipose tissue of the testis and epididymis. A fibro-granulomatous reaction with giant cells occurred and numerous encapsulated yeasts with budding and hyphal forms were detected. The most prominent finding was the lymphoid depletion in the spleen and in the lymph nodes. White pulps were mostly completely destroyed and lymph follicles were often absent in the lymph nodes. The normal architecture of these organs was necrotic and was replaced by fibro-granulomatous tissue reaction with many giant cells and many cryptococci. Within the necrotic centre of the cryptococcomas buddings and short hyphal forms were present (Fig. 4b).

ZYGOMYCOSIS

Systemic intravenous infection by *Rhizopus oryzae* and *Rh. microsporus* var. *rhizopodiformis* in guinea-pigs, rats and mice produced lesions and fungal invasion in various organs, including brain, eye, lungs, heart, kidneys, liver, spleen, intestine and skin. Blood vessels were seriously invaded. On the skin of guinea-pigs erythematous and inflammatory skin ulcers with crust formation were observed (Fig. 4c). Dermal mycotic granulomas produced inflammation and some hyphae migrated through the acanthotic epidermis into the hyperkeratotic crust, where hyphae developed in the keratin (Fig. 4d).[27,28]

Intravenous infection with these two strains of *Rhizopus* revealed thrombi and branching distorted, coenocytic hyphae in the brain, the sclera and choroid of the eye, in the kidneys, liver and white and red pulps of the spleen. The presence of many thrombi and fungi provoked several necrotic zones in these organs and the development of large mycotic granulomas with often large coenocytic hyphae. Meningitis, encephalitis and pyelitis were often seen. Cortical inflammation and degeneration was observed in the kidney. Intratubular leucocytic cylinders loaded with hyphae provoked papillary necrosis, pyelitis and ureteritis with fungal invasion of the pelvic cavity and ureter. Various forms of large swollen yeast-like

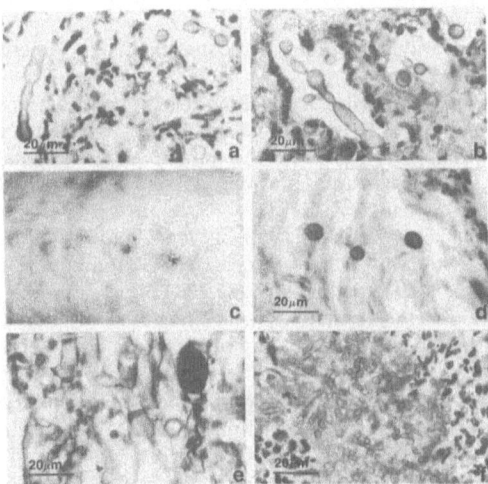

Fig. 4 Fungal IV infections in guinea-pigs.
   a. *C. neoformans*: yeasts and short hyphal form in the necrotic centre of a skin granuloma. GMSN.
   b. *C. neoformans*: lymph node with yeasts and a hypha in the centre of a granuloma. PAS.
   c. *Rh. oryzae*: Erythematous and inflammatory skin ulcers.
   d. *Rh. oryzae*: Skin: hyphae in hyperkeratotic crust. PAS.
   e. *Rh. microsporus*: swollen yeast and sporangium-like elements in ureteral lumen. GMSN.
   f. *T. mentagrophytes*: lung: trichophytoma with many hyphae and some budding yeasts. PAS.

elements, vesicles and sporangium-like elements were present in the pelvic cavity and ureteral lumen (Fig. 4e). Similar structures were also seen in the lungs and the spleen.

DERMATOPHYTOSIS

Dermatophytes are in general agents of superficial infection of the glabrous skin, the hairy skin, the folds and the nails. However, some other organs may be invaded. Systemic and disseminated trichophytosis was obtained in guinea-pigs by IV infection with a zoophilic strain of *Trichophyton mentagrophytes*.[55] Dissemination to the skin, the hairs and the nails occurred in the guinea-pig. In some internal organs such as the liver and the kidneys, but especially the lungs large mycotic granulomas with epithelioid- and giant cells were observed. These granulomas or trichophytomas had a necrotic centre filled with many hyphae and some budding yeasts (Fig. 4f).

REFERENCES

1. F.C. Odds, "*Candida* and Candidosis", ed., Baillière Tindall, London, (1988).
2. H.J. Scholer, Experimentelle Vaginal-Candidiasis der Ratte, *Pathol. Microbiol.* 23: 62 (1960).
3. M. Miyaji, "Animal Models in Medical Mycology", C.R.C. Press, Boca Raton (1987).
4. T.L. Ray, Animal models of experimental *Candida* infections of the skin, in: "Models in Dermatology", Vol. 1, H.I. Maibach and N.J. Lowe, eds., Karger, Basel (1985).
5. J. Van Cutsem and D. Thienpont, Experimental cutaneous *Candida albicans* infections in guinea-pigs, *Sabouraudia* 9: 17 (1971).
6. D. Thienpont, J. Van Cutsem and D.A. Gough, Treatment of gastrointestinal candidosis in predisposed guinea-pigs and in conventional mice with miconazole, *Mykosen* 21: 417 (1978).
7. D. Thienpont, J. Van Cutsem and M. Borgers, Ketoconazole in experimental candidosis, *Rev. Infect. Dis.* 2: 570 (1980).
8. M.N. Guentzel, G.T. Cole and L.M. Pope, Animal models for candidiasis, in: "Current Topics in Medical Mycology", Vol. 1, M.R. MacGinnis, ed., Springer-Verlag, New York (1985).
9. J. Van Cutsem, F. Van Gerven, J. Fransen and P.A.J. Janssen, Experimental candidosis in animals and chemotherapy, in: "*Candida* and Candidamycosis", E. Tümbay, H.P.R. Seeliger and Ö. Ang, eds., Plenum Press, New York (1991).
10. J. Van Cutsem, E. Drouhet, J. Fransen and F. Van Gerven, Candidose systémique et disseminée du cobaye avec manifestations cutanées et oculaires. Administration expérimentale de l'héroïne, *Soc. Fr. Mycol. Méd.* Abstract Nov. 22-23 (1991).
11. J. Van Cutsem, Animal models for dermatomycotic infections, in: "Current Topics in Medical Mycology", Vol. 3, M.R. MacGinnis and M. Borgers, eds., Springer-Verlag, New York (1990).
12. J. Van Cutsem, Fungal models in immunocompromised animals, in: "Mycoses in AIDS Patients", H. Vanden Bossche, D.W.R. MacKenzie, G. Cauwenbergh, J. Van Cutsem, E. Drouhet and B. Dupont, eds., Plenum Press, New York, (1990).
13. D.W. Williams, Drug therapy in animal models of histoplasmosis, in: "Experimental Models in Antimicrobial Chemotherapy", Vol. 3, O. Zak and M.A. Sande, eds., Academic Press, London, (1986).
14. J.R. Graybill, Animal models for treatment of cryptococcosis, in: "Experimental Models in Antimicrobial Chemotherapy", Vol. 3, O.Zak and M.A. Sande, eds., Academic Press, London (1986).
15. J. Van Cutsem, J. Fransen, F. Van Gerven and P.A.J. Janssen, Experimental cryptococcosis: dissemination of *Cryptococcus neoformans* and dermotropism in guinea-pigs, *Mykosen* 29: 561 (1986).
16. A. Polak, Experimental infection of mice by *Fonsecaea predrosoi* and *Wangiella dermatitidis*, *Sabouraudia* 22: 167 (1984).

17. J. Van Cutsem, Modelos experimentales de micosis profundas en animales para el estudio de antifúngicos nuevos, *Rev. Arg. Micología* 15: 10 (1992).

18. J. Van Cutsem, *In vitro* antifungal spectrum of itraconazole and treatment of systemic mycoses with old and new antimycotic agents, *Chemotherapy* 38 (S1): 3 (1992).

19. G. Cauwenbergh and J. Van Cutsem, Role of animal and human pharmacology in antifungal drug design, *Ann. N.Y. Acad. Sc.*, 544: 264 (1988).

20. J. Van Cutsem, F. Van Gerven and P.A.J. Janssen, Activity of orally, topically and parenterally administered itraconazole in the treatment of superficial and deep mycoses. Animal models, *Rev. Infect. dis.* 9 (S1): 15 (1987).

21. J. Van Cutsem, Oral, topical and parenteral antifungal treatment with itraconazole in normal and in immunocompromised animals, *Mycoses* 32 (S1): 14 (1989).

22. J. Van Cutsem, Oral, topical and parenteral treatment with itraconazole in various superficial and systemic experimental fungal infections. Comparison with other antifungals and combination therapy, *Br. J. Clin. Pract.* 44 (S71): 32 (1990).

23. A. Schaffner and P.G. Frick, The effect of ketoconazole on amphotericin B in a model of disseminated aspergillosis, *J. Infect. Dis.* 151: 902 (1985).

24. P.M. Laduron and P.F.M. Janssen, Impaired axonal transport of opiate and muscarinic receptors in streptozotocin-diabetic rats, *Brain Res.* 380: 359 (1986).

25. J.R. Boelaert, G.F. van Roost, P.L. Vergauwe, J.J. Verbanck, C. De Vroey and M.F. Segaert, The role of desferrioxamine in dialysis associated mucormycosis: report of three cases and review of the literature, *Clin. Nephrol.* 29: 261 (1988).

26. J.R. Boelaert, A.Z. Fenves and J.W. Coburn, Deferoxamine therapy and mucormycosis in dialysis patients: report of an international registry, *Am. J. Kidney Dis.* 18: 660 (1991).

27. J. Van Cutsem, J. Fransen and P.A.J. Janssen, Experimental zygomycosis due to *Rhizopus* spp. Infection by varios routes in guinea-pigs, rats and mice, *Mycoses* 31: 563 (1988).

28. J. Van Cutsem and J.R. Boelaert, Effects of deferoxamine, feroxamine and iron on experimental mucormycosis, *Kidney Int.* 36: 1061 (1989).

29. F. Abe, H. Inaba, T. Katoh and M. Hotchi, Effects of iron and desferrioxamine on *Rhizopus* infection, *Mycopathologia* 110: 87 (1990).

30. J. Fransen, J. Van Cutsem, R. Vandesteene and P.A.J. Janssen, Histopathology of experimental systemic candidosis in guinea-pigs, *Sabouraudia* 22: 455 (1984).

31. P.A. Thomas, D.J. Abraham, G.M. Kalavathy and J. Rajasekaran, Oral itraconazole therapy for mycotic keratitis, *Mycoses* 31: 271 (1989).

32. J. Van Cutsem and F. Van Gerven, Activité antifongique *in vitro* de l'itraconazole sur les champignons filamenteux opportunistes. Traitement de la kératomycose et de la pénicilliose expérimentales, *J. Myc. Méd.* 1: 10 (1991).

33. M. Baxter, The association of *Pityrosporum pachydermatis* with the normal external ear canal of dog and cats, *J. Small Anim. Pract.* 17: 231 (1976).

34. R. Dufait, Über die Bedeutung von *P. canis* bei Otitis externa und Dermatitis des Hundes, *Kleintier Praxis* 23: 29 (1978).

35. J. Van Cutsem, F. Van Gerven, J. Fransen, P. Schrooten and P.A.J. Janssen, The *in vitro* antifungal activity of ketoconazole, zinc pyrithione, and selenium sulfide against *Pityrosporum* and their efficacy as a shampoo in the treatment of experimental pityrosporosis in guinea-pigs, *J. Am. Ac. Dermatol.* 22: 993 (1990).

36. G. Segretain, *Penicillium marneffei*, agent d'une mycose du système réticulo-endothélial, *Mycopathol. Mycol. Appl.* 11: 327 (1959).

37. K.J. Kwon-Chung and J.E. Bennett, Penicilliosis marneffei, in: "Medical Mycology", Lea and Febiger, Malvern, USA, (1992).

38. M.A. Viviani and A.M. Tortorano, Unusual agents of mycoses in AIDS patients, in "Mycoses in AIDS Patients", H. Vanden Bossche, D.W.R. Mackenzie, G. Cauwenbergh, J. Van Cutsem, I. Drouhet and B. Dupont, eds., Plenum Press, New York (1990).

39. J.R. Perfect, D.V. Savani and D.T. Durack, Comparison of itraconazole and fluconazole in treatment of cryptococcal meningitis and *Candida* pyelonephritis in rabbits, *Antimicrob. Agents Chemother.* 29: 579 (1986).

40. A. Restrepo, J. Robledo, I. Gómez, A.M. Tabares and R. Gutierez, Itraconazole therapy in lymphangitic and cutaneous sporotrichosis, *Arch. Dermatol.* 122: 413 (1986).

41. R.J. Hay and J.M. Clayton, Treatment of dermatophytosis and chronic oral candidosis with itraconazole, *Rev. Infect. Dis.* 9 (S1): 114 (1987).

42. A. Restrepo, I. Gómez, J. Robledo, M.M. Patiño and L.E. Carro, Itraconazole in the treatment of paracoccidioidomycoses, *Rev. Infect. Dis.* 9 (S1): 51 (1987).

43. B. Dupont and E. Drouhet, The treatment of aspergillosis with azole derivatives, in: "*Aspergillus and aspergillosis*", H. Vanden Bossche, D.W.R. MacKenzie and G. Cauwenbergh, eds., Plenum Press, New York (1988).

44. R.M. Tucker, P.L. Williams, E.G. Arathoon and D.A. Stevens, Treatment of mycoses with itraconazole, *Ann. N.Y. Acad. Sc.*, 544: 451 (1988).

45. A. Restrepo, M.I. Múnera, I.D. Arteaga, I. Gómez, A.M. Tabares, M.M. Patiño and M. Arango, Itraconazole in the treatment of pulmonary aspergillosis and chronic pulmonary aspergillosis, in: "*Aspergillus* and aspergillosis", H. Vanden Bossche, D.W.R. MacKenzie and G. Cauwenberg, eds., Plenum Press, New York, (1988).

46. J.R. Graybill, Azole therapy of systemic fungal infections, in: "Diagnosis and therapy of systemic fungal infections", K. Holmberg and R. Meyer, eds., Raven Press, New York (1989).

47. J.R. Graybill, The modern revolution in antifungal drug therapy, in: "Mycoses in AIDS Patients", H. Vanden Bossche, D.W.R. MacKenzie, G. Cauwenbergh, J. Van Cutsem, E. Drouhet and B. Dupont, eds., Plenum Press, New York (1990).

48. D.W. Denning, R.M. Tucker, J.S. Hostetler, S. Gill and D.A. Stevens, Oral itraconazole therapy of cryptococcal meningitis and cryptococcosis in patients with AIDS, in: "Mycoses in AIDS Patients", H. Vanden Bossche, D.W.R. MacKenzie, G. Cauwenbergh, J. Van Cutsem, E. Drouhet and B. Dupont, eds., Plenum Press, New York (1990).

49. J.W. van 't Wout, Itraconazole in neutropenic patients, *Chemotherapy* 38 (S1): 23 (1992).

50. R. Negroni, Triazoles en el tratamiento de la paracoccidioidomicosis, *Rev. Arg. Micología* 15: 16 (1992).

51. M.A. Viviani, Opportunistic fungal infections in patients with acquired immune deficiency syndrome, *Chemotherapy* 38 (S1): 35 (1992).

52. J.S. Hostetler, D.W. Denning and D.A. Stevens, U.S. experience with itraconazole in *Aspergillus*, *Cryptococcus* and *Histoplasma* infections in the immunocompromised host, *Chemotherapy* 38 (S1): 12 (1992).

53. G.A. Sarosi, P.M. Silverbarb and F.E. Tosh, Cutaneous cryptococcosis: a sentinel of disseminated disease, *Arch. Dermatol.* 104: 1 (1971).

54. K.J. Kwon-Chung and J.E. Bennett, Cryptococcosis, "Medical Mycology", Lea and Febiger, Malvern, USA (1992).

55. J. Van Cutsem, J. Fransen, P.A.J. Janssen, Animal models for systemic dermatophyte and *Candida* infection with dissemination to the skin, in: "Models in Dermatology", Vol. 1, H.I. Maibach and N.J. Lowe, eds., Karger, Basel (1985).

MORPHOLOGICAL ASPECTS OF ANTIFUNGAL ACTION

Hideyo Yamaguchi, Yayoi Nishiyama, and Yukiyo Aoki

Research Center for Medical Mycology
Teikyo University
Hachioji-shi
Tokyo, Japan

ABSTRACT

To learn the cellular consequence of antifungal action of azoles and two other types of currently available ergosterol biosynthesis inhibiting antifungals (EBIA), terbinafine and amorolfine, the morphological changes induced by fungistatic concentrations of these drugs were examined electron microscopically using growing hyphae of *Trichophyton mentagrophytes* and yeast-form cells of *Candida albicans*, as well as reverting protoplasts of *C. albicans* , as the test organisms. All of these EBIA seemed to have a common morphological effect on both hyphal (*T. mentagrophytes*) and yeast-form (*C. albicans*) cells. The treatment of these fungi with any test drug induced intracellular and, more frequently, intra-cell wall deposition of electrondense vesicles, presumably resulting from degeneration of cell membanes primarily caused by inhibition of ergosterol biosynthesis. However, even more prominent were the morphological changes involving the cell wall. Characteristic of these changes were formation of fragile and deteriorated walls and aberrant thickening of cell wall at specific sites, particularly apical walls and septal walls, with the resultant blockage of hyphal elongation or proliferation of yeast-form cells. This adverse effect of azoles and other types of EBIA on the morphogenesis of fungal cell walls was studied in more detail in C. *albicans* protoplasts which reverted to osmotically stable cells. The experimental results provided evidence suggesting that all of these antifungals impair coordinated synthesis and/or assembly of glucans, chitin and other fungal wall-constituting materials.

INTRODUCTION

Recent progress in bioscience has allowed the primary target of most major antifungal drugs to be identified at the molecular level, and details of biochemical and physicochemical aspects of their mode of antifungal action are now understood. However, it is far more difficult to speculate on the basis of these biochemical data the subsequent cellular processes which ultimately lead to

Dimorphic Fungi in Biology and Medicine, Edited by
H. Vanden Bossche *et al.*, Plenum Press, New York, 1993

fungistasis or cell death, particularly those involved in cell wall formation, because of the complexity of the architecture and the morphogenesis of the fungal cell walls. Nevertheless, a number of papers from our laboratory and others show that morphological examination of antifungal action performed with growing cells or reverting protoplasts of susceptible fungi can provide information of particular use in understanding the drug-induced alterations of the cell envelope.

Until the 1970's when an antifungal imidazole derivative, miconazole, appeared on the market, amphotericin B and flucytosine were the only antifungal drugs available for the treatment of deep mycoses. Since then, three congeners, viz, one imidazole, ketoconazole and two triazoles, fluconazole and itraconazole, have been developed. All of these azole antifungals are now considered to constitute the major therapeutic armamentarium against deep mycoses. In addition, topical preparations of several new azoles as well as those of some other types of antifungal compounds including allylamine and morpholine derivatives have also been developed to treat superficial mycoses.

As reviewed by Vanden Bossche,[1] although these three types of antifungals have different targets of attack on the fungal ergosetrol biosynthesis pathway, all of them selectively inhibit ergosterol biosynthesis at low drug concentrations (10^{-6} M-10^{-8} M), eventually resulting in an accumulation of metabolic intermediates and decreased availability of ergosterol. Since ergosterol is a lipid component of fungal cell membranes essential to their retention of normal structure and function, primary inhibition of ergosterol biosynthesis by these drugs could induce impairment of the cell membrane. Such possible effects secondary to ergosterol biosynthesis inhibition have been most intensively studied with several azoles, and cytochemical and biochemical evidence suggests that depletion of ergosterol and/or accumulation of its metabolic precursors in the cell membrane of azole-treated fungal cells modifies the activity of membrane-bound enzymes, including those involved in synthesis of cell wall polysaccharides, particularly chitin.[2-6]

One of our major interests has been to collect electron microscopic data which may confirm this possibility and to delve deeper into the morphological aspect of antifungal action of EBIA. We began our studies using whole cells from growing cultures of dermatophytes and C. albicans. More recently, C. albicans protoplasts capable of reverting at relatively high frequency to osmotically stable and proliferative cells were successfully developed in our laboratory.[7] These protoplasts were also employed to see how inhibition of ergosterol biosynthesis influences cell wall regeneration.

EFFECT OF ERGOSTEROL BIOSYNTHESIS INHIBITING-ANTIFUNGALS (EBIA) ON WHOLE CELLS IN GROWING FUNGAL CULTURES

T. mentagrophytes TIMM 1189 and C. albicans TIMM 1046 were used throughout the studies presented in this paper. The EBIA tested were: an allylamine derivative, terbinafine; a morpholine derivative, amorolfine; two imidazole derivatives, neticonazole and omoconazole; and a triazole derivative, itraconazole. Appropriate specimens for examination by scanning electron microscopy (SEM) and transmission electron microscopy (TEM) were prepared from fungal cultures untreated or treated with drugs at growth-inhibiting concentrations for 12 to 24 h.

For SEM, harvested fungal cells were fixed with 2.5% glutaraldehyde in 0.1M cacodylate buffer (pH 7.4) at 4 °C for 2 h. After dehydration in a graded alcohol series, the samples were infiltrated with isoamyl acetate, and dried by the critical-point drying method with liquid carbon dioxide. The specimens were mounted on stubs by means of double-sided sticky tape, coated with gold-palladium in a Technics Hammer II Sputter coater, and observed under a Hitachi S-800 scanning electron microscope at an accelerating voltage of 7kV.

For TEM, fungal cells were fixed with 2.5% glutaraldehyde in 0.1 M cacodylate buffer (pH 7.2) at 4 °C for 2 h, washed with the same buffer, and post-fixed with a 1.5% aqueous solution of potassium permanganate at 4 °C for 18 h. The specimems were washed with chilled distilled water, dehydrated with a graded acetone series, and embedded in Spurr's resin mixture   Ultrathin sections were cut with an Ultracut N (Reichert-Nissei), stained with uranyl acetate and lead citrate, and then observed under a Hitachi H-700 electron microscope at 75 kV.

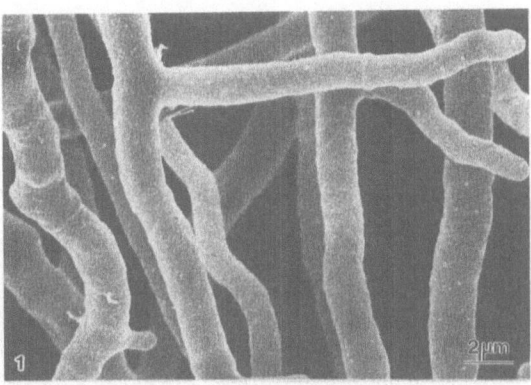

Fig. 1 SEM image of untreated control culture of *T mentagrophytes* grown for 24 h showing straight elongation of hyphae with smooth surfaces

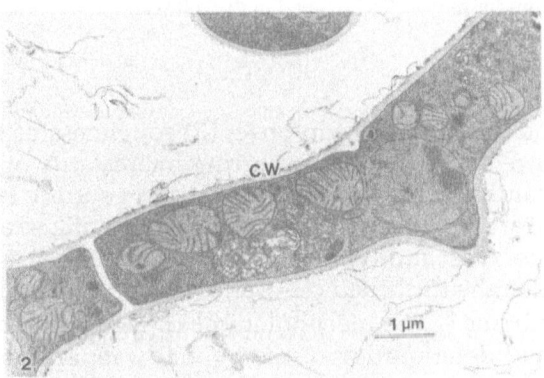

Fig. 2 Thin section of untreated control hyphae of *T mentagrophytes* grown for 24 h CW cell wall

Figures 1 and 2 give SEM and TEM images, respectively, of 24 h untreated control cultures of *T. mentagrophytes*. They show straight elongating hyphae with normal contour and smooth surface (Fig. 1) and the typical profile of the cell surface and cell interior (Fig. 2). In contrast, treatment with terbinafine at a fungistatic level (2 ng/ml) for 24 h produced substantial changes in the hyphal contour (Fig 3) and the fine structure of the peripheral region of cells, although the structures of cytoplasmic organelles and nuclei were relatively well preserved (Fig 4) The characteristic findings obtained with the treated cultures were as follows

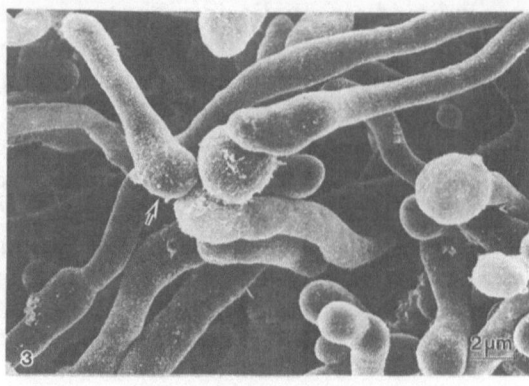

Fig. 3 SEM image of *T. mentagrophytes* cultures grown with terbinafine (2 ng/ml) for 24 h. Swelling of the hyphal tip (arrow) is common.

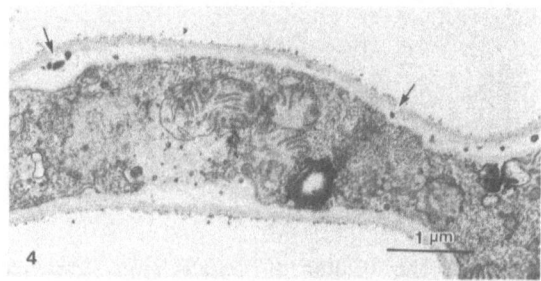

Fig. 4 Thin section of *T mentagrophytes* hyphae grown with terbinafine (2 ng/ml) for 12 h. Note accumulation of electron-dense vesicles (arrows) in the cell wall.

(i) deformation and/or collapse of hyphae; (ii) roughness of the hyphal surface: (iii) bulb-shaped swelling of hyphae occurring intercalarily or, more often, at the hyphal tip accompanied by breakage or exfoliation of surface layers of the swollen cell wall; (iv) aberrant thickening of cell walls at specific sites and retraction of opposing cell membranes; and (v) deposition of electrondense vesicles of various sizes scattered within the thickened cell wall and/or between the cell wall and the retracted cell membrane.[8] The morphological changes represented in (i) to (iii) suggest fragility and deterioration of the hyphal wall, and those in (iv) and (v) were more frequently observed at the apical wall than the lateral wall of hyphae. As illustrated in Figs. 5 and 6, virtually the same morphological changes were also induced by amorolfine (8 ng/ml) and neticonazole (0.3 μg/ml) at fungistatic concentrations.[9,10]

Figures 7 and 8 show TEM images of yeast-form cells of *C. albicans* treated for 24 h with fungistatic levels of omoconazole (4μg/ml) and itraconazole (1 ng/ml). It was demonstrated that these two azoles induced changes in the ultrastructure involving the cell wall and the cell membrane which were closely similar to those for *T.mentagrophytes* hyphae treated with azole or other types of EBIA. Other prominent features of the morphology of azole-treated *C. albicans* cultures are swelling of cells and defective separation of the daughter cell from the parent cell with the resultant formation of chains of interconnected yeast-form cells. These

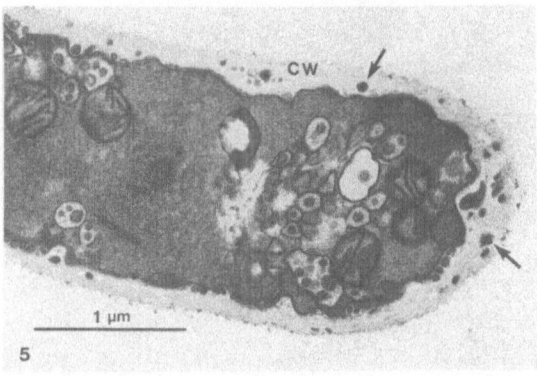

Fig. 5 Thin section of *T mentagrophytes* hyphae grown with amorolfine (8 ng/ml) for 12 h Numerous electron-dense vesicles (arrows) are seen within the cell wall (CW) of the hyphal tip

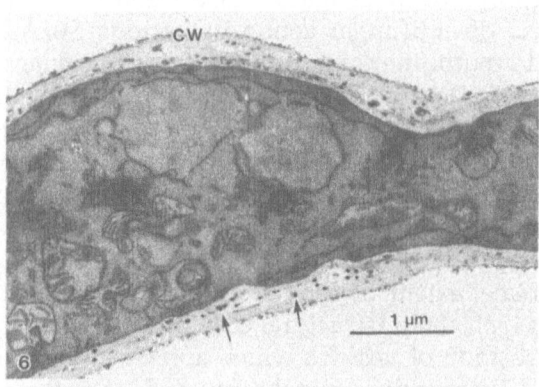

Fig. 6 Thin section of *T mentagrophytes* hyphae grown with neticonazole (0 3 μg/ml) for 24 h Numerous electron-dense vesicles (arrows) are observed within the thickened cell wall

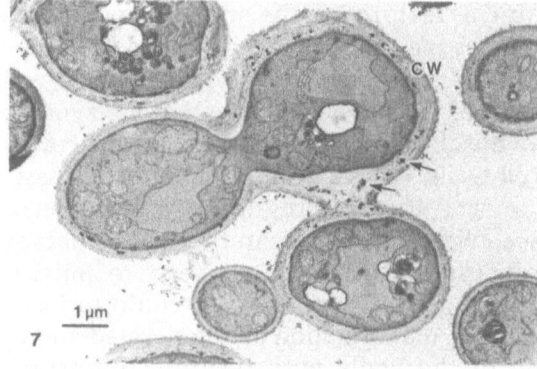

Fig. 7 Thin section of *C albicans* cells grown with omoconazole (4 μg/ml) for 24 h Note thickness of the cell wall and accumulation of electron dense vesicles (arrows) within it

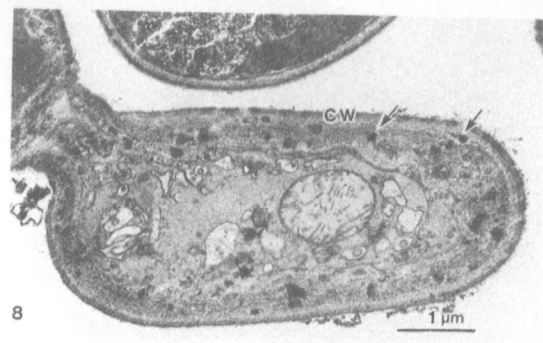

Fig. 8 Thin section of *C. albicans* grown with itraconazole (1 ng/ml) for 24 h. Numerous electron-dense vesicles (arrows) are present in the thickened wall.

findings are comparable to those obtained with bifonazole- and itraconazole-treated *Saccharomyces cerevisiae* in our previous study.[11]

Several other groups of investigators also published papers in which the morphological aspect of antifungal action of various EBIA including several azoles, naftifine and amorolfine toward *C. albicans*, *T. mentagrophytes* and some other yeast-like or filamentous fungi was individually studied by means of electron microscopic techniques.[2-4,6,12,13] In spite of the great diversity of test drugs and test fungi used in these studies, the drug-induced morphological changes documented were basically the same as found in our studies. It therefore appears possible that, irrespective of difference in the primary target of attack, all types of EBIA now available have a common morphological effect on the cell membrane and, to greater extent, on the cell wall of susceptible fungi.

As mentioned, a typical ultrastructural change induced by these antifungals is the intracellular deposition of vesicles which appear to be mainly composed of lipidic materials and surrounding membranes and, therefore, originating from degenerated cell membranes. Considering the biochemical actions of EBIA, it is possible that emergence of these membranous deposits in the drug-treated fungal cells reflects an abnormality of lipid composition and structure of the cell membrane resulting from the inhibition of ergosterol biosynthesis. This is true because ergosterol is the major sterol of the fungal cell membrane ultimately regulating membrane fluidity; therefore, its depletion and/or the accumulation of some unphysiological metabolic intermediates, such as 14α-methylsterols, might perturb the normal structure and function of the cell membrane.

A more prominent feature of the morphological changes induced by fungistatic EBIA involves the cell wall. There is a deforming swelling of hyphal cells and yeast-form cells and aberrant thickening of cell walls at specific sites, particularly septal and/or apical walls. Since those enzymes which are involved in synthesis of two major classes of polysaccharides constituting the fibrous skeleton of fungal cell walls, *i.e.*, β-glucans and chitin, are associated with the cell membrane, it is conceivable that the drug-induced depletion of ergosterol and/or accumulation of metabolic intermediates in the cell membrane can lead, through resultant alterations in its physicochemical property, to impairment of coordinated synthesis or assembly of glucan fibrils and/or chitin to form normal cell walls. This may relate causally to the formation of deteriorated and fragile cell walls and localized thickening of the walls with resulting interruption of the normal sequence of the cell separation following the completion of cell division in the case of yeast-form cells. In this connection, Barug et al.[3,4], De Nollin and Borgers[2] and other investigators[5,6] provided cytochemical and biochemical data suggesting that some azoles at fungistatic or subinhibitory concentrations concomitantly induced uncoordinated activation of chitin biosynthesis in fungi. It therefore appears likely that the excessive production of chitin at least partly explains the abnormal cell wall formation induced by azoles and probably other types of EBIA.

EFFECT OF EBIA ON REVERTING PROTOPLASTS OF *C. ALBICANS*

Protoplasts of *C. albicans* were produced by the following procedure.[7] Yeast-form cells of *C. albicans* 0146 grown in yeast extract-peptone-glucose (YPG) broth were harvested in the logarithmic phase of growth, washed with distilled water, and then resuspended in 50mM Na_2HPO_4/KH_2PO_4 buffer (pH 7.6) containing 0.6M KCl as an osmotic stabilizer. This cell suspension was incubated with Zymolyase 20-T (Kirin Breweries Co., Japan), 0.3mg/ml at 30 °C for 1 h and then with Novozyme (Novo Biolabs), 2 mg/ml for 30 min with gentle shaking. Completion of protoplast formation was confirmed by staining with a fluorescent dye, Fungifluora Y (Biomate, Japan) which sensitively stains all species of wall polysaccharides. This fluoromicroscopic technique was also used to examine the reverting protoplasts. Reversion of protoplasts to osmotically stable cells was performed at 30 °C in the medium of Lee et al.[15] supplemented with 0.6M KCl, with gentle shaking, in the absence and presence of test antifungal. Samples were removed at designated times from 0 h to 24 h and used for examination by negative-staining electron microscopy after staining with a 2% aqueous solution of uranyl acetate as well as for counts of viable cells or counts of osmotically fragile and stable cells were made by implanting samples on both 0.6M KCl-supplemented and unsupplemented YPG agar plates and incubating them at 30 °C for 72 h prior to counting the number of visible colonies. The viable counts measured on KCl-supplemented agar are the combined number of osmotically fragile and stable cells, and those measured on unsupplemented agar indicate only the number of the latter type of cells.

Reversion of Protoplasts in Antifungal-Free Medium

Fluoromicroscopic examination of reverting protoplasts revealed that fluorescent spots usually appeared at some regions of the cell surface after 20 min incubation and bright fluorescence was uniformly distributed over the entire cell surface after a further 40 min or more incubation. Consistent with this, negative-staining electron microscopy demonstrated that organized nets of glucan microfibrils 3-5 nm wide but of undetermined length, which looked similar to those observed to be formed by regenerated *C. albicans* spheroplasts or protoplasts,[16,17] developed on parts of the cell surface after 15 min of incubation (Fig. 9) and these fibrils with increasing width and length extended over the entire cell surface after 60 min or more (Fig. 10).
Correspondingly, TEM images of 20 min-incubated protoplasts showed the development of fluffy structures consisting of microfibrils 5 to 100 nm long (Fig. 11). It thus seems likely that, under the present experimental conditions, regeneration of the cell wall started and proceeded rapidly following the onset of incubation.
Figure 12 illustrates a typical result of experiments conducted to see what portion of starting protoplasts reverted to osmotically stable cells and what time

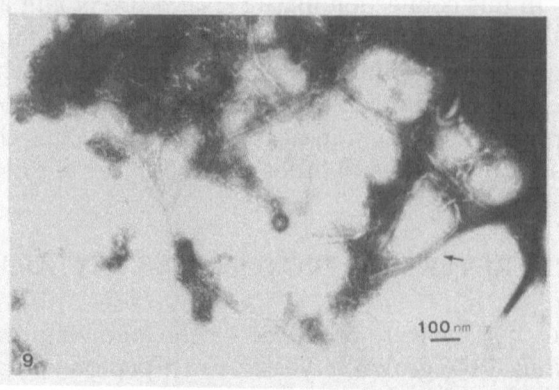

Fig. 9 Negatively stained image of reverting *C albicans* protoplast after 15 min incubation without itraconazole Note that growing protoplasts have synthesized long fibril structures (arrows) on their surface

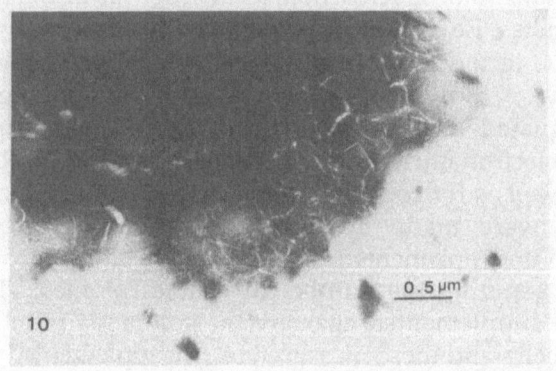

Fig. 10 Negatively stained image of reverting *C albicans* protoplast after 60 min incubation without itraconazole The entire surface of reverting protoplast is covered with many microfibril networks

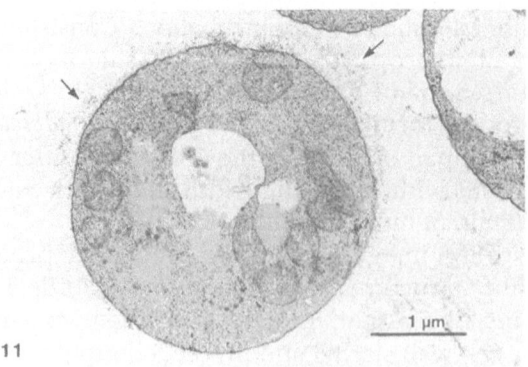

Fig. 11 Thin section of reverting *C albicans* protoplast after 20 min incubation without itraconazole Newly synthesized microfibrils have developed from their cell membrane

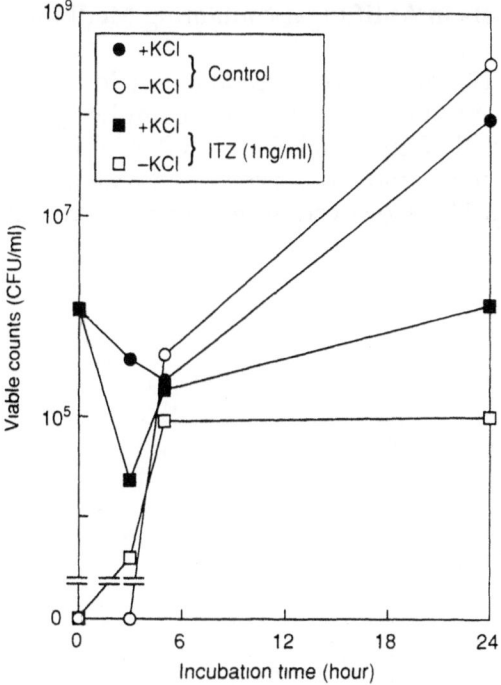

Fig. 12 Effect of itraconazole on regeneration of osmotically stable cells from *C. albicans* protoplasts in LBC broth. Nonsignificant data of CFU values (≤ 10) are plotted as zero.

was required to complete the reversion in the antifungal-free medium and the itraconazole (1 ng/ml)-containing medium. As shown, when incubated free of drug, usually 20 to 30% of inoculated protoplasts were able to revert to osmotically stable cells and the reversion was completed and the cells started to proliferate after 3 to 5 h. Comparable TEM findings demonstrated that after 5 h incubation, almost all cells with an intact interior were covered with a dense and compact fibrous layer approximately 100 to 200 nm thick and that some of them were in the budding phase (Fig. 13).

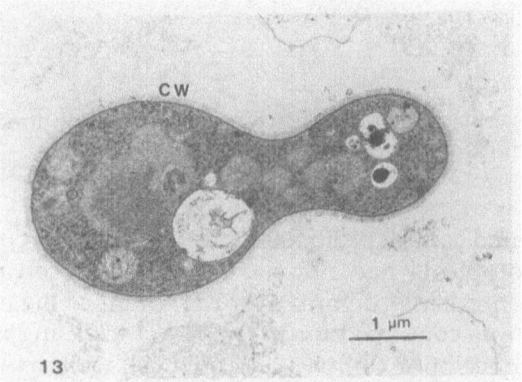

Fig. 13 Thin section of reverting protoplast after 5 h incubation without itraconazole. Intact cell wall structure is formed outside the cell membrane.

Reversion of Protoplasts in Antifungal-containing Medium

As shown in Fig. 12, itraconazole even at a very low concentration (1 ng/ml) reduced the relative number of protoplasts completing the reversion. TEM examination revealed that itraconazole-treated protoplasts developed fluffy fibrillar structures similar to those observed in untreated protoplasts in the early stage of reversion (Fig. 14). However, even after 5 or 8 h of incubation by which

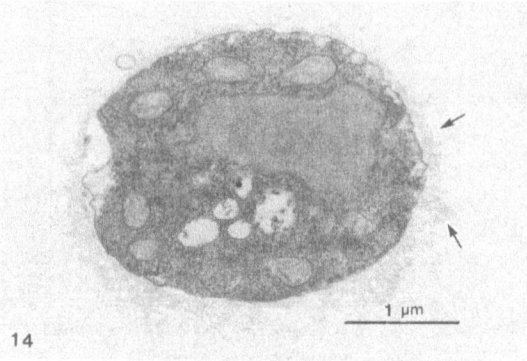

Fig. 14 Thin sections of reverting protoplast after 60 min incubation with itraconazole (1 ng/ml) Fibrillar structures have developed from the cell membrane.

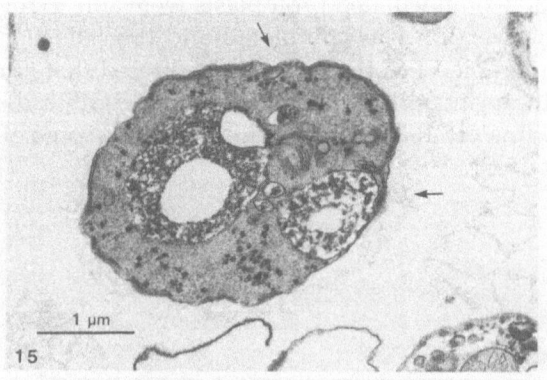

Fig. 15 Thin section of reverting protoplast after 8 h incubation with itraconazole (1 ng/ml) showing fluffy fibrillar structures on the cell membrane

time reversion of untreated control protoplasts has usually been completed, a large portion of treated protoplasts had not developed rigid cell walls but were surrounded with fluffy fibrillar structures over part or all of the cell surface (Fig. 15) When incubation was continued for up to 24 h, increasing numbers of cells were observed to have developed compact and rigid cell walls as shown in Fig 16 Interestingly, in those cells which apparently completed reversion, localized thickening of apical walls and deposition of electron-dense vesicles within the thickened walls were seen (Fig 16), as was the case for itraconazole-treated whole

370

cells of *C. albicans*. Virtually the same results were obtained when reverting protoplasts were treated with terbinafine and amorolfine.

All of these data suggest that all types of EBIA share a common adverse effect on cell wall formation: apart from aberrant thickening of apical walls, this involves impairment of formation of organized net structures principally consisting of glucan fibrils.

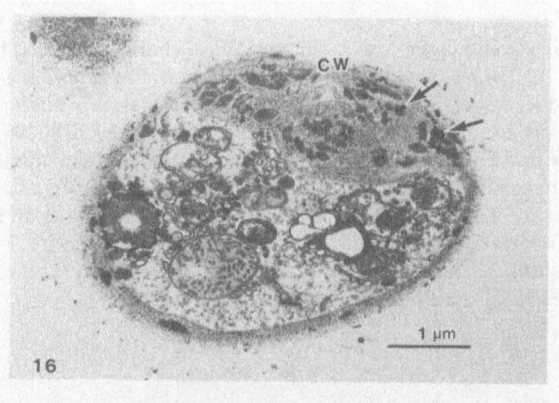

Fig. 16 Thin section of reverting protoplast after 24 h incubation with itraconazole (1 ng/ml). Note thickness of the regenerating cell wall and accumulation of electron-dense vesicles (arrows) within it.

REFERENCES

1. H. Vanden Bossche, Anti-*Candida* drugs - mechanisms of action, in:"*Candida* and Candidamycosis", E. Tümbay, H.P.R. Seeliger and Ö. Ang, eds., Plenum Press, New York (1991).
2. S. De Nollin and M. Borgers, Scanning electron microscopy of *Candida albicans* after *in-vitro* treatment with miconazole, *Antimicrob. Agents Chemother.* 7: 704 (1975).
3. D. Barug, R.A. Samson and A. Kerkenaar, Microscopic studies of *Candida albicans* and *Torulopsis glabrata* after *in-vitro* treatment with bifonazole, Light and scanning electron microscopy, *Arzneim. Forsch.*. 33: 528 (1983).
4. D. Barug, C. de Groot and R.A. Samson, Morphology and ultrastructure of *Candida albicans* after *in vitro* treatment with bifonazole, in:"Recent Trends in the Discovery, Development and Evaluation of Antifungal Agents", R.A. Fromtling, ed., J.R. Prous Science Publishers, S.A., Barcelona (1987).
5. R.F. Hector and P.C. Braun, The effects of bifonazole on chitin synthesis in *Candida albicans*, in:"Recent Trend in the Discovery, Development and Evaluation of Antifungal Agents", R.A. Fromtling, ed., J.R. Prous Science Publishers, S.A., Barcelona (1987).
6. H. Vanden Bossche, Biochemical targets for antifungal azole derivatives: hypothesis on the mode of action, in:"Current Topics in Medical Mycology", ed., M.K. McGinnis, Vol. 1, Springer Verslag, New York (1985).
7. Y. Nishiyama, Y. Aoki and H. Yamaguchi, Cytochemical and electron microscopic studies on the reversion of *Candida albicans* protoplasts to osmotically stable cells in a liquid medium, *Microbiol. Immunol.* (to be published).
8. Y. Nishiyama, Y. Asagi, T. Hiratani, H. Yamaguchi, N. Yamada and M. Osumi, Ultrastructural changes induced by terbinafine, a new antifungal agent, in *Trichophyton mentagrophytes*, *Jpn. J. Med. Mycol.* 32: 165 (1991)
9. Y. Nishiyama, Y. Asagi, T. Hiratani, H. Yamaguchi, N. Yamada and M. Osumi, Morphological changes with growth inhibition of *Trichophyton mentagrophytes* by amorolfine, *Clin. Expl. Dermatol.* (to be published).

10. Y. Nishyama, K. Maebashi, Y. Asagi, T. Hiratani, H. Yamaguchi, N. Yamada, A. Taki, J.I.X. Rong and M. Osumi, Effect of SS717 on the ultrastructure of *Trichophyton mentagrophytes* as observed by scanning and transmission electron microscopy, *Jpn. J. Med. Mycol.* 32: 43 (1991).

11. H. Yamaguchi and M. Osumi, Morphological aspect of azole action, in:"Sterol Biosynthesis Inhibitors. Pharmaceutical and Agrochemical Aspects", D. Berg and M. Plempel, eds., Ellis Horwood, Chichester (1988).

12. P. Marichal, J. Gorrens and H. Vanden Bossche, The action of itraconazole and ketoconazole on growth and sterol synthesis in *Aspergillus fumigatus* and *Aspergillus niger, J. Med. Vet. Mycol. 23: 13 (1985).*

13. J.G. Meingassner, U. Sleytr and G. Petranyi, Morphological changes induced by naftifine, a new antifungal agent, in *Trichophyton mentagrophytes, J. Invest. Dermatol.* 77: 444 (1981)

14. J. Müller, A. Polak and R. Jaeger, The effect of the morpholine derivative amorolfine (Roche 14-4767/002) on the ultrastructure of *Candida albicans, Mycosen.* 30: 528 (1987).

15. K.L. Lee, M.R. Buckley and C. Campbell, An amino acid liquid synthetic medium for development of mycelial and yeast forms of *Candida albicans, Sabouraudia* 13: 148 (1975).

16. P. Gopal, P.A. Sullivan and M.G. Shepherd, Isolation and structure of glucan from regenerating spheroplasts of *Candida albicans, J. Gen. Microbiol.* 130: 1217 (1984).

17. H. Yamaguchi, T. Hiratani, M. Baba and M. Osumi, Effect of aculeacin A on reverting protoplasts of *Candida albicans, Microbiol. Immunol.* 31: 625 (1987).

DIMORPHISM AND CANDIDOSIS: CLINICAL AND THERAPEUTIC IMPLICATIONS

Roderick J. Hay

St. John's Institute of Dermatology
Guys Hospital
London SE1 9RT
UK

ABSTRACT

Because dimorphism in *Candida,* as in other fungi, is an active process, to assign pathogenetic status to fungi in one phase but not in the other on the basis of morphology alone is not a valid approach to judging invasive capacity. Rather it is a screening exercise which, combined with other clinical and laboratory data, can be used to make a clinical decision - provided the limitations of this approach are fully recognized.

INTRODUCTION

Infections caused by *Candida albicans* are amongst the commonest of superficial mycoses as well the most frequently recognized of the systemic fungal infections in immunocompromised and postsurgical patients. Infections caused by related species such as *C. tropicalis, C. parapsilosis* and *C. glabrata* may also cause disease in similar groups of patients. The superficial infections caused by *Candida* species include mucous membrane infections involving the mouth or vagina (thrush) as well as those affecting the keratinized tissues of the body folds (intertrigo), nail folds (paronychia) and nail plates (onychomycosis). Deep *Candida* infections range from focal infections of sites such as the peritoneum or CSF and the urinary dissemination of yeasts (Table 1).

The range of clinical appearances seen in systemic *Candida* infections reinforces the view that systemic candidosis comprises a group of syndromes whose clinical expression depends on a number of factors such as the route of entry of the fungus and the underlying immune state of the host. For instance bloodstream dissemination in different groups of patients may be followed by the appearance of retinal deposits or hair, retinal and joint abscesses or hepatic and splenic granulomas. These represent, respectively, disseminated infections in patients receiving parenteral nutrition, in intravenous drug abusers and in leukemia patients receiving chemotherapy, generally after recovery of neutrophil counts. In addition the pattern of infection due to non-*albicans Candida* species may also vary in a similar manner. For instance 40-50% of cases of yeast endocarditis are caused by an organism other than *C. albicans*, most commonly *C. parapsilosis*, a yeast species which is an unusual cause of fungal infection in other circumstances.

Table 1. Clinical classification of *Candida* infections.

Superficial infections:
          Oropharyngeal candidosis
          Vaginal candidosis
          Paronychia, onychomycosis
          Intertrigo or interdigital candidosis
          Keratomycosis
          Otitis externa

Deep infections:
          Deep focal   *Candida* peritonitis
                          Candida oesophagitis
                          Candida pyelitis

          Candidemia
          Systemic   *Candida* endocarditis
                          Candida septicemia
                          Candida IV drug abuser syndrome
                          Hepatosplenic candidosis (chronic
                                    disseminated candidosis)
                          Systemic candidosis associated with
                                    hyperalimentation
                          Neonatal systemic candidosis

          Uncertain status – diarrhoea, asthma

Clinical dogma has it that disease due to *Candida* is always accompanied by yeast to mycelial conversion and tissue penetration by hyphae. Commensal carriage on the other hand is held to be associated with *Candida* in the yeast phase. In fact it is almost impossible to draw such a clear distinction between the behavior of the two phases; equally there are also a number of clinical examples where pathological processes may not be accompanied by tissue invasion. This is a source of confusion to the clinician who is frequently having to discriminate between genuine infection and saprophytic carriage. The status of these processes needs to be established more clearly. Two examples where considerable doubt exists as to the role of *Candida* species in the pathogenesis of pathological processes - nail disease and diarrhoea – will be considered.

THE ROLE OF *CANDIDA* IN NAIL DISEASE

An example of this dilemma is seen in considering the pathogenetic role of *Candida* species in nail disease. The frequency of the isolation of *Candida* species from nail material is very variable.[1] Whereas in the experience of many it is a rare event, in one study up to 66% of nail material was found to contain yeasts on culture.[2] The interpretation of these results is difficult in that in some cases the nature of the material may explain the differences; if, for example, nail clippings from all sources including nails with pre-existing nail abnormalities such as onycholysis are examined the frequency of isolation is higher. A second factor is climate. In one study from Saudi Arabia for instance there was a high isolation rate of *Candida* from nails[3] where over 83% of 243 cases of onychomycosis were caused by *C. albicans*, as well as foot infections. Biopsy evidence from the material suggests deep penetration of yeasts and mycelium into the nail plate.

A number of distinct clinical patterns of nail disease are associated with this organism.[4] In the first, paronychia, there is inflammatory swelling of the proximal nail fold with discharge of serous ooze or pus and discomfort on pressure. Organisms isolated from these lesions are generally either yeasts or *Staphylococcus aureus*. *Candida* paronychia may be acute or chronic although the latter pattern is much commoner. While in the early stages there is no involvement of the nail plate itself, with time there is progressive lateral onycholysis and buckling of the nail plate with the development of transverse grooves (proximal subungual onychomycosis or PSO). Hyperkeratosis is seldom seen. Cultures of a number of different species including *C. albicans* can be obtained from the nail fold and under the plate.[5]

In the second pattern of nail disease associated with *Candida* there is onycholysis without thickening of the nail plate usually originating from the distal margin. Patients frequently have another underlying disease, particularly psoriasis, but in some the nail changes are the sole manifestation of disease.[4] Once again a variety of different fungal species can be isolated from such nails with *C. albicans* being isolated in about half the cases. Other fungi present include *C. glabrata*, *C. guilliermondii* and *C. parapsilosis*. Thirdly there is a form in which *Candida* causes a severe and hyperkeratotic nail dystrophy with gross distortion of the nail plate in patients with chronic mucocutaneous candidosis (CMC).[6,7] These patients have severe nail disease in addition to oral and cutaneous candidosis and the nail involvement is very extensive, affecting the nail plate, the bed and the proximal nail fold as well as skin around the nail apparatus (total dystrophic onychomycosis or TDO). The fourth pattern of nail disease caused by *Candida* is distal and lateral subungual onychomycosis (DLSO) usually seen in women, particularly those who have developed peripheral vascular disease and those with severe Raynaud's phenomenon.[4,8] The main changes are as follows: distal erosion of the nail plate with irregular distal and lateral onycholysis, mild hyperkeratosis. A fifth and rare pattern is a superficial nail plate invasion, superficial white onychomycosis (SWO), which has on rare occasions been reported in infants.[9] Here there is a white powdery infection of the superficial nail plate which is covered by discrete or circular patches. These changes are summarized in Table 2.

The role of *Candida* in producing or contributing to these changes is not always entirely clear. In paronychia *Candida* spp may be difficult to isolate unless swabs or scrapings are taken from under the proximal nail fold. In one recent study from Malaysia 62% of patients with chronic paronychia had yeasts seen on direct microscopy.[10] This is not always the case and in a similar U.S. study only 4 of 28 patients with paronychia had yeasts (all *C. albicans*) recovered from the nail fold. In contrast, a mixture of aerobic and anaerobic bacteria was found in these patients, notably *Staph. aureus*, *Eikenella corrodens*, *Klebsiella pneumoniae*, *Bacteroides* and *Fusobacterium* species.[11] Nonetheless a high proportion of patients may respond to antifungal therapy. The duration of treatment may be long because, despite slow clinical improvement, *Candida* is continuously isolated from the site of infection or from the nail.[5] A study of the histology of onychomycosis associated with paronychia showed both hyphae and yeasts in the stratum corneum of the undersurface of the nail fold, in the epinychium, nail pocket and the epithelium of the proximal nailbed.[12] Most fungal elements were seen in the lunula area. Onycholysis spread to the distal nail margin in some cases and the author comments on the failure of 12 months oral therapy with ketoconazole to eradicate hyphae from the area. Interestingly there was almost no neutrophil infiltrate in the stratum corneum, unlike the pattern seen in dermatophytosis. In another study of *Candida* onychomycosis both mycelium and yeasts were found in four of five specimens examined both on the ventral nail plate surface and also extending a short distance into the nail, usually without deeper penetration.[4]

Table 2. Patterns of nail disease associated with *Candida*.

Type[a]	Primary target	Prediposition	Organism
Paronychia	Nail fold	Occupation	*Candida* spp.
(PSO)	Lateral nail	Women	
Onycholysis	Nail plate?	Any nail disease	*Candida* spp.
TDO	Nail fold	CMCC	*C. albicans*
DLSO	Nail plate	Vascular disease in women	*C. albicans*
SWO	Superficial nail	Infants	*C. albicans*

[a] PSO: proximal subungual onychomycosis; TDO: total dystrophic onychomycosis; DLSO: distal and lateral subungual onychomycosis; SWO: superficial white onychomycosis.

The role of the yeast in causing inflammation of the nail fold has been questioned. Is this due to invasion or, for instance, an immunologically mediated reaction? In a study of the immune responses of 12 patients with chronic *Candida* paronychia there was no difference in ELISA of specific antibody to *Candida* cytoplasmic antigen for IgG, IgM, IgE and IgA classes between patients and age-matched controls (14) compared to chronic vaginal condidosis patients (11) who had a significantly increased frequency of raised IgE and IgM antibody titres (Y. Jamil and R.J. Hay unpublished data). This contrasts with evidence that patients with chronic paronychia have a significantly increased chance of showing immediate type hypersensitivity on prick testing to food antigens.[13] The data described above, while not excluding a role for immune responses to non-fungal antigens in paronychia, suggest that at least a proportion of patients develop biopsy proven invasion of the proximal nailfold and the underface of the nail plate itself by *Candida*.

In patients with chronic mucocutaneous candidosis there appears to be no doubt that *Candida* species can invade the nail plate keratin. Nail material is grossly infiltrated by hyphae and yeasts. There is less certainty about the group of patients with onycholysis and distal and lateral onychomycosis from whom *C. albicans* is isolated.[8] One study showed that a number of risk factors were significantly associated with nail plate invasion in these circumstances. These included the presence of underlying disease (mainly peripheral vascular disease), Raynaud's phenomenon and Cushing's syndrome; both yeasts and mycelium were seen in nails on direct microscopy.[4] Those with these features were significantly more likely to have either biopsy evidence of mycelial penetration deep into the nail plate keratin[12] and/or response to oral antifungal therapy with either ketoconazole or itraconazole. The majority of these patients were female.

In patients either with preexisting nail disease (e.g. due to psoriasis) and those with onycholysis without erosive changes or underlying peripheral vascular disease, the presence of both yeasts and mycelium is uncommon. In addition at least 25% of the isolates are non-*albicans Candida* species. Biopsy changes are generally different and, for instance, large clusters of yeast without significant accumulation of mycelium can be found on the ventral surface of the nail without a significant degree of penetration into the nail plate itself.[12]

The most likely inference from these studies is that *Candida* can act as either a commensal or a true invader of the nail plate. Factors which favour invasion include immunological defects, such as those seen in CMC, or peripheral vascular disease. By contrast, if there is merely primary or secondary onycholysis, yeasts are generally carried without nail plate invasion. However, these rules are not absolute and it is clear that a minority of patients who present with onycholysis and have positive cultures for *Candida* may derive some benefit from antifungal therapy. Here the presence of both yeast and mycelium in direct microscopy of nail material is a very useful predictor of invasion – but not infallible. The ability of *Candida* species to metabolize keratin has been contested. *C. albicans*, as well as other *Candida* species, is said to be non-keratinolytic. However there are several pieces of evidence that confirm that *C. albicans*, if not other species, can attack nail keratin. Firstly there is biopsy evidence of nail penetration.[12] Secondly, in the presence of glucose *C. albicans* has been shown to pentrate nail plate *in vitro*[14] although experiments attempting to inoculate the fungus under nails of volunteers have failed to produce nail penetration.[15] Thirdly, *C. albicans* can liberate amino acids from keratin in the presence of glucose.[16] Fourthly, other keratinized sites such as hairs can be attacked by these organisms in systemic candidosis seen in heroin addicts.[17]

The main guidelines for the treatment of nails presumed to be infected with *Candida* is fairly straight forward. For paronychia there appears to be no benefit in most cases from use of an oral drug such as ketoconazole or itraconazole rather than an azole lotion (clotrimazole, econazole) although the former work perfectly well.[5] Where there is nail plate invasion with terminal erosion itraconazole is probably the best choice;[4] ketoconazole would be an alternative.[9] There is no published experience of fluconazole in this indication. Finally where there is onycholysis with positive cultures for *Candida* the primary cause of the onycholysis should be treated where possible, combined, if steroids are used, with a topical azole lotion applied under the nail. If this fails to produce a response and there is persistent isolation of *Candida* a nail biopsy should be taken and if this shows mycelium penetrating into the nail plate a trial of itraconazole therapy should be used.

CANDIDA AND DIARRHOEA

In deep candidosis hypha formation is once again regarded as a "sine qua non" of infection. However, there are also controversial disease states where there is evidence that *Candida* may contribute to infection by means other than invasion of tissue. For instance there is a putative association between *Candida* and persistent diarrhoea. Carriage rates for *Candida* species in stool samples in normal patients range from 8-20% and of yeasts from 8-60%.[18] High carriage rates have been reported from patients with either diarrhoea (83%)[19] or those on antibiotics (45%).[20] In recent years it has been suggested that certain patients with persistent stool cultures positive for *Candida* will respond to antifungal chemotherapy. One recent study, [21] for instance, reported 10 elderly patients with persistent diarrhoea, most of whom were receiving oral antibiotics. The stool cultures showed a heavy growth of a single or mixed *Candida* species – *C. albicans* (7), *C. tropicalis*(2), *C. glabrata* (2) and *C. parapsilosis* (1). The diarrhoea was profuse and watery and contained few leukocytes. All responded within 2-4 days to oral nystatin. A second report examined 24 elderly patients with diarrhoea, negative for *C. difficile* toxin. [22] They found that faecal colony counts for *Candida* were very high (up to 91 x 10^6 cfu/ml). Treatment with nystatin led to resolution with significant falls in *Candida* counts. One patient relapsed and there was an increase in stool *Candida* levels. The pathogenesis of this process is not clear. No biopsy data have been presented and paucity of faecal leucocytes is evidence against significant inflammation resulting from tissue invasion. One possible solution is inhibition of the reabsorption of water from the GI lumen. Whatever the explanation this phenomenon needs further investigation. Previous studies of *Candida* in the gastrointestinal tract have suggested that there are three other principle interactions – colonization, invasion and persorption. Colonization has been mentioned previously.[18] Invasion has generally been reported only in neutropenic patients where histopathologically proven invasion has been recorded in the jejunum, stomach and other areas.[23] The second route, persorption, is one which gives rapid access of fungal cells into the bloodstream and has been demonstrated in humans and mice.[24] In studies of this phenomenon the rapid transit of organisms from the stomach and jejunum has been demonstrated within 5 min of ingestion in mice.[25] In the GI tract this is associated with the presence of yeasts but not hyphae in the submucosa. The therapeutic consequences of this observation are potentially important. It is possible to block the adherence of *Candida* to gastrointestinal mucosal cells at relatively low concentrations of antifungal agents.[25] However, if uptake does not involve adherence or even a different adherence mechanism, use of drugs which do not affect yeast viability may be associated with a higher failure rate of antifungal chemoprophylaxis. The role of *Candida* in the pathogenesis of antibiotic-associated diarrhoea is amenable to further study and it should be possible once again to assess the significance of its presence.

The formation of mycelium is also variable in other forms of *Candida* infection, notably when the organism is slowly replicating on a tissue or foreign surface. This is typically seen in endocarditis where large masses of yeasts with minimal hypha formation can be seen adhering to heart valves. Pathology in this process is caused by slow destruction of the valve and by massive embolism from the clusters of yeasts. Species other than *C. albicans* which do not readily form hyphae *in vivo* are prominent amongst causes of yeast endocarditis. Once again, the response to antifungal drugs alone in this situation is slow.

REFERENCES

1. M. English, Nails and fungi, *Br. J. Dermatol.* 94: 697 (1976).

2. G. Achten and J. Wanet-Rouard, Onychomycosis in the laboratory, *Mykosen* 23: 125 (1978).

3. S.M. Al-Sogair, M.K. Moawad and Y.M. Al-Humaidan, Fungal infection as a cause of skin disease in the Eastern Province of Saudi Arabia: prevailing fungi and pattern of infection, *Mycoses* 34: 333 (1991).

4. R.J. Hay, R. Baran, M.K. Moore and J.D. Wilkinson, *Candida* onychomycosis – an evaluation of the role of *Candida* species in nail disease, *Br. J. Dermatol.* 118: 47 (1988).

5. E.S.M. Wong, R.J.Hay, Y.M. Clayton and W.C. Noble, Comparison of the therapeutic effect of ketoconazole tablets and econazole lotion in the treatment of chronic paronychia, *Clin. Exp. Dermatol.* 9: 489 (1984).

6. N. Zaias, Onychomycosis, *Arch. Dermatol.* 105: 263 (1974).

7. R.J. Hay and Y.M. Clayton, The treatment of patients with chronic mucocutaneous candidosis and *Candida* onychomycosis with ketoconazole, *Clin. Exp. Dermatol.* 7: 155 (1982).

8. S. Watanabe, Y. Seki, M. Shimozuma and K. Takizawa, Nail candidiasis, *J. Dermatol.* 10: 189 (1983).

9. N. Zaias, "The nail in health and disease",Spectrum Publications Inc., New York (1980).

10. E. Chow and C.L. Goh, Epidemiology of chronic paronychia in a skin hospital in Singapore, *Int. J. Dermatol.* 30: 795 (1991).

11. I. Brook, Aerobic and anaerobic microbiology of paronychia, *Ann. Emerg. Surg.* 19: 994 (1990).

12. E. Haneke, Nail biopsies in onychomycosis, *Mykosen* 28: 473 (1985).

13. N. Zaias, "The nail in health and disease", Spectrum Publications Inc., New York (1990).

14. L. Kapica and F. Blank, Growth of *Candida albicans* on keratin as sole source of nitrogen, *Dermatologica* 115: 81 (1957).

15. X. Vilanova, M. Cassanovas and J. Francino, Onychomycosis, and experimental study, *J. Invest. Dermatol.* 27: 77 (1956).

16. H. Remold, B. Fasold and F. Staib, Purification and characterization of a proteolytic enzyme from *Candida albicans, Biochim. Biophys. Acta* 167: 399 (1968).

17. B. Dupont and E. Drouhet, Cutaneous, ocular and osteoarticular candidiasis in heroin addicts. New clinical and therapeutic aspects in 38 patients, *J. Infect. Dis.* 152: 577 (1985).

18. F.C. Odds, "*Candida* and candidosis", Ballière Tindall, London (1988).

19. P. Soeparto, F.J. Irwantono, S. Isa, M.S. Makmuri, N. Rachman and M. Masduki, Enteric candidiasis in infantile gastroenteritis, *Paed. Indones.* 19: 129 (1979).

20. A.R. Chitale and Y.M. Bhende, The incidence of *Candida albicans* in throat and feces of healthy persons and patients on antibiotic therapy, *J. Postgrad. Med.* 11: 30 (1965).

21. P.L. Danna, C. Urban, E. Bellinb and J.J. Raha, Role of *Candida* in pathogenesis of antibiotic associated diarrhoea in elderly patients, *Lancet* 337: 551 (1991).

22. J.P. Gupta and M.N. Ehrinpreis, *Candida* associated diarrhoea in hospitalized patients, *Gastroenterology* 98: 780 (1990).

23. A.W. Maksymiuk, S. Thongprasert, R. Hopfer, M. Luna, V. Fainstein and G.P. Bodey, Systemic candidiasis in cancer patients, *Am. J. Med.* 77: 20 (1985).

24. L.H. Field, L.M. Pope, G.T. Cole, M.N. Guentzel and L.J. Berry, Persistence and spread of *Candida albicans* after intragastric inoculation of infant mice, *Infect. Immun.* 31: 783 (1981).

25. J. Mehentee and R.J. Hay, Effect of antifungal drugs on the *in vitro* adherence of *Candida albicans* to gastrointestinal mucosal cells, *J. Gen. Microbiol.* 25: 111 (1990).

TREATMENT OF SPOROTRICHOSIS, BLASTOMYCOSIS, AND HISTOPLASMOSIS

John R. Graybill, and Patricia Kay Sharkey

Infectious Diseases Section
Audie Murphy Memorial Veterans Hospital
7400 Merton Minter Blvd
San Antonio, Texas 78284, USA

INTRODUCTION

Sporotrichosis, blastomycosis, and histoplasmosis are all fungal pathogens of man, and all are controlled in large part by cell mediated immunity; beyond this they share relatively few characteristics. Sporotrichosis usually follows percutaneous inoculation, while the others infect by inhalation of conidia. Sporotrichosis is a widely scattered illness associated with outdoor plant contact, such as nursery work, and is found both in temperate climates and the tropics. Blastomycosis and histoplasmosis both have restricted ranges in which they are most commonly seen. Most cases of blastomycosis occur in North America. Histoplasmosis is seen in North and South America. In their widely disseminated forms, both sporotrichosis and blastomycosis are characterized by focal cutaneous lesions, and less commonly bone and joint disease. In its widely disseminated form, histoplasmosis causes generalized reticuloendothelial organ enlargement, invades the bone marrow, and occasionally causes focal mucous membrane lesions. Adrenal gland invasion is a common complication only of histoplasmosis. Meningitis is occasionally seen with histoplasmosis, and is rare in the others. Finally, histoplasmosis is the only one of these three mycoses which is highly associated with HIV infection.

For most of their known history, these three pathogens have caused disease sporadically or in limited outbreaks, with insufficient numbers of patients to generate the interest of major pharmaceutical houses. For years, the treatment of lymphocutaneous sporotrichosis has been potassium iodide. For disseminated sporotrichosis, blastomycosis, and histoplasmosis, amphotericin B has been used. In the late 1970's the Mycoses Study Group (MSG) was formed to define optimal treatment for another uncommon systemic mycosis, cryptococcosis. At about the same time, Janssen Pharmaceutica entered the field of systemic azole antifungal therapy. Miconazole first appeared, but was limited by toxicity and the need for intravenous administration. This was followed by ketoconazole, the first orally administered azole, a drug which is still widely used. Because of poor central nervous system penetration, and the finding that very high doses were required to treat coccidioidal meningitis, the MSG did not pursue ketoconazole for cryptococcal meningitis.

However, in seeking to broaden its mandate from cryptococcosis to other mycoses, the MSG conducted an open pilot study of ketoconazole in sporotrichosis, histoplasmosis, blastomycosis, and coccidioidomycosis.[1] The results of these studies were in general encouraging, and the era of azole antifungal therapy was launched. The initial studies were conducted with the majority of patients having blastomycosis, and the fewest with sporotrichosis.

In addition to the availability of a major antifungal "new player" on the field of mycoses, the field itself has undergone considerable redefinition. Immunosuppressive agents and AIDS have brought systemic mycoses from relative obscurity to prominence. These factors have not impacted much on sporotrichosis and blastomycosis, but histoplasmosis has undergone a dramatic shift from "endemic mycosis" to "opportunist". The first evidence of this was noted by the appearance of disseminated histoplasmosis in residents of east and west coast cities of the USA.[2,3] The patients were residents of Carribean Islands, normally endemic for *H. capsulatum*. Presumably they had become infected in their homelands; when they later developed AIDS, their *H. capsulatum* escaped its confines and produced disseminated disease.

This importation of histoplasmosis anticipated a second event, which was the explosion of indigenous cases that occured as AIDS moved to the endemic regions of the United States. Here, primary disseminated histoplasmosis replaced reactivation disease, and became very common in AIDS patients.[4-8] As many as 20% of AIDS patients in Indianapolis or Kansas City developed histoplasmosis during the course of their AIDS...making this one of the most common HIV associated opportunistic infections.[4-6] The disease is aggressive, may present with manifestations indistinguishable from the "sepsis syndrome", and may run a very acute course of days to weeks.

It was soon apparent that histoplasmosis in AIDS patients did not respond well to ketoconazole.[6-8] This is in part because of the relatively poor absorption of the drug in these rather achlorhydric patients, but low potency may also be a factor. Pressure for new therapeutic approaches mounted rapidly, and ongoing itraconazole and fluconazole studies in histoplasmosis in non-HIV infected patients were expanded to include these with AIDS.

At present, the majority of experience with histoplasmosis is occuring in patients with AIDS, and cases far outnumber those with blastomycosis and sporotrichosis together. The remainder of this discussion will focus on current approaches to each of these diseases.

SPOROTRICHOSIS

For many years iodides have been the basis of antifungal therapy. Iodide mechanisms of action are not totally clear. Iodides are administered orally as a saturated solution, taking 5 drops 3 times daily, and increasing the dose until 40 drops 3 times daily or maximal tolerance is reached. This is usually manifested by increased salivation, salivary gland swelling, or gastrointestinal intolerance, but allergic rashes also occur.[9] Iodides are useful only for lymphocutaneous disease, and not for extracutaneous disease, although there are a few reports of success in osteoarticular disease (Table 1).

Table 1. Response of osteoarticular sporotrichosis to antifungal drugs. (Selected studies)[a].

Refs. No.		Drug	Response/Relapse		Summary
10	13	KI	7/0		Total KI:
11	2	KI	0/0		N = 19;
14	4	KI	4/0		RESPONSE = 58%;
					RELAPSE = 0%
10	5	AMB	4/0		Total AMB:
13	3	AMB	3/0		N = 16;
11	5	AMB	5/1		Response = 94%;
16	1	AMB	1/1		Relapse = 12%
15	2	AMB	2/0		
12	8	KETO	6/1		Total KETO:
1	4	KETO	2/1		N = 12;
					Response = 67%;
					Relapse = 16%
17*	15	ITRA	11/4		Total ITRA:
18	1	ITRA	1/0		N = 16;
					Response = 75%
					Relapse = 25%

[a] Antifungal therapy was combined with synovectomy in some patients. Patients who had a successful response and later relapsed are included with successful initial responses, and then independently determined as percent of the total patients. AMB=Amphotericin B; KI=Potassium iodide (results of treatment biased by small numbers and one unusually optimistic report); KETO=ketoconazole; ITRA=itraconazole. Relapses are also counted as responders and relapses are % of total number.
*Includes updated patients not in original abstract.

Amphotericin B has not been considered necessary for most patients with lymphocutaneous disease, and has been reserved for those with extracutaneous disease, where it is effective in about 2/3 of patients.[9] The experience is irregularly reported from small series, some of which are summarized in Table 1. Unfortunately, not all reported patients are evaluable for response. Gladstone and Littman[10] report 4 of 6 responding to amphotericin B, while Crout et al.[11] had all of 5 with osteoarticular disease who responded to amphotericin B, with 1 later relapse. Pulmonary disease also responds irregularly to amphotericin B. Bennett suggests that only about one third of such patients respond.[9] One problem with assessing responses is the small number of patients. Long courses of amphotericin B, known toxicities, and the tendency to relapse after therapy, particularly in pulmonary disease, have prompted a search for less noxious and more effective agents.

Ketoconazole was the first such agent tried. It is clearly less noxious than amphotericin B. There are a few reports and there are differences of opinion regarding efficacy. In the MSG open-label series there were 4 failures among 7 patients treated for up to 6 months.[1]

A subsequent study by Calhoun et al. was more optimistic, reporting 8 of 11 patients treated as successes, including 6 of 8 with osteoarticular disease.[12]

However, in sum this is still only 67% of 12 patients, and 16% relapse post treatment (Table 1). With doses given up to 800 mg per day, and marked increase of gastrointestinal intolerance at that dose, ketoconazole was still not ideal.

Fluconazole has been recently evaluated in treatment of sporotrichosis. Diaz et al.[19] found that 13 of 19 patients treated with 200 to 400 mg per day responded. Five responders received 400 mg per day from outset, while three responded to 200 mg per day and were raised to 400 mg per day for response. Two failures received only 200 mg per day, but others were increased from 200 to 400 mg per day, without response. There were no relapses in 10 responders followed for 8 months or more post therapy. These patients all had lymphocutaneous disease, and there was no experience with the more serious osteoarticular or pulmonary forms of sporotrichosis. Fluconazole is clearly effective, and a modest advance over ketoconazole. Because fluconazole is well tolerated, it may be that higher doses of 600 or 800 mg per day would be useful for failures at lower doses.

The largest experience with triazoles for treatment of lymphocutaneous sporotrichosis has been accumulated with itraconazole and its difluorinated analogue, saperconazole.[17,18,20-23] The first published experiences with itraconazole were a few scattered cases in a 1987 symposium.[18,21,22] All had lymphocutaneous disease and responded well to itraconazole. A larger more systematic experience was reported by Restrepo et al.,[20] a series of 17 patients treated with 100 mg per day for up to 6 months. Restrepo et al. convincingly showed a 100% response rate, with lesions generally clearing by 3 months, cultures negative where done, and no relapses post treatment (but with a limited post treatment observation). All patients had lymphocutaneous disease. This experience was expanded upon by Sharkey et al.[17] with a Mycoses Study Group report of itraconazole therapy for 26 patients given 29 courses of drug. All of 10 patients with lymphocutaneous disease responded, with only 2 relapses. Thus the earlier observations of Restrepo et al. [20] were confirmed.

However, in addition this study demonstrated a role for itraconazole in osteoarticular disease (Table 1). Of 15 patients treated, with doses up to 400 mg per day, 11 had clinical resolution of disease. Four patients failed. One who had initially responded and relapsed responded to a second course of treatment. Therefore, in osteoarticular disease, a former requirement for amphotericin B has been replaced by itraconazole as a well tolerated and highly effective agent. There were only four patients with pulmonary disease in this study. One failed outright, and another patient initially responded and later relapsed. A third patient remains on itraconazole and a fourth is free of disease after completing therapy.

Finally, according to a report by Franco et al.,[23] saperconazole appears even more rapidly effective in lymphocutaneous disease than itraconazole. All of 13 patients with lymphocutaneous disease responded to 100 mg per day for an average of 3.5 months of treatment. However, numbers of patients are small and post treatment follow up is still short.

Present treatment recommendations for lymphocutaneous sporotrichosis include potassium iodide as the drug of choice, based largely on economic considerations and long proven efficacy. Itraconazole and saperconazole appear highly effective, and well tolerated but much more costly. For pulmonary disease our information with azoles is too sketchy to permit therapeutic recommendations relative to amphotericin B, though no agent appears excellent. For osteoarticular and other extracutaneous forms of disease itraconazole appears to be the drug of choice, with response rates comparable to amphotericin B and much better tolerance. Fluconazole may be a reasonable alternative if an optimal dose is more effective than those used thus far.

BLASTOMYCOSIS

Untreated, blastomycosis carries a mortality as high as 78%, but the disease responds very well to amphotericin B, with up to 97% cures in responses to doses of 2 grams or more.[24,25] Relapse is rare following amphotericin B. Both pulmonary and disseminated blastomycosis also respond well to ketoconazole. In the first report of 16 patients, by the NIAID Mycoses Study Group, 12 of 16 patients responded to ketoconazole, but 5 patients later relapsed.[1]

In a follow up study comparing 400 and 800 mg per day ketoconazole, the overall response rate for 80 patients was 78%, and was 89% of the 65 patients treated 6 months or longer.[26]

All of the 32 patients given 800 mg per day for 6 or more months responded, but gastrointestinal intolerance was considerable. Only 6% of the 65 patients relapsed. In an independent confirmation of this study, Bradsher et al. [27] gave 400 mg ketoconazole per day to 46 patients with blastomycosis. Similarly, 89% of the patients responded, but 6 patients (13%) later relapsed, largely due to noncompliance with treatment. The 3 patients who succumbed were all treated less than 2 weeks. Therefore, ketoconazole is highly effective for most patients with blastomycosis. However, patients with blastomycosis occasionally develop meningeal disease, and ketoconazole at 400 mg per day is insufficient for this.[28]

Because ketoconazole is less well tolerated than the triazoles, these were explored in a search for alternatives with similar efficacy but improved tolerance. In a randomized comparison of fluconazole, 200 and 400 mg per day, a total of 12/24 patients were cured. Five had preogressive disease, and 1 had persistent disease (Mycoces study group annual reports). Fluconazole at doses up to 400 mg per day is not highly effective in blastomycosis.

The other widely available triazole is itraconazole. Itraconazole has been highly effective. Of 48 patients treated by the Mycoses Study Group 90% were successfully treated ([30], Mycoses study group annual reports). On the basis of the Mycoses Study Group trials, itraconazole appears to be the optimal drug from viewpoints of both efficacy (equal or superior to ketoconazole) and toxicity (clearly better tolerated).

There have been very few patients with AIDS-associated blastomycosis, with only 15 cases reported.[31] CD4 counts were <200/mm^3 in 14 patients. Seven patients had local pulmonary disease, and 8 had disseminated disease, with central nervous system involvement in 6 patients. Forty percent of the patients died within 3 weeks. Amphotericin B was given to 12 patients, of whom 9 survived more than 1 month, and 8 of these continued suppressive therapy with ketoconazole at 400 mg per day (7 patients) or 800 mg per day (1 patient). Of those who survived more than 1 month, 2 remained on ketoconazole, and 7 died a mean of 11 months after presentation. Two patients had active blastomycosis, contributing to their deaths. Thus the disease is considerably more severe in patients with AIDS.

HISTOPLASMOSIS

In the immunocompetent host treatment with ketoconazole at 400 mg per day is effective in more than 77% of patients, and well tolerated, and when treatment was continued for 6 or more months, response rate was 92% of 25 patients.[26] Other studies support a high rate of response for ketoconazole in treatment of histoplasmosis in the non-compromised patient.[1,32] However, at 200-400 mg per day for 6 months, itraconazole has the same range of efficacy and is even better tolerated, making it the drug of choice for normal people with histoplasmosis.[30] Of 37 patients with histoplasmosis treated by the Mycoses Study Group, 81% responded. This excellent rate of response was confirmed by Negroni et al., [33] who had responses in 16 of 17 patients treated. One patient later relapsed, but was treated successfully in a second course.

One exception to initial antifungal azole therapy might be the patient with overwhelming disease who may benefit from a short course of 500 mg of amphotericin B over 1-2 weeks, with later conversion to azole treatment. In normal patients this has not been explored, although it is one approach used in cases of histoplasmosis in patients with AIDS.

Of these 3 mycoses, histoplasmosis is the only one which has undergone dramatic evolution in this past decade. This may be best illustrated by the great increase in histoplasmosis in patients with AIDS. The aggressiveness of the disease is inversely related to the vigor of host defenses. These patients commonly have fulminating disease with widespread involvement of the lungs, lymph nodes, bone marrow, liver, intestines and spleen.[5,8,34] The course may be chronic over months, or acute within weeks, but untreated patients do not recover, and death is usual. Patients treated with amphotericin B do recover promptly, but after completion of treatment they often relapse (Table 2).[5,7] Chronic suppression has

Table 2. Treatment of histoplasmosis in AIDS patients.

Primary Treatment:	Refs.	No.	% Failure
Amphotericin B	5	82	16
Ketoconazole	5	7	86
	36	7	86
Itraconazole	37*	61	17
	38, Sharkey**	8	12

Suppressive Treatment	Refs.	No.	% Failure
Amphotericin B	5	21	19
	35	40	12
Itraconazole	39	42	5
Fluconazole	35	6	16

* = AIDS Clinical Treatment Group Study, LJ Wheat, personal communication
** = Sharkey, PK, Graybill, JR unpublished observations

been reported by McKinsey et al.,[7] involving an initial course of 1 or 2 grams of amphotericin B followed by weekly suppressive treatment at a dose of 1 mg/kg.[7,35] Chronic suppression is accompanied by the usual problems of gradually increasing toxicity and bacterial sepsis caused by secondary infections of the chronic indwelling catheters.

In this situation, oral alternatives seem particularly attractive. Ketoconazole was tried by Johnson *et al.*, and by Wheat *et al.*[5,8,36] Unfortunately, because of either poor absorption or efficacy, ketoconazole has been effective in fewer than 50% of patients, as either primary therapy or as chronic maintenance.

Itraconazole, on the other hand, appears highly effective in patients with the more severe forms of disease.[34,37-39] Patients with widely disseminated disease associated with HIV infection have responded well to itraconazole 400 mg per day. An initial report by Graybill indicated that >75% of patients treated achieved remission.[37] More recently, the AIDS Clinical Treatment Group (ACTG) has conducted 2 studies, one of primary treatment and the other of suppressive maintenance therapy for histoplasmosis. For the suppressive study 42 patients were initially treated with ≥15 mg/kg amphotericin B and then transferred to itraconazole at 400 mg per day.[39] Forty two patients were entered in this study. Ten patients succumbed to other diseases and 2 withdrew during the course of the study. One withdrew because of toxicity. There were no relapses or failures while receiving itraconazole. Buoyed by this encouraging result, a followup study was commenced with primary treatment involving 800 mg per day for 3 days, and then 400 mg per day for 12 weeks, then 200 mg/day[37] indefinitely (Wheat, J., unpublished observations). Of the 62 patients entered in this study, there were only 18% failures. Only in 5 patients was itraconazole inefficacious.

The most realistic alternative to itraconazole at present is saperconazole, at 200 mg per day.[23] Only 3 patients were reported, all with HIV-associated histoplasmosis. All responded well to saperconazole. The role of saperconazole is unclear, due to limited experience.

Therefore, at present it appears that for all but the most acutely lethal forms of histoplasmosis, itraconazole is the drug of choice, at 400 mg per day for 6 months in patients with intact immune responses, or a short loading dose followed by 400 mg per day indefinitely in the setting of AIDS. There is insufficient information available for treatment of meningitis, but itraconazole has been successful in cocciodioidal and cryptococal meningitis, and it may be effective in central nervous histoplasmosis as well.[40,41]

REFERENCES

1. W.E. Dismukes, A. Stamm, J.R. Graybill, P.C. Craven, *et al.*, Treatment of systemic mycoses with ketoconazole: emphasis on toxicity and efficacy in 52 patients. National Institute of Allergy and Infectious Diseases Collaborative Antifungal Study, *Ann. Intern. Med.* 98: 13 (1983).
2. W. Mandell, D.M. Goldberg, and H.C. Neu, Histoplasmosis in patients with the acquired imunodeficiency syndrome, *Am. J. Med.* 81: 974 (1986).
3. C.M. Haggerety, M.C. Britton, J.M. Dorman and F.A. Marzoni, Gastrointestinal histoplasmosis in suspected acquired immunodeficiency syndrome, *Western. J. Med.* 143: 244 (1985).
4. L.J. Wheat, and C.B. Small, Disseminated histoplasmosis in the acquired immunodeficiency syndrome, *Arch. Intern. Med.* 144: 2147 (1984).
5. L.J. Wheat, P.A. Connoly-Springfield, R.L. Backer, *et al.*, Disseminated histoplasmosis in the acquired immunodeficiency syndrome: clinical findings, diagnosis, and treatment, and review of the literature, *Medicine* 69: 361 (1990).
6. L.J. Wheat, Histoplasmosis in Indianapolis. *Clin. Infect. Dis.* 14: 591 (1992).
7. D.S. McKinsey, M.R. Gupta, S.A. Ridler and M.R. Driks, Long -term amphotericin B therapy for disseminated histoplasmosis in patients with the acquired immunodeficiency syndrome (AIDS), *Ann. Intern. Med.* 111: 655 (1989).
8. P.C. Johnson, N. Khardori, A.F. Najjar, F. Butt, P.W.A. Mansell and G.A. Sarosi, Progressive disseminated histoplasmosis in patients with acquired immunodeficiency syndrome, *Am. J. Med.* 85: 152 (1988).

9. J.E. Bennett, *Sporothrix schenckii*, in: "Principles and Practice of Infectious Diseases", 3rd edition. G. Mandell, G. Douglas and J.E. Bennett, eds., Churchill Livingston, New York (1990).

10. J.L. Gladstone and M.L. Littman, Osseous sporotrichosis, *Amer. J. Med.* 51: 121 (1971).

11. J.E. Crout, N.S. Brewer and R.B. Tompkins, Sporotrichous arthritis, *Ann. Intern. Med.* 86: 294 (1977).

12. D.L. Calhoun, H. Waskin, M.P. White, J.R. Bonner, J.H. Mulholland, L.W. Rumans, D.A. Stevens and J.N. Galgiani, Treatment of systemic sporotrichosis with ketoconazole, *Rev. Infect. Dis.* 13: 47 (1991).

13. P.J. Lynch, J.J. Voorhees and R. Harrell, Systemic sporotrichosis, *Ann. Intern. Med.* 73: 23 (1970).

14. S. Govender, M.N. Rasool and M. Ngcelwane, Osseous sporotrichosis, *J. Infection* 19: 273 (1989).

15. C.W. Stratton, K.A. Lichtenstein, S.R. Lowenstein, D.B. Phelps and L.B. Reller. Granulomatous tenosynovitis and carpal tunnel syndrome caused by *Sporothrix schenckii*, *Amer. J. Med.* 71: 161 (1981).

16. R.M. Gullberg, A. Quintanilla, M.L. Levin, J. Williams and J.P. Pahir, Sporotrichosis: recurrent cutaneous, articular, and central nervous system infection in a renal transplant recipient, *Rev. Inf. Dis.* 9: 369 (1987).

17. P.K. Sharkey, C.A. Kauffman, J.R. Graybill, W.E. Dismukes and Members, NIAID Mycosis Study Group, Itraconazole treatment of sporotrichosis. Eleventh Congress of the International Society for Human and Animal Mycoses. Montreal, 1991, Abstract PS3.100.

18. A. Ganer, E. Arathgoon and D.A. Stevens, Initial experience in therapy for progressive mycoses with itraconazole, the first clinically studied triazole, *Rev. Infect. Dis.* 9 (Suppl 1): S77 (1987).

19. M. Diaz, R. Negronii, F. Montero-Gei, L.G. Castro, S. Sampaio, D. Borelli, A. Restrepo, L. Franco, J. Bran, E. Arathoon, D.A. Stevens *et al.*, A Pan-American 5-year study of fluconazole therapy for deep mycoses in the immunocompetent host, *Clinc. Inf. Dis.* 14 (Suppl 1): S68 (1992).

20. A. Restrepo, J. Robledo, I. Gomez, A.M. Tabares and R. Gutierrez, Itraconazole therapy in lymphangitic and cutaneous sporotrichosis, *Arch. Derm.* 122: 413 (1986).

21. D. Borelli, A clinical trial of itraconazole in the treatment of deep mycoses and leishmaniasis, *Rev. Infect. Dis.* 9 (Suppl 1): S57 (1987).

22. P. LaValle, P. Suchil, F. deOvando and S. Reynoso, Itraconazole for deep mycoses:preliminary experience in Mexico, *Rev. Infect. Dis.* 9 (Suppl 1): S64 (1987).

23. L. Franco, I. Gomez and A. Restrepo, Treatment of subcutaneous and systemic mycoses with saperconazole. Eleventh Congress of the International Society for Human and Animal Mycology. Montreal, 1991, Abstract PS 3.99.

24. D.S. Martin and D.T. Smith, Blastomycosis: I. A review of the literature, *Amer. Rev. Tuberc.* 39: 275 (1939).

25. R. Bradsher, Blastomycosis, *Clin. Inf. Dis.* 14 (Suppl 1): S82 (1992).

26. National Institute of Allergy and Infectious Diseases Mycoses Study Group. Treatment of blastomycosis and histoplasmosis with ketoconazole, *Ann. Intern. Med.* 103: 861 (1985).

27. R.W. Bradsher, D.C. Rice and R.S. Abernathy, Ketoconazole therapy for endemic blastomycosis, *Ann. Intern. Med.* 103: 872 (1985).

28. R.W. Yancy, C.A. Perlino and L. Kaufman, Asymptomatic blastomycosis of the central nervous system with progression in patients given ketoconazole therapy, *J. Inf. Dis.* 164: 807 (1991).

29. P.G. Pappas, R.W. Bradsher, S.W. Chapman, C. Kaufmann, A. Feldman, G.A. Cloud and W.E. Dismukes, Fluconazole (Flu) in the treatment of blastomycosis (B). Thirty-first Interscience Conference on Antimicrobial Agents and Chemotherapy, Chicago (1991).

30. M. Saag, R. Bradsher, S. Chapman, *et al.* and the NIAID Mycoses Study Group, Itraconazole (I) therapy (Rx) for blastomycosis (B) and histoplasmosis (H) and sporotrichosis (S). Abstract 574. Twenty-eighth Interscience Conference on Antimicrobial Agents and Chemotherapy, Los Angeles (1988).

31. P.G. Pappas, J.C. Pottage, W.G. Powderly, V.J. Fraser, C.W. Stratton, S. McKenzie, M.L. Tapper, H. Chmel, F.C. Bonebrake, R. Blum, R.W. Shafer, C. King and W.E. Dismukes, Blastomycosis in patients with the acquired immunodeficiency syndrome, *Ann. Intern. Med.* 116: 847 (1992).

32. R. Negroni, A.M. Robles, A. Arechavala, M.A. Tuculet and R. Galimberti, Ketoconazole in the treatment of paracoccidioidomycosis and histoplasmosis, *Rev. Infect. Dis.* 2: 643 (1980).

33. R. Negroni, O. Palmieri, F. Koren, I.N. Tirabischi and R.L. Galimbereti, Oral treatment of paracoccidioidomycosis and histoplasmosis with itraconazole in humans, *Rev. Infect. Dis.* 9 (Suppl 1): S47 (1987).

34. J.R. Graybill, P.K. Sharkey, P. Johnsdon and S. Nightingale, The major endemic mycoses in the setting of AIDS, in: " Mycoses in AIDS Patients", H. Vanden Bossche, D.W.R. Mackenzie, G. Cauwenbergh, J. Van Cutsem, E. Drouhet and B. Dupont, eds., Plenum Press, New York, (1990).

35. D.S. McKinsey, M.R. Gupta, M.R. Driks, D.L. Smith and M. O'Connor, Histoplasmosis in patients with AIDS: efficacy of maintenance amphotericin B therapy, *Amer. J. Med.* 92: 225 (1992).

36. G.A. Sarosi, P.C. Johnson, Disseminated histoplasmnosis in patients infected with the human immunodeficiency virus, *Clin. Inf. Dis.* 14 (Suppl 1): S60 (1992).

37. L.J. Wheat, R.E. Hafner, M. Ritchie and D. Schneider, Itraconazole (ITRA) is effective treatment for histoplasmosis in AIDS, Abstract 1206, 32nd Interscience Conference on Antimicrobial Agents and Chemotherapy, Anaheim (1992).

38. J.R. Graybill, Histoplasmosis and AIDS, J. Infect. Dis. 158: 623 (1988).

39. L.J. Wheat, R.E. Hafner, M. Wulfsohn, J. Johnson and S. Owens, Itraconazole is effective maintenance treatment for prevention of relapse of histoplasmosis in AIDS: prospective multicenter noncomparative trial. Abstract 290. Thirty First Interscience Conference on Antimicrobial Agents and Chemotherapy. Chicago (1991).

40. R.M. Tucker, D.W. Denning, B. Dupont, D.A. Stevens, Itraconazole therapy for chronic coccidioidal meningitis. *Ann. Intern. Med.* 112: 108 (1990).

41. D.W. Denning, R.M Tucker, J.S. Hostetler, S. Gill and D.A. Stevens, Oral itraconazole therapy of cryptococcal meningitis and cryptococcosis in patients with AIDS, in: "Mycoses in AIDS Patients." H. Vanden Bossche, D.W.R. Mackensie, G. Cauwenbergh, J. Van Cutsem, E. Drouhet, B. Dupont, eds., Plenum Press, New York (1990).

AZOLE COMPOUNDS IN THE TREATMENT OF PARACOCCIDIOIDO-MYCOSIS

Ricardo Negroni

Centro de Micologia del Departamento de Microbiologia
Facultad de Medicina de Buenos Aires, UBA
Buenos Aires, Argentina

The arrival of the azole-derivatives marked an important advance in the treatment of paracoccidiodomycosis.

The etiologic agent of this mycosis, named *Paracoccidioides brasiliensis*, turned out to be very sensitive to these drugs, both *in vivo* and *in vitro*.[1-3] This characteristic, plus the possibility of using azoles by oral route for long periods with tolerable side effects, made these chemical derivatives an ideal option for the treatment of paracoccidioidomycosis.

The first compounds used were miconazole nitrate and econazole nitrate. Both were effective in the control of paracoccidioidomycosis, but their investigation was interrupted because of the side effects observed: diarrhea, vein thrombosis, anemia, pruritus, hyponatremia and tachycardia.[3]

Ketoconazole was the first azole widely employed and whose efficacy is now beyond question.[2-5] This imidazole derivative should be administered at doses of 200-400 mg/day for almost one year. Its global efficacy is around 95% and post therapeutic relapses have been detected in approximately 10% of cases.[4-6] These relapses have generally been successfully treated with the same drug.[2,3] The 5% therapeutic failures has been related to the use of antiacidic agents, the presence of gastric achylia, gastrectomy or mesenteric adenopathies impairing absorption and also to the interaction with rifampin.[4,6]

Few side effects have been described with the use of ketoconazole in paracoccidioidomycosis. This may be attributed to the relatively low doses, which result in a low probability of male sexual impotence or gynecomastia. Another possibility could also be that most patients are men and that severe toxic hepatopathies are less frequent among them.[3] It remained necessary, however, to search for safer drugs causing less side effects without reducing efficacy.

The triazole antifungals show a high selective activity towards cytochrome P450-dependent ergosterol synthesis which makes them less toxic for mammalian cells.[26,27]

To date, four triazoles have been tried in paracoccidioidomycosis: itraconazole, fluconazole, Sch 39304 and saperconazole. They have all been shown to be very active in experimental models of paracoccidioidomycosis.[6-9] However, clinical studies with Sch 39304 and saperconazole have been temporarily interrupted due to the incidence of hepatomas and hepatocarcinomas with the former and ovarian tumors with the latter in long-term toxicity studies in rats.

Itraconazole and fluconazole are already being marketed and are currently used for the treatment of human paracoccidioidomycosis. Both triazolic compounds possess complementary pharmacokinetic properties. Itraconazole is highly lipophilic, showing an excellent digestive absorption, although its blood levels are low: 0.16 µg/ml at peak. It has a high tissue affinity, reaching concentrations about 3- to 30- fold higher than those found in blood. Itraconazole binds to plasma proteins, especially albumin, by 99.8%. Concentrations in cerebrospinal fluid are practically neglectable. However, itraconazole has been shown to be active in some mycoses of the central nervous system, both experimentally[9,10] and in humans,[10] probably because of its high affinity for the cerebral parenchyma. Maximum blood levels and half-life increase during the first two weeks of administration; once this period is over, steady-state is achieved. In these conditions, the half-life of itraconazole ranges between 17 and 21 hours. Excretion occurs via the digestive route, particularly in the bile; urinary excretion of the active form of this drug is very low. Most of the drug is metabolized in the liver, where inactive compounds are formed. Concurrent H_2-receptor antagonist (cimetidine) treatment does not alter single-dose kinetics of itraconazole.[10] Both rifampin and, to a lesser extent, isoniazid accelerate its metabolism reducing blood levels and half-life.[6]

Itraconazole's major advantage compared to other azoles is the absence of endocrine side effects, at least at the usual therapeutical doses.[25] Digestive disturbances and asymptomatic and transitory increase in hepatic enzymes have been documented, although no cases of severe hepatotoxicity have been reported.[6,14]

Fluconazole is water soluble. It has an excellent and rapid digestive absorption reaching its peak blood concentration of 6-10 µg/ml after 2 hours. Bioavailability of this drug is over 80%. It may also be administered intravenously. Plasma levels are higher according to the dose administered. It has a low binding to plasma proteins, only about 11-12%, which allows for free circulation of the drug and for easy passage through the hematoencephalic barrier. Fifty to 90% of plasma concentrations are reached in cerebrospinal fluid.[11,12] In patients with fungal meningitis fluconazole levels in the CSF during therapy have been over 80% of corresponding serum levels.[12] Fluconazole penetrates well in all tissues although its tissue affinity is lower than that of itraconazole. Therefore, tissue concentrations correlate with those found in plasma.[11,12] Over 62% of the drug is eliminated in its active form in urine and less than 20% is metabolized in the liver forming inactive compounds. Half-life and blood levels are stable 3-4 days after drug administration. Half-elimination time is influenced by renal function and in normal conditions it may be 31 hours.[11,12] Digestive absorption is scarcely altered by gastric hypochylia and, due to the low hepatic metabolism of this drug, interaction with rifampin is low.[12] Although the clinical use of fluconazole has not been as wide as itraconazole's, a lesser number of side effects have been registered up to the present.[11] The increase in hepatic enzymes is very infrequent.

In summary, fluconazole offers advantages over intraconazole: a higher concentration in cerebrospinal fluid and in urine; the possibility of parenteral use; its more consistent digestive absorption; lower drug interaction and reduced side effects. However, since this drug is not lipophilic, tissue concentrations are lower than itraconazole's.

Pre-clinical studies have shown that itraconazole strongly inhibits the development of *P. brasiliensis in vitro*. Most of the yeast-form cells of this fungus die in the presence of just 70 ng itraconazole/ml of culture medium.[1] Ultrastructural studies have shown severe and irreversible alterations with 10- to 100-fold lower concentrations than are needed with ketoconazole.[1] Minimum inhibitory concentrations range between 0.0005 and 0.03 µg/ml, while fungicidal concentrations are around 0.1 µg/ml.[14,26] In animal models with experimental paracoccidioidomycosis, the efficacy of itraconazole has been shown in mice[7], rats[8] and guinea pigs. In these two latter experimental models, biological cure was reached with doses of 8-10 mg/kg/day during 3 weeks.

In a comparitive study of the antifungal action of the triazole derivatives in experimental paracoccidioidomycosis of Wistar rats, itraconazole proved to be significantly more active than fluconazole.[8] The latter drug only caused a marked decrease in the lesions at doses of 25 mg/kg/day.[8] In experimental trials with mice, fluconazole was clearly effective at doses of 100 mg/kg/day.[13] This bistriazolic compound is known to be scarcely active *in vitro* against fungi for which its *in vivo* antifungal action is very marked. Therefore, minimum inhibitory concentrations should not be taken into account for evaluating the efficacy of this drug. However, both the minimum inhibitory concentration and the fungicidal concentration were around 1.6 µg/ml in Sabouraud glucose medium at pH 7.[8]

The clinical use of itraconazole for the treatment of paracoccidioidomycosis started in 1982, so now 10 years of use are already over. Many papers have been published showing the benefits of this drug in the control of paracoccidioidomycosis.[6,14-20] Up to the time of this report, data on the evolution of 278 patients treated with this drug had already been published. Although the elements considered for evaluating its therapautic action have been different in all studies, the following main conclusions may be drawn: most of the cases were treated with 50-200 mg/day during 2 to 12 months; doses of 150-200 mg/day were employed only exceptionally. All cases but one showed clinical improvement: there were 6 clinical relapses (2.1%) after successful treatments; these relapses were multiple in 2 patients and all cases showed favorable responses to renewed treatment with itraconazole. The 22 patients with acute forms of the juvenile type and marked ganglion involvement showed therapeutical responses which were similar to those observed in the chronic adult form.[6,15,16,18-20] Among side effects, only the asymptomatic and transitory increase in hepatic enzymes should be noted, which were observed in 14.2% of cases.[6,15] These data include those patients studied by several research groups in Venezuela, Colombia, Brazil, Bolivia and Argentina. Table 1 summarizes the experience of the Mycology Unit at the "F.J. Muñiz" Infectious Diseases Hospital in Buenos Aires, Argentina using ketoconazole, itraconazole and fluconazole in paracoccidioidomycosis.

The only therapeutic failure was found by Rodriguez *et al.*[6] in a patient with the acute disseminated form of the juvenile type treated with sulphonamides, amphotericin B, miconazole, econazole, ketoconazole and finally, itraconazole. During a period of 11 years this patient developed resistance to all drugs until his death. The authors concluded that the therapeutic failures had been caused by irregular use of the medications.

Thirty-seven cases of paracoccidioidomycosis treated with fluconazole were studied in 5 countries.[21-24] The drug was given at 200-400 mg/day. Nineteen patients were treated during periods of over 6 months. Evident clinical improvement or total cure were seen in 34 patients and there was one early death.

In one case with involvement of mucocutaneous and renal lesions, active pulmonary and ganglion lesions were seen to persist even though the patient had been given treatment for 3 months. The therapeutic trial was concluded as a failure and the medication was changed to itraconazole 200 mg/day, resulting in a rapid involution of the lesions.[24] The remaining patient was considered non-assessable due to the irregular use of medications. Favorable clinical changes were observed between days 15 and 30 of therapy. Oral, cutaneous and ganglion lesions were the ones with more rapid improvement; pulmonary lesions improved more slowly and the changes in serologic test titers were to be seen only at very late stages. A favorable clinical response was observed, even among patients with paracoccidioidomycosis of chronic evolution who had been treated with other antifungal agents. Eight patients recieved cotrimoxazole (a trimethoprim/sulfamethoxazole combination) as a suppresive treatment after achieving clinical cure with fluconazole.[23] Favovrable clinical changes were noted after 2 to 4 weeks, but healing of pulmonary lesions and decrease of serological titers required more than 3 months. Twenty-five patients were followed-up for more than 6 months after interruption of treatment and only one relapse was observed.[21] Neither clinical side effects nor changes in the laboratory folluw-up could be determined in the cases treated.

Table 1. Experience of the Mycology Unit at the "F.J. Muñiz" Infectious Diseases Hospital in Buenos Aires, Argentina, using ketoconazole, itraconazole and fluconazole in paracoccidioidomycosis.

	KETOCONAZOLE	ITRACONAZOLE	FLUCONAZOLE
Daily dose (mg)	200-400	50-100	200-400
Duration of treatment (months)	12	6	6
Number of cases	104	70	8
- Failures	5 (4.8%)	0	1 (12.5%)
- Relapses	9%	5.6%	0%
Early interruption of treatment	52%	18%	12.5%
Increase in heaptic enzymes	10%	12.5%	0%
Endocrinal side-effects	3%	0%	0%
Pruritus	2%	0%	0%

CONCLUSIONS

a. Both triazole compounds are active against *P. brasiliensis*.
b. Both the *in vitro* studies as well as those with infected animals show that itraconazole is more active than fluconazole against this microorganism.

c. The pharmacokinetic properties of these drugs show that they are complementary agents which do not exclude each other.

d. Itraconazole, at doses of 100 mg/day given for 6 months resulted in the clinical cure of over 99% of the patients treated.

e. The relapse rate was low (2.1%). All patients relapsing responded to a renewed treatment with itraconazole.

f. The data available up to the present allow us to state that 18% of the patients started on itraconazole treatment give up their therapy before a period of 6 months for non-medical reasons.[15] These early interruptions of therapy are attributable to the low socio-economic and cultural level of these patients. However, the rate of therapy interruptions is significantly lower than that seen with ketoconazole, which reaches 52%.[15]

g. Tolerance was good and only the appearance of asymptomatic and transitory increases in hepatic enzymes should be noted.

h. The therapeutic action of itraconazole in human paracoccidioidomycosis could be measured by several parameters: 1) clinical and radiological improvement of lesions; 2) significant decreases in serologic tests titers in 86% of cases; 3) restoration of positive cutaneous test reactivity in 75% of initially anergic patients and 4) normalization of globular sedimentation rates in 76.5% of cases with high initial figures.[15]

i. Fluconazole at 200-400 mg/day doses was effective in 34/37 cases (91.8%). The 6 months treatment period appears to be adequate for avoiding relapses; only 1/34 patients (2.9%) died early. Treatment tolerance was excellent.

j. The clinical experience with fluconazole is not so wide and this drug is more expensive than itraconazole. Therefore, it does not seem likely that fluconazole could replace itraconazole as the drug of choice, but it may be used when parenteral administration is necessary or when effective antifungal concentrations are required in cerebrospinal fluid or urine.

REFERENCES

1. M. Borgers and M.-A. Van de Ven, Degenerative changes in fungi after intraconazole treatment, *Rev. Inf. Dis.* 9 (Suppl.1): S33 (1987).

2. R. Negroni, Estado actual del empleo del ketoconazol en paracoccidioidomicosis (ketoconazol: 6 anos despues), *Rev. Arg. Micologia*, 10 (Suppl.): 21 (1987).

3. R. Negroni, Azole derivatives in the treatment of paracoccidioidomycosis, *Ann. N.Y. Acad. Sciences*, 544: 497 (1988).

4. A. Resptrepo-Moreno, Paracoccidioidomycosis, South American blastomycosis, in: "Antifungal Drug Therapy", P.H. Jacobs and L. Nall, eds., Marcel Dekker, Inc., New York (1990).

5. A. Restrepo, D.A. Stevens, E. Leiderman, J. Fuentes, A. Arana, R. Angel, G. Mejia and I. Gomez, Ketoconazole in paracoccidioidomycosis: efficacy of prolonged oral therapy, *Mycopathologia*, 27: 35 (1980).

6. C. Marcano and R. Negroni, Paracoccicioidomicosis aspectos terapeuticos, *Interciencias* (Venezuela) 15: 227 (1990).

7. J.G. McEwen, G.R. Peters, T.F. Blaschke, E. Brummer, A.M. Perlman, A. Restrepo and D.A. Stevens, Treatment of Paracoccidioidomycosis with itraconazole in a murine model, *J. Trop. Med. Hyg.*, 88: 295 (1985).

8. R. Negroni, M.R.I. de Elias Costa, J.L. Finquelievich, C. Iovannitti, I.L. Agorio, I.N. Tiraboschi, Three triazoles in the treatment of experimental paracoccidioidomycosis, *Rev. Iberoamericana Micol.*, 8: 8 (1991).

9. J. Van Cutsem, F. Van Gerven, P.A.J. Janssen, Saperconazole, a new potent antifungal triazole: in vitro activity spectrum and therapeutic efficacy, *Drug of the Future*, 14: 1187 (1989).

10. J. Heykants, A. Van Peer, V. Van de Velde, P. Van Rooy, W. Meuldermans, K. Laurijsen, R. Woestenborghs, J. Van Cutsem and G. Cauwenbergh, The clinical pharmacokinetics of itraconazole: an overview, *Mycoses*, 32 (Suppl. 1): 67 (1989).

11. J. Bennet and S. Grant "Fluconazole. An overview", ADIS Press International Inc., Langhorne, Pa. (1989).
12. J.M. Feczko, Overview of fluconazole, in: "Recent Progress in Antifungal Chemotherapy", H. Yamaguchi, G.S. Kobayashi and H. Takahashi, eds., Marcel Dekker Inc., New York (1992).
13. D.A. Stevens, E. Brummer, J.G. McEwen and A. Perlman, Efficacy of fluconazole, a new oral triazole, in blastomycosis and paracoccidioidomycosis and in comparison with ketoconazole, *Rev. Iberica Micol.*, 5 (Suppl. 1): 26 (abstract) 1988.
14. M.S. Naranjo, M. Trujillo, M.I. Múnera, P. Restrepo, I. Gomez and A. Restrepo, Treatment of paracoccidioidomycosis with itraconazole, *J. Med. Vet. Mycol.*, 28: 67 (1990).
15. R. Negroni, A. Taborda, A. Arechavala and A.M. Robles, Experiencia con itraconazol en el tratamiento de la paracoccidioidomicosis, *Rev. Arg. Micologia*, 15: 83 (abstract) 1992.
16. M.A. Shikanai-Yasuda, Y. Higaki, G.B. del Negro, S. Ho Joo, E.H. Vaccari, G. Bernard, R.C.B. Gryschek, A.A. Segurado, J.P. Bueno, A.A. Barone, D.R. Andrade and V. Amato Neto, Randomized therapeutic trial with itraconazole, ketoconazole and sulfadiazine in paracoccidioidomycosis, *Rev. Arg. Micologia*, 15: 83 (abstract) 1992.
17. F.J. Vargas, El itraconazol en la paracoccidioidomicosis (experincia en 40 casos Bolivianos), *Rev. Arg. Micologia*, 15: 85 (abstract) 1992.
18. F. Queiros Telles, L. Bendhack, N.T. Hagi, K.S. Purim, R.P. Lameira and G.F. Bordignon, Itraconazole in the therapy of paracoccicioidomycosis, *Rev. Arg. Micologia*, 15: 85 (abstract) 1992.
19. S. Marques, R.P. Camargo, J.C. Lastoria, N.L. Machado and N.L. Dillon, Itraconazol: resultados terapeuticos na paracoccidioidomicose, *Rev. Arg. Micologia*, 15: 86 (abstract) 1992.
20. R.P. Mendes, B. Barraviera, L.R. Souza, P.C.M. Pereira, J. Marcondes-Machado, M.F. Franco and D.A. Meira, Evaluation of itraconazole in the treatment of paracoccidioidomycosis, *Rev. Arg. Micologia*, 15:86 (abstract) 1992.
21. M. Diaz, R. Negroni, F. Montero-Gei, L.G.M. Castro *et al.*, A Panamerican five years study of fluconazole therapy of deep mycoses in the noncompromised host, *Clin. Infect. Dis.*, 14 (Suppl. 1): 68 (1992).
22. L.G.M. Castro, G. del Negro, W. Belda, L.C. Cucé and S.A.P. Sampaio, Fluconazole no tratamento da paracoccicioidomicose, Resultados preliminares, *Rev. Arg. Micologia*, 15: 85 (1992).
23. R.P. Mendes, L.R. Souza, J. Marcondes-Machado, D.A. Meira *et al.*, Evaluation of fluconazole in the initial treatment of paracoccidioidomycosis (PBM) preliminary results. *Rev. Arg. Micologia*, 15: 84 (1992).
24. R. Negroni, A.M. Robles, A. Arechavala and A. Taborda, Experiencia terapeutica con el fluconazol en las micosis profundas. *Rev. Arg. Micologia*, 13: 26 (1990).
25. H. Vanden Bossche and P. Marichal, Azole antifungals: mode of action in: "Recent Progress in Antifungal Chemotherapy", H. Yamaguchi, G.S. Kobayashi and H. Takahashi, eds., Marcel Dekker Inc., New York (1992).
26. H. Van Cauteren, A. Lampo, J. Vandenberghe, Ph. Vanparys, W. Coussement, R. De Coster and R. Marsboom, Toxicological profile and safety evaluation of antifungal azole derivatives, *Mycoses*, 32 (Suppl. 1): 60 (1989).

CURRENT THERAPY FOR COCCIDIOIDOMYCOSIS

John N. Galgiani

Section of Infectious Diseases
Medical and Research Services
Veterans Affairs Medical Center
and
Department of Medicine
University of Arizona College of Medicine
Tucson, Arizona, USA

ABSTRACT

Coccidioidomycosis produces a spectrum of disease from inapparent or self-limited pulmonary symptoms to widespread lesions throughout the body. Immunosuppressed patients such as those with AIDS are at particular risk of complications. Treatment options have increased during the past decade and recent studies have better defined a role for azole antifungal agents in comparison to traditional amphotericin B therapy. Despite these advances, treatment failures still occur, and the need for new approaches to therapy remains.

INTRODUCTION

Coccidioidomycosis was first recognized exactly 100 years ago in an Argentinean soldier as a disfiguring chronic disease.[1] Two years later, Emmet Rixford, a surgeon at Stanford University in San Francisco, reported to his local medical society, first one and then a second similar infection[2,3] and suggested from epidemiologic details of the cases that the source of the infection might be the San Joaquin Valley of California. Examination of the lesions revealed multicellular spherical microbes, thought to be protozoa, but by 1896 Ophuls delineated the disease's true fungal nature.[4,5]

Gradually, it was appreciated that these unusual cases were only a small part of the spectrum of disease caused by *Coccidioides immitis*.[6-12] Current estimates indicate that 50,000 to 100,000 infections occur each year. Fully two-thirds of coccidioidal infections are mild and are not brought to medical attention. Most other infections cause symptoms of a subacute pneumonia. Since the primary illness usually develops from one to three weeks after exposure, it is only seen outside of the regions endemic for the infection if the patient is in transit. Fortunately, most patients will recover without specific therapy although it may take several weeks. However, approximately 5% of patients may develop pulmonary complications such as residual nodules, cavities, ruptured cavities, or unremitting pneumonia. In an even smaller percentage, the infection will spread outside of the chest to the skin, other soft tissues, bones, joints, and meninges. These problems may develop months to years after the initial infection and not be diagnosed properly, especially where the infection is less commonly encountered. Even worse, treatment may be only partially effective and sometimes difficult to administer. With this great diversity of complications, it is of little surprise that the optimal approach for individual patients varies widely.

The first report of coccidioidomycosis treated with amphotericin B was in 1957[13] and over the ensuing years this agent became accepted as the standard therapy. The first alternative to amphotericin B was miconazole which became available nearly two decades later.[14] Despite its effectiveness, miconazole has never rivaled amphotericin B because of an array of practical difficulties of administration. However, miconazole stimulated interest in related drugs such as ketoconazole, fluconazole, itraconazole, and others, all of which could be administered orally.[15-19]

TREATMENT OF COCCIDIOIDOMYCOSIS

Primary Coccidioidal Pneumonia

General Recommendations. For most patients with uncomplicated initial infection with *C. immitis*, no specific therapy is needed for the illness to resolve.[20] Furthermore, no antifungal agent has been shown to either shorten the course of the presenting illness, or to reduce the likelihood of future complications. Therefore, it is reasonable to refrain from treating patients with primary pneumonia if the illness appears to be following a benign course.

Withholding treatment was easily done when amphotericin B was the only available therapy. The morbidity of the primary infection had to be unusually severe to offset the untoward effects of this treatment. Even today, such patients are usually treated with amphotericin B despite the availability of other antifungal agents. However, with the availability of oral and less toxic alternatives, it has become increasingly difficult not to intervene for milder infections. As a result, physicians who commonly treat coccidioidal pneumonia now vary widely in their practices.

I still see little reason to treat the initial infection if there is little or no remaining disability by the time the etiology is established. On the other hand, treatment may benefit some patients despite the lack of available documentation. Where treatment seems appropriate, I usually use ketoconazole at a dose of 400 mg/day for a period of two to six months. In the United States at present, ketoconazole is about one tenth the cost of fluconazole and is just as likely to be effective in this situation. On this dose of ketoconazole, some patients will develop symptoms such as nausea, dizziness, or lightheadedness. These symptoms are usually easily attributable to the drug since they disappear promptly if ketoconazole is discontinued and recur on rechallenge.[21] If this occurs, switching to fluconazole or itraconazole may not show the same side effects. It would also be reasonable to reassess the original decision to proceed with therapy at all.

Immunocompromised Patients. Special considerations arise in the setting of immunodeficiency, either as the result of HIV infection or from immunosuppressive therapy. HIV-infected patients, especially those whose peripheral blood CD4 cell counts have decreased below 250 cells/mm,[3] frequently develop serious complications of coccidioidomycosis.[22-27] Therefore, if symptoms or serologic evidence of coccidioidal infection develop, HIV-infected patients should be treated. On the other hand, patients with a reactive coccidioidal skin test as the only evidence of infection are not at unusual risk and therefore this should not be used as an indication for treatment.

With the increasing use of organ transplantation, the risk of reactivating prior coccidioidal infection has become a matter of concern.[28-32] From the renal and cardiac transplantation programs at our centers, we identified 9 patients with lesions or serologic evidence of coccidioidomycosis prior to transplantation.[33] Three patients had unilateral pulmonary lesions, four others had coccidioidal antibodies but no lesions, and two had a lung cavity or coccidioidal antibodies in the past. Infection recurred in two of four patients who did not receive post-operative antifungal drugs, whereas none recurred in the other five who did receive such therapy. Both infections, associated with periods of heightened immunosuppression for acute rejection, were fatal. Additionally, 13 cardiac recipients demonstrated reactivity to coccidioidal skin testing antigens, only two received post-operative antifungal therapy, and none showed recurrence of their infection after transplantation. This experience indicates that some patients with prior coccidioidal infections can successfully undergo organ transplantation. However, they may require antifungal therapy after engraftment to prevent recurrence, especially during periods of intense immunosuppression.

Progressive Forms of Coccidioidomycosis

Therapy is appropriate for pulmonary infections if the primary pneumonia does not resolve, if pneumonia recurs months or years later, or if cavitation develops and results in hemoptysis or other problems. Similarly, *Coccidioides immitis* which spreads beyond the lungs should be treated with antifungal drugs. In some cases, surgical resection of infected tissue may help to control extensive lesions.

Amphotericin B is usually selected for patients with rapidly progressive illness, especially those patients that are immunocompromised. It is administered parenterally for many weeks or months until cumulative doses reach 1.5 to 3.0 grams.[34] Azoles such as ketoconazole, itraconazole, and fluconazole are particularly useful alternatives in patients with subacute manifestations of infection. Increasingly, patients whose initial therapy was with amphotericin B are being switched to one or another azole for continuation treatment once the progression of their coccidioidal infections has been arrested. The usual dose for ketoconazole and itraconazole is 400 mg/d.[16,35] For fluconazole, 400 mg/day also appears to be an effective dose, but whether larger doses will achieve better results is presently unknown.

Many infections can be controlled with these approaches. Table 1 summarizes results for these therapies. Amphotericin B results are derived from a review of 35 separate reports.[36] Studies of ketoconazole, itraconazole, fluconazole, and SCH-39304 were all performed by the Mycoses Study Group (MSG). As the table shows, all antifungals showed some degree of efficacy. The rates of response with ketoconazole can not be compared with those of the other azoles since an entirely different scoring system was used. Moreover, all of these studies were performed at different times and small differences in response rates may not be significant. In a study now under way, the MSG is comparing itraconazole and fluconazole in a blinded, randomized study. It should also be noted that for all drugs, relapse after discontinuing treatment has been a significant problem.

Table 1. Therapy of non-meningeal coccidioidomycosis.

| Drug | Dose | Number of patients | Outcome | | | | Toxicity |
			% response	% relapse	% 1 year remission	% occurence	% stopping therapy
AMB[36]	≈ 3.0 g	91	71			*	
KET[35]	400 mg/d	56	23	9	21	39	6
	800 mg/d	56	32	44	18	59	17
ITR[16]	400 mg/d	51	57	16	49	47	6
FLU**	200 mg/d	73	34	39	21	36	4
	400 mg/d	25	61	36	39		
SCH 39304**	100 or 200mg/d	54	66		*	44	4

Abbreviations: AMB = amphotericin B; KET = ketoconazole; ITR = itraconazole; FLU = fluconazole.
* not available.
** unpublished NIAID-Mycoses Study Group results.

Coccidioidal meningitis is nearly always a fatal complication unless treated.[37-39] In the past, the only effective therapy has been frequent intrathecal administrations of amphotericin B. However, recent studies have demonstrated that meningitis also responds to either fluconazole or itraconazole.[40,41] In recent studies by the MSG, the likelihood of response to 400 mg/day of fluconazole was 70% or greater. Patients with AIDS may have meningitis as their only manifestation of coccidioidal infection, and their response to fluconazole therapy has been equivalent to that in non-HIV-infected patients. Whether azole therapy is as effective as intrathecal amphotericin B is not known. It is also unclear whether responses to fluconazole will be sustained indefinitely, even while continuing therapy. This remains an important question since most evidence suggests that fluconazole does not eradicate meningeal infection and that discontinuation of azole therapy frequently results in clinical relapses.

CONCLUSIONS

From the above summary, it is evident that while the therapeutic options for coccidioidomycosis have expanded there remain serious limitations. Treatment failures are still all too common, and patients who demonstrate a good response while under treatment may have infection recur once treatment is stopped. Furthermore, certain groups of immunosuppressed patients are at unusually high risk of infection. Early findings suggest that it is possible to prevent infection in patients who undergo organ transplantation. However, it is not known if a similar strategy would be effective for patients with AIDS residing within areas that are endemic for *C. immitis*.

Fortunately, interest in evaluating new therapies for this infection continues, partly because of the increasing number of immunosuppressed patients at risk and partly because of the well-deserved reputation of coccidioidomycosis as a therapeutic challenge. Lipid-carrier formulations of amphotericin B offer the hope of reduced toxicity which in turn may permit the use of higher and more effective doses.[42] Another approach has been to identify classes of antifungal agents active against the fungal cell wall. The chitin synthase inhibitor, nikkomycin, is one such drug. After oral administration, this agent has proved to be very effective in treatment of murine coccidioidomycosis,[43] and at present there appears to be no technical reason for not proceeding to clinical trials. Thus, it is most distressing that Bayer AG, the pharmaceutical firm responsible for nikkomycins' development, has chosen not to do so. Finally, a therapeutic role for cytokines such as γ-interferon has both theoretical and experimental support.[44,45] Whether it will be useful for patients deserves further investigation. Thus, in the coming years it is hoped that further progress will be made in the treatment of this difficult disease.

ACKNOWLEDGEMENTS

This work was supported in part by the Department of Veterans Affairs and by the NIAID-Mycoses Study Group (Contract NO1-AI-15082).

REFERENCES

1. A. Posada, Ensayo Anatomo-Patologico sobre una neoplasia cutánea considerada como micosis fungoidea, *Anales del Circulo Médico Argentino* 15: 481 (1892).
2. E. Rixford, Case for diagnosis presented before the San Francisco Medico-Chirurgical Society, Mar. 5, 1894, *Occidental M. Times* 8: 326 (1894).
3. E. Rixford, A case of protozoic dermatitis, *Occidental M. Times* 8: 704 (1894).
4. W. Ophuls, Further observations on a pathogenic mold formerly described as a protozoan (Coccidioides immitis, coccidioides pyogenes), *J. Exper. Med.* 6: 443 (1905).
5. W. Ophuls, Coccidioidal granuloma, *JAMA* 45: 1291 (1905).
6. C.E. Smith, Epidemiology of acute Coccidioidomycosis with erythema nodosum, *Am. J. Public Health* 30: 600 (1940).
7. C.E. Smith, M.T. Saito, and S.A. Simons, Pattern of 39,500 serologic tests in coccidioidomycosis, *JAMA* 160: 546 (1956).
8. C.E. Smith, E.G. Whiting, E.E. Baker, H.G. Rosenberger, R.R. Beard and M.T. Saito, The use of coccidioidin, *Am. Rev. Tuberc. Pulm. Dis.* 57: 330 (1948).
9. M.A. Gifford, San Joaquin Fever, *Kern County Dept. Pub. Health, Ann. Rep.* 22 (1936).
10. E.C. Dickson and M.A. Gifford, Coccidioides infection (Coccidioidomycosis). II. The primary type of infection, *Arch. Intern. Med.* 62: 853 (1938).
11. M.J. Fiese, Coccidioidomycosis, Charles C. Thomas, Springfield, Ill., pp. 164 (1958).
12. D.J. Drutz and A. Catanzaro, Coccidioidomycosis. Part II, *Am. Rev. Respir. Dis.* 117: 727 (1978).

13. M.J. Fiese, Treatment of disseminated coccidioidomycosis with amphotericin B. Report of a Case, *Calif. Med.* 86: 119 (1957).

14. D.A. Stevens, Miconazole in the treatment of coccidioidomycosis, *Drugs* 26: 347 (1983).

15. J.N. Galgiani, Ketoconazole in the treatment of coccidioidomycosis, *Drugs* 26: 355 (1983).

16. J.R. Graybill, D.A. Stevens, J.N. Galgiani, W.E. Dismukes, G.A. Cloud and NAIAD Mycoses Study Group, Itraconazole treatment of coccidioidomycosis, *Am. J. Med.* 89:282 (1990).

17. R.M. Tucker, D.W. Denning, E.G. Arathoon, M.G. Rinaldi, and D.A. Stevens, Itraconazole therapy for nonmeningeal coccidioidomycosis: Clinical and laboratory observations, *J. Am. Acad. Dermatol.* 23 Suppl.: 593 (1990).

18. M. Diaz, R. Puente, L.A. De Hoyos and S. Cruz, Itraconazole in the treatment of coccidioidomycosis, *Chest* 100: 682 (1991).

19. A. Catanzaro, J. Fierer and P.J. Friedman, Fluconazole in the treatment of persistent coccidioidomycosis, *Chest* 97: 666 (1990).

20. S.S. Kerrick, L.L. Lundergan and J.N. Galgiani, Coccidioidomycosis at a university health service, *Am. Rev. Respir. Dis.* 131: 100 (1985).

21. A.M. Sugar, S. Alsip, J.N. Galgiani, J.R. Graybill, W.E. Dismukes, G.A. Cloud, P.C. Craven and D.A. Stevens, Pharmacology and toxicity of high-dose ketoconazole, *Antimicrob. Agents Chemother.* 31: 1874 (1987).

22. D.A. Bronnimann, R.D. Adam, J.N. Galgiani, M.P. Habib, E.A. Petersen, B. Porter and J.W. Bloom, Coccidioidomycosis in the acquired immunodeficiency syndrome, *Ann. Intern. Med.* 106: 372 (1987).

23. D.G. Fish, N.M. Ampel, J.N. Galgiani, C.L. Dols, P.C. Kelly, C.H. Johnson, D. Pappagianis, J.E. Edwards, R.B. Wasserman, R.J. Clark, D. Antoniskis, R.A. Larsen, S.J. Englender and E.A. Petersen, Coccidioidomycosis during human immunodeficiency virus infection. A review of 77 patients, *Medicine (Baltimore)* 69: 384 (1990).

24. N.M. Ampel, C.L. Dols, J.N. Galgiani, L. Hood and J.J. Rohwedder, Coccidioidomycosis (Coccy) during HIV infection. A continuing prospective study in the coccidioidal endemic area, *Seventh International Conference on AIDS. Program and Abstracts* (1991).

25. J.N. Galgiani and N.M. Ampel, Coccidioidomycosis in Human Immunodeficiency Virus-Infected Patients, *J. Infect. Dis.* 162: 1165 (1990).

26. J.N. Galgiani, Coccidioidomycosis: changes in clinical expression, serological diagnosis, and therapeutic options, *Clin. Infect. Dis.* 14 Suppl. 1: S100 (1992).

27. N.M. Ampel, C.L. Dols and J.N. Galgiani, Coccidioidomycosis during human immunodeficiency virus infection. Results of a prospective study in a coccidioidal endemic area, *Am. J. Med.* in press (1993).

28. S.E. Vartivarian, P.E. Coudron and S.M. Markowitz, Disseminated coccidioidomycosis. Unusual manifestations in a cardiac transplantation patient, *Am. J. Med.* 83: 949 (1987).

29. R.H. Britt, D.R. Enzmann and J.S. Remington, Intracranial infection in cardiac transplant recipients, *Ann. Neurol.* 9: 107 (1981).

30. L.G. Dodd and S.D. Nelson, Disseminated coccidioidomycosis detected by percutaneous liver biopsy in a liver transplant recipient, *Am. J. Clin. Pathol.* 93: 141 (1990).

31. I.M. Cohen, J.N. Galgiani, D. Potter and D.A. Ogden, Coccidioidomycosis in renal replacement therapy, *Arch. Intern. Med.* 142: 489 (1982).

32. D. L. Calhoun, J. N. Galgiani, C. Cukowski and J. G. Copeland, Coccidioidomycosis in recent renal or cardiac transplant recipients, Coccidioidomycosis. Proceedings of the 4th international conference on coccidioidomycosis, in: H. E. Einstein and A. Catanzaro, eds., National Foundation for Infectious Diseases, Washington,D.C., pp. 312 (1985).

33. K.A. Hall, J.G. Copeland, C.F. Zukoski, G.K. Sethi and J.N. Galgiani, Markers of coccidioidomycosis prior to cardiac or renal transplantation and the risk of recurrent infection, *Transplantation* (in press) (1993).

34. D.J. Drutz, Amphotericin B in the treatment of coccidioidomycosis, *Drugs* 26: 337 (1983).

35. J.N. Galgiani, D.A. Stevens, J.R. Graybill, W.E. Dismukes and G.A. Cloud, Ketoconazole therapy of progressive coccidioidomycosis. Comparison of 400- and 800-mg doses and observations at higher doses, *Am. J. Med.* 84: 603 (1988).

36. M.H. Hardenbrook and S.L. Barriere, Coccidioidomycosis: Evaluation of parameters used to predict outcome with amphotericin B therapy, *Mycopathologia* 78: 65 (1982).

37. T. Vincent, J.N. Galgiani, M. Huppert and D. Salkin, The natural history of coccidioidal meningitis. VA-Armed Forces Cooperative Studies, 1955 -1958, *Clin. Infect. Dis.* (in press) (1993).

38. P. C. Kelly, Coccidioidal meningitis, in: "Coccidioidomycosis", D. A. Stevens, ed., Plenum Press, New York, (1980).

39. H.E. Einstein, C.W.J. Holeman, L.L. Sandidge and D.H. Holden, Coccidioidal meningitis. The use of amphotericin B in treatment, *Calif. Med.* 94: 339 (1961).

40. R.M. Tucker, D.W. Denning, B. Dupont and D.A. Stevens, Itraconazole therapy for chronic coccidioidal meningitis, *Ann. Intern. Med.* 112: 108 (1990).

41. R.M. Tucker, J.N. Galgiani, D.W. Denning, L.H. Hanson, J.R. Graybill, K. Sharkey, M.R. Eckman, C. Salemi, R. Libke, R.A. Klein and D.A. Stevens, Treatment of coccidioidal meningitis with fluconazole, *Rev. Infect. Dis.* 12 Suppl. 3: S380 (1990).

42. P. K. Sharkey, R. Lipke, A. Renteria, J. Galgiani, A. Catanzaro, M. Diaz, M. Kramer, R. Whitney and R. Gupta, Amphotericin B lipid complex (ABLC) in treatment (Rx) of coccidioidomycosis, Program and Abstracts of the 31st Interscience Conference on Antimicrobial Agents and Chemotherapy, American Society for Microbiology, Washington, D.C., pp. 742A (1991).

43. R.F. Hector, B.L. Zimmer and D. Pappagianis, Evaluation of nikkomycins X and Z in murine models of coccidioidomycosis, histoplasmosis, and blastomycosis, *Antimicrob. Agents Chemother.* 34: 587 (1990).

44. L. Beaman, Effects of recombinant gamma interferon and tumor necrosis factor on in vitro interactions of human mononuclear phagocytes with *Coccidioides immitis, Infect. Immun.* 59: 4227 (1991).

45. N.M. Ampel, G.C. Bejarano, S.D. Salas and J.N. Galgiani, In vitro assessment of cellular immunity in human coccidioidomycosis: Relationship between dermal hypersensitivity, lymphocyte transformation, and lymphokine production by peripheral blood mononuclear cells from healthy adults, *J. Infect. Dis.* 165: 710 (1992).

TREATMENT OF PHAEOHYPHOMYCOSIS AND PITYRIASIS VERSICOLOR

Geert Cauwenbergh

Department of Clinical Research and Development
Janssen Research Foundation
B-2340 Beerse, Belgium

ABSTRACT

Phaeohyphomycosis and pityriasis versicolor are 2 fungal infections which are at completely opposite sides of the spectrum of mycoses. Pityriasis versicolor is considered by some to be nothing more than a very common, almost asymptomatic cosmetic burden, while phaeohyphomycosis is a less frequent, progressively disfiguring and sometimes fatal infection. Although the diseases may be quite opposite, their drug therapy has become a common feature for both infections.

In the past, treatment of phaeohyphomycosis has been frustrating with options that included surgery and chemotherapy. Resection of the lesions is of course not always possible. Amphotericin B treatment has never given satisfactory results. Failure, relapse and toxicity were common events with this drug. Similar disappointments were obtained with miconazole IV and ketoconazole, or with combinations of these agents. When itraconazole became available, patients with phaeohyphomycosis were also exposed to this new drug. The first patients treated showed a rapid and dramatic response to this triazole derivative. Thus far, 26 patients with this infection have been treated with itraconazole. No therapeutic effect was seen in 8 of them. They continued to worsen during therapy. Thirteen patients showed clear improvement or complete remission, and stabilization of the disease was observed in 5 patients. These results suggest that itraconazole to date is the treatment of choice for phaeohyphomycosis, although further improvement of efficacy would be most welcome.

Therapy for pityriasis versicolor has been much easier. Classical approaches include topical selenium sulphide, zinc pyrithione, tar and of course topical therapy with imidazole derivatives. The major problem with local treatment has always been the extent of the lesions, and the inability of the patients to treat the whole body area where the infection may occur. Treatment of pytiriasis versicolor has been revolutionized by the discovery of ketoconazole. This orally active imidazole derivative is excreted through the sweat very rapidly and at high concentrations. Short therapies of 200 mg daily for 5 to 10 days have resulted in more than 90% cure rates. Even the ultra-short treatment with 400 mg as a single

Dimorphic Fungi in Biology and Medicine, Edited by
H. Vanden Bossche *et al.*, Plenum Press, New York, 1993

dose has given cure rates of more than 70%. Needless to say, these very short treatment courses with ketoconazole are usually well tolerated. Itraconazole appears to be equally active as compared to ketoconazole; the more water-soluble triazole fluconazole and the allylamines are substantially less effective in the treatment of pityriasis versicolor.

The development of ketoconazole in a shampoo formulation has added another treatment modality with this potent pityrosporucidal agent to the therapeutic armamentarium.

Although the clinical picture and the course of phaeohyphomycosis and pityriasis versicolor are quite different, and although in the past this clinical difference was supplemented by a therapeutic difference since virtually no treatment existed for phaeohyphomycosis while acute episodes of pityriasis versicolor were relatively easy to manage, the availability of newer agents such as ketoconazole, but more importantly itraconazole, has created at least one common aspect between these 2 opposites of the spectrum of fungal infection. The availability of these newer azoles does not mean in any way that therapy of phaeohyphomycosis can be considered as optimal; and this in contrast to treatment of pityriasis versicolor. Further efforts have to be made to improve the therapeutic options for phaeohyphomycosis.

INTRODUCTION

From the clinical point of view phaeohyphomycosis and pityriasis versicolor are entities that have very little in common. Phaeohyphomycosis is an opportunistic infection caused by dematiaceous fungi.[1] The clinical classification of the infection is according to the level of invasion of the fungus. Phaeohyphomycosis includes superficial, cutaneous, corneal, subcutaneous and systemic infections. The infection can also be present under the form of mycotic cysts.[2]

The cutaneous forms of phaeohyphomycosis are most commonly caused by *Hendersonula toruloidea* and *Scytalidium hyalinum*. The clinical features of the infection resemble the dry type *Trichophyton rubrum* infection on the skin.[3] The infection of the nails resembles a classical dermatophytic onychomycosis. The infection starts at the distal or lateral nail borders and this leads to thickening and a yellowish-brown discoloration of the nail plate and subungeal hyperkeratosis. Onycholysis occurs and spreads to the whole nail plate.

The subcutaneous forms of phaeohyphomycosis invade the deep dermis and the subcutaneous tissue by traumatic or post traumatic inoculation with the organism. The most common etiologic agents are *Exophiala jeanselmei* and *Wangiella dermatitidis*. Other agents include *Phialophora richardsiae, Exophiala spinifera, Fonsecaea pedrosoi*, etc.[4]

The clinical features can start as a dark scaly patch with dark red shiny soft papules or also as a verrucous keloid-like presentation. These lesions will gradually become confluent and larger, and their hardness will increase. This leads to abscesses and subcutaneous masses. Scratching breaks the cysts and this leads to auto-inoculation. In the tissues one can find dark yeast-like or hypha-like elements.[5]

Finally, the invasive systemic infection or the cerebral infection can be localized or disseminated to other organs. This infection is rare and occurs in general in immunocompromised patients. Its evolution is often fatal.[6]

In contrast to these more severe infections stands the very common and benign infection pityriasis versicolor. It is a chronic superficial fungal infection usually located on the upper trunk, neck or upper-arms. Lesions are slightly scaly and may involve larger parts of the body. A dimorphic lipophilic yeast, *Malassezia furfur*, is responsible for this infection.[7] The yeast form of this organism, *Pityrosporum ovale*, has been associated with other dermatological conditions such as folliculitis, seborrheic dermatitis, dandruff and atopic dermatitis.[8]

TREATMENT OF PHAEOHYPHOMYCOSIS AND PITYRIASIS VERSICOLOR

Some years ago, treatment of phaeohyphomycosis was as different from that of pityriasis versicolor as the difference between the clinical manifestations between the two infections. In recent years, with the availability of ketoconazole and itraconazole, the two infections have started to show at least some therapeutic parallels.

Cutaneous Phaeohyphomycosis

Treatment of cutaneous phaeohyphomycosis is difficult. *Hendersonula* and *Scytalidium* are both resistant to griseofulvin (MIC of more than 0.2 mg/ml). The drug has been used at daily doses of 750 mg for as long as 3 years without any objective improvement. Also systemic ketoconazole does not seem to give significant improvement, even after prolonged courses of therapy. The MIC values of both organisms for miconazole, clotrimazole, amphotericin B and nystatin are within the range of what can be achieved on the skin after topical administration; however the penetration of these topically applied agents is probably to limited to result in any therapeutic effect. Six weeks treatment with miconazole and clotrimazole has not given acceptable results.[9] The tioconazole nail solution has been tried but the results here also were not very encouraging. As for the topical preparations, probably Whitfield's ointment is still the most efficacious antifungal readily available.[10]

Itraconazole, the more lipophilic triazole derivative with a good tissue distribution, may be a somewhat more useful agent in the treatment of cutaneous phaeohyphomycosis. Indeed, three clinical isolates of *Scytalidium hyalinum* have been tested for their sensitivity to itraconazole, and all 3 were completely inhibited at a concentration of 100 ng/ml. *Hendersonula toruloidea* on the other hand was clearly less sensitive. One of the three clinical isolates tested was inhibited at 0.01 mg/ml, and the two others required 0.1 mg/ml.[11]

At present it seems that the best remedy against cutaneous phaeohyphomycosis is still to avoid contact with the plants on which these organisms live as plant pathogens. Trinidad, Tobago and Jamaica are places where these pathogens are highly prevalent.

Subcutaneous Phaeohyphomycosis

As for the cutaneous manifestations of this infection, treatment remains rather difficult. First of all, and apart from an accurate diagnosis, it is important to obtain some anamnestic facts about the patients: immune status, predisposing factors (diabetes, tuberculosis, steroids or antimitotic treatments). Correction of these predisposing factors can already help substantially to improve the chances for a favourable therapeutic outcome.

Two large therapeutic options exist: surgery and chemotherapy. Localized cystic lesions can easily be treated surgically. A generously made excision of the cyst is usually curative. Special attention should be given as to not incise the lesion itself because this can cause contamination of the surrounding tissues and lead to relapse.[1]

Chemotherapy to date gives 5 options: flucytosine, amphotericin B, intravenous miconazole, ketoconazole, itraconazole and thiabendazole. The latter substance has *in vitro* activity against the fungi causing chromomycosis at 750 ng/ml. The clinical evidence of activity however is not overwhelming. Flucytosine has a broad range of sensitivities for each different organism that can cause subcutaneous phaeohyphomycosis. Some organisms are sensitive to concentrations of 30 ng/ml; but the majority of the organisms require more than 0.1 mg/ml. With amphotericin B, the sensitivity range is much smaller. For most organisms, sensitivities range between 1 and 32 µg/ml. The only exception is *Ph. richardsiae* which is sensitive at concentrations above 0.1 mg/ml. The imidazoles show a completely different pattern of sensitivity. Drugs like miconazole and ketoconazole are fairly active *in vitro* against *Ph. richardsiae* and also against *F. pedrosoi*. The activity against *E. jeanselmi* is somewhat lower. Itraconazole, the triazole derivative has a very high *in vitro* activity against most of these organisms. Many strains and species were inhibited at concentrations between 1 and 100 ng/ml. This is substantially less than what is required for the other agents.[1,12,13]

Some general guidelines can be given for the treatment of phaeohyphomycosis with the above mentioned agents. Flucytosine is generally given at daily doses of 50 to 150 mg/kg divided over 4 daily intakes. This results in serum levels between 50 and 75 µg/ml. If the MIC of flucytosine for the organism is greater than 15 µg/ml, the organism is considered resistant to the drug. When improvement occurs, it can already be noted within 2 to 4 weeks after starting therapy. Toxicity with flucytosine is acceptable. It usually occurs when the drug is given at high dose. A plasma level of 0.1 mg/ml is sort of a target concentration from which toxicity may occur. Resistance development has been described in a number of cases during prolonged therapy. Because of this, flucytosine is not frequently given anymore as a monotherapy. Usually it is combined with amphotericin B. This latter drug achieves serum concentrations between 0.2 and 2 µg/ml when daily doses between 0.25 and 1 mg/kg are administered intravenously. If *in vitro* the MIC value for the isolated fungus is above 2 µg/ml, the fungus is considered resistant to amphotericin B. The toxicity and side effects of amphotericin B are well known. It is essential to monitor renal function, hepatic function, electrolyte balance and hemogram very regularly in order to avoid major toxicity.

When thiabendazole is used, daily doses of 25 to 50 mg/kg are given divided over the different meals. At doses below 2 g daily the drug is well tolerated. Nausea, anorexia and vomiting occur at doses of 3 g daily. Response occurs slowly, within 6 to 8 weeks after starting treatment, which can take up to 2 years. Development of resistance during therapy has been reported with thiabendazole.[1,12]

Experience with intravenous miconazole in phaeohyphomycosis is limited. Ketoconazole has been used somewhat more frequently with variable degrees of success. In general the daily dose of ketoconazole in this infection is 400 to 600 mg resulting in peak plasma levels between 4 and 6 µg/ml. An organism can be considered resistant to ketoconazole if the MIC is higher than 6 µg/ml. The toxicity of both miconazole and ketoconazole has been documented extensively. Intravenous miconazole should not be given as a bolus since this can cause cardiac arrythmias. Because of the cremophore in the excipient, intravenous miconazole has caused anaphylactic reactions in a number of patients. Ketoconazole is known to cause liver toxicity in some patients.[14] In addition it interferes at high doses with cortisol and testosterone production.[15] These latter effects are reversible within 24 h after stopping therapy. To prevent patients from developing liver damage, liver function tests should be performed regularly.

In view of the difficult-to-treat nature of this infection, a lot of emphasis has been given to combination treatments. Combinations that have been suggested include amphotericin B plus ketoconazole for *Drechslera rostrata* and *F. pedrosoi*; amphotericin B plus flucytosine for *F. pedrosoi* and *Ph. richardsiae*; miconazole plus rifampicin for *W. dermatitidis*; amphotericin B plus rifampicin for *F. pedrosoi*, *Ph. richardsiae* and *Phialophora verrucosa*; ketoconazole plus rifampicin for *E. jeanselmi* and *F. pedrosoi* and finally flucytosine plus ketoconazole for *E. jeanselmi*.[1]

The most promising treatment for phaeohyphomycosis appears to be the newer triazole agent itraconazole.[6] As already indicated, this molecule has a good *in vitro* activity against many of the fungi causing this infection. The best documented data fot the use of itraconazole in phaeohyphomycosis come from the group of Graybill and co-workers, who treated 19 patients with this infection. Seventeen of these patients were clinically evaluable, and of these 17 only 2 had received no prior therapy. Five patients had shown no response to amphotericin B, 4 failed ketoconazole or miconazole therapy and 5 had failed amphotericin and imidazole treatment. In his case series, Graybill had fungi of 7 diferent genera causing disease of the skin in 9 patients, soft tissue in 9, simuses in 8, bone in 5, joints in 2 and lung in 2. Itraconazole doses ranged between 50 and 600 mg per day for a maximum of 48 months. Nine patients showed clinical improvement or remission and in 2 patients disease progression was halted. Six cases were treatment failures, while one patient relapsed after an initial succesful treatment course. Considering the refractory nature of these cases to other treatments, the results obtained in that study indicate that itraconazole could be one of the first treatment options to be used in these patients. Toxicity and side effects are in general not significant during itraconazole therapy. Liver test monitoring is suggested during prolonged treatment. No information is available to date on the use of itraconazole in combination with other agents. Some reports suggesting additive effects of itraconazole with thiabendazole in chromoblastomycosis may be an indication that this combination could be useful in phaeohyphomycosis as well.

Systemic Phaeohyphomycosis

Basically the treatment options in this often life-threatening condition are the same as for the subcutaneous forms. The combinations as indicated earlier can be used and apparently, itraconazole is a promising new drug that may well become a treatment of choice. Patients with the disseminated systemic infection are often severely immunocompromised. This has to be taken into account when deciding for one or another treatment. In some cases it may be necessary to have an intravenous therapy available, at which moment a combination of itraconazole with amphotericin B should be considered.

Treatment of phaeohyphomycosis, even with itraconazole, remains a significant problem. Often the patient and the treating physician will have to accept that just stopping the further progression of the disease should be regarded as a therapeutic success. A problem remains the unpredictable response of the different body sites affected by the disease. Some areas or organs will respond well while others remain apparently unaffected despite high drug concentrations in the tissues, and sensitivity of the fungus to the drug at low concentrations.

Pityriasis versicolor

Pityriasis versicolor can be treated topically or systemically. In a substantial number of patients, recurrence after topical therapy is a significant problem, reaching 60% within 1 year after treatment. For this reason topical treatment should not just cover the infected patches, but the skin as a whole should be treated. In view of these large body sites to be treated topically, a logical preference exists for the use of lotions or shampoos, since these formulations are easier to apply than creams or ointments.

Topicals that have proven effective are propylene glycol 50% applied for 1 to 2 weeks, selenium sulphide 2.5 or 5% for 1 week, zinc pyrithione 1% for 1 to 2 weeks and finally, the topical imidazole agents miconazole, econazole, etc.[7] All these agents have their inconveniences. The first inconvenience is the necessity to treat large body areas, including the back. This is a problem with all the topical agents. Propylene glycol 50% is cheap, but irritation and contact dermatitis are seen in a number of patients. Selenium sulphide and zinc pyrithione have to remain on the skin for 5 to 10 min before they can be showered off. In addition to this inconvenience, both products have a bad odour and cause irritation in a fair number of people. Topical imidazoles give no major problems with tolerance.

Systemic treatment for pityriasis versicolor was not possible until ketoconazole[16] became available. Initially ketoconazole was given at daily doses of 200 mg for 4 weeks. The occurrence of rare but serious hepatic toxicity with longer treatments of ketoconazole has made this treatment regimen of little use. The finding by Borelli, Sampaio and Jacobs[17] that ketoconazole, as a single dose of 400 mg, resulted in about 80% response has renewed the interest of the medical profession in ketoconazole as a systemic agent for treatment of pityriasis versicolor. It is now considered to be one of the standard treatment regimens.

In more recent years some newer systemically active antifungals have become available. Terbinafine, fluconazole and itraconazole have been evaluated in the treatment of pityriasis versicolor. Terbinafine has only a limited therapeutic potential in this infection. Fluconazole is active, but higher doses than those recommended are required, and short courses as in the case of ketoconazole are not very effective. Itraconazole is active in pityriasis versicolor.[18] A daily dose of 200 mg given for 5 to 7 days yields 80% response rates. The final cure rates 1 month after treatment are similar for this treatment schedule compared with the single dose ketoconazole therapy. Ketoconazole seems however to give a faster improvement of symptoms. Itraconazole is better tolerated than ketoconazole.

During the last 5 years, the work by several researchers has indicated a link between the presence on the skin of the lipophylic yeasts *P. ovale* and the occurrence of dermatoses such as head and neck dermatitis, seborrheic dermatitis, dandruff and pityrosporum folliculitis. Because of the cosmetic nature of diseases such as dandruff, and to a lesser degree seborrheic dermatitis, ketoconazole has also been formulated as a cream and as a shampoo.[19] Both forms have shown to be well active in these *P. ovale* related illnesses.

CONCLUSION

The infections discussed in this chapter clearly belong to two completely opposite sides of the mycotic universe. One is debilitating, occasionally fatal, and the other is considered by many to be a cosmetic burden. In the past, the treatments for both infections have been completely different. Phaeohyphomycosis has been refractory to most treatments given, while pityriasis versicolor responded easily to well known topical remedies. The development of the systemically active agents, especially ketoconazole and itraconazole, has made that these two opposite sides of the mycotic universe have encountered eachother at least on a therapeutic level. Itraconazole, but especially ketoconazole are very useful drugs for pityriasis, while ketoconazole, to a limited extend, and itraconazole especially, are of use in phaeohyphomycosis.

REFERENCES

1. B. Kotrajaras, Phaeohyphomycosis, in: "Antifungal Drug Therapy - A Complete Guide for the Practitioner", P.H. Jacobs, L. Nall, eds., Marcel Dekker, Inc., New York and Basel (1990).
2. J. Rippon, Medical Mycology: The pathogenic fungi and the pathogenic antinomycetes, 2nd edition, W.B. Saunders, Philadelphia (1982).
3. R.J. Hay and M.K. Moore, Clinical feature of superficial fungal infection caused by *Hendersonula toruloidea* and *Scytalidium*, *Br. J. Dermatol.* 110: 677 (1984).
4. Y.A. Doory, Chromomycosis, in: "Occupational Mycosis", A.F. Disalia, ed., Lea & Febiger, Philadelphia (1983).
5. Z. Herschil, Phaeohyphomycotic cyst of the skin caused by *E. jeanselmei*, *J. Am. Acad. Dermatol.* 12: 207 (1985).
6. P.K. Sharkey, J.R. Graybill, M.G. Rinaldi, D.A. Stevens, R.M. Tucker, et al., Itraconazole treatment of phaeohyphomycosis, *J. Am. Acad. Dermatol.* 23 (3) part 2: 577 (1990).
7. J.N. Faergeman, Pityriasis Versicolor, Tinea Nigra and Piedra, in: "Antifungal Drug Therapy - A Complete Guide for the Practitioner", P.H. Jacobs, L. Nall, eds., Marcel Dekker, Inc., New York and Basel (1990).
8. S. Shuster, The aetiology of dandrufff and the mode of action of therapeutic agents, *Br. J. Dermatol.* 111: 235 (1984).
9. C.K. Campbell, Studies on Hendersonula toruloidea isolated from skin and nails, *Sabouraudia* 12: 150 (1971).
10. R.J. Hay, R.M. Mackie, Y.M. Clayton, Tioconazole nail solution: an open study of its efficacy in onychomycosis, *Clin. Exp. Dermatol.* 10: 111 (1985).
11. J. Van Cutsem, F. Van Gerven and P. Janssen, Activity of orally, topically and parenterally administered itraconazole in the treatment of superficial and deep mycoses, *Rev. Infect. Dis.* 9 (1) Jan.-Feb.: S15 (1987).
12. W.B. Pratt and R. Fekety, Antimicrobial Drugs, Oxford University Press, New York (1986).
13. M.R. McGinnis, Chromoblastomycosis and phaeohyphomycosis: new concepts, diagnosis and mycology, *J. Am. Acad. Dermatol.* 8: 1 (1983).
14. P.A.J. Janssen and J. Symoens, Hepatic reactions during ketoconazole treatment, *Am. J. Med.* 74: (1B): 80 (1983).
15. D.A. Stevens, P.L. Williams, A.M. Sugar, and A. Pont, Ketoconazole effects, *Ann. Intern. Med.* 97 (2): 284 (1982).
16. J. Faergeman and L. Djärv, Tinea versicolor: treatment and prophylaxis with ketoconazole, *Cutis* 30: 542 (1982).
17. D. Borelli, Treatment of pityriasis versicolor with ketoconazole, *Rev. Infect. Dis.* 2: 592 (1980).
18. J. Delescluse, G. Cauwenbergh, and H. Degreef, Itraconazole, a new orally active antifungal in the treatment of pityriasis versicolor, *Br. J. Dermatol.* 114: 701 (1986).
19. G. Cauwenbergh, P. De Doncker, P. Schrooten, and H. Degreef, Treatment of dandruff with a 2% ketoconazole scalp gel: a double blind placebo-controlled study, *Int. J. Dermatol.* 25 (8): 541 (1986).

PENICILLIUM MARNEFFEI: DIMORPHISM AND TREATMENT

M.A. Viviani[1], J.O. Hill[2], and D.M. Dixon[3*]

[1] Istituto di Igiene e Medicina Preventiva
    Università degli Studi
    Milano 20122, Italy
[2] Trudeau Institute
    Saranac Lake, New York 12983, USA
[3] Laboratories for Mycology
    Wadsworth Center for Labs and Research
    New York State Department of Health
    Albany, NY 12201-0509, USA

ABSTRACT

Penicillium marneffei grows at 25 °C as a mould typical of the genus, but is atypical by its ability to convert to a parasitic, yeast-like phase of growth at 37 °C on the appropriate media *in vitro* or in infected animals and humans. Natural infections occur in bamboo rats and in humans living in or having travelled in Indonesia or Southest Asian countries. The fungus infects cells of the reticuloendothelial system and is most often found in patients with diseases or disorders involving cell mediated immunity (Hodgkin's disease, tuberculosis, corticosteroid therapy, and more recently, AIDS). The first natural human infection was reported in 1973. The total number of reported cases rose from 18 through 1987 to 73 as of June 1992, with 71% involving AIDS patients. To investigate the role of T cells in limiting *P. marneffei* disease, a mouse model was developed. Mice were depleted of CD4+ T cells, CD8+ T cells, or both. The results show that CD4+ T cells and, to a lesser extent CD8+ T cells play a role in the protective host response against *P. marneffei*. Because of the enormous increase of HIV infection in the endemic area, *P. marneffei* has the potential to emerge as a major opportunistic pathogen in AIDS.

* Present address: Division of Microbiology and Infectious Diseases, NIH/NIAID, Bethesda, Maryland 20892, USA

Dimorphic Fungi in Biology and Medicine, Edited by
H. Vanden Bossche *et al.*, Plenum Press, New York, 1993

INTRODUCTION

Penicillium marneffei as a Dimorphic Fungus

Penicillium marneffei grows in culture at 25 °C on Sabouraud agar as a mycelial fungus typical of the genus by producing rapidly growing, greenish-yellow sporulating colonies. A characteristic red pigment is elaborated into the medium.

Microscopically, the mycelial phase of growth is characterized by divaricate penicilli, phialides in verticils, and chains of smooth conidia with prominent extensions on either end (disjunctors).[1] The fungus can be considered dimorphic on the basis of its conversion to cylindrical to ovoid yeast-like cells that divide by transverse septations when growing at 37 °C *in vitro* on the appropriate media, or *in vivo* in infected animals and humans. It is the only species of *Penicillium* known to be dimorphic.[2-4]

First Description of the Organism

The organism was first described by Capponi *et al.* in 1956 as an etiologic agent infecting the reticuloendothelial system in Vietnamese bamboo rats (*Rhizomys sinensis*).[2] Segretain named the fungus *Penicillium marneffei* and gave the formal mycological description in 1959.[3] In that same year, Segretain also described the first human infection which resulted as a consequence of accidental inoculation of the investigator's finger.[4]

Natural Infection

The first natural infection of a human was repoted by Di Salvo *et al.* in 1973.[5] The patient had Hodgkin's disease an had travelled in Southeast Asia. All subsequently reported cases have involved subjects living in or visiting Indonesia or Southeast Asian countries such as the People's Republic of China, Vietnam and Thailand. The infection has been associated with Hodgkin's disease, tuberculosis, and corticosteroid-treated lymphoproliferative disorders, but also occured in previously healthy subjects.[6-10]

The AIDS epidemic has resulted in an increase in reports of *P. marneffei* infection, and within the last five years the total number of cases rose from 18 through 1987 to a total of 73 as of June 1992.[10-30] This represents a tenfold increase, with 71% involving patients with AIDS. Of the 18 cases diagnosed outside of Asia, 13 occurred in AIDS patients and most were reported in Europe (Table 1).

Table 1. Distribution of *P. marneffei* cases* in relationship to the emergence of the AIDS epidemic.

Period	N° of Cases	% in AIDS	Fatality
1973-1987	18	0	14/18 (78%)
1988-June 1992	55	71	21/55 (38%)

* Naturally acquired infection

Penicilliosis marneffei was the terminology applied by Jayanetra *et al.* to the disease.[6] More recently, the suggestion has been made to name individual mycoses in the form "infection or pathological entity caused by the specific etiologic agent". It was recommended that disease terminology for the *Penicillium* species be approached in this manner, for example, disseminated fungal disease due to *P. marneffei*.[31]

The route of acquisition of the infection has not been established, but a respiratory portal of entry would be consistent with infections caused by other dimorphic fungal pathogens that produce conidia in the saprophytic phase of growth.

With the exception of a few cases of localized infections in which the fungus was isolated only from the spleen or lung, *P. marneffei* infection has been diagnosed as an invasive, disseminated disease. The sites most frequently involved were lung, lymph node, liver, skin, soft tissue, and bone.

The high mortality rate in the first cases of *P. marneffei* infection can be ascribed to the failure to make an initial accurate diagnosis. This mycosis was often misdiagnosed as tuberculosis and treated accordingly. But recently, mortality has been reduced by early treatment, due to the increased awareness by clinicians and microbiologists.

The clinical presentation is relatively non specific. Symptoms may be suggestive of tuberculosis or other systemic infections. In addition, coexisting opportunistic infections, frequently occurring in AIDS, may mask *P. marneffei* disease. Intensive laboratory surveillance, however, will uncover the infection since the fungus is easily isolated in culture and is readily distinguished as described above.

Histology is also helpful in diagnosis. The yeast-like cells can be identified in smears or tissue sections stained with PAS or methenamine-silver and discriminated from *Leishmania, Pneumocystis* and *Histoplasma* which also frequently cause disease in patients with AIDS. Even though *P. marneffei* superficially resembles *Histoplasma capsulatum* when yeast-like cells are clustered within macrophages or histiocytes, the two organisms multiply differently. *P. marneffei* replicates by fission, instead of budding, and the considerable variation in size and shape of the extracellular yeasts (enlarged, occasionally septate, tubular or sausage-shaped fungi) make *P. marneffei* easily distinguishable from *H. capsulatum*.

Transmission and scanning electron microscopy have been used by Garrison and Boyd[32] and Deng and coworkers[10] to study the morphology of the saprophytic hyphae and the temperature-dependent transformation to the yeast-like tissue form. Hyphae first become shorter, develop more septa and branches and stop producing conidia. After 2 weeks at 37 °C, there is a gradual shift to spherical or ellipsoidal yeast-like cells, that are 2-6 µm across and divide by fission.

Antifungal treatment is usually successful when started early when tissue burden is low. The fungus is very sensitive to antimycotics used in systemic treatment, as has been shown *in vitro* and in animal models.[33,34] Amphotericin B with or without flucytosine is used most often. Successful results have also been obtained with itraconazole. Nine cases were successfully treated with itraconazole at 400 mg/day.

It is generally believed that the host generates an immune response against *P. marneffei*[10] and that the failure of a CD4+ T cell-dependent immunity in people with AIDS contributes to the development of disseminated systemic infection.[29] Yet we know very little about how this T cell subpopulation, or other components of immunity function against this opportunistic pathogen.

Because of the relative ease with which T lymphocyte populations can be manipulated in mice, we have begun to investigate the importance of T cell subsets in protective immunity in a mouse model of disseminated infection.

MATERIALS AND METHODS

Fungi

Two isolates of *P. marneffei* were studied. SC325 was from the Culture Collection, Wadsworth Center for Laboratories and Research, New York State Department of Health; it was a subculture derived from the human isolate reported by Di Salvo *et al.*[5] and deposited in the ATCC under accession number ATCC24100. M134 was isolated from bronchoalveolar lavage from a patient with AIDS described by Romana *et al.*[17]

Inoculum Preparation

Fungi were grown on Sabouraud's dextrose agar (SDa) slants for 10-14 days at 26 °C. Suspensions of conidia were prepared by flooding cultures with phosphate-buffered saline (PBS). Conidial suspensions were qualified with a hemocytometer and adjusted to contain 10^6 conidia/ml.

Experimental Design

C57BL/6 mice were obtained from the Trudeau Institute Animal Breeding Facility, which is fully accreditated by the American Association of Laboratory Animal Care. Animals were maintained in accordance with the Institutional Animal Care and Use Committee. Mice were housed in standard wire-topped cages and provided food and water *ad libitum*. To deplete T lymphocytes subsets, mice were thymectomized and then treated with T cell monoclonal antibodies (mAbs), as previously described.[35] After mAb treatment, mice were held one week to allow the targeted T cell subpopulation to be eliminated from the hosts. T cells remaining were enumerated by flow cytofluorometry (FAC Scan, Becton Dickinson, Sunnyvale, CA, USA), using fluorescein conjugated $F(ab')_2$ fragments of GK1,5 (anti-CD4) TIB-210 (anti-CD8) and 3OH12 (anti-THY) mAbs, as described in an earlier publication.[35] A cocktail containing FITC-conjugated goat anti-mouse IgG, IgA and IgM was obtained from Cappel Laboratories (West Chester, PA, USA) to enumerate Ig+ B cells.

Groups of mice were infected intravenously with 10^5 *P. marneffei* conidia of one of 2 different strains. One day after injection, representative mice were killed. The liver and spleen were homogenized and plated on SDA. The number of colony forming units (CFU) was then used to estimate the initial tissue burden.

In addition, flow cytofluorometric analysis was performed on spleen cells from these mice to determine if the targeted T cell subpopulation had been depleted.

Over the next 3 weeks, mice were observed for signs of morbidity, and any deaths were noted. At the end of the third week, all the remaining mice were killed and *P. marneffei* in the livers and spleens was quantified by determination of CFU.

RESULTS AND DISCUSSION

The results in Table 2 show that one day after i.v. injection virtually the entire inoculum could be recovered in the liver; approximately 10% of the inoculum (10^4 organisms) was found in the spleen. This indicates that most of the organisms are removed from the blood within these organs. The table also shows that systemic treatment with T cell mAbs depleted the animal's corresponding T cell subpopulation. For example, control spleen contained 33 million Thy+ cells (mostly T lymphocytes) and 55 million Ig+ B lymphocytes. The ratio of CD4 to CD8 cells was approximately 2: / 1. In mice treated with anti-CD4 mAbs, no CD4+ cells could be detected in the spleens and there was a concomitant reduction in the total number of Thy+ cells. Analysis of the spleen for CD8+ T cells showed a similar pattern of depletion.

Table 2. Status of normal and T cell-deficient C57BL/6 mice one day after an intravenous inoculation of 10^5 *P. marneffei*.*

P.marneffei strain	Group	mAb Treatment	Log$_{10}$ Viable *P. marneffei* Liver	Spleen	n	Number (x10^6) in spleen Thy$^+$	CD4$^+$	CD8$^+$	Ig$^+$
SC325	A	rat IgG	5.14	3.63	1	43.0	19.0	16.0	49.0
M134	E	rat IgG	5.10	3.69	1	33.0	20.0	10.0	55.0
	F	CD4	2.09	3.60	1	15.0	<0.3	11.0	42.0
	G	CD8	5.08	3.61	1	9.5	4.2	<0.4	83.0
	H	CD4 + CD8	5.04	3.61	1	5.4	<0.2	<0.6	49.0

* To deplete T cell subpopulations, groups of mice were thymectomized and treated with CD4, CD8, or both mAbs. One week after treatment, mice were inoculated intravenously with 10^5 *P. marneffeii* in 0.2 ml PBS. The following day, one mouse from each group was killed and viable organisms enumerated by planting samples of homogenized liver and spleen on SDA. To determine the T cell status of the mice, lymphocytes obtained from their spleens were analyzed by flow cytofluorometry.

No mortality or morbidity was observed in mice infected with *P. marneffei* strain SC325 (Table 3). At sacrifice, the livers of normal immunocompetent mice contained $1\text{-}2\times10^5$ viable fungi, the same range found in the organ three weeks earlier. By comparison, the livers of mice depleted of CD4+ T cells, either alone or together with CD8+ T cells, yielded 10-30 times more organisms than the controls.

P. marneffei strain M134, on the other hand, was more virulent for mice than strain SC325. The original 10^5 spores implanted in the liver increased 5000-fold to over 10^8 during three weeks. Depletion of CD4+ and CD8+ T cells rendered mice even more susceptible to the M134 strain. Three out of the four CD4 T cell-deficient mice died during the third week of the infection with liver burderns of $>10^{10}$ organisms. A role for CD8 T cells in immunity against the M134 strain was shown by the fact that also 2/4 mice in this group died during the third week.

To summarize briefly the results of the histological studies: in the section of liver of normal mice infected with SC325 strain typical yeast-like organisms were confined by mononuclear cells within remarkably discrete granulomas, each containing up to 3 intact organisms (Fig. 1A).

Table 3. Status of systemic disease in normal and T cell-deficient C57BL/6 mice 3 weeks after an intravenous inoculation of 10^5 *P. marneffei*.

P. marneffei strain	Group	mAb Treatment	Deaths	Log$_{10}$ Viable *P. marneffei** Liver	Spleen	n
SC325	A	rat IgG	0/4	5.12 ± 0.12	4.16 ± 0.15	4
	B	CD4	0/5	6.18 ± 0.16	4.23 ± 0.10	5
	C	CD8	0/5	4.68 ± 0.57	3.57 ± 0.17	5
	D	CD4 + CD8	0/5	6.67 ± 0.12	4.50 ± 0.21	5
M134	E	rat IgG	0/4	8.69 ± 0.70	5.26 ± 0.36	4
	F	CD4	3/4	9.96	8.80	1
	G	CD8	2/4	9.17 ± 0.31	6.75 ± 0.18	2
	H	CD4 + CD8	3/4	10.00	8.55	1

* Viable organisms (CFU) are expressed as the mean (± SD).

Similar numbers of microabscesses were observed in mice depleted of CD4+ T cells, but the granulomas were 3-4 times larger than in normal mice and contained numerous septate fungal cells (Fig. 1B).

In mice depleted of CD4+ and CD8+ T cells the microabscesses were numerous and ranged in size. Aggregates of microabscesses occurred, and fungal cells were abundant.

In livers and spleen of normal mice infected with strain M134 microabscesses were numerous with an enormous number of yeast cells per lesion (Fig. 2A). Depletion of CD4+ T cells allowed the lesions to increase in size to confluency. There were many fungal cells that individual organism was difficult to identify, except at the periphery of the lesions where internal septations were evident (Fig. 2B).

CONCLUSIONS

These preliminary studies of *P. marneffei* infection in mice suggest that CD4+ T cells, and to a lesser extent CD8+ T cells, play a role in the protective host response against the fungus. *P. marneffei* strains can differ in their virulence for mice. Depending on this virulence, CD4+ T cell mediated protective immunity functions to slow or halt multiplication of the fungus in the visceral organs. It is therefore likely that AIDS patients are susceptible to systemic *Penicillium* infection because they are unable to generate a CD4+ T cell response. Furthermore CD8+ T cells that are spared by the virus may be insufficient to eliminate the organisms within developing granulomas. It is not known, however, whether these T cells function primarily in cell-mediated immunity or as helper cells for antibody production.

This mouse model can be used to functionally identify the CD4+ and CD8+ T cell components of anti-*Penicillium* immunity. In addition it should be possible to examine the efficacy of antibiotic chemotherapy in the immunodeficient host, and to aid in the development of molecular biology-based methods of diagnosis.

Because of the enormous increase of HIV infection in the endemic area, *P. marneffei* could emerge as a major opportunistic infection in AIDS. Thus the

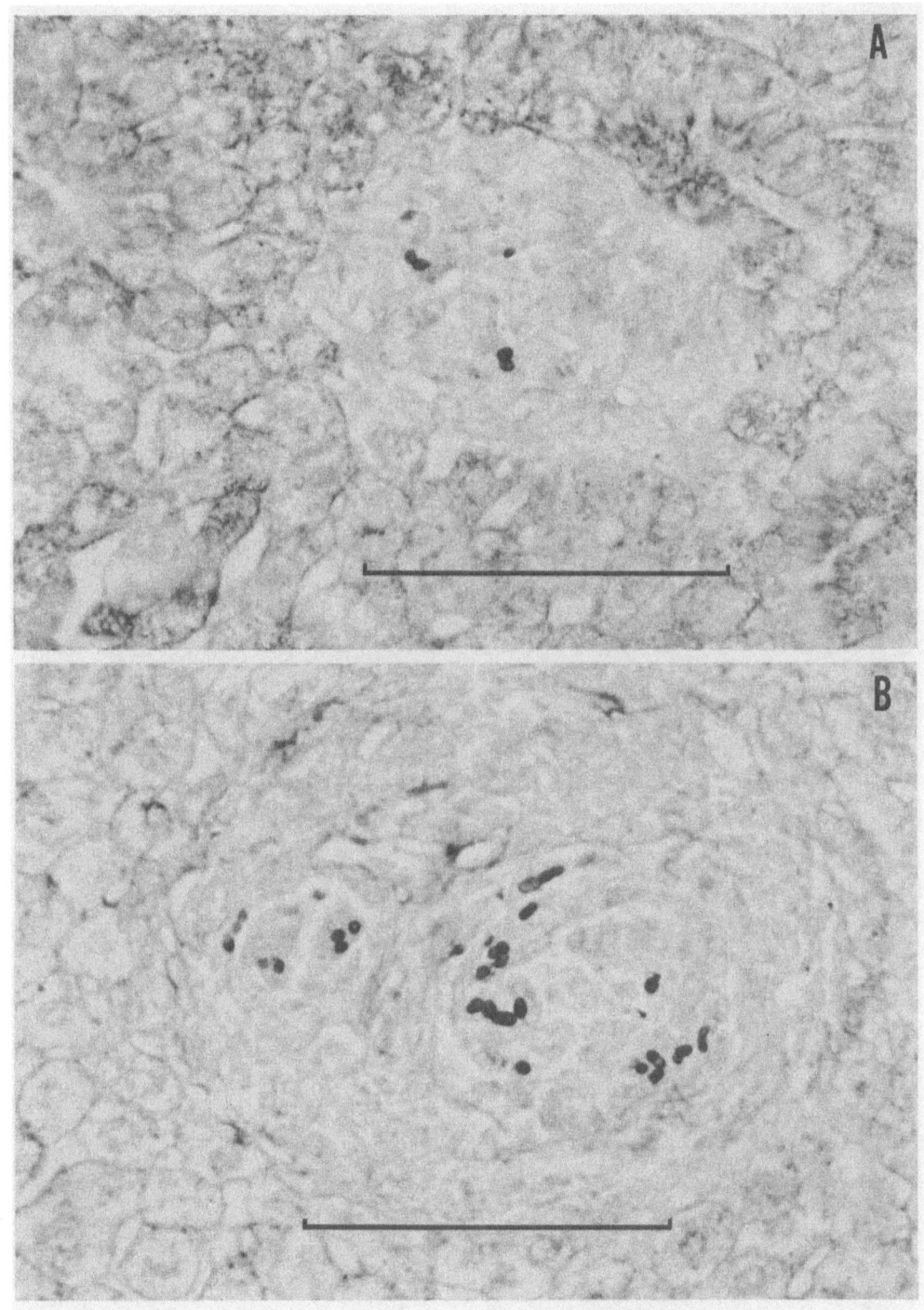

Fig. 1 Photomicrographs of foci in the liver of normal mice (A) and CD4+ T cell-depleted mice (B) at 3 weeks of a systemic *P marneffei* SC325 infection initiated by injecting 10^5 conidiospores intravenously (Gomori-Grocott methenamine silver stain, Bar = 100 μM)

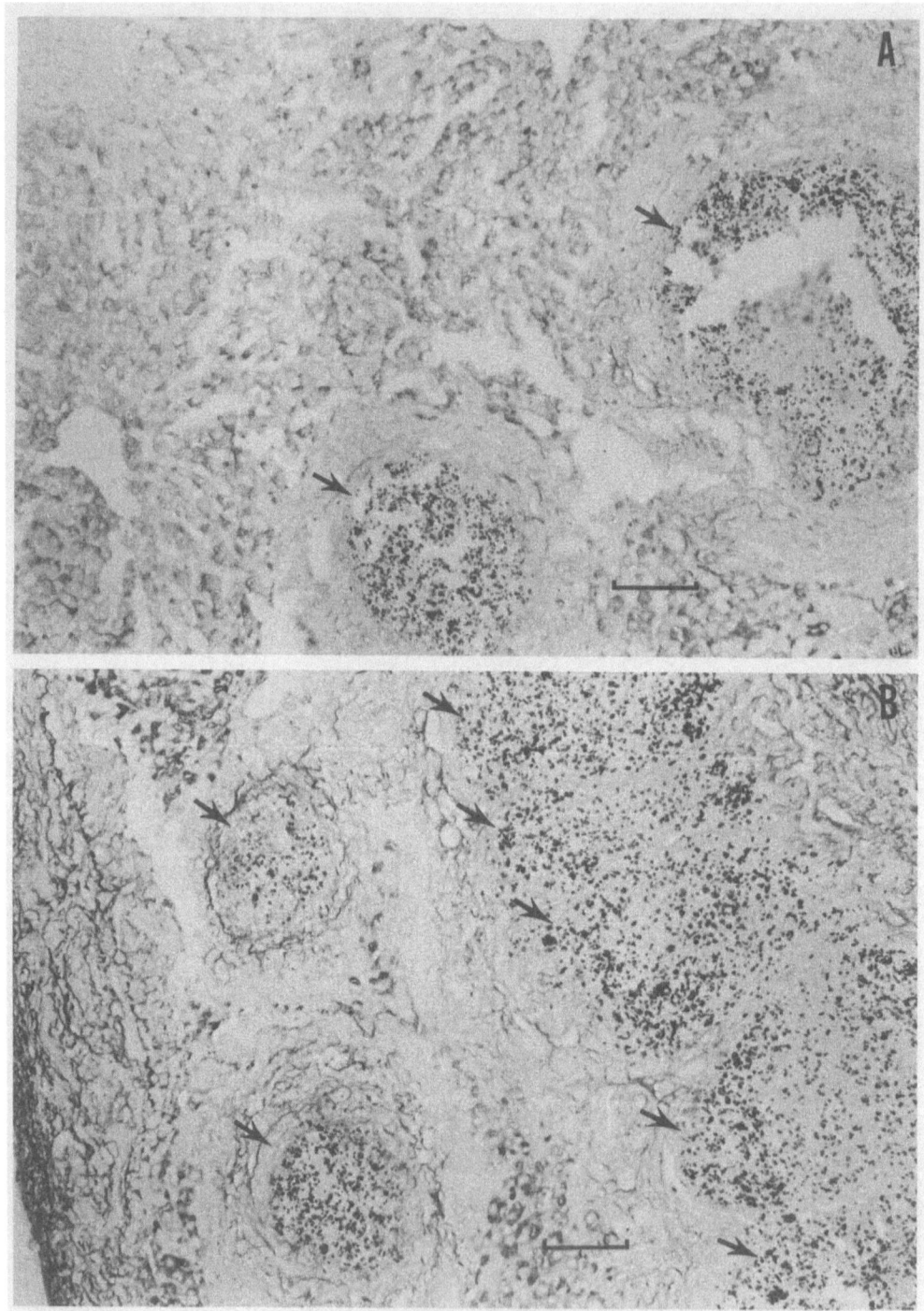

Fig. 2 Photomicrographs of foci in the liver of normal mice (A) and CD4+ T cell-depleted mice (B) at 3 weeks of a systemic *P marneffei* M134 infection initiated by injecting 10^5 conidiospores intravenously Arrows point to the margins of individual and coalescing granulomas (Gomori-Grocott methenamine silver stain, Bar = 100 μM)

medical community should be made aware of the need for early diagnosis and proper management of this disease.

The inclusion of *P. marneffei* disseminated infection among the infections defining AIDS would alert clinicians to consider this disease in differential diagnosis and to differentiate this mycosis from tuberculosis and *P. carinii* infection, which require completely defferent treatment.

REFERENCES

1. J.I. Pitt, The *genus Penicillium* and its *teleomorphic states Eupenicillium* and*Talaromyces*, Academic Press, New York (1979).
2. M. Capponi, P. Sureau, G. Segretain, Pénicillose de *Rhizomys sinensis*, *Bull. Soc. Pathol. Exot.* 49: 418 (1956).
3. G. Segretain, Description d'une nouvelle espèce de penicillium: *Penicillium marneffei* n. sp., *Bull. Soc. Mycol. Fr.* 75: 412 (1959).
4. G. Segretain, *Penicillium marneffei* n. sp., agent d'une mycose du système reticuloendothèlial, *Mycopathol. Mycol. Appl.* 11: 327 (1959).
5. A.F. Di Salvo, A.M. Fickling and L. Ajello, Infection caused by *Penicillium marneffei*: description of first natural infection in man, *Am. J. Clin. Pathol.* 60: 259 (1973).
6. P. Jayanetra, P. Nitiyanant, L. Ajello, A.A. Padhye, S. Lolekha, V. Atichartakarn, P. Vathesatogit, B. Sathaphatayavongs and R. Prajaktam, Penicilliosis in Thailand: report of five human cases, *Am. J. Trop. Med. Hyg.* 33: 637 (1984).
7. Z. Deng and D.H. Connor, Progressive disseminated penicilliosis caused by *Penicillium marneffei*: report of eight cases and differentiation of the causative organism from *Histoplasma capsulatum*, *Am. J. Clin. Pathol.* 84: 323 (1985).
8. X.G. Wei, L.T. Zhou, Q.S. Zhang, S.Y. Song and O.G. Nong, Report of the first case of penicilliosis marneffei in China, *Natl. Med. J. China* 65: 533 (1985).
9. W.C. Yuen, Y.F. Chan, S.L. Loke, W.H. Seto, G.P. Poon and K.K. Wong, Chronic lymphadenopathy caused by *Penicillium marneffei*: a condition mimiking tuberculous lymphadenopathy, *Br. J. Surg.* 73: 1007 (1986).
10. Z. Deng, J.L. Ribas, D.W. Gibson and D.H. Connor, Infection caused by *Penicillium marneffei* in China and Southeast Asia: review of eighteen published cases and report of four more Chinese cases, *Rev. Infect. Dis.* 10: 640 (1988).
11. T.E.A. Peto, R. Bull, P.R. Millard, D.W.R. Mackenzie, C.K. Campbell, M.E. Haines and R.G. Mitchell, Systemic mycosis due to *Penicillium marneffei* in a patient with antibody to human immunodeficiency virus, *J. Infect.* 16: 285 (1988).
12. T. Ancelle, J. Dupouy-Camet, F. Pujol, L. Ferradini et J. Lapierre, Un cas de pénicilliose disseminée à *Penicillium marneffei* chez un malade atteint de SIDA, *Bull. Soc. Fr. Mycol. Méd.* 17: 73 (1988).
13. D.N.C. Tsang, J.K.C. Chang, Y.T. Lau, W. Lim, C.H.Tse and N.K. Chang, *Penicillium marneffei* infection: an underdiagnosed disease?, *Histopathology* 13: 311 (1988).
14. M.K. Piehl, R.L. Kaplan and M.H. Haber: Disseminated penicilliosis in a patient with acquired immunodeficiency syndrome, *Arch. Pathol. Lab. Med.* 112: 1262 (1988).
15. M. Coen, M.A. Viviani, G. Rizzardini, A.M. Tortorano, C. Bonaccorso and T. Quirino, Disseminated infection due to *Penicillium marneffei* in a HIV positive patient, 5th International Conference on AIDS, Montreal (Quebec) Canada June 4-9, (1989) Abstract M.B.P. 94.
16. B. Sathapatayavongs, S. Damrongkitchaiporn, P. Saengditha, S. Kiatboonsri and P. Jayanetra, Disseminated penicilliosis associated with HIV infection, *J. Infect.* 19: 84 (1989).
17. C.A. Romana, M. Stern, S. Chovin, E. Drouhet et J.F. Pays, Pénicilliose pulmonaire à *Penicillium marneffei* chez un patient atteint d'une syndrome immunodeficitaire acquis, Deuxième cas français, *Bull. Soc. Fr. Mycol. Méd.* 18: 311 (1989).
18. J.K.C. Chan, D.N.C. Tsang and D.K.K. Wong, *Penicillium marneffei* in bronchoalveolar lavage fluid, *Acta Cytol.* 33: 523 (1989).
19. I.L. Wang, H.P. Yeh, S.C. Chang and J.S. Chen, Penicilliosis caused by *Penicillium marneffei*, A case report, *Derm. Sinica* 7: 19 (1989).
20. Y.F. Chan and T.C. Chow, Ultrastructural observations on *Penicillium marneffei* in natural human infection, *Ultrastruct. Pathol.* 14: 439 (1990).

21. C.M.J. Hulshof, R.A.A. van Zanten, J.F. Sluiters, M.E. van der Ende, R.S. Samson, P.E. Zondervan and J.H.T. Wagenvoort, *Penicillium marneffei* infection in an AIDS patient, *Eur. J. Clin. Microbiol. Infect. Dis.* 9: 370 (1990).

22. S. Chienchanvit, P. Mahanupab, P. Hirunsri and N. Vanittanakom, Cutaneous manifestations of disseminated *Penicillium marneffei* mycosis in five HIV-infected patients, *Mycoses* 34: 245 (1991).

23. I. Hilmarsdottir, A. Datry, J.L. Meynard, O. Rogeaux, C. Katlana, M. Chalot, G. Guermonprez, G. Brucker, M. Danis and M. Gentilini, Infection disseminée à *Penicillium marneffei* observée chez deux patients infectés par le VIH, Réunion de la Societé Française de Mycologie Medicale, Novembre 22-23, 1991, Abstract p. 62.

24. D.N.C. Tsang, P.C.K. Li, M.S. Tsui, Y.T. Lau, K.F. Ma and E.K. Yeoh, *Penicillium marneffei*: another pathogen to consider in patients infected with human immunodeficiency virus, *Rev. Infect. Dis.* 13: 76 (1991).

25. A.S. Sekhon, J. Galbrait, B.W. Mielke, W.A. Black, A.K. Garg, G. Sheehan, L.Stein, J.D. Glezes and C. Wong, Esoteric mycoses: cerebral phaeohyphomycosis caused by *Xylohypha bantiana* and penicilliosis marneffei, XI Congress of the International Society for Human and Animal Mycology, Montreal (Quebec) Canada, June 24-28, 1991, Abstract S13.4, p. 36.

26. P. de Truchis, M.E. Bonioux, J. Roussi, F. Paraire, P. Nordmann et E. Dournon, Septicemie à *Penicillium marneffei* au cours du SIDA, Réunion interdisciplinaire de chimioantobioticothèrapie, Paris, Décember 5, 1991, Abstract 35/C4.

27. D.J. Manion, F. Auclair and R. Saginur, *Penicillium marneffei* mycosis: the first canadian case report, 59th Conjoint Meeting on Infectious Diseases, Québec (Québec). Canada December 2-5, 1991, Abstract C-3.

28. I. Kok, H. Boot, P.G.J.M. Rietra and H.M. Weigel, Successful treatment of *Penicillium marneffei* infection, 3rd European Conference on Clinical Aspects and Treatment of HIV Infections, Paris March 12-13, 1992, Abstract P18O.

29. K. Supparatpinyo, S. Chiewchanvit, P. Hirunsri, C. Uthammachai, K.E. Nelson and T. Sirisanthana, *Penicillium marneffei* infection in patients infected with human immunodeficiency virus, *Clin. Infect. Dis.* 14: 871 (1992).

30. W.M.S. Tsui, K.F. Ma and D.N.C. Tsang, Disseminated *Penicillium marneffei* infection in HIV-infected subjects, *Histopathology* 20: 287 (1992).

31. F.C. Odds, T. Arai, A.F. Di Salvo, E.G.V. Evans, R.J. Hay, H.S. Randhawa, M.G. Rinaldi and T.J. Walsh, Nomenclature of fungal disease: a report and recomendations from a Sub-Committee of the International Society for Human and Animal Mycology (ISHAM), *J. Med. Vet. Mycol.* 30: 1 (1992).

32. G.R. Garrison and K.S. Boyd, Dimorphism of *Penicillium marneffei* as observed by electron microscopy, *Can. J. Microbiol.* 19: 1305 (1973).

33. M.A. Viviani, unpublished data.

34. J. Van Cutsem and F. Van Gerven, Activité antifongique *in vitro* de l'itraconazole sur les champignons filamenteux opportunistes, Traitement de la kératomycose et de la pénicilliose éxperimentales, *J. Mycol. Méd.* 118: 10 (1991).

35. J.O. Hill, M. Awwad and R.J. North, Elimination of CD4+ suppressor T cells from susceptible BALB/c mice releases CD8+ T lymphocytes to mediate protective immunity against *Leishmania*, *J. Exp. Med.* 169: 1819 (1989).

INDEX

Tunicamycin, 173

Ustilago maydis
  chitin synthase
    genes, 55
  pathogenicity, 100
  sterols, 180
  terbinafine-resistance, 180

Virulence, 3, 5-7, 23, 33, 40, 49, 73, 106, 117, 170,
      192, 214, 217, 242, 250, 292, 307-311,
      418
VSC (vesicle supply center), 133, 135-143

Wangiella dermatitidis
  actin, 247
  aculeacin A, 109
  albino auxotrophs, 105, 114
  β-glucan, 109
  brain, 307-308
  calcium, 241-242, 247-253
  chitin, 108-109, 246
  chitin synthase
    genes, 55, 246-247
    polyoxin, 246
  fusion products, 111-113
  genes, 245

  glucan, 104, 246
  mannoprotein, 104
  melanin, 105, 117, 242, 246, 307
  mutants, 105, 108, 112, 114
  parasexual genetic analysis, 110-116
  pathogenicity, 308-310
  pentaketide, 307
  phase transition, 105-116
  phenotypes, 112
  polymorphism, 107, 243-245, 247
  proton-calcium antiport, 248
  virulence, 308-310
  Woronin-bodies, 243
  zinc, 247

Xylohypha bantiana
  chitin synthase, 55

Yersinia pestis, 217

Zinc pyrithione, 405, 410
Zygomycosis
  animal models, 357-358
  deferoxamine, 341, 343